FUNDAMENTAL STELLAR PROPERTIES:
THE INTERACTION BETWEEN OBSERVATION AND THEORY

INTERNATIONAL ASTRONOMICAL UNION

UNION ASTRONOMIQUE INTERNATIONALE

FUNDAMENTAL STELLAR PROPERTIES: THE INTERACTION BETWEEN OBSERVATION AND THEORY

PROCEEDINGS OF THE 189TH SYMPOSIUM OF THE INTERNATIONAL ASTRONOMICAL UNION, HELD AT THE WOMEN'S COLLEGE, UNIVERSITY OF SYDNEY, AUSTRALIA, 13–17 JANUARY 1997

EDITED BY

T. R. BEDDING

A. J. BOOTH

and

J. DAVIS

School of Physics,
University of Sydney, Australia

KLUWER ACADEMIC PUBLISHERS

DORDRECHT / BOSTON / LONDON

A C.I.P. Catalogue record for this book is available from the Library of Congress

ISBN 0-7923-4651-3 (HB)
ISBN 0-7923-4652-1 (PB)

Published on behalf of
the International Astronomical Union
by
Kluwer Academic Publishers, P.O. Box 17, 3300 AA Dordrecht, The Netherlands.

Sold and distributed in the U.S.A. and Canada
by Kluwer Academic Publishers,
101 Philip Drive, Norwell, MA 02061, U.S.A.

In all other countries, sold and distributed
by Kluwer Academic Publishers,
P.O. Box 322, 3300 AH Dordrecht, The Netherlands.

Printed on acid-free paper

Printed in the Netherlands

Contents

6. THE STELLAR EFFECTIVE TEMPERATURE SCALE

7. STELLAR ABUNDANCES

DEDICATION

These Proceedings are dedicated to

Emeritus Professor Robert Hanbury Brown, AC, FRS, FAA

to mark his 80th birthday and in recognition of the contributions made by the Narrabri Stellar Intensity Interferometer program to the determination of fundamental stellar properties.

Robert Hanbury Brown has made major contributions to many fields including radar, radio astronomy and quantum optics. His discovery and development of intensity interferometry, and particularly the design and operation of the Narrabri Stellar Intensity Interferometer which was used under his leadership for the determination of fundamental stellar properties, made it appropriate to dedicate the meeting to him. A highlight of the meeting was the recognition that the results obtained with the Narrabri Stellar Intensity Interferometer in the 1960s and early 1970s have stood the test of time and have not yet been superseded.

PREFACE

This Symposium began with a proposal for a meeting to honour Emeritus Professor Robert Hanbury Brown on the occasion of his 80th birthday. He requested that any such meeting should be on a topic that would be of benefit to the Sydney University Stellar Interferometer (SUSI) program. With SUSI and several other high angular resolution instruments either in operation or coming on line within the next decade, and with advances in astrometry, spectroscopy and in theoretical models of stellar atmospheres and interiors, it appeared to be both appropriate and timely to hold a symposium on "Fundamental Stellar Properties: the Interaction between Observation and Theory."

The emphasis of the meeting was on the critical assessment of the quality, accuracy, and prospects for improvement of the observational data and theoretical models, on the outstanding problems in stellar astrophysics, and on the feasibility of achieving the observational and theoretical advances required for their solution. Invited papers comprised the major part of the oral program and the speakers responded to the challenge issued by the Scientific Organising Committee to critically review the current status and prospects for their area of expertise.

The Symposium was opened by the Chancellor of the University of Sydney, Emeritus Professor Dame Leonie Kramer, who welcomed the 126 participants from 22 countries on behalf of the University. The oral program included 52 invited reviews and papers and 10 contributed papers. In addition, 60 poster papers were displayed for the duration of the meeting and a session was set aside for their viewing.

The social program included a reception, an afternoon cruise on Sydney Harbour and evening visits to the Sydney Observatory—now part of Sydney's Powerhouse Museum—and to a performance of "La clemenza di Tito" at the Sydney Opera House. The Symposium dinner, which was attended by three generations of the Hanbury Brown family, was held at Sydney's Taronga Park Zoo, which has magnificent views across Sydney Harbour to the city centre.

The Editors, on behalf of the Chairs of the Scientific and Local Organising Committees, thank everyone associated with organising the symposium. While the organisation was very much a team effort, we would particularly like to thank the postgraduate students of the Chatterton Astronomy Department for their help both before and during the meeting; Bill Tango for ensuring that the Symposium budget was balanced; and Sioe-Gek Chew for transcribing much of the discussion. Special thanks are due to Johannes

Andersen, IAU Assistant General Secretary, for his advice and assistance. Finally, we thank all the participants for their contributions to the Symposium, which ensured its success.

T. R. BEDDING, A. J. BOOTH & J. DAVIS (Editors)

SCIENTIFIC ORGANISING COMMITTEE

J. Andersen (Denmark)
Y. Balega (Russia)
B. Barbuy (Brazil)
M. Bessell (Australia)
C. Chiosi (Italy)
J. Christensen-Dalsgaard (Denmark)
J. Davis (Australia; Chair)
R-P. Kudritzki (Germany)
D. Lambert (USA)
M. Spite (France)
D. Vandenberg (Canada)

LOCAL ORGANISING COMMITTEE

T. R. Bedding
A. J. Booth (Chair)
S-G. Chew
W. A. Lawson
J. W. O'Byrne
W. J. Tango (Treasurer)
F. Watson
A. E. Vaughan

SPONSORS

The Organising Committees thank the following organisations for their generous financial support which made the Symposium possible:

- the International Astronomical Union
- the Anglo-Australian Observatory
- the Mount Stromlo and Siding Spring Observatories
- the Science Foundation for Physics within the University of Sydney
- the Research Centre for Theoretical Astrophysics, University of Sydney

SPONSORING IAU DIVISION AND COMMISSIONS

Division IV: Stars
Commission 29: Stellar Spectra
Commission 35: Stellar Constitution
Commission 36: Theory of Stellar Atmospheres

LIST OF PARTICIPANTS

Johannes ANDERSEN	Copenhagen University, Denmark ja@astro.ku.dk
David ARNETT	Steward Observatory, USA darnett@as.arizona.edu
Suchitra C. BALACHANDRAN	Univ. Maryland, USA suchitra@astro.umd.edu
Ivan Karim BALDRY	University of Sydney, Australia baldry@physics.usyd.edu.au
John BALDWIN	MRAO Cambridge, UK jeb@mrao.cam.ac.uk
Isabelle BARAFFE	Ecole Normal Superieure de Lyon, France ibaraffe@cral.ens-lyon.fr
Beatriz BARBUY	Univ. Sao Paulo, Brazil barbuy@atmos.iagusp.usp.br
Paul Stuart BARKLEM	University of Queensland, Australia barklem@kepler.physics.uq.oz.au
Jean Philippe BEAULIEU	Kapteyn Laboratorium, The Netherlands beaulieu@astro.rug.nl
Timothy Russell BEDDING	University of Sydney, Australia bedding@physics.usyd.edu.au
Peter BEHRENS	cordula@magna.com.au
Roger Alistair BELL	Univ. of Maryland, USA roger@astro.umd.edu
James A. BENSON	USNO, NPOI, USA jbenson@nofs.navy.mil
David BERSIER	Mt Stromlo Observatory, Australia bersier@mso.anu.edu.au
Michael BESSELL	MSSSO, Australia bessell@mso.anu.edu.au
Robert C. BLESS	University of Wisconsin, USA bless@sal.wisc.edu
Giuseppe Quinto BONO	Trieste Astronomical Observatory, Italy bono@oat.ts.astro.it
Andrew John BOOTH	University of Sydney, Australia booth@physics.usyd.edu.au

Arnold I. BOOTHROYD
Monash University, Australia
boothroy@vulcan.maths.monash.edu.au

Maxwell Howard BRENNAN
Australian Research Council, Australia
BrennanMH@msn.com

Theo ten BRUMMELAAR
Georgia State University, USA
theo@chara.gsu.edu

Gilles CHABRIER
ENS - Lyon, France
chabrier@cral.ens-lyon.fr

Corinne CHARBONNEL
Lab. d'Astrophys. de Toulouse - CNRS, France
corinne@obs-mip.fr

Cesare CHIOSI
Dept. Astronomy - Univ. Padova, Italy
chiosi@astrpd.pd.astro.it

Jørgen CHRISTENSEN-DALSGAARD
Theoretical Astrophysics Center, Denmark
jcd@obs.aau.dk

Jens Viggo CLAUSEN
Copenhagen University Observatory, Denmark
jvc@astro.ku.dk

Martin COHEN
UC-Berkeley, USA
mcohen@astro.berkeley.edu

Kem Holland COOK
Lawrence Livermore National Laboratory, USA
kcook@llnl.gov

Peter COTTRELL
University of Canterbury, New Zealand
p.cottrell@phys.canterbury.ac.nz

Lawrence CRAM
University of Sydney, Australia
l.cram@physics.usyd.edu.au

Paul Adrian CROWTHER
University College London, England
pac@star.ucl.ac.uk

Gary Stewart DA COSTA
MSSSO, Australia
gdc@mso.anu.edu.au

Conard Curtis DAHN
USNO Flagstaff, USA
dahn@nofs.navy.mil

John DAVIS
University of Sydney, Australia
davis@physics.usyd.edu.au

Leen DECIN
K.U.Leuven, Belgium
leen@ster.kuleuven.ac.be

G. Paolo DI BENEDETTO
CNR-Fisica Cosmica, Italy
pdibene@ifctr.mi.cnr.it

Jadzia DONATOWICZ
Inst. for Astronomy, Univ. Vienna, Austria
jadzia@shakti.edvz.tuwien.ac.at

Stefan DREIZLER University Kiel, Germany
dreizler@astrophysik.uni-kiel.de

Nicholas ELIAS USNO/NPOI, USA
nme@fornax.usno.navy.mil

Francis FEKEL Tennessee St. Univ., USA
fekel@coe.tnstate.edu

Masayuki FUJIMOTO Hokkaido Univesity, Japan
fujimoto@phys.hokudai.ac.jp

Flavio FUSI PECCI Osservatorio Astronomico, Italy
flavio@astbo3.bo.astro.it

Wolfgang GIEREN Universidad de Concepcion, Chile
wgieren@astro.puc.cl

Yuri GNEDIN Pulkovo Observatory, Russia
gnedin@pulkovo.spb.su

Jean-Franois GONZALEZ Centre de Recherche Astron. de Lyon, France
jfgonzal@cral.ens-lyon.fr

Bengt GUSTAFSSON Uppsala Astronomical Observatory, Sweden
bengt.gustafsson@astro.uu.se

Robert Hanbury BROWN

Mohammad Ridwan HIDAYAT National University of Malaysia, Malaysia
baksa@po.jaring.my

Desmond John HILLIER University of Pittsburgh, USA
jdh@magpie.phyast.pitt.edu

Michael HRYNEVYCH University of Sydney, Australia
mick@physics.usyd.edu.au

Christian HUMMEL USRA, USA
cah@fornax.usno.navy.mil

Donald J. HUTTER U.S. Naval Observatory, USA
djh@fornax.usno.navy.mil

Philip A. IANNA Univ. of Virginia, USA
pai@mso.anu.edu.au

Andrew JACOB University of Sydney, Australia
ande@physics.usyd.edu.au

Carme JORDI Univ. Barcelona, Spain
carme@facjn0.am.ub.es

Steve KAWALER Iowa State University, USA
sdk@iastate.edu

Andreas KELZ	University of Sydney, Australia kelz@physics.usyd.edu.au
Hans KJELDSEN	Aarhus University, Denmark hans@obs.aau.dk
Jeffrey KRUK	Johns Hopkins University, USA kruk@pha.jhu.edu
Robert L. KURUCZ	Center for Astrophysics, USA rkurucz@cfa.harvard.edu
Norbert LANGER	MPA Garching / Univ. Potsdam, Germany ntl@astro.physik.uni-potsdam.de
Wendie Susan LANGER	Melbourne University, Australia wsl@mozart.ph.unimelb.edu.au
John LATTANZIO	Monash Uni, Australia johnl@flash.maths.monash.edu.au
Warrick LAWSON	Australian Defence Force Academy, Australia wal@phadfa.ph.adfa.oz.au
Sandy K. LEGGETT	UKIRT, Hawaii, USA skl@jach.hawaii.edu
Zaixiong LI	Shaanxi Astronomical Observatory, China zxli63@physics.usyd.edu.au
Robert E. LUCAS	University of Sydney, Australia lucas@physics.usyd.edu.au
Andre MAEDER	Geneva Observatory, Switzerland andre.maeder@obs.unige.ch
Pierre MAXTED	Southampton, UK pflm@astro.soton.ac.uk
Michel MAYOR	Geneva Observatory, Switzerland michel.mayor@obs.unige.ch
Harold A. McALISTER	CHARA/Georgia State Univ., USA hal@chara.gsu.edu
Claude MEGESSIER	Obs de Meudon, France claude.megessier@obspm.fr
Alberto MENDEZ	Sydney Uni, Australia alberto@physics.usyd.edu.au
Ben Bryan MESSENGER	Monash University, Australia bbmes1@jeep.maths.monash.edu.au
Dante MINNITI	Lawrence Livermore Nat'l Lab., USA dminniti@llnl.gov

Anna Marie MOORE — University of Sydney, Australia
anna@physics.usyd.edu.au

Carlo MOROSSI — Astronomical Observatory of Trieste, Italy
morossi@oat.ts.astro.it

David MOZURKEWICH — NRL, USA
mozurk@rira.nrl.navy.mil

Jennifer NICHOLLS — University of Sydney, Australia
j.nicholls@physics.usyd.edu.au

Poul Erik NISSEN — Institute of Physics and Astronomy, Denmark
pen@obs.aau.dk

Birgitta NORDSTROM — Copenhagen University, Denmark
birgitta@astro.ku.dk

John Edward NORRIS — Mount Stromlo Observatory, Australia
jen@mso.anu.edu.au

John O'BYRNE — University of Sydney, Australia
j.obyrne@physics.usyd.edu.au

Bernard James O'MARA — University of Queensland, Australia
omara@kepler.physics.uq.oz.au

Sergio ORTOLANI — Dept. of Astronomy Padova, Italy
ortolani@astrpd.pd.astro.it

Guy PERRIN — Observatoire de Meudon, France
perrin@hplyot.obspm.fr

Orlon PETTERSON — University of Canterbury, New Zealand
o.petterson@phys.canterbury.ac.nz

Daniel M. POPPER — UCLA, USA
popper@bonnie.astro.ucla.edu

Martin John PORTER — Univ. of Queensland, Australia
porter@physics.uq.oz.au

Nigel Patrick PRESTAGE — University of Sydney, Australia
nigel@ips.oz.au

Charles R. PROFFITT — Computer Sciences Corp
Catholic Univ. of America, USA
hrsproffitt@hrs.gsfc.nasa.gov

Prabaharan RAGLAND — Physical Research Laboratory, India
sam@prl.ernet.in

Sofia RANDICH — ESO, Germany
srandich@eso.org

Andrea RICHICHI	Osservatorio di Arcetri, Italy richichi@arcetri.astro.it
Robert T. ROOD	University of Virginia, USA rtr@virginia.edu
Maitreyee ROY	University of Sydney, Australia m.roy@astron.physics.usyd.edu.au
Frédéric ROYER	DASGAL - Obs. de Paris-Meudon, France frederic.royer@obspm.fr
Goedele RUYMAEKERS	Royal Observatory of Belgium, Belgium goedele.ruymaekers@oma.be
Sean RYAN	Anglo-Australian Obs., Australia sgr@aaoepp.aao.gov.au
Elaine Margaret SADLER	University of Sydney, Australia ems@physics.usyd.edu.au
Stefano SANDRELLI	Dept. di Astronomia, Univ. Bologna, Italy sandrelli@astbo4.bo.astro.it
Dimitar D. SASSELOV	Harvard-Smithsonian CfA, USA dsasselov@cfa.harvard.edu
Michael SCHOLZ	Univ. Heidelberg, Germany scholz@ita.uni-heidelberg.de
William John SCHUSTER	Univ. Nacional Autonoma, Mexico schuster@bufadora.astrosen.unam.mx
Matthew David SHETRONE	ESO, Chile mshetron@eso.org
Lindsey SMITH	Univ. Sydney, Australia lsmith@physics.usyd.edu.au
Monique SPITE	Obs. de Paris-Meudon, France Monique.Spite@obspm.fr
Michelle Claire STOREY	University of Sydney, Australia michelle@physics.usyd.edu.au
William TANGO	University of Sydney, Australia tango@physics.usyd.edu.au
Rosaria TANTALO	Department of Astronomy, Padua, Italy tantalo@astrpd.pd.astro.it
Benjamin J. TAYLOR	Brigham Young University, USA taylorb@astro.byu.edu
Melinda TAYLOR	Univ. of Sydney, Australia melinda@physics.usyd.edu.au

Catherine TURON	Observatoire de Paris, France catherine.turon@obspm.fr
Antonella VALLENARI	Observatory of Padova, Italy vallenari@astrpd.pd.astro.it
Nicole S. VAN DER BLIEK	Sterrewacht Leiden, The Netherlands nvdbliek@strw.LeidenUniv.nl
Hans VAN WINCKEL	K.U.Leuven, Belgium hans@ster.kuleuven.ac.be
Don Allan VANDENBERG	Univ. of Victoria, Canada davb@uvvm.uvic.ca
Myriam VRANCKEN	Royal Observatory of Belgium, Belgium Myriam.Vrancken@oma.be
Rainer WEHRSE	Inst. Theoret. Astrophys. Heidelberg, Germany wehrse@rw.iwr.uni-heidelberg.de
Nathaniel WHITE	Lowell Observatory, USA nmw@lowell.edu
Walter WINDSTEIG	Institute for Astronomy, Vienna, Austria windsteig@auro.ast.univie.ac.at
Peter R. WOOD	MSSSO, Australia wood@mso.anu.edu.au
Charles WORLEY	U.S. Naval Observatory, USA eqb@draco.usno.navy.mil
Kinwah WU	University of Sydney, Australia kinwah@physics.usyd.edu.au

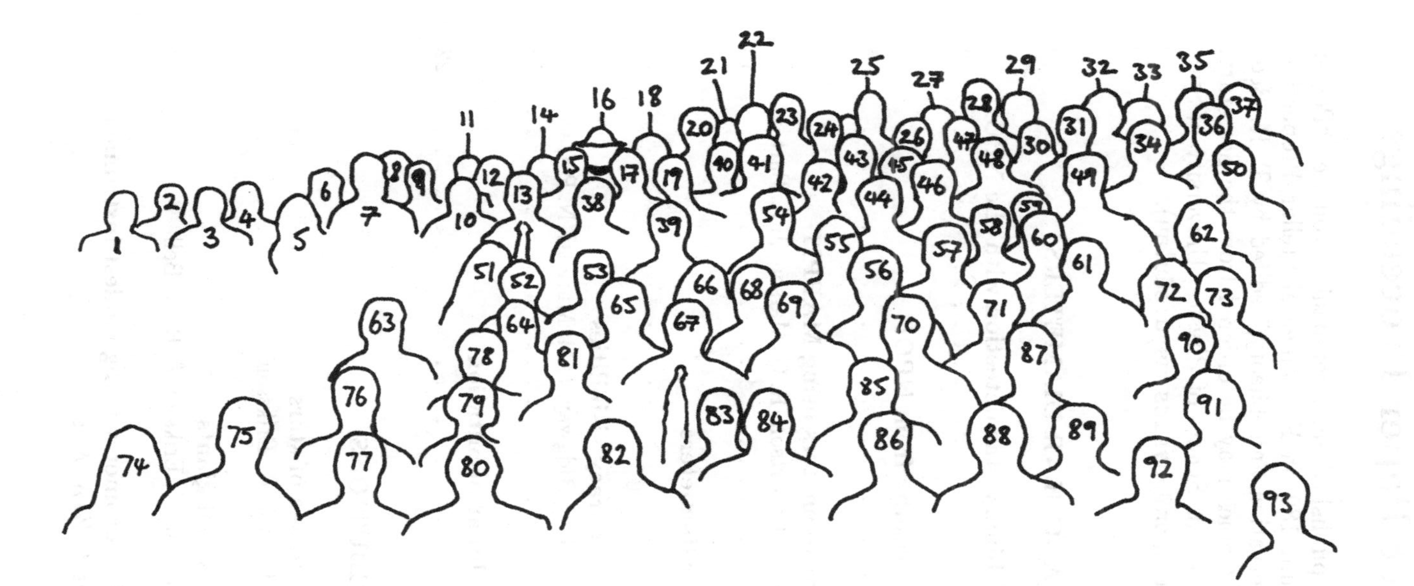

Andersen 12, Balachandran 51, Baldry 76, Baldwin 71, Barbuy 32, Barklem 4, Beaulieu 49, Bedding 36, Bell 21, Bersier 60, Bessell 69, Bless 42, Bono 31, Booth 75, Boothroyd 88, ten Brummelaar 68, Charbonnel 5, Christensen-Dalsgaard 8, Clausen 9, Cook 57, Cottrell 16, Cram 67, Crowther 84, Da Costa 48, Dahn 50, Davis 80, Decin 59, Di Benedetto 53, Dreizler 77, Elias 70, Fujimoto 1, Gnedin 26, Gonzalez 25, Gustafsson 6, Hanbury Brown 82, Hidayat 24, Hillier 56, Hrynevych 65, Hummel 54, Hutter 41, Jacob 14, Jordi 52, Kawaler 11, Kelz 81, Kjeldsen 87, Kurucz 33, Langer 43, Lawson 86, Leggett 62, McAlister 7, Megessier 73, Mendez 78, Minniti 61, Moore 74, Morossi 47, Mozurkewich 39, Nissen 23, Nordstrom 79, Norris 91, O'Mara 20, Petterson 27, Popper 3, Porter 64, Proffitt 89, Richichi 19, Rood 22, Royer 45, Ruymaekers 30, Sadler 72, Sasselov 34, Scholz 17, Schuster 37, Tango 13, Tantalo 40, M.M. Taylor 83, B.J. Taylor 85, Turon 93, Vallenari 66, van der Bliek 15, Van Winckel 46, VandenBerg 38, Vrancken 58, White 63, Windsteig 55, Wood 18, Worley 10.

Contents of Poster Paper Proceedings

The poster papers are being published in a companion volume, "Poster Proceedings of IAU Symposium 189 on Fundamental Stellar Properties: The Interaction between Observation and Theory," edited by T. R. Bedding (University of Sydney). Copies may be ordered from the Secretary, Chatterton Astronomy Department, School of Physics, University of Sydney 2006, Australia (e-mail: `astron@physics.usyd.edu.au`).

3. THE EFFECTIVE TEMPERATURE SCALE

4. BINARY & MULTIPLE STARS

5. STELLAR ABUNDANCES & SPECTRA

1. INTRODUCTION

INTRODUCTORY OVERVIEW

J. ANDERSEN
Astronomical Observatory; Niels Bohr Institute for Astronomy, Physics, and Geophysics; Juliane Maries Vej 30, DK-2100 Copenhagen, Denmark

Abstract. The subject of this meeting is the confrontation between precise observational and theoretical determinations of fundamental stellar properties. Its goal is to better define the limits of our present understanding of the structure and evolution of stars, and of our own and other galaxies. That goal is not approached by keeping to the "safe" side of the border: The areas where significant *dis*agreement is found between the best observations and the best theories also show the directions where progress is to be made.

1. Introduction

It is a privilege to introduce this IAU Symposium in honour of Professor Hanbury Brown's 80th birthday, on a subject also close to my own heart. His own work with the Narrabri Intensity Interferometer is, of course, an established classic in the field, and the many references to it throughout this meeting show it to be still very much alive today. Moreover, his group – our hosts – has remained in the forefront of stellar interferometry ever since, and results presented at this meeting show the field to be poised for another breakthrough in the next few years. No doubt, fascinating results from optical interferometers – some of them fed by 8-10 m telescopes – will feature prominently on the programme when we meet here again in 2007 in honour of Professor Hanbury Brown's 90th birthday!

The subtitle of the meeting reminds us that the observational determination of fundamental stellar properties is not a goal in itself: The results only become really interesting when placed in a larger astrophysical context. Keeping in mind what today's burning questions are helps us to focus

T.R. Bedding et al. (eds.),
Fundamental Stellar Properties: The Interaction between Observation and Theory, 3–6.

our efforts on obtaining the data that will be needed to answer them: *Which* fundamental stellar parameters are needed to answer this question? *How accurate* must the results be to be useful? *Which other types of data* will be needed to apply the results in the most meaningful way? *How are they analysed* to produce the most interesting astrophysical results?

While in the heat of the battle, it is natural to focus on how to best carry out the work – how to find "the right answer" to the initial question. But in lucky cases, when the observations are sufficiently accurate, what one finds is that, in fact, the *question* was wrong, based on an inadequate theoretical framework. And one is reminded that the most interesting progress is made by looking for the right questions rather than for the right answers.

2. A Few Examples ...

A few examples, immodestly drawn from my own experience, will serve to illustrate the point. As is natural for a meeting on fundamental stellar properties, they will pertain directly to such aspects of stellar astrophysics as stellar atmospheres, compositions, structure, end evolution. Yet, as the interaction between observations and theory has advanced our understanding of these subjects, the ramifications to the larger picture of the formation and evolution of galaxies and the synthesis of the chemical elements has become ever clearer and more important.

The first review at the very first IAU meeting I ever attended (Popper, 1970) discussed new fundamental results on stellar masses and radii in the context of the then vigorously developing theory for the evolution of single and binary stars. An immediate result from these accurate data was the demise of the long-cherished notion that mass and radius are strict functions of spectral type or colour for main-sequence stars, a lesson that took another 20 years to be fully appreciated (Nordström, 1989). More significantly, the general feeling at the time, that the "Algol paradox" had been solved by the theory of mass exchange in close binaries, was shattered shortly after when Refsdal et al. (1974) found no models which could account for the first precise fundamental data for an actual Algol system (Popper, 1973), highlighting the shortcomings of the theory at the time.

The chemical evolution of our own Milky Way galaxy provides another set of examples. The key fundamental stellar properties in this context are ages, detailed element abundances, and galactic orbital parameters derived from observed positions and space motions. These were determined for a number of disk F dwarfs by Edvardsson et al. (1993) with sufficient accuracy that "the" age-metallicity and other key relations predicted by chemical evolution models for the local disk could be defined with superior precision. The results, however, showed that the basic concept of a well-defined age-metallicity relation for the solar neighbourhood, and with it the basic paradigm for standard models of its chemical evolution, is inadequate in the real galaxy. Yet, individual element ratios were found to be surprisingly tight functions of overall metallicity and galactocentric distance, in ways that will strongly constrain more realistic physical models of nucleosynthesis in the galactic disk. This result was found later to apply also to nearby stars of metallicities down to those of the most metal-poor globular clusters. Yet, through very careful spectroscopy, Nissen & Schuster (1997) have recently uncovered subtle abundance anomalies in one group of halo stars, characterised by outlying orbits, which may hold important clues to the processes by which the halo was assembled. Similarly, the highly unusual abundance patterns recently found in the most metal-poor halo stars (Sneden et al. 1996, Barbuy et al. 1997) tell us that nucleosynthesis in our galaxy proceeded very differently before and after the stage when globular clusters were formed.

Finally, noting that the precise sequence in which the major components of the Milky Way galaxy were assembled – and how – remains uncertain even today, the value of precise stellar data in establishing a reliable age scale for galactic stars is worth recalling. That precise knowledge on such fundamental stellar properties as cluster membership, duplicity, and mass is important when testing stellar models for this purpose is, in retrospect, not really surprising. However, the example presented at this meeting (Nordström et al., 1997) shows not only how imprecise broad-band photometry, neglect of interstellar reddening, and subjective membership assignments may conspire to make the age of a supposedly "classical" open cluster uncertain by a factor of four, but also provides a striking reminder that the stellar content of a present-day cluster may be but a faint shadow of the original population.

3. ... and Let's Get Started!

Over the next few days we will hear many new and interesting results, and some long-standing discrepancies between observations and theory will no doubt be shown to have been reduced or eliminated. However, let us not get swept away with contentment, but keep the border to the unknown within sight, not only within the field of stellar astrophysics *per se*, but also in those other major fields of contemporary astrophysics where our results are perhaps less directly visible, but nonetheless crucial. I look forward to five days of pleasant, constructive, and harmonious disagreement!

Acknowledgements I am delighted that Dan Popper can be with us at this meeting: My own presence here is a direct consequence of his own review

28 years ago. Financial support from the Danish Natural Science Research Council and the Carlsberg Foundation for the research discussed here is gratefully acknowledged.

References

Barbuy B., Cayrel R., Spite M., Beers T.C., Spite F., Nordström B., Nissen P.E. 1997, A&A 317, L63

Edvardsson B., Andersen J., Gustafsson B., Lambert D.L., Nissen P.E., Tomkin J. 1993, A&A 275, 101

Nissen P.E., Schuster W.J. 1997, A&A, in press

Nordström B. 1989, ApJ, 341, 934

Nordström B., Andersen J., Andersen M.I. 1997, this meeting; poster paper no. 64

Popper D.M. 1970, in Mass Loss and Evolution in Close Binaries, (IAU Colloquium No. 6), eds. K. Gyldenkerne and R.M. West, Univ. of Copenhagen, Copenhagen, p. 13

Popper D.M. 1973, ApJ 185, 265

Refsdal S., Roth M.L., Weigert A. 1974, A&A 36, 113

Sneden C., McWilliam A., Preston G.W., Cowan J.J., Burris D., Armosky B.J. 1996, ApJ 467, 819

2. STELLAR DISTANCES

THE HIPPARCOS RESULTS [1]

C. TURON [2]
Dasgal / URA CNRS 335, Observatoire de Paris-Meudon
92 195 MEUDON Cedex, France

Abstract. After a brief presentation of the Hipparcos mission, and an overview of the astrometric and photometric results obtained from the main mission and from Tycho, more details are given on the parallaxes. Absolute parallaxes have been obtained for 117 955 entries of the Hipparcos Catalogue, out of a total of 118 218, with a median precision of 0.97 mas for stars brighter than 9. This precision varies with apparent magnitude and ecliptic latitude. The estimated systematic error is smaller than 0.1 mas. The distances of more than 20 000 stars are determined to better than 10 %. Some more statistics are presented.

A few applications of this fantastic amount of new and accurate data are presented, in the fields of absolute magnitude calibrations, stellar physics, distance scale determination.

1. Introduction

The ESA Hipparcos satellite, included in the European Space Agency scientific programme in 1980, was launched by Ariane in August 1989. High quality scientific data were obtained during 37 months, from November 1989 to March 1993. Operations were terminated on 15 August 1993. Comprehensive description of the Hipparcos mission is given in Perryman et al. (1992). The whole project was supervised by ESA; the satellite and the instrument were constructed under the leadership of Matra Marconi Space and Alenia Spazio; the scientific aspects, supervised by the Hipparcos Science Team, were conducted by four scientific consortia.

[1] Based on observations made with the ESA Hipparcos satellite, and on work performed within the INCA, FAST, NDAC and TDAC Consortia.
[2] For the Hipparcos Science Team

T.R. Bedding et al. (eds.),
Fundamental Stellar Properties: The Interaction between Observation and Theory, 9–18.

The INCA Consortium was responsible for the preparation of the observing programme, published as the Hipparcos Input Catalogue (Turon et al. 1992a and 1992b). The two consortia, FAST and NDAC, were responsible for the global treatment of the 1 000 Gbit of the main mission data (Lindegren et al. 1992, Kovalevsky et al. 1992). The TDAC consortium was responsible for the analysis of the star mapper data (Høg et al. 1992).

The final products of the mission are

- the Hipparcos Catalogue, resulting from the merging of the two independent solutions obtained by FAST and NDAC, and including 118 218 preselected entries, brighter than $V = 12.5$, with a median precision at the level of the milliarcsec for each of the five astrometric parameters, and of a few millimagnitudes for the mean photometric parameters,
- annexes to the Hipparcos Catalogue, including details on double and multiple systems, and on variable stars,
- the Hipparcos Epoch Photometry Annex, including calibrated epoch photometry for each programme star, at an average of some 110 different epochs of observation,
- the Tycho Catalogue, including 1 052 031 entries observed in survey mode, brighter than $V = 11.5$, with a median precision at the level of 3 to 50 milli-arcsec (depending on magnitude), for each of the five astrometric parameters, and of 0.003 to 0.12 magnitude for the mean photometric parameters,
- the Tycho Epoch Photometry Annex, including epoch photometry for the brighter stars, at each epoch of observation,

About 115 000 stars are common to the Hipparcos and Tycho Catalogues. 935 000 stars are in the Tycho Catalogue only, not observed by Hipparcos due to limited observing time. 6 300 stars are contained in the Hipparcos Catalogue only, not observed by Tycho due to the crowding of some zones of the sky, due to their faintness (about 2 300 Hipparcos stars were below Tycho detection limits), or to their brightness (very few bright stars were not observed by Tycho). These entries are flagged in the Tycho Catalogue.

2. The Astrometric Parameters

2.1. SUCCESSIVE SPHERE SOLUTIONS

The overall data analysis was decomposed in five major steps: the treatment of the raw photon counts, the attitude determination, the great circle reduction, the sphere solution and the determination of the astrometric parameters. The whole process is described in detail in Volume 3 of 'The Hipparcos and Tycho Catalogues' (ESA 1997). The last four steps were repeated several times, independently by the FAST and NDAC consortia, as successively larger data sets were available from satellite observations. Finally, the results obtained for each star by each consortium, were merged in order to provide a unique set of astrometric parameters together with their associated covariance matrix.

Figures 1 and 2 are illustrations of the improvement obtained in the determination of the astrometric parameters by including successively larger sets of data in the solution, and then by merging the two independent solutions obtained by the two data reduction consortia. Distributions of formal standard errors in parallax for the successive NDAC sphere solutions, compared with that of the final Hipparcos Catalogue are shown in Figure 1. The temporal evolution of the fraction of negative parallaxes, for the solutions obtained by each consortium, compared to the successive merged solutions, is illustrated in Figure 2. The parallaxes improve, as expected, roughly as $t^{-1/2}$, if t is the total duration of the observations, the position at the same rate, and the proper motions slightly slower than the expected $t^{-3/2}$. Empirically, the fraction of negative parallaxes improves roughly as $t^{-1.0}$.

Finally, the Hipparcos reference frame was linked to the International Celestial Reference System (ICRS), defined by radio VLBI observations of extragalactic sources, by means of several dedicated programmes (Lindegren and Kovalevsky 1995; Kovalevsky, Lindegren and Perryman 1997): VLBI, Merlin, and VLA observations of radio-emitting stars observed by Hipparcos, Hubble observations of separations of Hipparcos stars next to compact extragalactic objects, wide field photographic plate measurements of Hipparcos stars with respect to compact extragalactic objects, use of Earth orientation parameters. The accuracy of the extragalactic link is estimated to be ± 0.6 mas in the orientation and ± 0.25 mas/yr in the spin.

2.2. THE PARALLAXES

The number of observations per star was conditioned by the scanning motion of the satellite all over the sky. The number of astrometric observations (abscissae on reference great circles) varies from about 10 to about 60, depending principally on the ecliptic latitude. This is directly reflected in the standard errors of the parallaxes. The other parameter of importance is the star magnitude. Median standard errors in parallax, σ_π, as function of the Hp magnitude and of the ecliptic latitude are given in Table 1.

Hipparcos parallaxes have been compared with the best available ground-based parallaxes (Arenou et al. 1997). Twelve optically bright radio-emitting stars were observed by VLBI as part of the Hipparcos link programme (Lestrade et al. 1995). The precision of VLBI parallaxes is 0.2 to 1 mas, and an excellent agreement is found between the two sets of measurements.

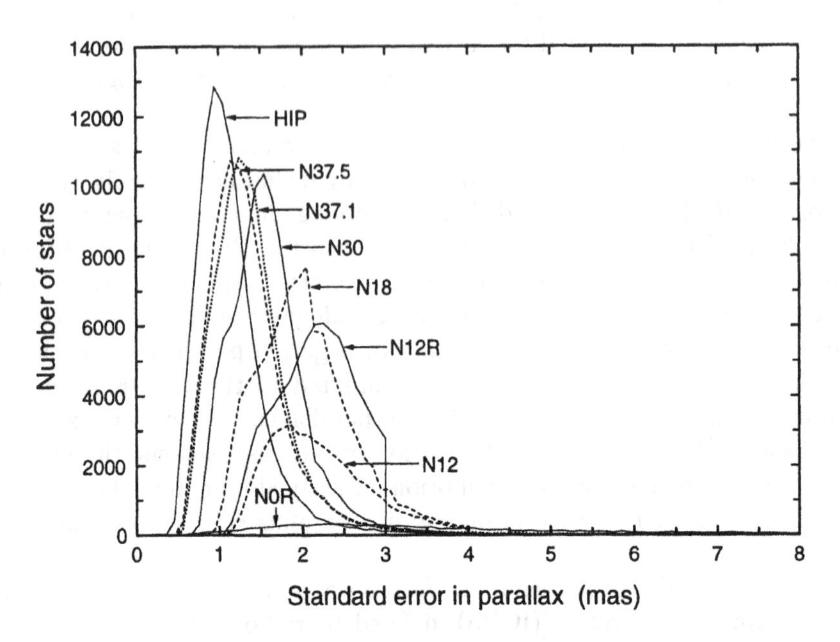

Figure 1. Distribution of formal standard errors in parallax for the successive NDAC sphere solutions (NOR to N37.5) and for the Hipparcos Catalogue (HIP). Courtesy L. Lindegren, Lindegren et al. 1997, Figure 16.13.

TABLE 1. Median standard errors in parallax, σ_π, as function of the Hp magnitude and of the ecliptic latitude

	Ecliptic latitude ($\|\beta\|$, deg)								Nb of
Hp	0-20	20-30	30-40	40-50	50-60	60-70	70-90	all	stars
< 6	0.86	0.81	0.75	0.67	0.56	0.52	0.49	0.75	4559
6-7	0.91	0.86	0.81	0.73	0.61	0.55	0.52	0.80	9381
7-8	1.02	0.97	0.92	0.86	0.72	0.63	0.59	0.91	23679
8-9	1.20	1.16	1.12	1.07	0.89	0.77	0.71	1.09	40729
9-10	1.48	1.43	1.37	1.35	1.11	0.94	0.86	1.36	27913
10-11	1.99	1.93	1.82	1.84	1.54	1.24	1.13	1.85	8563
11-12	2.97	2.81	2.74	2.73	2.26	1.88	1.73	2.72	2501
≥12	4.26	4.11	4.10	4.00	3.67	3.49	2.51	4.11	630
all	1.23	1.20	1.15	1.11	0.89	0.77	0.71	1.10	
Nb	34775	18286	17354	16878	13954	9728	6980		117955

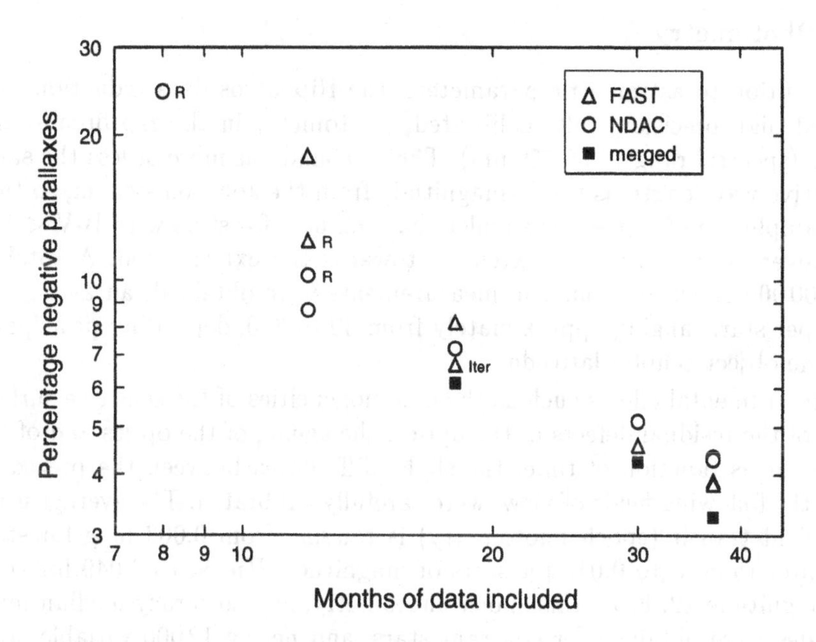

Figure 2. Fraction of negative parallaxes in the various solutions as function of the length of the data set, compared with the merged solutions. Note that the points for the merged catalogues H18, H30 and HIP (filled squares) are significantly below the corresponding points for the FAST and NDAC solutions. Courtesy L. Lindegren, Lindegren et al. 1997, Figure 16.22.

Comparison with the US Naval Observatory photographic trigonometric parallax programme with formal mean errors in the range 1 to 4 mas (Harrington et al. 1993) shows a median difference of 0.2 ±0.35 mas for 88 common stars. Comparison with the last edition of the General Catalogue of Trigonometric parallaxes (GCTP, van Altena et al. 1995), shows a median difference of 1.8 ±0.2 mas for 4292 common stars (2.6 ±0.3 for stars farther than 50 parsecs, 0.5 ±0.4 for stars nearer than 20 parsecs; 1.2 ±0.3 mas for northern stars farther than 50 parsecs). Systematic differences up to 7 mas are found at δ = - 30 deg. Apart from these systematic differences, the standard width of the differences divided by the combined standard errors is 1.04 ±0.01.

Finally the zero-point and unit-weight error of the Hipparcos parallaxes have been investigated using Hipparcos stars for which the distance can be estimated by other methods: stars in the Magellanic Clouds and in open clusters, stars with $uvby\beta$ photometry (Arenou et al. 1997). These external comparisons shows that the zero-point can be safely assumed to be smaller than 0.1 mas and that the standard errors of the parallaxes are not underestimated by more than 10 per cent.

3. Photometry

In addition to astrometric parameters, the Hipparcos data reduction produced high precision, fully calibrated, photometry in the Hp broad-band filter (spectral range: 375-750 nm). The Hp band has more or less the same effective wavelength as the V-magnitude from the Johnson system, so that the amplitude of $Hp - V$ is smaller than 0.2 mag for stars with B-V < 1.5. However, its sensitivity is extended towards the extreme red. A total of 13 000 000 epoch photometric measurements were obtained, an average of 110 per star, ranging approximately from 30 to 380, depending principally on the object ecliptic latitude.

Instrumental effects such as the inhomogeneities of the sensitive surface and of the residual defects of the optics, the ageing of the optics and of the detector as function of time, the slight differences between the preceding and the following fields of view, were carefully calibrated. The average error per field transit (epoch photometry) is ranging from 0.003 mag for stars brighter than 4, to 0.015 for stars of magnitude Hp=9, to 0.049 for stars of magnitude 12. From this epoch photometry, high accuracy median magnitudes were obtained for constant stars, and nearly 12 000 variable stars (8200 new) were investigated. Average error on median Hp magnitudes as function of Hp magnitude are given in Table 2.

TABLE 2. Average errors, in milli-magnitudes, on the median Hp magnitude as function of Hp, for stars found to be constant.

Hp	2	3	4	5	6	7	8	9	10	11	12
error	0.4	0.4	0.5	0.6	0.7	0.9	1.3	1.9	2.8	4.4	7.2

Among the variable stars, 2 712 were found to be periodic (970 new), 5 542 non-periodic or unsolved (4 145 new). Among the periodic variables, there are 273 Cepheids (2 new), 186 RR Lyrae (9 new), 108 Delta Scuti and SX Phoenicis (35 new), 917 eclipsing binaries (347 new), and many other types such as Mirae, SR, RV Tau, etc. (1 238, 576 new).

In addition to the photometry obtained from the main detector, photometry in two pass-bands was recorded by the star mapper detectors. These photon counts were processed in the Tycho reduction chain, and B_T and V_T magnitudes obtained for stars brighter than $V_T \sim 11.5$ mag, with median standard errors ranging from 0.003 mag for stars brighter than V_T = 6.0 mag to 0.014 mag for stars in the range 8-9 mag, to 0.12 mag for stars fainter than $V_T = 11.0$ mag.

4. Some statistics on the results

4.1. THE OBSERVING PROGRAMME

The Hipparcos observing programme has been defined, in successive steps, over the period 1982–1991, on the basis of scientific proposals submitted to the European Space Agency, while taking into account the observing possibilities of the satellite. Much attention has been paid to the selection of stars in order to enhance the scientific return expected from the mission (Turon et al. 1992c, Turon et al. 1995). In parallel, extensive ground-based programmes were organised to obtain, before launch, good positional and photometric data about programme stars in order to optimise the observation with the satellite (Jahreiß et al. 1992, Grenon et al. 1992).

As a result, 118 000 stars, one quasar (3C273), 48 minor planets, and three satellites of major planets were selected. The stellar part of the catalogue is essentially complete for stars brighter than $V = 7.9 + 1.1\ |sin\ b|$ for spectral types earlier than G5 (included), and for stars brighter than $V = 7.3 + 1.1\ |sin\ b|$ for spectral types later than G5. The 60 000 fainter stars have been selected from the proposals on grounds of scientific merit. As a results, most stars are in the range 7-10 in Hp, with a selection of fainter stars down to $Hp = 12.7$. Due to the pre-selection, most programme stars are closer than 500 parsecs.

4.2. SOME STATISTICS ON PARALLAXES

The Hipparcos data are providing a dramatic increase, qualitatively and quantitatively, of the basic distance information available for any application. The numbers of stars for which a relative accuracy better than 1, 5, 10 and 20 % is available from Hipparcos data and from ground-based data are given in Table 3.

TABLE 3. Number of stars per range of relative accuracy of trigonometric parallaxes

	Hipparcos	ground-based
$\sigma_\pi\ /\ \pi\ <\ 1\%$	442	a few tens
$\sigma_\pi\ /\ \pi\ <\ 5\%$	7 388	100 ?
$\sigma_\pi\ /\ \pi\ < 10\%$	22 396	1 000 ?
$\sigma_\pi\ /\ \pi\ < 20\%$	50 643	

Moreover, the range of spectral types and luminosity classes for which precise parallaxes are available is considerably enlarged, especially towards the upper part of the main sequence and towards the giant branch, where the 'clump' is already clearly marked when considering stars with relative accuracy better than 5 %. The bottom of the main sequence is populated down to absolute Hp magnitude 14. The ranges of absolute Hp magnitude and $B-V$ covered by Hipparcos observations per range of relative accuracy on trigonometric parallaxes are shown in Table 4.

TABLE 4. Ranges of absolute Hp magnitude and $B-V$ per range of relative accuracy of trigonometric parallaxes

	Hp	$B-V$
$\sigma_\pi / \pi < 1\%$	0.5-13.5	0.0-1.8
$\sigma_\pi / \pi < 5\%$	-3.0-14.0	-0.2-1.8
$\sigma_\pi / \pi < 10\%$	-5.5-14.5	-0.3-1.8

The comparison with the last edition of the Catalogue of Nearby Stars (CNS3, Gliese & Jahreiß 1991) shows the outstanding improvement expected from the Hipparcos data: in CNS3, which is the best available compilation of nearby stars, there are 851 stars with a relative accuracy of trigonometric parallaxes better than 10% and a measured $B-V$. The main sequence is well populated for absolute V magnitudes between 10 and 15, very sparsely between 5 and 10, with a few stars at brighter magnitudes (including a few subgiants). The white dwarf sequence is well represented. In contrast, a similar HR diagram drawn from Hipparcos observations includes 21 739 stars, a very dense main sequence between absolute Hp magnitudes 0 and 8, very dense subgiant and giant sequences, and a few white dwarfs (Perryman et al. 1995).

The new CCD ground-based parallax programmes (Dahn 1997), reaching precisions of the milliarcsecond level, are very much complementary of the Hipparcos data as they include stars in the faintest parts of the HR diagram.

4.3. SOME APPLICATIONS

4.3.1. *Open clusters*

Stars in galactic open clusters have been carefully selected for Hipparcos observations (Mermilliod & Turon 1989), and some of the closest clusters are nicely sampled: 210 stars observed in the Hyades, 120 in Coma Ber, 80 in the Pleiades, 130 in α Per, etc. In addition to detailed studies of these very nearby clusters, Hipparcos observations allow to determine the mean distances with relative precision better than 10% for about 10 clusters of various ages and metallicities, and for 10 other clusters with relative precision between 10 and 20%.

4.3.2. *Pulsating variable stars*

The Hipparcos observation of variable stars raised specific problems: the observing time for each type of variable stars had to be chosen according to their light curve, and ephemerides had to be predicted for large amplitude variable stars (Mennessier & Figueras 1989, Mennessier at al. 1992). Both the astrometric and the photometric results obtained from the satellite will greatly improve our knowledge of the physics of these stars and the use which can be made of some of these in distance scale determinations (Turon & van Leeuwen 1995).

Parallax results obtained by Hipparcos for five types of pulsating variables are given in Table 5. About twenty Cepheids can be used to rediscuss the zero-point of the period-luminosity relation, with periods in the range 2 to 36 days. In addition cluster membership will be greatly improved, and it will be possible to reassess the effect of coulour and metallicity on the period-luminosity relation.

TABLE 5. Pulsating variables in the Hipparcos Catalogue. Relative precisions on parallax

Type	Observed stars	Stars in clusters	(σ_π / π) < 0.05	0.05 – 0.10	0.10 – 0.20	0.20 – 0.30
δ Ceph	273	6	0	1	2	7
RR Lyr	186	0	1	0	8	3
δ Scu	87	14	12	31	22	11
SX Phe	21	0	1	2	1	4
Mirae	233	0	0	0	5	8

Accurate parallaxes are obtained for a dozen of RR Lyrae, half of these being newly detected as variables by Hipparcos. In addition, many objects of various metallicity have been measured very accurately by Hipparcos, especially stars with low metallicity such as subdwarf stars (Pont et al. 1997). These observations are leading to the revision of the calibration of the position of the main sequence in the HR diagram as function of the metallicity, and to a re-discussion of the dependence of RR Lyrae absolute magnitudes with metallicity.

5. Conclusion

Hipparcos is providing a dramatic improvement in distance determination which will lead to a reassessment of many topics of astrophysics such as stellar physics and evolution; galactic structure, kinematics, dynamics and evolution; distance scale zero points through the study of open clusters, pulsating variables, sub-dwarf stars. Accurate direct distance determinations are now available within a sphere of 100 to 150 parsecs centred on the sun. The next step forward, leading to direct distance determination throughout our Galaxy, would be obtained from a space mission such as Gaia (Lindegren & Perryman 1996), estimated to lead to parallaxes for some 50 million objects, with an accuracy better than 10 microarcsec. This has been recommended as a future ESA cornerstone mission, and detailed studies of this mission are now beginning.

References

Arenou F., Mignard F., Palasi J., 1997, in ESA SP-1200, Volume 3, Chapter 20.

Dahn C.C., 1997, this Symposium

ESA, 1997, *The Hipparcos and Tycho Catalogues*, ESA SP-1200

Gliese W., Jahreiß H., 1991, Astron. Rechen-Institut, Heidelberg

Grenon M., Mermilliod, J.C., Mermilliod, M., 1992, *A&A*, **258**, 88

Harrington R.S. et al., 1993, *AJ*, **105**, 1571

Jahreiß H., Réquième Y., Argue A.N., Dommanget J., et al., 1992, *A&A*,**258**, 82

Høg E., Bastian U., Egret D., et al., 1992, *A&A*, **258**, 177

Kovalevsky J., Falin J.L., Pieplu J.L., Bernacca P.L., Donati F., Frœschlé M., Galligani I., Mignard F., Morando B., Perryman M.A.C. et al., 1992, *A&A*, **258**, 7

Kovalevsky J., Lindegren L. and Perryman M.A.C., 1997, in ESA SP-1200, Volume 3, Chapter 18.

Lestrade J.F., Jones D.L., Preston R.A et al., 1995, *A&A*, **304**, 182

Lindegren L., Frœschlé M., Mignard F., 1997, in ESA SP-1200, Volume 3, Chapter 16.

Lindegren L., Perryman M.A.C., 1996, *A&A Suppl.*, **116**, 579

Lindegren L., Kovalevsky J., 1995, *A&A*, **304**, 189

Lindegren L., Høg E., van Leeuwen F., Murray C.A., Evans D.W., Penston M.J., Perryman M.A.C., Petersen C., Ramamani N., et al., 1992, *A&A*, **258**, 18

Mennessier M.O., Figueras F., 1989, ESA SP-1111, Vol. II, 177

Mennessier M.O., Barthès D., Boughaleb H, Figueras F., Mattei J., 1992, *A&A*, **258**, 99

Mermilliod J.C., Turon C., 1989, ESA SP-1111, Vol. II, 177

Perryman M.A.C., Lindegren L., Kovalevsky J., Turon C., Høg E., Grenon M., Schrijver H., et al., 1995, *A&A*, **304**, 69

Perryman M.A.C., Høg E., Kovalevsky J., Lindegren L., Turon C., Bernacca P.L., Crézé M., Donati F., Grenon M., Grewing M., van Leeuwen F., et al., 1992, *A&A*, **258**, 1

Pont F., Mayor M., Turon C., 1997, this Symposium

Turon C., van Leeuwen F., 1995, in *Astrophysical Applications of Stellar Pulsation*, R.S. Stobie and P.A. Whitelock eds, p 241

Turon C., Réquième Y., Grenon M., *et al.*, 1995, *A&A*, **304**, 82

Turon C., *et al.*, 1992a, *The Hipparcos Input Catalogue*, ESA SP-1136.

Turon C. *et al.*, 1992b, Bull. Inform. CDS, **41**, 9

Turon C., Gómez, A. Crifo F., Crézé M., Perryman M.A.C., Morin D., Arenou F., Nicolet B., Chareton M., Egret D., 1992c, *A&A* **258**, 74

van Altena W.F. et al., 1995, The General Catalogue of Trigonometric parallaxes, 4th edition, Yale University Observatory

REVIEW OF CCD PARALLAX MEASUREMENTS

C.C. DAHN
U.S. Naval Observatory, Flagstaff, AZ USA

Abstract. Several groups, using 1-m to 2-m telescopes on 'visitor' arrangements, have employed CCDs to measure stellar trigonometric parallaxes with precisions in the range ± 2–5 mas. More intensive observing efforts now routinely achieve sub-mas results with the best obtained to date in the ± 0.3–0.5 mas range. Selective dimming of bright target stars using neutral density spots now permit CCD parallax measures of stars as bright as R ∼ 4.

1. Introduction

Until recently, the determination of trigonometric stellar parallaxes has remained the province of a handful of institutions with telescopes (primarily long-focal-length refractors) and plate measuring equipment essentially dedicated to this work. The demonstration that a general purpose reflecting telescope equipped with a CCD detector could produce high quality parallax determinations (Monet & Dahn 1983), along with the widespread availability of direct imaging CCD cameras at a majority of observatories, has dramatically altered this arrangement over the past decade. Today there are at least seven groups – varying in size from a single astronomer to collaborations of 10 to 15 investigators – engaged in CCD parallax measures. The present report summarizes the current activity in CCD parallax work throughout the astronomical community.

2. Synopses of Six Non-USNO CCD Parallax Efforts

All groups known to be making CCD parallax measures were queried for information regarding their efforts – including the focus/scope of the program (e.g., numbers of and types of stars targeted), the telescope and instrumen-

T.R. Bedding et al. (eds.),
Fundamental Stellar Properties: The Interaction between Observation and Theory, 19–24.

tation (CCD camera) employed, the access to observing time, the formal internal precision of the relative parallaxes measured, and what, in their opinions, prevented them from obtaining even better results. All kindly responded and were more than generous in sharing information about their work. Very abbreviated summaries of these responses follow.

2.1. UNIV. VIRGINIA & SIDING SPRING OBS. CCD PARALLAX PROG.

Collaborators P. Ianna (PI), R. Patterson, and M. Begam are employing the Siding Spring 1-m (scale: 25.8 arcsec mm^{-1}) along with an EEV88530 2K x 1K CCD (22.5 μ pixels, giving 0.58 arcsec pix^{-1} image sampling). Normally only a 700 x 700 portion of the chip is employed providing a $\sim$6.8 x 6.8 arcmin field of view. The Cousins R or I bandpasses are employed. Telescope access is by quarterly application and an average of 1 week per month of gray/bright time has been received. Targets include 130 stars from the Third Catalogue of Nearby Stars (CNS3) plus 100 secondary, fainter stars from the LHS Catalogue. Based on experience from 1985 to the present, Pi(rel) precisions of $\pm$ 1–2 mas are being achieved for brighter stars (V $\sim$ 11.4–15.0; R $\sim$ 10.5–13) with > 50 frames spanning > 1.5 years. For fainter stars (V $\sim$ 15–21; I $\sim$ 14–17), precisions of $\pm$ 2–5 mas are obtained with > 30 frames spanning > 2.0 years. (See also Ianna 1993.)

2.2. UNIV. OF CHILE CCD PARALLAX PROGRAM AT CTIO

Collaborators C. Anguita (PI), P. Loyola, and M.T. Ruiz employ the CTIO 1.5-m (scale: 10.0 arcsec mm^{-1}) and CCD cameras with either Tek1K or Tek2K chips (both thinned, back-side illuminated; 24.0 μ pixels, giving 0.24 arcsec pix^{-1} image sampling). Telescope access is by semester application and 3 or 4 nights every 3 months (generally close to New Moon) have been received. A "Broad R" bandpass is used and the field of view employed is $\sim$4.1 x 4.1 arcmin (dictated by the Tek1K field). Targets include 40–60 cool WDs (from Ruiz's proper motion survey) for improving the WD LF, 3 PN central stars and 18 hot or late-type subdwarfs. All targets are faint (15 < R < 19). Based on experience from 1985 to the present, Pi(rel) precisions of $\pm$ 1–2 mas are being achieved for well-configured, 8+ star reference star frames; otherwise, $\pm$ 2–3.5 mas. A limitation mentioned is that telescope scheduling constraints necessitate their mixing data from two different CCD cameras. (See also Ruiz et al. 1990.)

2.3. TORINO OBSERVATORY CCD PARALLAX PROGRAM

This collaboration includes M. Lattanzi and R. Smart (Co-PIs), along with B. Bucciarelli, R. Casalegno, G. Chiumiento, R. Drimmel, H. Eichhorn, F.

Massone, F. Morale, A. Spagna; plus Program Support from A. Ferrari, L. Lanteri, R. Morbidelli, F. Porcu, and F. Racioppi. The Torino 1.05-m (scale: 20.7 arcsec mm^{-1}) is employed with an EEV 1246 x 1155 CCD (thick, front-side illuminated; 22.5 μ pixels giving 0.47 arcsec pix^{-1} image sampling). A Cousins I bandpass is used and the field of view is 10.0 x 9.0 arcmin. Allotment of observing time consists of all evening and morning hours of $\sim$ 50% of the nights – plus "some midnight hours." Of the $\sim$ 140 targets, 40 are bright ($I < 11$). The program stars include: a 10% overlap with the USNO Parallax Program; 30% of specific interest to the Torino collaborators (Miras, PN central stars, WDs); 20% from CNS3; 25% T Tauri and/or flare stars; 15% misc. Since observations only started in 1995, the achievable Pi(rel) precision is still unknown; but $\pm$ 2 mas from 30–40 frames is anticipated. Experiments with a neutral density spotted I band filter to provide $\sim$ 5 mag of selective target star dimming look promising.

2.4. CURRENT CCD PARALLAX EFFORTS AT LA SILLA

Two groups are currently carrying out CCD parallax determinations using the Danish 1.54-m telescope at La Silla (scale: 18.8 arcsec mm^{-1}). Access to this telescope is by semester application and both groups are employing a Gunn i bandpass.

The first collaboration – involving M. Hawkins (PI), C. Ducourant, H. Jones, and M. Rapaport – is employing a Tek1K CCD (24.0 μ pixels, giving 0.45 arcsec pix^{-1} and a $\sim$ 7.7 x 7.7 arcmin field of view). Targets are $\sim$ 30+ very red, low-mass stars and brown dwarf candidates identified from the deep UK Schmidt plate stack in ESO/SERC field 287 at 21^h28^m,–45 (1950). Pi(rel) precisions of $\sim \pm 4$ mas are anticipated but there is concern about inadequate observing time. At present the project is receiving 3 hrs (evening or morning) on 10 nights every 6 months. Preliminary results have been submitted for publication (Hawkins et al. 1996).

The second collaboration – involving T. Forveille (PI), F. Delfosse, F. Crifo, and F. Epchtein – is employing the DFOSC focal reducing camera which provides $\sim$ 0.4 arcsec pix^{-1} image sampling. The targets are $\sim$ 20 very red stars identified in the DENIS survey. Allotted observing time amounts to 3 runs of 2 nights each per semester near Quarter Moon. Pi(rel) precisions of $\sim$ 3–5 mas are anticipated but there is concern about possible flexures in the reimaging system.

2.5. TINNEY CCD PARALLAX MEASURES AT PALOMAR & LA SILLA

An important example of a single investigator carrying out CCD parallax determinations targeted at a very specific scientific problem – that of elucidating the nature of very-low-mass star and brown dwarf candidates – is the work of C. Tinney. Although no longer active, these efforts illustrate the important contributions that "limited" projects carried out at 'visitor' facilities can accomplish. In both instances, telescope access was by semester applications and the Gunn i bandpass was employed.

The Palomar 1.5-m (scale: 15.5 arcsec mm−1) was used with two different Tek1K CCDs (the first thick, front-side illuminated; the second thinned, back-side; both 24 μ pixels, giving 0.37 arcsec pix^{-1} image sampling and a $\sim$6.3 x 6.3 arcmin field of view – the latter noted as only "sufficient"). The project received 5 runs of gray-dark observing time a year between September 1990 and December 1993. Parallax precisions in the range $\pm$4–5 mas were achieved on targets typically I $\sim$ 14.5–17.5 from 3–4 years of data. (See Tinney et al. 1995.)

The La Silla work employed the ESO 2.2-m (scale: 17.5 arcsec mm^{-1}) and the EFOSC2 reimaging system which uses a Thomson 1K x 1K CCD (19.0 μ pixels, giving 0.33 arcsec pix^{-1} image sampling and a 5.7 x 5.7 arcmin field of view – again, only "sufficient"). This project received 4 observing runs a year, typically at gray-dark lunar phases between May 1993 and February 1995. Typical Pi(rel) precisions were in the range $\pm$3–7+ mas (median: $\pm$4.5 mas) from 2 years of data. Concerns were expressed about flexures in the reimaging system. (See Tinney 1996.)

3. Status of the USNO, Flagstaff CCD Parallax Program

The USNO, Flagstaff efforts are an outgrowth of work carried out with the KPNO 4-m in 1980–1983 (Monet & Dahn 1983). This program has evolved considerably over the ensuing years and presently can be summarized best as a set of sub-programs, each defined by a specific CCD camera employed for the observations. Current in-house collaborators include: C. Dahn (PI), H. Guetter, F. Harris, H. Harris, A. Henden, C. Luginbuhl, A. Monet, D. Monet, J. Pier, R. Stone, F. Vrba, and R. Walker. The telescope used is the USNO 1.55-m Astrometric Reflector (scale: 13.55 arcsec mm^{-1}) and access for CCD astrometry (primarily parallax work) has remained at $\sim$20 nights per lunation, centered around New Moon.

3.1. THE USNO TI800 PARALLAX PROGRAM

The TI800 program, which operated from March 1983 through October 1995, has been described by Monet et al. (1992) where the results for 72 stars were presented. A total of $\sim$175 parallax determinations have been completed with this camera and publication of the remaining results for the additional 100+ stars is in preparation. Due to the small field of view of the TI 800 x 800 CCD ($\sim$2.7 x 2.7 arcmin at the USNO 1.55-m) many marginal quality reference star frames were used. Nevertheless, the formal precisions of the Pi(rel) determinations ranged from ± 1.0–2+ mas for average-to-poorer reference frames, to ± 0.7–1.0 mas for better reference frames, to ± 0.5–0.7 mas for the best cases.

3.2. THE USNO TEK2K PARALLAX PROGRAM

Initiated in February 1992, a program employing a Tek2K CCD (thinned, back-side illuminated; 24.0 μ pixels giving 0.325 arcsec pix^{-1} and an 11 x 11 arcmin field of view) has now become the mainstay of USNO parallax efforts. The large full-well capacity of this chip (> 300k e$^-$) provides a working astrometric dynamic range of 4.5–5 magnitudes and assures good-to-excellent reference star frames for > 95% of the targets attempted to date. Note, however, that without selective magnitude compensation – that is, the ability to selectively dim a target star while simultaneously exposing fainter reference stars – the Tek2k program, like all other CCD parallax efforts discussed thus far, is restricted to relatively faint stars; in this case to those primarily with R > 12. The ~ 200 stars under observation (or already completed) have been selected primarily from Luyten's Palomar Proper Motion Survey (a mixture of degenerates, late-type dwarfs and subdwarfs) but includes special interest targets such as PN central stars and very low mass stars from the work of Tinney and others. Based on solutions for fields with epoch ranges spanning 2–4+ years, Pi(rel) precisions of < ± 1.0 mas are routinely being obtained. The median Pi(rel) precision achieved for 67 completed solutions derived from 3.1–4.4 years of data (median: 4.2 years) is ± 0.55 mas. For the best cases (30 completed solutions), the values range between ± 0.28 mas and ± 0.54 mas.

3.3. THE USNO ND9 PARALLAX PROGRAM

In an attempt to address the inability to target stars brighter than R ≈ 12, a second Tek2K CCD camera has been outfitted with a neutral density attenuation spot to selectively dim a target star near the center of the field. The spot is a highly uniform deposit of Inconel on an optically flat quartz substrate and is mounted ~ 1 mm in front of the CCD. This spot is circular, roughly 3 mm in diameter (≈ 40 arcsec) and provides about 9.0 magnitudes of attenuation. (Hence, the name "ND9.") The particular Tek2K CCD employed in this camera is only an 'Engineering Grade' device and, consequently, has several blocked columns, a number of 'charge traps' and a higher read-noise than the chip employed in the previously described system. The substrate mounting arrangement also diminishes the available field of view to approximately 9.0 x 9.0 arcmin. Placed in operation in December 1995, this camera is being used on 68 targets, mostly in the 5.2 < V < 9.5 brightness range. Included are: 27 sdF-G stars, 9 Solar

Analogs, 5 Periodic Variables (SU Cas, VZ Cnc, RR Lyr, η Aql, δ Cep), 4 Field Horizontal Branch stars, 4 Hyades Cluster members, and 2 Mira Variables (o Cet, R Cas). Sixty-five of the 68 present targets are also in Hipparcos Input Catalogue. Several potential concerns – such as excessive scattered light and flattening of the chip response in the area behind the spot – have proved not to be serious. Preliminary reductions have been carried out on the 40 fields which have sufficient epoch coverage (354–360 days) and enough variation in parallax factor to support separation of proper motion from parallax. The results are very encouraging in that: (1) the internal Pi(rel) precisions are already down to $\pm$ 0.9–2.8 mas (median: $\pm$ 1.7 mas); and (2) where good quality independent parallax determinations are available from the Yale Parallax Catalogue, the agreement is excellent.

4. Summary and Conclusions

Relative trigonometric parallaxes with formal internal errors in the range $\pm$ 2–5 mas are routinely being measured under 'visitor' arrangements by investigators employing 1–2-m aperture telescopes and a variety of CCD cameras. With more intensive observing programs, sub-mas results are routinely acheived and the best results to date are in the $\pm$ 0.3–0.5 mas range. The absence of a facility equipped (and willing) to provide such intensive observations for Southern Hemisphere targets is still a significant deficiency for the astronomical community. On the positive side, successful magnitude compensation has now been demonstrated, permitting CCD parallax measures for stars as bright as fifth magnitude.

References

Hawkins, M.R.S., Ducourant, C., Jones, H.R.A., and Rapaport, M. 1996, MNRAS, submitted.

Ianna, P.A. 1993, in *Developments in Astrometry and Their Impact on Astrophysics and Geodynamics*, IAU Symp. No. 156, ed. I.I. Mueller and B. Kolaczek, 75.

Monet, D.G. and Dahn, C.C. 1983, AJ, 88, 1489.

Monet, D.G, Dahn, C.C., Vrba, F.J., Harris, H.C., Pier, J.R., Luginbuhl, C.B., and Ables, H.D. 1992, AJ, 103, 638.

Ruiz, M.T., Anguita, C., and Maza, J. 1990, AJ, 100, 1270.

Tinney, C.G. 1996, MNRAS, 281, 644.

Tinney, C.G., Reid, I.N., Gizis, J., and Mould, J.R. 1995, AJ, 110, 3014.

STELLAR DISTANCES BY THE BAADE-WESSELINK METHOD

G. P. DI BENEDETTO
Istituto di Fisica Cosmica del CNR, Milano, Italy

Abstract. Galactic and extragalactic distances to Cepheid variable stars according to a modern realization of the geometric Baade-Wesselink method are presented. Reliable results as accurate as few percent are currently achievable, allowing the cosmic distance scale calibration to be significantly improved with respect to previous calibrations.

1. Introduction

Modern calibrations of stellar surface brightness enable reliable and accurate apparent angular diameters to be inferred for stars from A to M spectral type by applying high-precision photometry in the magnitude-color combination $(V, V-K)$ suitable to minimize biasing effects such as reddening and metallicity. These distance-independent angular sizes can be matched with the distance-dependent linear radii achievable from radial velocity measurements to provide stellar distances according to the geometric BW method. The color domain sampled by the surface brightness scale includes several important distance markers of major cosmological concern. In this contribution I shall outline the implementation of the BW method for Cepheid variables, being these stars the most relevant standard candles for the cosmic distance scale calibration.

2. Calibrations by interferometric and spectroscopic techniques

BW results are known to depend *crucially* on the quality of the empirical surface brightness-color SC correlations available for inferring stellar sizes and then on the choice of the magnitude-color combination adopted *ab initio* to calibrate most of potential biasing effects. This is notably important

T.R. Bedding et al. (eds.),
Fundamental Stellar Properties: The Interaction between Observation and Theory, 25–28.

for Cepheid variable stars, since variations in gravity and/or microturbulence along their pulsation cycle can destroy the well defined single-valued SC correlation required by any BW realization (Gautschy 1987). The modern Michelson stellar interferometry has recently provided angular diameters of non variable stars as accurate as 5% or less (Davis and Tango 1986; Di Benedetto and Rabbia 1987; Mozurkewich et al. 1991) enabling now a deeper investigation and fundamental calibration of the SC correlations. By using these data, a much tighter correlation can be found for the visual surface brightness S_V as a function of the infrared color (V-K), rather than of other current optical colors, and reliable and accurate angular sizes have been predicted through the almost *ideal* magnitude-color combination $(V, V-K)$ (Di Benedetto 1993). The same calibration has been also applied for representing Cepheid angular sizes (Di Benedetto 1994). Fig. 1 shows the most recent zero-point calibration of the SC correlation over the Cepheid color domain. A slope of (1.30 ± 0.01) has been derived according to radial displacement data of the cluster Cepheid U Sgr (Welch 1994) and found to be in fairly good agreement with that of non variable supergiants, giving evidence for a likely constant slope representing the correlation of *all Cepheids.* The zero-point is obtained by averaging the most accurate data of practically unreddened dwarfs and giants. But there appears clear evidence for several other stars, notably 6 supergiants, to be well represented by the same absolute calibration. Then, the SC correlation representing photospheric size variations of galactic Cepheids is likely to be:

$$S_V = V_0 + 5\,\log(\Phi/\mathrm{mas}) = 2.762 + 1.30\,(V-K)_0, \tag{1}$$

where the unreddened photometry is in the Johnson magnitude system. The Eq. (1) also allows average angular diameters to be derived according to the intensity mean magnitudes of only two phase points out of the whole cycle with results largely insensitive to reddening and metallicity (Di Benedetto 1994, 1995).

The correlation currently applied for inferring linear radii of Cepheids takes the form of a *suitably* calibrated period-radius PR relation, where the potentially very accurate period P is the current observable. The most recent realization of the spectroscopic BW technique has provided calibrating linear radii of galactic Cepheids by using radial displacement data along with the infrared photometry less sensitive to limiting factors critically affecting optical BW radii (Laney and Stobie 1995). The corresponding PR relation will be adopted below as the most reliable ridge-line correlation. It allows the mean linear radius (in solar units) of a galactic Cepheid to be derived according to:

$$5\,\log\langle R\rangle = 5.355 + 3.75\,\log(P/\mathrm{days}) \tag{2}$$

Because of its geometrical nature, the PR relation is expected to be largely independent of metallicity effects.

3. BW distances to Cepheid variable stars

The angular and linear sizes achievable by Eq.s (1) and (2), respectively, can be combined to yield BW distance moduli to any Cepheid of known V, K photometry and period P according to the following observational relationship:

$$\mu_0 = \langle V \rangle + 3.75 \log P - 1.30 \left(\langle V \rangle - \langle K \rangle \right) + 2.434 + 0.18\, A_V \qquad (3)$$

where intensity mean magnitudes are reported now along with an overall absorption term due to interstellar reddening. The small formal errors affecting the above calibrations would lead to individual BW distance moduli as accurate as 0.04 mag or 2% in distance. However, both the reliability and uncertainty in using Eq. (3) should be critically checked by comparing the results with the calibrating distances to cluster Cepheids well determined by the ZAMS fitting approach. Fig. 2 shows such comparison for 25 ZAMS calibrators with currently published P, V, K and absorption data. I find that 20 of these stars show an average residual (BW − ZAMS) = $-(0.01 \pm 0.03)$ mag with a scatter SD = ± 0.12 mag, whereas the remaining 5 stars deviate by up to 5 SD. By removing these 5 discordant stars inducing significant systematic errors, the overall uncertainty on the absolute galactic distance scale can be set to the value of $\pm$ 0.04 mag, i.e. a factor two smaller than that currently quoted for the Pleiades distance modulus which limits the ZAMS calibration.

The Eq. (3) might also be applied in a straightforward manner to obtain distances to any set of extragalactic Cepheids with P, V, K and absorption data. However, for more accurate individual distances, a composite *ad hoc* PR relation has to be recalibrated by using the available sets of extragalactic Cepheids in the Magellanic Clouds (LMC and SMC) (Di Benedetto 1994). According to the most recent composite fit, the overall PR slope in Eq.s (2) and (3) decreases to the value 3.65; their zero-points increase to 5.472 and 2.551, respectively, leading to the BW distances of $d(LMC) = (52.0 \pm 0.6)$ kpc and $d(SMC) = (63.2 \pm 0.7)$ kpc with uncertainties *not* including the above contribution of 2% due to the absolute distance scale calibration. Strong support to reliability of the actual BW results comes from the distance to LMC itself. Its value compares remarkably well with that measured by the geometric expansion parallax of the SN1987A in LMC via HST observations, given by $d(SN1987A) = (51.1 \pm 1.5)$ kpc (Panagia et al. 1996).

The actual BW approach can be readily extended for including either the Johnson-Cousins color relevant in the HST observations of Cepheids,

References

Davis, J., Tango, W. J.: 1986, *Nature* 323, 234
Di Benedetto, G. P., Rabbia Y.: 1987, *Astron. Astrophys.* 188, 114
Di Benedetto, G. P.: 1993, *Astron. Astrophys.* 270, 315
Di Benedetto, G. P.: 1994, *Astron. Astrophys.* 285, 819
Di Benedetto, G. P. : 1995, *Ap. J.* 452, 195
Gautschy, A.: 1987, *Vistas in Astron.* 30, 197
Laney, C. D., Stobie, R. S.: 1995, *Mon. Not. R. Astr. Soc.* 274, 337
Mozurkewich et al.: 1991, *Astron. J.* 101, 2207
Panagia et al.: 1996, in *The Extragalactic Distance Scale*, STScI, May Symp., Baltimore (in press)
Welch D. L.: 1994, *Astron. J.* 108, 1421

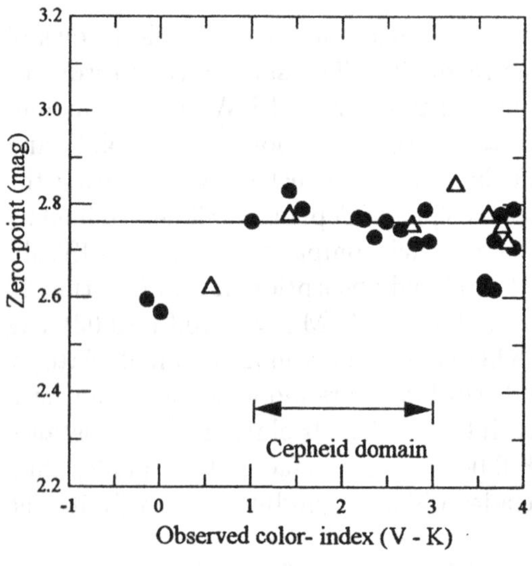

Fig. 1. Zero-point calibration of visual surface brightness by angular diameter measurements of non variable stars. Slope of linear SC correlation: 1.30. Solid circles: dwarfs and giants. Open triangles: supergiants

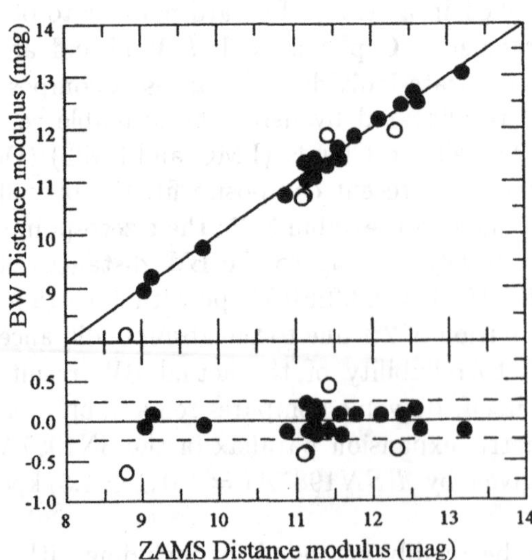

Fig. 2. BW vs. ZAMS distances of Cepheids in clusters. Solid line: locus of equal distances. Dashed lines are drawn at ± 0.24 mag (2-sigma scatter). Open circles are stars falling more than 3 sigmas away from solid line.

Discussion of this paper appears at the end of these Proceedings.

3. STELLAR ANGULAR DIAMETERS AND RADII

STELLAR ANGULAR DIAMETER MEASUREMENTS BY INTERFEROMETRY

JOHN DAVIS
Chatterton Astronomy Department, School of Physics
University of Sydney, NSW 2006, Australia

Abstract. Stellar angular diameter measurements have been made with a range of interferometric techniques including speckle, aperture masking and long baseline optical/infrared interferometry. The current status of these measurements are summarised in terms of the range of spectral types and luminosity classes measured, the accuracies achieved, the wavelengths used for observations, and the reliability of the results. A number of major long-baseline interferometers are coming on-line, or are under development, and their potential is assessed in terms of wavelength cover, accuracy, angular resolution, and the range of spectral type and luminosity class cover.

1. Introduction

The angular diameters of stars are important for the determination of fundamental stellar properties including emergent fluxes, effective temperatures, radii and absolute luminosities, and for providing constraints on theoretical stellar models. In this contribution the emphasis is on their determination by means of interferometry.

The first angular diameter of a star (α Ori) was measured interferometrically by Michelson and Pease in 1921. Although they succeeded in measuring 6 late type giants and supergiants, the accuracy was poor, and the technique was limited by the technology available and the deleterious effects of atmospheric turbulence. The field did not progress significantly until Hanbury Brown and Twiss developed the technique of intensity interferometry in the 1950s. This overcame the problems posed by atmospheric turbulence and led to the Narrabri Stellar Intensity Interferometer (NSII). Since this meeting marks Hanbury Brown's 80th birthday, a brief list of the achievements made with the NSII under his leadership is included:

T.R. Bedding et al. (eds.),
Fundamental Stellar Properties: The Interaction between Observation and Theory, 31–38.

- Angular diameters for 32 early type stars, including main-sequence stars, which led to an effective temperature scale for stars hotter than the Sun. These are still the only measurements of early-type stars.
- The distance, the mass, radius and luminosity of the primary, and the mass of the secondary for a double-lined spectroscopic binary (α Vir).
- Performance of a number of exploratory experiments including:
  - Measurement of the angular size of the emission envelope surrounding a Wolf-Rayet star (γ^2 Vel).
  - An attempt to measure limb-darkening (α CMa).
  - Measurement of an extended corona (β Ori).
  - An attempt to measure the distortion of a rapidly rotating star (α Aql).

Although the intensity interferometer has now been superseded, because of its inherent low sensitivity, by various forms of amplitude interferometer, the NSII demonstrated that angular diameters of stars can be determined accurately through the earth's atmosphere.

As the achievements of the NSII illustrate, interferometry can do a great deal more than measure the angular diameters of single stars, but this will be the subject of other contributions to this meeting.

2. Techniques and Instruments

Several interferometric techniques have been used for the determination of stellar angular diameters and, very briefly, they are:

- Speckle - limited by the diameter of the largest telescope apertures to resolve only a small number of stars for angular diameter measurements.
- Aperture Masking (Non-Redundant Masking or NRM) - has some advantages over speckle but is also limited in resolution by telescope aperture diameters. Both speckle and NRM complement long-baseline interferometers by providing observations at the short baselines generally inaccessible to multi-aperture instruments.
- Intensity - overcomes the effects of atmospheric turbulence but limited in sensitivity. The only instrument (now closed) was the NSII.
- Amplitude (Modern Michelson) - the very high angular resolution, long-baseline optical/infrared instruments of today are of this type.

Table 1 lists the amplitude interferometers which have been developed, are under construction, or are being planned. These are all ground-based instruments. Space based interferometer projects are under development but generally with objectives other than stellar angular diameter determinations.

TABLE 1. Long-baseline optical/infrared amplitude interferometers

Instrument	Location	Aperture Diameter (m)	Max. B'line (m)	λ Range (μm)	Status
SUSI Prototype	Australia	2 x 0.10	~13	0.4–0.5	Closed
Mark III	USA	2 x 0.05	~32	0.45–0.8	Closed
I2T	France	2 x 0.26	144	Visible	Working
GI2T	France	2 x 1.5	65	Visible	Working
SUSI	Australia	2 x 0.14	640	0.4–0.9	Working to 80 m
IOTA	USA	3 x 0.45	38	Vis/IR	2 apert. working
COAST	UK	4 x 0.4	100	Red/IR	3 apert. working
Palomar Interf.	USA	2 x 0.4	100	2.2	Commissioning
NPOI (Astrom.)	USA	4 x 0.125	38	0.45–0.9	Commissioning
NPOI (Imaging)	USA	6 x 0.35	437	0.45–0.9	Construction
CHARA Array	USA	7 x 1	354	0.55-0.9	Construction
				2.1–2.5	
VLTI	Chile	4 x 8 plus	200	0.45–20	Construction
		3 x 1.8			Finalising design
Keck	USA	2 x 10 plus	165	2.2–10	
		SideKecks			Planned

3. The Measurement of Angular Diameters by Interferometry

In principle, the response of an interferometer to a star, as a function of baseline, is the Fourier transform of the brightness distribution across the equivalent strip source. It is not appropriate to go into details but there are certain aspects important in the context of angular diameter determinations. If both the phase and visibility of the interference fringes could be measured, a unique solution for the brightness distribution could be determined which would include asymmetries due to spots etc. In practice, for measures made with a single baseline at a time, which has generally been the case, only the fringe visibility is meaningful and symmetry in the brightness distribution has to be assumed. In most cases this is a reasonable assumption.

Assuming symmetry in the brightness distribution, the response of an interferometer depends on the angular size and the limb-darkening of the stellar disk. In principle, from measurements of fringe visibility as a function of baseline, one should be able to determine the limb-darkened angular diameter of the star. Unfortunately this turns out to be extremely difficult. The effect of limb darkening on the shape of the response of an interferometer is very small—too small to measure with the accuracy required to distinguish between different limb-darkening laws (Hanbury Brown *et al.*, 1974b). The main difference is in the scale of the responses which varies from 1.0 for a uniformly bright disk to $\sim$1.13 for a fully darkened disk. Unfortunately, this cannot be determined but it has been shown, both from accurate measurements of visibility around the first zero in the response and from the variation of angular diameter with wavelength, that observations are consistent with the predictions of model atmospheres.

The general practice has been to fit the response for a uniform disk to determine an angular diameter and then to convert it, using data from atmospheric models, to obtain an estimate of the limb-darkened angular diameter. The magnitude of the correction is up to $\sim$10% with an uncertainty of the order of 10%. The result is a systematic uncertainty in the limb-darkened angular diameter of the order of 1% in addition to the measurement uncertainty of the uniform disk angular diameter. This must be borne in mind in any discussion of the final accuracies of stellar angular diameters.

In summary, ideally the limb-darkened diameter would be measured but the vast majority of angular diameter measurements in the literature have been determined as uniform disk diameters and subsequently converted to limb-darkened diameters.

4. Stellar Angular Diameters

4.1. THE DATA

A database of 490 interferometric measurements of the angular diameters of 156 stars has been assembled. These data have been obtained using speckle, NRM, intensity and amplitude interferometry and much of the data have been taken from the literature (space limitations do not permit a listing of the sources). I am grateful to David Mozurkewich for allowing me to include unpublished results for 78 stars obtained with the Mark III interferometer. More than 98% of the angular diameters are uniform disk determinations.

Figure 1(a) shows, in a fairly coarse grid, the distribution of 145 measured stars as a function of spectral type and luminosity class. All stars, irrespective of the accuracy of the measured angular diameter have been included. The entries for the early-type stars (O and B) are all from the NSII as are most of the A-type stars. There are no measurements for main-sequence stars later than A7. In Fig. 1(b) the entries have been restricted to measurements with an uncertainty $\leq$ 5%. The range of spectral types and luminosity classes covered is poorer with only late-type giants and supergiants well represented. This trend is emphasised if entries are restricted to measurements with smaller uncertainties.

The major bodies of data come from the NSII and Mark III instruments. As a first step in examining the accuracies of angular diameter measure-

Range in Spectral Type	Luminosity Class				
	I	II	III	IV	V
O	3	0	0	0	1
B0-B4	2	2	3	2	2
B5-B8	2	0	1	1	1
A0-A3	1	0	0	2	5
A5-A7	0	0	1	0	1
F0-F5	3	0	0	1	0
F8	2	0	0	0	0
G0-G5	2	1	2	3	0
G7-G9.5	2	1	12	0	0
K0-K3.5	3	10	17	0	0
K4-K7	1	1	7	0	0
M0-M4	5	6	18	0	0
M5-M8	1	2	15	0	0
TOTAL:	27	23	76	9	10

(a)

Range in Spectral Type	Luminosity Class				
	I	II	III	IV	V
O	0	0	0	0	0
B0-B4	0	0	1	1	1
B5-B8	1	0	0	0	1
A0-A3	1	0	0	1	3
A5-A7	0	0	0	0	1
F0-F5	2	0	0	1	0
F8	2	0	0	0	0
G0-G5	2	1	0	3	0
G7-G9.5	2	1	11	0	0
K0-K3.5	3	7	11	0	0
K4-K7	1	0	6	0	0
M0-M4	5	6	15	0	0
M5-M8	1	0	9	0	0
TOTAL:	20	15	53	6	6

(b)

Figure 1. The distribution of interferometric angular diameter measurements as a function of spectral type and luminsity class: (a) all measurements regardless of accuracy; (b) measurements with an uncertainty in angular diameter $\leq \pm 5\%$.

ments, the uncertainties for these two groups of data are plotted against visual magnitude in Figs. 2 and 3.

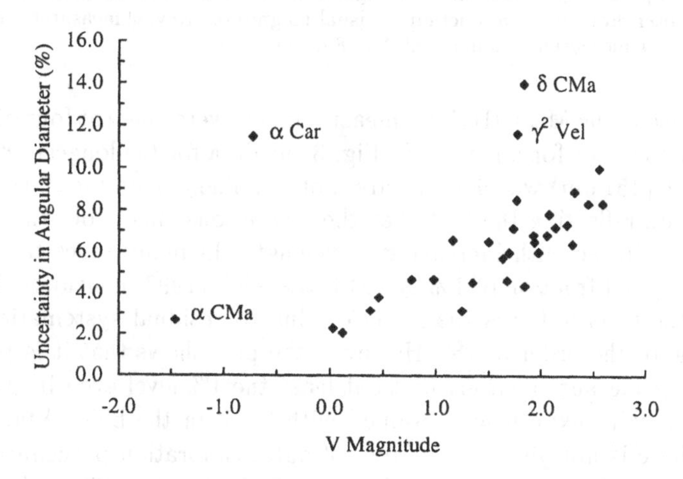

Figure 2. The percentage uncertainty in angular diameter measurements for 32 stars measured at $\lambda = 443$ nm with the Narrabri Stellar Intensity Interferometer as a function of visual magnitude.

For the NSII data the uncertainties range from $\sim 2\%$ at $V = 0$ to $\sim 10\%$ at $V = 2.5$. There are four stars which lie off the main trend. The plotted uncertainty for α CMa is the published value (Hanbury Brown *et al.*, 1974a) which was arbitrarily double the formal uncertainty because it was felt at the time that unidentified systematic errors of the order of $\sim 1\%$ might be present. Subsequent measurements with the SUSI prototype and with SUSI agree with the NSII value within the formal error so the uncertainty plotted in Fig: 2 should be halved. α Car was measured low down on the visibility curve resulting in low accuracy, γ^2 Vel is a binary and δ CMa was the faintest star at λ443 nm measured with the NSII.

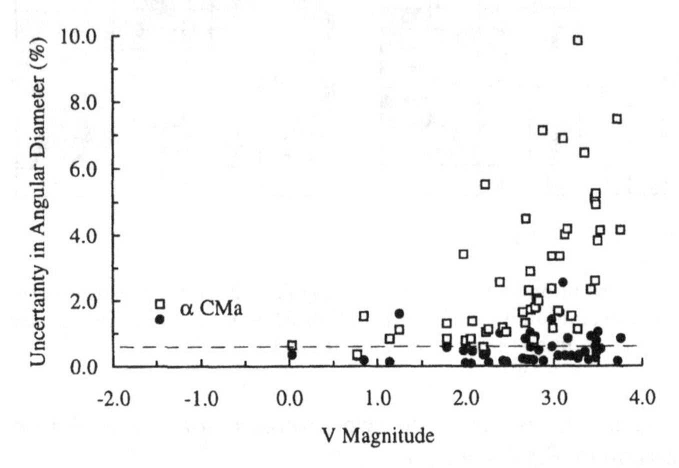

Figure 3. The percentage uncertainty in angular diameter measurements measured with the Mark III interferometer as a function of visual magnitude. Key: □ measurements made at $\lambda = 451$ nm; • measurements made at $\lambda = 800$ nm.

In the case of the Mark III data, measurements were made at four different wavelengths but not for all stars. In Fig. 3 the data for the longest (800 nm) and shortest (451 nm) wavelengths are plotted. The results for 800 nm have smaller errors reflecting the fact that the deleterious effects of atmospheric turbulence decrease with increasing wavelength. In many cases the formal errors are a small fraction of 1% but Mozurkewich (1997) considers that the real uncertainty is not less than $\sim$0.5% due to residual systematic errors and may be of the order of 1%. However, the plot shows that it is possible to reduce the measurement errors to at least the 1% level even in the blue. This accords with experience obtained with SUSI in the blue (λ442 nm)—although there is not yet a large body of data, calibration procedures have been developed that give angular diameters to better than 2% and in some cases $\leq$ 1%. Angular diameters with uncertainties $\leq$ 1% have also been obtained with the I2T and IOTA instruments at $\lambda 2.2\,\mu$m.

4.2. THE RELIABILITY OF ANGULAR DIAMETER MEASUREMENTS

Formal uncertainties and estimates of systematic errors are indicators of reliability but the real test is how well measurements made with different instruments and techniques agree. With this in mind, stars in the database measured by more than one instrument or technique have been identified. In addition, stars in the interferometric database that have lunar occultation angular diameter measurements listed in the catalog published by White and Feierman (1987) have also been identified. The resulting list of stars with more than one angular diameter determination was analysed as follows:

1. Measurements with uncertainties $> 10\%$ were removed.
2. Formal uncertainties less than 0.5% for the Mark III data were increased to 0.5%. This is the best estimate of the uncertainty when systematics are taken into account (Mozurkewich, 1997).
3. In each comparison only measurements at the same wavelength or within 50 nm of each other were retained in order to avoid significant wavelength effects. TiO band measurements were omitted.
4. For each remaining star with 2 or more measurements a weighted mean for the angular diameter was calculated.
5. The difference between each measurement and its associated weighted mean was expressed in terms of the measurement's uncertainty.
6. The results are plotted in Fig. 4.

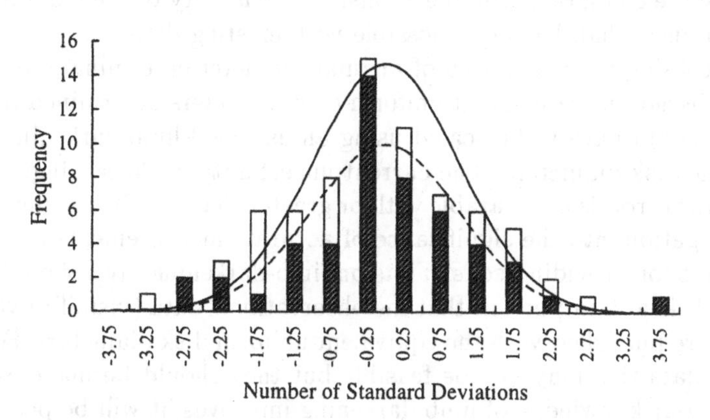

Figure 4. The distribution of angular diameters determined by different instruments or techniques expressed in terms of their standard deviations from the weighted mean angular diameter for each star. The hatched areas are for stars observed by two or more interferometric groups and the clear areas for stars observed by lunar occultation and interferometry. The normal error curves have been fitted to the data: the full curve is for the entire data set and the dashed curve for the interferometric data alone.

The distribution for interferometric measurements alone in Fig. 4 is in reasonable agreement with the fitted normal error curve indicating that there are no major systematic differences between the results for different instruments and techniques. The agreement in the case of interferometric and lunar occultation measurements with the normal error curve is not so good, suggesting that there may be some systematic effects in the occultation data which have not been included in the assessment of the uncertainties.

5. The Prospects

There are several high angular resolution interferometers coming into operation, under construction, or being planned. These are summarised in Table 1. The wavelengths covered range from ~ 440 nm to 20 μm and planned baselines range up to 640 m. It should be noted that a combination of the longest baselines and shortest wavelengths will be necessary to measure a sample of the hottest stars. Conversely, larger apertures operating in the red or near infrared will be required to measure stars at the cool end of the main sequence. Referring to Fig. 1, we can expect measurements of essentially all spectral types and classes with the instruments listed in Table 1 except for the coolest main-sequence stars but, even here, it should be possible to fill in down to at least the M6 dwarfs with, for example, the VLTI.

Current data suggest that formal fitting errors can be reduced to $< 1\%$ in the blue and $< 0.5\%$ at red wavelengths and that systematic errors can be reduced to this level. Nevertheless, it is highly desirable that the new instruments observe common stars to establish the reliability of the measurements more rigorously than has been possible with existing data.

In discussing the accuracies of angular diameter determinations it must be remembered that equivalent uniform disk diameters determined from observational data have to be scaled using an assumed limb-darkening law to obtain true disk diameters. The current uncertainty in the scaling, which is a systematic error, is comparable with or greater than the fitting uncertainty. An investigation into the significance of accurate measurements at multiple wavelengths for providing constraints on limb-darkening laws should be undertaken. It is also important that the observational data (visibility values v. baseline) are published with the equivalent uniform disk diameters. For large bodies of data this may not be feasible but they should be made available so that, as our knowledge of limb darkening improves, it will be possible for limb-darkened models to be fitted directly to the observational data.

References

Hanbury Brown, R., Davis, J. and Allen, L.R. (1974a) The angular diameters of 32 stars, *Mon. Not. R. Astron. Soc.*, **167**, pp. 121–136

Hanbury Brown, R., Davis, J. and Allen, L.R. (1974b) The effects of limb darkening on measurements of angular size with an intensity interferometer, *Mon. Not. R. Astron. Soc.*, **167**, pp. 475–484

Mozurkewich, D. (1997) *Private communication*

White, N.M. and Feierman, B.H. (1987) A catalog of stellar angular diameters measured by lunar occultations, *Astron. J.*, **94**, pp. 751–770

ASTROMETRY USING INTERFEROMETRY AT OPTICAL WAVELENGTHS

K. J. JOHNSTON, D. J. HUTTER, J. A. BENSON AND N. M. ELIAS II
US Naval Observatory, USA

J. T. ARMSTRONG, D. MOZURKEWICH AND T. A. PAULS
Naval Research Laboratory, USA

AND

C. A. HUMMEL
Universities Space Research Association, USA

Abstract. Interferometry offers an improvement in the accuracy with which astrometric measurements can be made. Using this technique, radio astronomers together with geodeticists have established a global inertial reference frame that is accurate to 0.1 milliarcseconds. At optical wavelengths, interferometry was first developed by Michelson at the turn of the twentieth century, but due to the complexities of precise beam combination at high speeds, it has lagged in its development. Now, with the availability of lasers, detectors and computers that allow path length compensation on millisecond time scales and distance determination between light collectors with a precision of 0.01 μm, interferometry at optical wavelengths will achieve the results in astrometry comparable to those at radio wavelengths.

1. Introduction

Interferometry at radio wavelengths has allowed a global inertial reference frame to be established with a precision of 0.1 milliarcseconds (mas) (Johnston *et al.* 1995). Relative positions of radio sources may be determined with even greater precision over small angles. After the demonstration of interferometry by Michelson & Pease (1921) in measuring the diameters of supergiant stars such as Betelgeuse in the 1920s, progress was slow. These early measurements were made with a twenty foot beam attached to the 100

T.R. Bedding et al. (eds.),
***Fundamental Stellar Properties: The Interaction between Observation and Theory*, 39–44.**

inch Hooker telescope. Later, a fifty foot interferometer was constructed on Mt. Wilson, which measured the diameters of a few more supergiant stars as well the separation of bright close binary stars such as Mizar. However, difficulties in maintaining mechanical stability and the inability to address fringe motions due to atmospheric turbulence severely limited this interferometer. In the 1960s, the diameters of 32 bright stars were measured by a group led by Hanbury Brown and John Davis (1974) using amplitude interferometry. In the 1970s, Shao & Staelin (1980) demonstrated that one could recover the complex amplitude and phase of the correlated signal by using a dithered delay line. These developments led the Office of Naval Research (ONR) to initiate a program in 1982 to develop this technology for astrometry and imaging. This program resulted in the Mark III optical interferometer and the Navy Prototype Optical Interferometer (NPOI), which will be described here. The accuracy of these ground-based instruments is limited by the atmosphere. Space-based interferometers will overcome the limitations of the atmosphere, and will also be briefly described here.

2. Ground-based Astrometry

The principles of radio astrometry are well developed (Thompson, Moran, & Swenson, 1986). Precise positions of stars have been determined using interferometry at radio wavelengths (Johnston *et al.* 1985). In these observations the Very Large Array observed the stars at several hour angles, solving for their positions from the observed phases. In Very Long Baseline Interferometry, the observed delays are used to solve for the positions. The geometry of the antennas must allow for sufficient spatial frequencies in the east-west and north-south directions to obtain an optimum solution for the celestial position. Corrections must be made for differential path delays in the instrument, atmosphere and ionosphere.

Very little has been accomplished at optical wavelengths using interferometry for astrometry. The Mark II interferometer, located on Mt. Wilson, was the first instrument capable of tracking phase and measuring fringe visibility and to report the measurement of stellar positions over large angles. The interferometer was mounted on a 5×12 foot optical table. The light collectors were siderostats with effective apertures of 3 inches, separated by 3.1 m along a north-south baseline. A variable delay line in the south arm was used to maintain path length equality in the arms of the interferometer. A delay line in the north arm of the interferometer induced a one-wavelength 500 Hz path length modulation to track and measure the parameters of the white light fringe. This allowed measurements of delay to be made over a large range of hour angles. Observations of the delay of four stars is sufficient to solve for the instrumental baseline, a constant delay offset, and the relative stellar positions. The 3.1 m baseline was measured to an accuracy of 50 μm (Shao *et al.* 1987). This corresponds to a celestial positional accuracy of (50×10^{-6} m/3.1 m) radians, or three arcseconds.

The Mark III interferometer, also located on Mt. Wilson, incorporated the developments of the Mark II. It was specifically designed to demonstrate fundamental interferometric astrometry. Star tracking and fringe tracking were completely automated to allow for rapid switching between stars. The siderostats were mounted on massive concrete piers. It had three fixed siderostat locations, giving a 12 m north-south baseline and a 12 m east-south baseline. Only one baseline could be used at a time. Light from the siderostats was directed along vacuum pipes into a temperature controlled beam combining building that contained vacuum delay lines. The stellar fringes were detected in a wide-band channel near $\lambda 0.7$ μm as well as narrow-band channels at 0.5 and 0.8 μm. Preliminary measurements with this instrument using a wide 0.3 μm band centered at 0.7 μm measured the 12 m north-south baseline to an accuracy of 6 μm, corresponding to a celestial accuracy of 22 mas in declination (Mozurkewich *et al.* 1988).

The major limitations in the accuracy of these measurements were twofold. First, turbulence in the atmosphere caused the atmospheric delay to fluctuate as $f^{-2/3}$, which implies that the S/N increases as $t^{1/6}$, where t is the integration time. By observing at two or more wavelengths, the dispersion of the atmosphere may be used to estimate the error caused by atmospheric turbulence. This two-color technique will correct for temperature microfluctuations, which are the major contribution to this error, but will not correct for turbulent water vapor fluctuations. Applying the two-color correction to all of the astrometric wide angle data obtained by the Mark III interferometer indicated an accuracy of 13 mas in declination and 23 mas in right ascension (Hummel *et al.* 1994). Simultaneous measurements at several wavelengths are necessary to improve the elimination of atmospheric microturbulent effects from the measurements.

The second major limitation of the Mark III was rapid variations in the baseline length (caused by imperfect manufacture of the mechanical components such as the siderostat bearings) on scales of a micron. A laser metrology system to precisely measure the positions of the siderostats relative to their massive piers was not successfully developed.

Parallel to the measurements for astrometry, a program was undertaken with the Mark III interferometer to study the spatial structure of stars. To accomplish this, a variable north-south baseline with lengths of 2 to 38 m was added. This instrument determined the angular diameters of over 70 stars, the orbits of 26 binary stars with separations of 3 to 120 mas, limb darkening for α Bootis, and the disk size of Nova Cygni, as well as other interesting astrophysical results.

The astrometric and astrophysical results of the Mark III interferometer led to the design of a two-dimensional interferometer, the Navy Prototype Interferometer (NPOI). This instrument consists of six 50 cm siderostats for imaging and four 50 cm siderostats for astrometry. The siderostats are arranged in a Y shaped configuration. The astrometric siderostats are located at the center of the array and $\approx$ 20 m out along each arm. The imaging siderostats have variable spacings from 2 to 252 m from the center of the array along the arms of the Y. (As of January 1997, the spacings available from the center of the Y are 2 to 34 m. Construction of the remaining Y is underway.) There are six vacuum delay lines to equalize the pathlengths between the siderostats. Vacuum delay lines are used to ensure that fringes can be simultaneously tracked and to determine the effect of atmospheric longitudinal dispersion on the fringe position. Data are taken in 32 channels distributed approximately evenly in wavenumber between $\lambda\lambda$450 and 850 nm. A complex metrology system consisting of over fifty lasers will measure the motions of the pivot points of the four astrometric siderostats with respect to bedrock. The expected accuracy is 0.01 μm.

With the array metrology system, the improved observing geometry with four simultaneous siderostats, and the extension of the two-color dispersion correction method to use several of the 32 spectral channels, it is expected that the baselines will be measured to an accuracy of 0.1 μm, resulting in an accuracy in celestial position, with a 20 m baseline, of 1 mas.

All of the previous discussion deals with measurements over wide angles, for which the astrometric accuracy σ_θ improves as $t^{-1/6}$. For narrow-angle astrometry (separations smaller than $\sim 10'$), $\sigma_\theta \propto t^{-1/2}$ (Lindegren 1980). Narrow-angle astrometry is itself divided into two regimes. For $\theta < B/h$, where h is the height of the dominant atmospheric turbulence, θ is the angle between the positions of the two stars, and B is the baseline length, $\sigma_\theta \propto B^{-2/3}\theta t^{-1/2}$. For the $\theta > B/h$ case, $\sigma_\theta \propto \theta^{1/3}t^{-1/2}$. These dependencies have been verified by Han (1991) for the $\theta > B/h$ case and Colavita (1994) for the $\theta < B/h$ case, for which the error is consistent with 21 μas/$\sqrt{t}$ for a 12 m baseline with the Mark III interferometer. The Palomar Testbed Interferometer (PTI) now under fabrication by JPL is specifically designed for the $\theta < B/h$ case in that it can simultaneously observe two very closely separated stars with a long baseline. This system operates at λ2 μm and should come on line in 1998.

3. Space Astrometry

The advantage of space interferometry for astrometry can be clearly seen from the previous discussion. Absence of the atmosphere means that the astrometric accuracy is limited only by the knowledge of the interferometer geometry. In addition, larger apertures can be used so that very faint objects can be observed. There have been many systems proposed. The two most prominent systems in the early 1990s were POINTS, proposed by SAO, and OSI, proposed by JPL. The leading contender is now the Space Interferometry Mission (SIM), an offshoot of OSI, that will be capable of 4 μas accuracy. This instrument consists of a linear array of seven apertures with spacings from 1 to 10 m making up three interferometers, two for pointing and one for scientific data. Expected launch is 2004 or later.

4. Future Prospects

Future prospects look excellent for both ground and space based astrometric interferometers. The previous discussion has shown that very little has been accomplished thus far. The instruments are very complex, and a large amount of resources is needed to develop them. The author list on this paper gives testament to these facts. The ONR support in this area has been crucial in developing the technology in this area. As a result of this support, the NPOI is nearing completion. NASA, through its interest in extra-solar planets, is developing the PTI, which also is nearing completion, and is also beginning to seriously support a space based mission, SIM.

The NPOI as of January 1997 is very near completion. Three of the astrometric siderostats, one in the center of the array and those on the east and west arm, are in operation nightly. The system is being debugged for automated operation in order to achieve the large number of observations needed for astrometry, say 200 per night. At this time, about 100 observations per night have been realized. The metrology system for the astrometric siderostats is nearing completion. The fourth siderostat is expected to be in operation by the fall of 1997. With this, the instrument will have full capability for astrometric observations.

The NPOI has achieved some significant milestones. Using simultaneous multichannel closure phase and visibility amplitude data, it has imaged the spectroscopic binary star Mizar A (Benson *et al.* 1997) with three siderostats and baselines of 19, 22, and 38 m. These are the first phase closure (ϕ_C) and squared visibility amplitude (V^2) measurements made simultaneously in multiple spectral channels. Figures 1c and 1d of Hummel & Benson (1997 [this volume]) display V^2 and ϕ_C versus wavelength for an observation obtained on 1996 May 1. The closure phases show a 180° phase jump at the minimum in V^2, confirming earlier results that show that the components of Mizar A are almost identical (Hummel *et al.* 1995). Data from several scans on 1996 May 1 were processed using the standard techniques of radio astronomy to form an image of Mizar A, displayed in Fig. 1a of Hummel & Benson (1997). This image has a dynamic range of approximately 100:1 and a spatial resolution of 3 mas.

Another significant result in obtaining precise values for stellar parameters has also been demonstrated. Multichannel observations of α Cas have been obtained at several hour angles. Figure 2 of Hummel & Benson (1997) displays the amplitude of the triple product (the product of the complex visibilities on the three baselines). These data allow a precise model for the stellar size and limb darkening to be made. For α Cas, Hajian *et al.* (1997) have determined a linear limb-darkening coefficient of 0.4 and an angular diameter of 5.58 mas, with a formal error of 5 μas. This result demonstrates the capability of optical interferometry to determine stellar diameters at the 0.1% level.

These two results clearly demonstrate that optical interferometry has matured, substantially overcoming many of the technical problems of operating at optical wavelengths. By the year 2000, the ground-based systems will produce significant results. The addition of space-based instruments in the 21st century will further these results, especially in giving sub-milliarcsecond positions over wide angles.

References

Benson, J. A., Hutter, D. J., Elias, N. M. II, Bowers, P. F., Johnston, K. J., Hajian, A. R., Armstrong, J. T., Mozurkewich, D., Pauls, T. A., Rickard, L. J, Hummel, C. A., White, N. M., Black, D., & Denison, C. S. 1997, AJ, submitted

Brown, R. H., Davis, J., & Allen, R. L. 1974, MNRAS, 167, 121

Colavita, M. M. 1994 A&A 283, 1027

Hajian, A. R., *et al.* 1997, to be submitted to ApJ

Han, I. 1989, AJ, 97, 607

Hummel, C. A., Armstrong, J. T., Buscher, D. F., Mozurkewich, D., Quirrenbach, A., & Vivekanand, M. 1995, AJ, 119, 376

Hummel, C. A., & Benson, J. A. 1997, this volume

Hummel, C. A., Mozurkewich, D., Elias, N. M. II, Quirrenbach, A., Buscher, D. F., Armstrong, J. T., Johnston, K. J., Simon, R. S. & Hutter, D. J. 1994, AJ, 108, 326

Johnston, K. J., deVegt, C., Florkowski, D. M., & Wade, C. M. 1985, AJ, 90, 2390

Johnston, K. J., Fey, A. L., Zacharias, N., Russell, J. L., Ma, C., deVegt, C., Reynolds, J. L., Jauncey, D. L., Archinal, B. A., Carter, M. S., Corbin, T. E., Eubanks, T. M., Florkowski, D. M., Hall, D. M., McCarthy, D. D., McCulloch, D. M., King, E. A., Nicholson, G., & Shaffer, D. B. 1995, AJ, 110, 880

Lindegren, L. 1980, A&A, 89, 41

Michelson, A. A. & Pease, F. G. 1921, ApJ, 53, 249

Mozurkewich, D., Hutter, D. J., Johnston, K. J., Simon, R. S., Shao, M., Colavita, M. M., Staelin, D. H., Hines, B., Hershey, J. L., Hughes, J. A., & Kaplan, G. H. 1988, AJ, 95, 1296

Shao, M. & Staelin, D. H. 1980, J. Opt. Soc. Am., 19, 1519

Shao, M., Colavita, M. M., Staelin, D. H., Johnston, K. J., Simon, R. S., Hughes, J. A. & Hershey, J. L. 1987, AJ, 93, 1280

Thompson, A. R., Moran, J. M. & Swenson, G. W. 1986, Interferometry and Synthesis in Radio Astronomy (Wiley, New York)

LUNAR OCCULTATION MEASUREMENTS OF STELLAR ANGULAR DIAMETERS

A. RICHICHI
Osservatorio Astrofisico di Arcetri
Largo E. Fermi, 5 - 50125 Firenze, Italy

1. Introduction

Offering an angular resolution which has remained unattained by any other technique for decades, lunar occultations have traditionally been the most productive method for the measurement of stellar angular diameters. Unlike interferometric methods, which are limited in resolution by the size of the aperture or of the baseline between apertures, in a lunar occultation the key to high angular resolution is the phenomenon of diffraction by a straight edge, that occurs at the Moon's limb in a turbulence–free environment. For the reader not familiar with the physics and technical aspects of the lunar occultation (LO) technique, it is sufficient here to show in Fig. 1 some practical examples of occultation lightcurves for sources with different angular diameters. It can be noted that the contrast of the fringes is maximum for a point–like source; it then decreases with the angular diameter, and eventually reaches the regime of a monotonic drop in the signal –as predicted by simple geometrical optics– when the angular extent of the source is large. In practice, the LO method is well suited to measure angular diameters in the range 1 to 50 milliarcseconds (mas). There is no real limitation concerning the wavelength of observation, although at present the near–IR is the region of choice for several different reasons (Richichi 1994).

Several hundred measurements have been collected so far by this method. A primary application has been the calibration of the effective temperature (T_{eff}) of the cooler spectral types, a fundamental parameter of great importance because it allows us to compare directly the theoretical models of stellar atmospheres with the observations. In order to derive T_{eff}, it is necessary to measure the bolometric flux F and the angular diameter ϕ, as

T.R. Bedding et al. (eds.),
Fundamental Stellar Properties: The Interaction between Observation and Theory, 45–50.

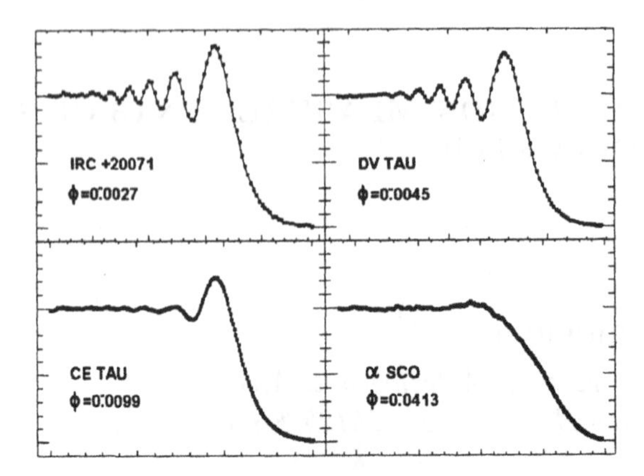

Figure 1. Occultation data (dots) and corresponding fits (solid lines) for four late–type stars with increasing angular diameter (data of the Arcetri LO group). The axes have been rescaled for each data set, bringing the time of occultation and the rate of the event to a common value and renormalizing the intensity values, in order to provide a more direct illustration of the phenomenon of fringe smoothing as a function of the angular size.

follows from the definition

$$F = \sigma \left(\frac{\phi}{2}\right)^2 \mathrm{T}_{\mathrm{eff}}^4 \tag{1}$$

Traditionally, our knowledge of $\mathrm{T}_{\mathrm{eff}}$ for cool stars has been poor, and as a result the contribution of LO in this area has been of particular interest. Lunar occultation observations have led also to important results in related areas, such as the study of surface structure, circumstellar matter, and close binaries. While there will not be sufficient space here to deal with these other applications of the method, the interested reader can find further reading in Richichi et al. (1996). Detailed accounts of the technical aspects of the method and of different aspects of data analysis are also available in the literature (see for instance Richichi 1994).

2. An overview of recent results

Ten years ago, White and Feierman (1987, WF hereafter) published a compilation of all LO angular diameters available through the end of 1986. They listed 348 determinations for 124 stars with spectral types from A to M (including several carbon stars). This large database has constituted a precious reservoir for people working in this field: both for those interested in effective temperatures, and more recently for those working with Michelson interferometry (MI) and interested in a comparison of LO versus MI

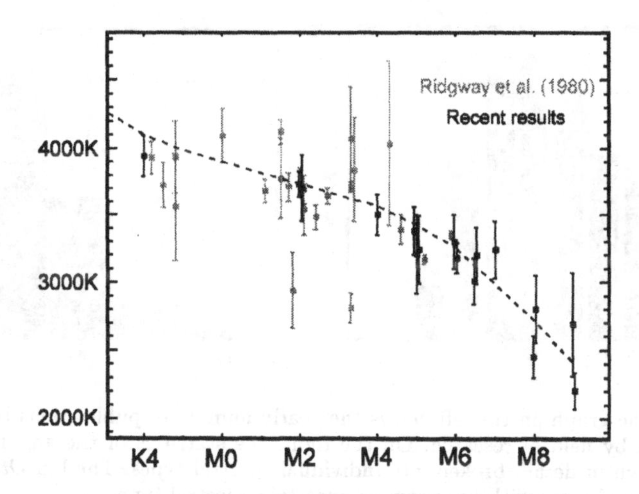

Figure 2. The data used by RJWW (shown here only below K4, in light gray), plotted as T_{eff} vs. spectral type. The dark gray points are the most recent determinations obtained by the Arcetri LO group. The dashed line is the RJWW calibration for spectra warmer than M5, and a tentative best–fit to the most recent results for spectra below M5 (see also Fabbroni and Richichi, these proceedings).

results. However, one should not overlook the fact that the WF compilation includes results from many different groups, often obtained in an early era of LO research when data treatment was not as advanced as today (in particular with respect to the understanding of the formal errors and of the instrumental and atmospehric effects). As a result, it is inevitable that the data show a large scatter which is not, in fact, a true limitation of the method as it has sometimes been assumed. A good example of this is the T_{eff} calibration obtained by Ridgway et al. (1980, RJWW hereafter), shown in Fig. 2, where it can be appreciated how many of the stars show a much wider scatter around the mean calibration, than their formal errors would imply. Nevertheless, the RJWW calibration has proven basically correct over the range K0-M5, as shown by independent MI results (Di Benedetto and Rabbia 1987, Dyck et al. 1996).

What can be said today, looking back at these results? First of all, it can be appreciated how the research in the area is still quite active, in spite of the fact that interferometric methods have undergone a huge technical development and are now clearly the choice for the future (see the several contributions in these proceedings). In Fig. 3, it is shown how the production of papers based on LO results has remained essentially constant over the past decade. There are only few observatories where LO are observed on a regular basis, the most important ones being those at Calar Alto[1] (Spain)

[1]Run by the German–Spanish Astronomical Center, Heidelberg

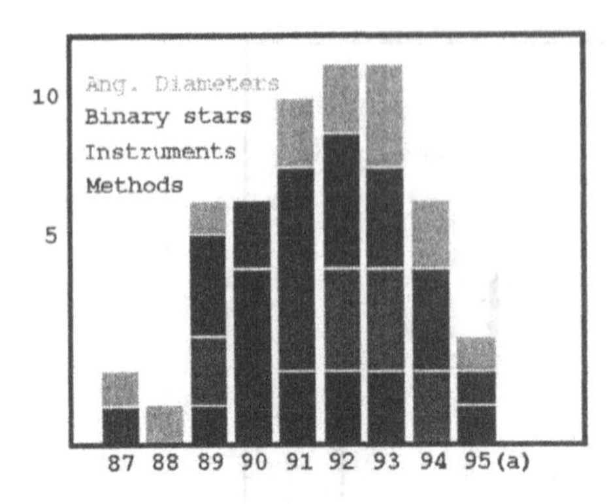

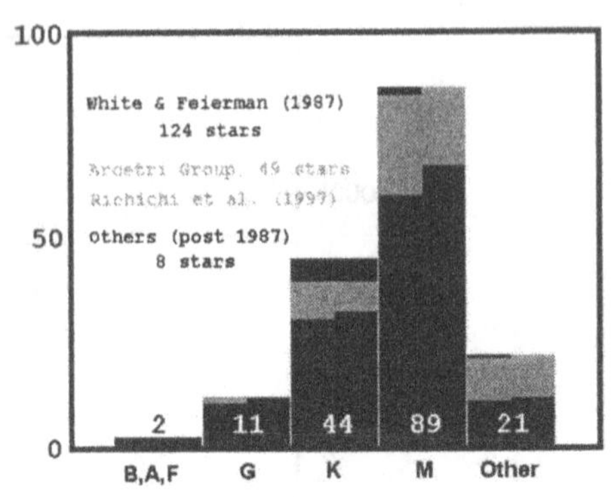

Figure 3. The graph on the left shows the yearly number of publications based on LO results, listed by field of research. On the right, the statistics of the angular diameter results is shown in detail, broken into individual spectral types. The bin *Other* includes carbon stars, and stars with unknown or uncertain spectral type.

and Mt. Gornergrat[2] (Switzerland). Activity on a smaller scale is present also at sites in the USA, Chile, India, Russia and FSU, South Africa, China, Mexico.

Almost 60 new measurements have been added in the past decade to the database of the WF compilation. Allowing for some overlap as shown in Fig. 3, angular diameters have been determined for the first time for about 50 stars. Of these, about 30 new stars belong to the M class and can be used to improve the T_{eff} calibration. At this stage, we are able to use only 17 of the stars in this sample, because for the remaining ones we are still in the process of obtaining the bolometric flux F or of assigning an accurate spectral type. While this number is only a small fraction with respect to the database already available for the M class, one should consider that the intrinsic accuracy of the results is now higher than before. Even more importantly, it should be noted that the LO–derived T_{eff} extends now from M5 to M9, in a range of temperatures for which no calibration at all was available (more recently, Dyck et al. have somewhat extended the RJWW calibration to reach M6). Figure 2 shows these more recent results, along with a tentative extension of the T_{eff} calibration. This preliminary result seems in good agreement with very recent work in the same temperature range by means of MI (Perrin et al. , this Symposium).

3. Conclusions and a look into the future

We can conclude that the measurement of angular diameters by LO is a field which is still producing competitive results, in terms of volume of data

[2]TIRGO, run by CAISMI, Florence, Italy

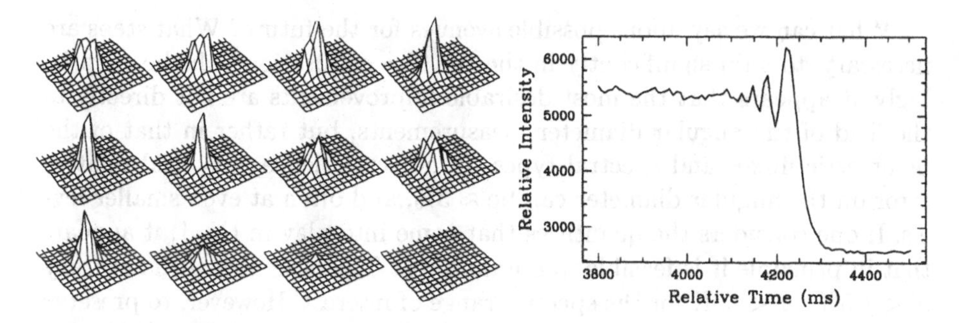

Figure 4. The disappearance of V Ari observed at TIRGO with a Nicmos3 detector operated in subarray mode. The frames on the left show a time sequence of ≈ 0.3 s in a 16X16 subarray, each representing a 7 ms integration (only a subset of frames is shown). The reconstructed lightcurve is shown to the right. An angular diameter of about 3.5 mas is measured. From Richichi et al. (1996).

and accuracy, with respect to more moderm methods such as long–baseline interferometry. If one is to look into the reasons of this prolonged vitality of LO, at least two main factors can be found:

- the method is easy. In spite of severe limitations concerning the *when* and *what* can be observed by LO, the instrumentation is relatively cheap and the data reduction simple, making the method attractive also for small observatories where the expensive technology required for other high angular resolution methods cannot be afforded. Moreover, a telescope in the 1.5m class is sufficient to observe LO at least for what concerns angular diameters.
- The improvements in the data reduction methods and in the understanding of systematic errors have been conspicuous over the past decade. This has made possible to achieve a better consistency of the results, with a much smaller scatter in the T_{eff} data for stars with the same spectral type (see Fig. 2).

There is one additional aspect that is worth mentioning, and it is the technological breakthrough constituted by the use of 2–D detectors used in fast mode on a subarray, to record LO events. Four years ago this author gave a talk in this same venue (Richichi 1994), in which the predicted perfomance of this novel approach was investigated and the implications discussed. Now, this has become a reality (see Fig. 4), with a fast LO mode implemented in several instruments in the near and mid–IR ranges (Richichi et al. 1996, Stecklum et al. 1996). At least 2 magnitudes can be gained, thanks to the drastic reduction in the background intensity allowed by these detectors. The advantages are expecially important for the study of fainter sources than those discussed here, but they are nevertheless considerable also for angular diameter applications.

What can we say about possible avenues for the future? What steps are necessary, to gain significantly in the the T_{eff} calibration? Maybe surprisingly, it appears that the most desirable improvements are not directly in the field of the angular diameter measurements, but rather in that of the bolometric fluxes and spectral types. In our experience, a typical relative error on the angular diameter can be $\approx 3\%$, and often at even smaller levels. If one considers the quantities that come into play in Eq. 1, it appears that in principle it is feasible to obtain direct T_{eff} estimates with accuracy of $< 1.5\%$, or < 50K for the spectral range of interest. However, to preserve this accuracy, it would be necessary to obtain bolometric fluxes with accuracies of $<$ 5-10% and in our experience this is very difficult in practice, expecially if one considers that the cooler stars are usually variable, and relatively faint in the visual bands. For the same reasons, also the spectral type can often be a non–negligible source of uncertainty. Not to mention, that the next step would be temporal studies of diameter and temperature in pulsating variable stars, where these problems are even more serious.

Therefore, it seems that LO (as well as MI) angular diameter measurements should be coupled to a systematic effort to obtain photometric coverage in a large number of bands from the visual to the IR, as close as possible to the date of the diameter determination, as well as to obtain accurate spectral types. While these measurements are simple from an experimental point of view, they require rapid and prolonged access to well equipped telescopes, and in fact the key to serious advancement in the field of direct T_{eff} determinations is probably in a coordinated effort across different astronomical disciplines.

Acknowledgments. The author has been partially supported in his work by a Chretien Grant awarded by the American Astronomical Society.

References

Di Benedetto, G.P., Rabbia, Y. (1987), A&A, 188, 114
Dyck, H.M., Benson, J.A., van Belle, G.T., Ridgway, S.T. (1996), AJ, 111, 1705
Perrin, G., Coude du Foresto, V., Ridgway, S.T., Mariotti, J.-M. (1996), these proceedings.
Richichi, A. (1994), IAU Symposium 158 *Very High Angular Resolution Imaging*, Tango, W.J., Robertson, J.G. (eds.), p. 71
Richichi, A., Baffa, C., Calamai, G., Lisi, F. (1996), AJ, 112, 2786
Stecklum, B., Käufl, U., Richichi, A. (1996), ESO Workshop *Science with the VLT Interferometer*, Paresce, F. (ed.)
Ridgway S.T., Joyce, R.R., White, N.M., Wing, R.F. (1980), ApJ, 235, 126
White, N.M., Feierman, B.H. (1987), AJ, 94, 751

STELLAR RADII

M. SCHOLZ
Institut für Theoretische Astrophysik, Universität Heidelberg, Tiergartenstr. 15, D-69121 Heidelberg, Germany

Abstract. Observing a stellar radius basically means observing a center-to-limb intensity variation. The significance and properties of center-to-limb variations, common approximations, the correlation with optical-depth radii in extended-photophere stars, and direct measurements of angular (interferometry, lunar occultation) and absolute diameters (binary eclipses) are discussed. Spectrophotometric and doppler techniques of diameter determination are also briefly outlined.

1. Introduction

Mass M, luminosity L and radius R are the three fundamental parameters of a (spherical) star where two of these are may be replaced by the surface gravity $g_s = GM/R^2$ and the effective temperature $\sigma_{SB}T_{eff}^4 = L/(4\pi R^2)$. Both the mass and the luminosity are measurable physical quantities whereas the star's radius is a fictitious quantity because a star is a gaseous sphere and does not have a sharp edge. The relevant observable quantity ist the center-to-limb variation (= clv) of intensity or limb-darkening $I_\Delta(r)/I_\Delta(0)$ (r: distance from the star's (disk) center; Δ: observational bandpass (monochromatic, filter, bolometric)). In case of a compact photosphere the variable r may be replaced by $\cos\theta = \mu = (1 - r^2/R^2)^{1/2}$ (θ: angle between the radius vector and the line-of-sight). As there are always tiny light contributions from outermost layers, the clv has an inflection point whose position is used to define the photospheric radius of the Sun. This definition may in principle be transferred to other stars though present observational techniques are still far from yielding details of stellar clv curves. This article summarizes the state of the art of direct (i.e. clv-based) as well as spectrophotometric and doppler diameter observations.

T.R. Bedding et al. (eds.),
Fundamental Stellar Properties: The Interaction between Observation and Theory, 51–58.

2. Direct diameter determinations

The Sun is the only star whose clv we can observed directly. Any stellar clv received by an observer only produces an interference pattern (interferometry = i), a diffraction pattern (lunar occultation = lo) or a specific light curve (binary eclipse). For reconstructing the star's clv from an interferometric or lunar occultation observation, a clv is assumed which yields a predicted pattern to which the observed pattern has to be fitted. A fit at only one (significant) point does not provide any information about the clv shape. This is the most common situation. A 2- or 3-point-fit, however, would yield in principle the full shape of a clv that can be described by a 1- or 2-parameter representation. There are so far about a dozen stars for which clv reconstructions have been attempted which, however, are in practice just sufficient to decide whether a model-predicted clv is roughly correct or grossly incorrect. These limb-darkening studies include Sirius (A1V, Hanbury Brown et al 1974 (i)), Arcturus (K1III, Quirrenbach et al 1996 (i)), Betelgeuze (M1-2I, Roddier & Roddier 1985 (i), Cheng et al 1986 (i), Wilson et al 1992 (i), Gilliland & Dupree 1996 (HST imaging)), Antares (M1.5I, Richichi & Lisi 1990 (lo)), 7 non-Mira and Mira M giants (Bogdanov & Cherepashchuk 1984, 1990, 1991 (lo), Di Giacomo et al 1991 (i), Wilson et al 1992 (i)), 3 C stars (Richichi et al 1991, 1995 (lo)), as well as the WN5 component of V444 Cyg (Cherepashchuk et al 1994 (light curve analysis)).

The standard procedure of evaluating interferometric, lunar occultation and binary eclipse data only determines a radius position on the basis of a parametrized approximation of a model-predicted limb-darkening curve. Common representations are

$$\begin{aligned}
I_\Delta(\mu)/I_\Delta(1) &= 1 - u_1(1-\mu), \quad \mathrm{UD}: u_1 = 0, \quad \mathrm{FDD}: u_1 = 1 \\
I_\Delta(\mu)/I_\Delta(1) &= 1 - u_1(1-\mu) - u_k(1-\mu)^k, \quad k \geq 2 \\
I_\Delta(\mu)/I_\Delta(1) &= 1 - u_1(1-\mu) - v_{nm}\mu^m(\ln\mu)^n, \quad n, m \geq 1 \\
I_\Delta(\mu)/I_\Delta(1) &= 1 - u_1(1-\mu) - w(1-\mu^{1/2})
\end{aligned}$$

As the variable is μ, they may only be used for compact photospheres. Data fits on the basis of this type of limb-darkening identify the $\mu = 0$ point as the position of the star's radius. Published limb-darkening coefficients of the past two decades cover a wide range of stellar parameters and bandpasses: Manduca et al (1977), Al-Naimy (1978), Manduca (1978), Wade & Rucinski (1985), Claret & Giminez (1990), Rubashevskii (1991a), Van Hamme (1993), Diaz-Cordoves et al (1995), Claret et al (1995). One should be aware, however, that these parameterized forms have historical roots and that the permanent discussion about their adequacy (Rubashevskii 1991b,

Van Hamme 1993, Diaz-Cordoves et al 1995) would become superfluous if model-predicted electronic I_Δ tabulations were provided instead.

Since clv curves depend noticeably on wavelength filter-independence of diameters measured in different bandpasses is a robust test of the quality of the adopted limb-darkening in the case of a compact photosphere. In contrast, diameters of extended-photosphere stars may and often do depend on wavelength. Baschek et al (1991) have discussed stellar parameter combinations leading to extended configurations and have summarized various radius definitions. One has to *choose* a specific layer whose distance from the star s center *shall be called* the stellar radius.

One of the most common definitions uses the layer $\tau_\lambda = 1$ for defining a radius $R_\lambda = r(\tau_\lambda = 1)$ that depends of course on the extinction coefficient k_λ at this wavelength. Since the optical-depth interval $\mathrm{d}\tau_\lambda = -k_\lambda(r)\rho(r)\mathrm{d}r = -\mathrm{d}r/l_\lambda(r)$ (ρ: density) measures the local distance interval $\mathrm{d}r$ in units of the local photon mean free path $l_\lambda(r)$, this choice of radius definition means choosing the layer which is just one integrated (radial) photon mean free path below the surface. If any, this type of radius is expected to be well related to the shape of the clv. It turns out, however, that *(i)* the photons collected by the observer often originate from a wide range of depths around the selected $\tau_\lambda = 1$ layer, and that *(ii)* there is no trivial correlation between the clv shape and the position of the $\tau_\lambda = 1$ radius on that curve.

Figure 1 shows normalized intensity contribution functions for four different source functions (from left to right: depth-independent, slightly, moderately, strongly increasing with τ_λ). In the case of a hydrostatic stratification and of depth-independent extinction k_λ the $\log \tau_\lambda$ scale corresponds roughly to the geometric r scale. Then, the $\Delta \log \tau_\lambda \approx 1 \ldots 2$ intensity contribution range seen in Fig. 1 may comprise a substantial part of the total thickness of the photosphere typically extending over 6 to 8 powers of optical depth. An important exception from this rule occurs where k_λ is a strongly increasing function of $T(\tau_\lambda)$ resulting in a steep $\Delta \log \tau_\lambda / \Delta r$ gradient and a dramatically shrinking geometric contribution range. This happens for continuous absorption of H, H^- and H_2^- in middle- to late-type (super)giants which, therefore, use to have a fairly compact continuum-forming region and well-defined continuum radii.

Figure 2 shows illustrative examples of clv curves predicted by models of extended photospheres. The U350 M giant model is almost compact and all $\tau_\lambda = 1$ positions are close to the clv end points. In contrast, the Mira model and the supernova model demonstrate clearly that no straight correlation exists between the clv shape and the position of the $\tau_\lambda = 1$ layer. The inflection point has no special meaning, and even Gauss-type limb-darkening is often found. Thus, parameters of clv approximations (UD,

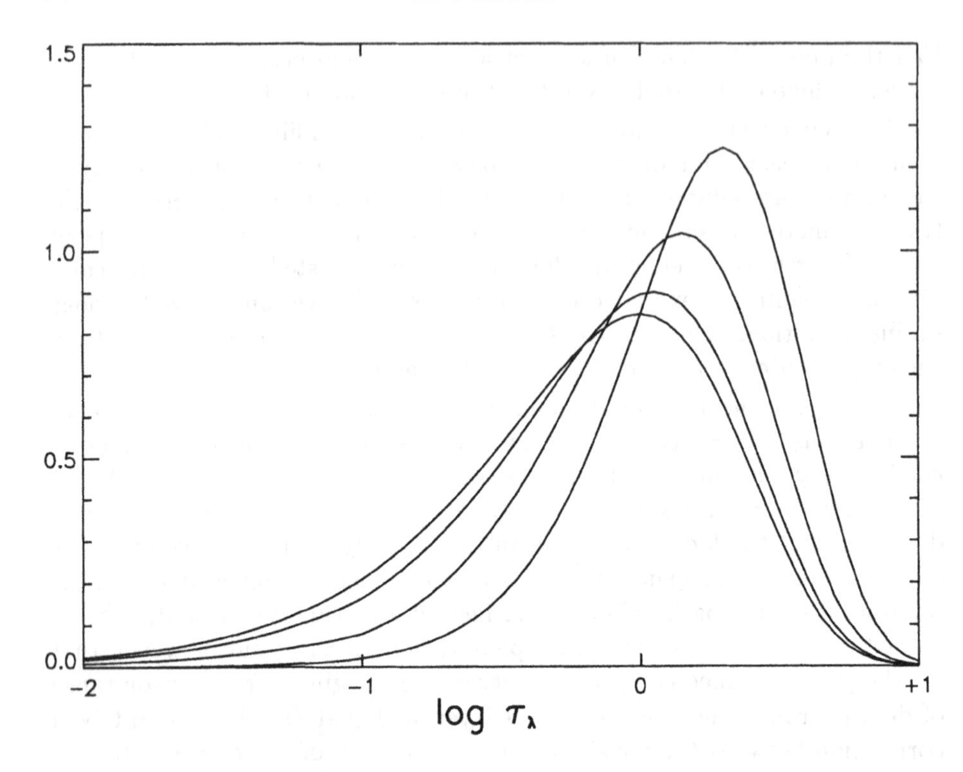

Figure 1. Intensity contribution functions for 4 different source functions.

FDD, Gauss) have no direct physical relevance with respect to the $\tau_\lambda = 1$ radius, and the position of this radius on a model-predicted clv has to be taken from this specific model. If continuum or Rosseland radii are to be determined, a "scaling" procedure has to be performed which first converts the $\tau_\lambda = 1$ near-continuum radius of an observed contaminated bandpass into the corresponding real-continuum radius if necessary and thereafter converts the continuum radius into the $\tau_{\rm Ross} = 1$ radius of the model. Though both steps usually involve only small scaling factors there is a substantial risk of mis-scaling if inadequate models are used for this procedure. Particular problems occur when impure filters assemble both deep-layer and high-layer photons.

Good direct measurements of *angular* diameters by interferometry or lunar occultation and of *absolute* diameters from binary eclipses have internal accuracies of the order of 5% in favorable cases. Anderson (1991) even quotes $\leq$ 2% for 2×44 selected binary components. One should realize, however, that errors introduced by inadequate limb-darkening are systematic errors and that measurements of the same star by different observers often differ more than expected from quoted error bars.

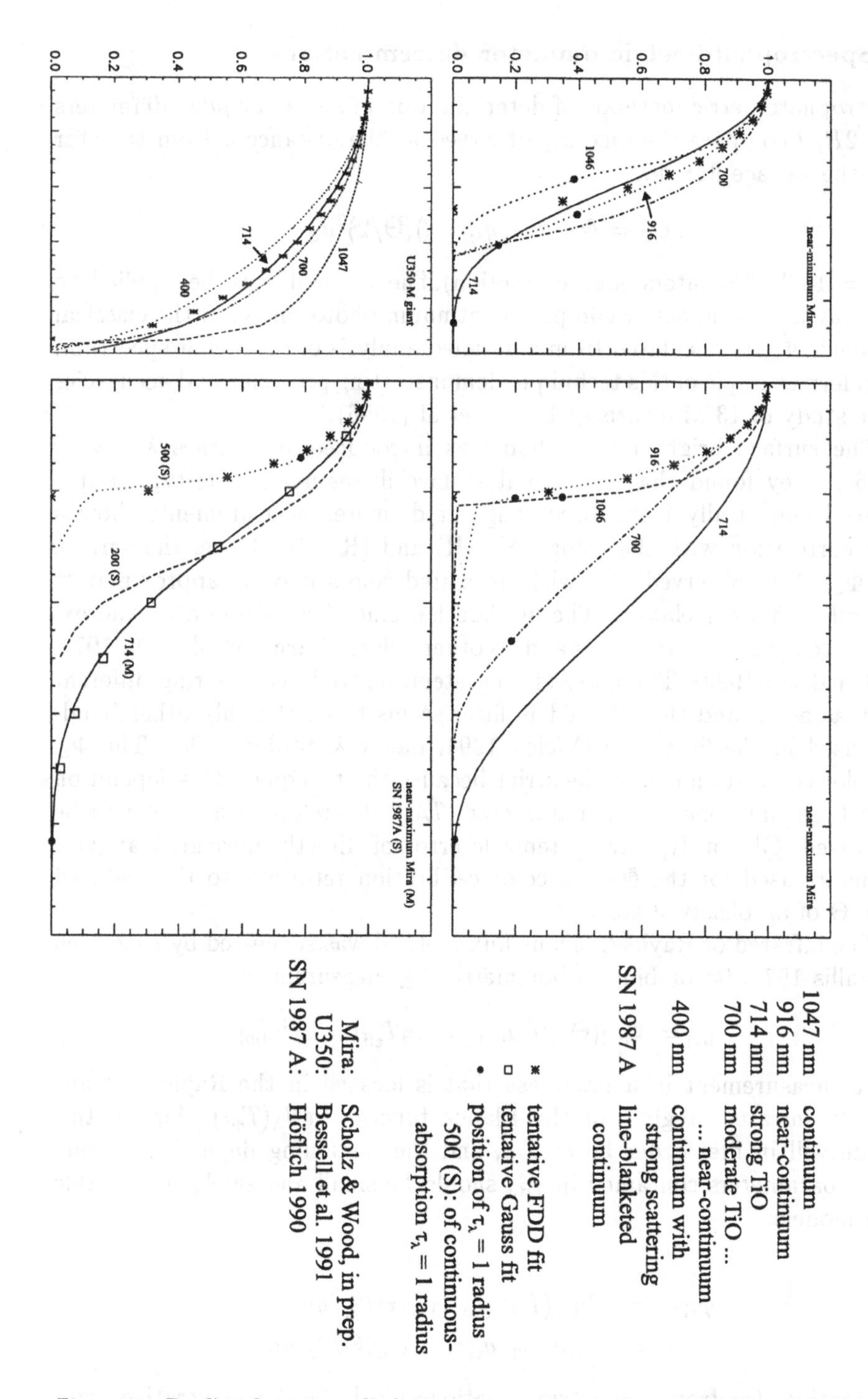

Figure 2. Predicted center-to-limb intensity variations of extended-photosphere stars.

3. Spectrophotometric diameter determinations

Spectrophotometric methods of determination of stellar *angular* diameters $\Theta = 2R/d$ compare the flux φ_Δ observed at the distance d from the star with the surface flux Φ_Δ:

$$\varphi_\Delta = \Phi_\Delta(T_{\mathrm{eff}}, g_{\mathrm{s}}, \ldots)(\Theta/2)^2 a_\Delta$$

($a_\Delta = 10^{-0.4A_\Delta}$: interstellar extinction).The method may be applied to stars having a compact or compact-continuum photosphere. In the classical approach, Φ_Δ is calculated from a detailed analysis of the stellar spectrum. A modern example of this technique demonstrating problems and accuracies is the study of 13 M dwarfs of Legget et al (1996).

The surface brightness method was introduced by Barnes & Evans (1976). They found that the visual surface fluxes Φ_{V} of late-type giants deduced empirically from direct angular diameter measurements show a tight correlation with the colors (V – R) and (R – I). Hence, the surface flux Φ_Δ of an observed star with measured colors may be approximately read off such a correlation. The method has since been elaborated and extended to other types of stars and to other colors (Barnes et al 1976, 1978, Di Benedetto 1993). There exists no systematic study considering different bandpasses Δ, and the infrared K filter seems to be the only other bandpass used in the literature (Welch 1994, Laney & Stobie 1995). The Φ_Δ vs. color correlation cannot be strict because the two quantities depend on more than only one stellar parameter (T_{eff}). Therefore, stars have to be pre-sorted. Obviously, any systematic error of directly measured angular diameters used for the Φ_Δ vs. color calibration returns into the deduced value Θ of an observed star.

The infrared or Rayleigh-Jeans flux method was suggested by Blackwell & Shallis 1977. It combines a bolometric Φ_Δ measurement

$$\varphi_{\mathrm{bol}} = \Phi_{\mathrm{bol}}(\Theta/2)^2 a_{\mathrm{bol}} = \sigma_{\mathrm{SB}} T_{\mathrm{eff}}^4 (\Theta/2)^2 a_{\mathrm{bol}}$$

and a measurement in a bandpass that is located in the Rayleigh-Jeans (usually infrared) regime of the Planck function $\pi B_\lambda(T_{\mathrm{eff}})$. Φ_{RJ} in this regime will in first order be $\propto T_{\mathrm{eff}}$, and the remaining dependence upon stellar parameters contained in f_{RJ} should be small and safely predictable from models:

$$\begin{aligned} \varphi_{\mathrm{RJ}} &= \Phi_{\mathrm{RJ}}(T_{\mathrm{eff}}, g_{\mathrm{s}}, \ldots)(\Theta/2)^2 a_{\mathrm{RJ}} \\ &= f_{\mathrm{RJ}}(T_{\mathrm{eff}}, g_{\mathrm{s}}, \ldots) T_{\mathrm{eff}} (\Theta/2)^2 a_{\mathrm{RJ}} \end{aligned}$$

Eliminating T_{eff} from these two equations yields Θ. An illustrative error assessment may be obtained by eliminating only the explicit T_{eff} terms:

$$\Theta = 2\sigma_{\rm SB}^{1/6}(a_{\rm bol}/\varphi_{\rm bol})^{1/6}(\varphi_{\rm RJ}/a_{\rm RJ})^{2/3}f_{\rm RJ}^{-2/3}$$

It shows that the bolometric quantities $\varphi_{\rm bol}$ and $a_{\rm bol}$ which are notoriously prone to errors only enter in the 6th root, and that the influence of stellar parameters and modelling inaccuracies (Blackwell et al 1991, Blackwell & Lynas-Gray 1994, Megessier 1994) enter with $f_{\rm RJ}^{2/3}$. Errors below 5% may be achieved in favorable cases, whereas accuracies of good spectral analysis or surface brightness diameters rather are in the 5 to 10% range. Some caution is recommended when the method is to be applied to O and Wolf-Rayet stars (Underhill 1982, 1983) because of complications arising from the the scattering-dominated continuum.

4. Doppler diameter determinations

Under special circumstances observed doppler shifts of lines originating in moving photospheres may be used to derive *absolute* stellar diameters $2R$. The Baade-Wesselink method is applied to expanding or pulsating photospheres. The photospheric motion leads to a change of the radius between the time t_1 and the time t_2,

$$R_2 - R_1 = \int_{t_1}^{t_2} (v_{\rm ph}(t) - v_{\rm c}){\rm d}t$$

where $v_{\rm ph}$ is the (spherically symmetric) expansion or pulsation velocity and $v_{\rm c}$ is the center-of-mass velocity. The velocity $v_{\rm ph}(t)$ must be determined from disk-integrated doppler profiles of selected lines. This procedure, called "conversion" or "projection" of an artificially defined "observed radial velocity" into $v_{\rm ph}$, is a critical step of the method. Combining the equation with two direct or spectrophotometric angular diameter measurements yields

$$R_2 = (R_2 - R_1)/(1 - R_1/R_2) = (R_2 - R_1)/(1 - \Theta_1/\Theta_2)$$

Recent studies of photospheres treated by the Baade-Wesselink technique indicate that most (or all?) of them show $v_{\rm ph}$ gradients and are not compact (supernovae: Eastman et al 1996; δ Cep and RR Lyr stars: Sasselov & Karovska 1994, Bono et al 1994, Butler et al 1996). These findings complicate the "conversion" process and imply that some "scaling" procedure has to be carried out that transfers $v_{\rm ph}$ of line-forming into $v_{\rm ph}$ of continuum-forming layers so that R_i and Θ_i refer to identical layers. Under these circumstances published error bars below 5% appear over-optimistic. Gautschy (1987) has written an excellent review on the method.

Schmutz et al (1994) have suggested to derive the diameter $2R = v_{\rm rot}P/\pi$ from the rotationally broadened line profiles of a co-rotating eclipsing binary component seen equator-on ($v_{\rm rot}$: rotational velocity; P: Period). Accuracies of this method are of the order of 15% .

References

Al-Naimy H.M. 1978, Ap&SS 53, 181
Andersen J. 1991, A&A Rev. 3, 91
Barnes T.G., Evans D.S. 1976, MNRAS 174, 489
Barnes T.G., Evans D.S., Parsons S.B. 1976, MNRAS 174, 503
Barnes T.G., Evans D.S., Moffett T.J. 1978, MNRAS 183, 285
Baschek B., Scholz M., Wehrse R. 1991, A&A 246, 374
Bessell M.S., Brett J.M., Scholz, M., Wood P.R. 1991, A&AS 89, 335
Blackwell D.E., Lynas-Gray A.E. 1994, A&A 282, 899
Blackwell D.E., Lynas-Gray A.E., Petford A.D. 1991, A&A 245, 567
Blackwell D.E., Shallis M.J. 1977, MNRAS 180, 177
Bogdanov M.B., Cherepashchuk A.M. 1984, SvA 28, 549
Bogdanov M.B., Cherepashchuk A.M. 1990, SvA 34, 393
Bogdanov M.B., Cherepashchuk A.M. 1991, SvA 35, 392
Bono G., Caputo F., Stellingwerf R.F. 1994, ApJ 432, L51
Butler R.P., Bell R.A., Hindsley R.B. 1996, ApJ 461, 362
Cheng A.Y.S., Hege E.K., Hubbard E.N., et al. 1986, ApJ 309, 737
Cherepashchuk A.M., Eaton J.A., Khaliullin Kh.F. 1984, ApJ 281, 774
Claret A., Diaz-Cordoves J., Giminez A. 1995, A&A 114, 247
Claret A., Giminez A. 1990, A&A 230, 412
Diaz-Cordoves J., Claret A., Giminez A. 1995, A&AS 110, 329
Di Benedetto G.P. 1993, A&A 270, 315
Di Giacomo A., Richichi A., Lisi F., Calamai G. 1991, A&A 249, 397
Eastman R.G., Schmidt B.P., Kirshner R. 1996, ApJ 466, 911
Gautschy A. 1987, Vistas Astron. 30, 197
Gilliland R.L., Dupree A.K. 1996, ApJ 463, L29
Hanbury Brown R., Davis J., Lake R.J.W., Thompson R.J. 1974, MNRAS 167, 475
Höflich P. 1990, A&A 229, 191
Laney C.D., Stobie R.S. 1995, MNRAS 274, 337
Leggett S.K., Allard F., Berriman G., Dahn C.C., Hauschildt P.H. 1996, A&AS 104, 117
Manduca A. 1979, A&AS 36, 411
Manduca A., Bell R.A., Gustafsson B. 1977, A&A 61, 809
Megessier C. 1994, A&A 289, 202
Quirrenbach A., Mozurkewich D., Buscher D.F., Hummel, C.A., Armstrong J.T. 1996, A&A 312, 160
Richichi A., Chandrasekhar T., Lisi F. et al 1995, A&A 301, 439
Richichi A., Lisi F. 1990, A&A 230, 355
Richichi A., Lisi F., Calamai G. 1991, A&A 241, 131
Roddier F., Roddier C. 1985, ApJ 295, L21
Rubashevskii A.A. 1991a, SvA 35, 396
Rubashevskii A.A. 1991b, SvA 35, 626
Sasselov D.D., Karovska M. 1994, ApJ 432, 367
Schmutz W., Schild H., Mürset U., Schmid H.M. 1994, A&A 288, 819
Underhill A.B. 1982, ApJ 263, 741
Underhill A.B. 1983, ApJ 266, 718
Van Hamme W. 1991, AJ 106, 2096
Wade R.A., Rucinski S.M. 1985, A&AS 60, 471
Welch D.L. 1994, AJ 108, 1421
Wilson R.W., Baldwin J.E., Buscher D.F., Warner P.J. 1992, MNRAS 257, 369

4. STELLAR FLUX DISTRIBUTIONS

THE VISUAL AND INFRARED FLUX CALIBRATIONS

C. MEGESSIER
Observatoire de Paris-Meudon
F-92195 Meudon Cedex France

Abstract. We present a critical review of the available visible and near infrared flux calibrations. In the visible, the accuracy and the good consistency of three independent determinations of Vega monochromatic flux allow one to recommend with confidence $f_{5556-\mathrm{Vega}} = 3.46\ 10^{-11}\mathrm{Wm^{-2}nm^{-1}}$ within 0.7%. In the near infrared, the possible flux excess of Vega, as compared to that derived from the atmosphere models fitting the visible, does not allow such a good accuracy. The agreement between the calibrations, either from a comparison of Vega to blackbodies or from solar analog stars, would question the calibrations relying on models. More work is necessary to conclude with confidence and then to reduce the uncertainty on the near infrared calibrations.

1. Introduction

Significant progress is made possible in our knowledge of the fundamental stellar properties because of the recent improvements both in the models (atmospheres, internal structure, evolution) and in the accuracy of the direct measurements (Hipparcos, IRAS, Hubble, photometry and spectrophotometry). Then it is important to constrain the accuracy of the absolute stellar fluxes required for comparison to the models and for the stellar parameter determinations, *which depends strongly on the absolute astrophysical flux calibrations.* In this review, I shall present successively:

- The principle of measurement of the absolute astrophysical fluxes.
- The calibrations available in the visible, which rely mostly on Vega.
- The calibrations available in the near infrared, which rely on Vega and on solar analog stars.
- A discussion on the calibration relying on model atmospheres.

T.R. Bedding et al. (eds.),
Fundamental Stellar Properties: The Interaction between Observation and Theory, 61–66.

- A conclusive discussion on the present limitations, and what is necessary to improve the calibrations.

2. Principle of the absolute astrophysical flux measurements

The two methods relying on observations that have been used are presented.

2.1. DIRECT COMPARISON OF THE STELLAR FLUX TO A REFERENCE SOURCE

The stellar flux and the standard source, placed in the vicinity of the observatory, are observed with the same telescope, equipped with the same receptor and photometric filter. The resulting accuracy depends on two important factors:

- the intrinsic quality of the standard sources and their own calibrations
- the atmospheric transmission and the correction for the atmospheric extinction.

The reference sources used are Cu or Pt blackbodies, Tu lamps or standard furnaces. They have to be calibrated in laboratory against gold blackbodies. Indeed the gold melting temperature, T = 1337.58 K, is the highest on the International Practical Temperature Scale, but the Au blackbodies can be operated in laboratory only. Moreover, it is preferable to work with a source which temperature is as high as possible, to be compared to stars which are far hotter than the blackbodies. The melting temperature of Cu is T = 1357.8 K and that of Pt is T = 2042.1 K. The Tu strip lamps can be worked at higher temperatures, up to 2900K, however they are less reliable since their emissivity depends strongly on the intensity of the current and on their temperature. The intensity radiated by each source, which is given by the Planck law, is accordingly a function of the wavelength. The emission of Tu lamps and Pt-BB are strong enough from λ=3000 to 9000 Å, whereas the Cu-BB emission is faint shorter than 6000 Å. (for more details see Mégessier 1995 and references therein)

2.2. CALIBRATION FROM THE SOLAR ABSOLUTE FLUX THROUGH ANALOG STARS

This method has been used in the near infrared domain. The energy distributions of the solar analog stars are supposed to be identical to that of the sun, so that they have the same color indices as it, in the Johnson photometric system (V, J, H, K, L, M). The solar energy distribution is known (Neckel and Labs 1981). The solar flux density in each photometric band $I_{V_\odot}$, $I_{J_\odot}$,..., is computed by convolving the solar energy distribution with the filter transmission functions. These solar fluxes are raccorded to the stellar's by means of Vega, for which the flux density trough the filters are obtained by the convolution of Vega flux models, calibrated according to the monochromatic flux $f_{5556-\mathrm{Vega}}$, with the filters transmission functions.

Then, in each band, the solar flux density is scaled in proportion to the differences $(V_\star - V_\odot)$, ..., i.e.: $I_{V\star} = I_{V\odot}\ 10^{-0.4(V_\star - V_\odot)}$, ..., and the absolute flux for a null magnitude is given by $I_{m=0.0} = I_{m\star}\ 10^{-0.4\ m\star}$, where $m_\star$ is the stellar magnitude in the color considered.

3. The visible flux calibration

From the beginning, the A0V star Vega has been chosen as the reference standard star in photometry as well as for the absolute flux measurements. The absolute visual calibrations are given as its monochromatic flux at λ 5556 Å. The improvement of the techniques allowed an increase of the accuracy. We discuss the calibrations obtained by various groups since that performed at Palomar 5m telescope by Oke and Schild (1970). Table 1 gives, for each group, the value of f_{5556} for Vega, the reference sources and the observatory where they worked.

TABLE 1. Visible flux calibrations f_{5556} for Vega

Authors	f_{5556} x 10^{-11} $Wm^{-2}nm^{-1}$	sources	Observatory
Oke and Schild (1970)	3.36 ± 2%	Tu, Cu	Palomar
Hayes et al. (1975)	3.45 ± 1.9%	Tu, Cu	Lick, Mt Hopkins
Tüg et al. (1977)	3.47 ± 1%	Cu, Pt	Lowell
Terez (1985)	3.44 ± 1.2%	Tu	Mt Ararat, Armenia

An extensive discussion of the accuracy of the visual calibration is given in Mégessier (1995). The main points are reported here.

In the visible domain, the correction for the extinction due to earth atmosphere is constrained satisfactorily and it contributes for less than 0.5% to the calibration uncertainty (see by ex. Hayes and Latham 1975).

The role of the standard reference source is crucial and the blackbodies are more reliable than the Tu lamps, as mentioned above. The calibration against the Au blackbody is important, as well as an inter-comparison of the Au blackbodies developed in the various laboratories over the world.

One has to remark the progressive improvement of the intrinsic accuracy obtained on each calibration (Table 1). Clearly, the Palomar calibration is lower than the three others for which the internal consistency is as good as ± 0.4%. Mégessier (1995) showed that the low Palomar value is due to the

Tu lamp worked by Oke and Schild (1970). Indeed the comparison of the Tu lamps performed by Hayes, Oke and Schild (1970) shows a systematic difference in the emissivities, that of the Palomar lamp being the largest. This leads to an underestimation of the stellar flux at Palomar. Then the best absolute Vega flux is given by the mean of the values obtained by the three latest groups listed in Table 1, weighted by the uncertainties.

The most reliable visible calibration is:

$\mathbf{f_{5556-Vega} = 3.46\ 10^{-11} Wm^{-2} nm^{-1}}$ **with** $\sigma = \mathbf{0.025}$**, i.e. 0.7%**

One has to notice that the mean values of $f_{5556-Vega}$ given successively by Hayes and Latham (1975) and Hayes (1985), widely used up to recently, were lowered by the Palomar value.

4. Near infrared flux calibrations

Three methods have been used, if one excepts that relying on the assumption that the stellar flux is given by the Planck law.

- The direct comparison of the observed Vega stellar flux to that of blackbodies or standard furnaces, as in the visible. This has been done by Blackwell's group between 1980 and 1989 (Blackwell et al. (1983), Selby et al. (1983), Mountain et al. (1985), Petford et al. (1985), Booth et al. (1989))

- Photometry of solar analog stars calibrated through the solar absolute flux. Two works have been done by Wamsteker (1981) and Campins et al. (1985) respectively, the photometric data given by Wamsteker being included in Campins (1985) work. The principle of the procedure is recalled above in sect. 2.

- The comparison of Vega near infrared photometry to the atmosphere model energy distribution fitting the visible data. The most recent works are those of Bessell and Brett (1988), Blackwell et al. (1994), Alonso et al. (1994).

The comparison of the calibrations obtained through the three procedures shows that a systematic difference exists. The calibrations from Vega flux models are lower than those derived either from direct comparison to furnaces or through the solar analog stars.

4.1. CALIBRATIONS RELYING ON MODEL ATMOSPHERES - QUESTION OF THE VEGA NEAR INFRARED EXCESS

The discrepancy between the observed Vega near infrared flux and the models has been reported in several works: Campins (1985), Blackwell et al. (1983) and Mountains et al. (1985), who compared their absolute fluxes to Dreiling and Bell (1980) models, and recently Castelli and Kurucz (1993), comparing ATLAS9 Vega energy distribution to the observed one constructed by Hayes (1985). The discrepancy increases with the wavelength, from 1 or 2 % at 1.25 μm to 4% at 2.2 μm and more than 6% around 3.7 μm (Blackwell et al. 1983 and Mountains et al. 1985). This is of importance since, if the star presents a near infrared excess, logically the calibrations from models yield lower calibrations (fluxes for a null magnitude) than the direct comparison to a blackbody. Indeed, one assumes the model, which has a lower flux than the star, represents the actual stellar flux. To be confident in the validity and in the accuracy of the calibrations, it is important to confirm whether Vega presents or not a flux excess as compared to the models.

The agreement between two independent works, by two different procedures relying on observations (Vega or solar analogs i.e. the first and second ones mentioned above) questions the third procedure relying on models.

If Vega near infrared flux is really larger than that of the models, one has to check the incidence on the works using absolute fluxes derived from model calibrations, specially the calibrations of new astrophysical measurements. Underestimated fluxes will be derived from such calibrations.

This effect is not an uncertainty but a systematic effect which has to be included in the uncertainty on the derived fluxes until this point is not cleared up. A more detailed study of that question will be given in Mégessier (1997).

5. Conclusion - Discussion - Requirements

In the visible, the values of the Vega monochromatic flux at λ5556 Å obtained totally independently by three groups using different absolute reference sources in different observatories are in excellent agreement. Then one can be confident and conclude that now the astrophysical visible flux calibration is satisfactorily determined. It is given by:

$f_{5556-\mathrm{Vega}} = 3.46\ 10^{-11}\mathrm{Wm}^{-2}\mathrm{nm}^{-1}$ with $\sigma = 0.025$, i.e. 0.7%

In the near infrared domain the only series of direct absolute flux measurements agrees with that obtained from solar analog star measurements, but disagrees significantly from that obtained assuming Vega's energy distribution is given by atmosphere models. Then to choose between them and so increase the accuracy one has to answer the questions:

- Do the models represent actually Vega's energy distribution?
- Is Vega's energy distribution similar to that of the A0 V stars?
- Does Vega exhibit any flux excess in the near infrared?

What is required to go further?

- More absolute flux measurements in the infrared
- Observations of another A0V star?
- Improvements of the models?

- "Definitive" absolute solar energy distribution?
- Comparison of the standard reference sources to laboratory Au-blackbodies and an inter-comparison of the Au-BB from the different laboratories.

References

Bessell M.S., Brett M.J., 1988, PASP 100, 1134

Blackwell D.E., Lynas-Gray A.E., 1994, A&AS 282, 899

Blackwell D.E., Leggett S.K., Petford A.D., Mountain C.M., Selby M.J., 1983, MNRAS 205, 897

Booth A.J., Selby M.J., Blackwell D.E., Petford A.D., Arribas S., 1989, A&AS 218, 167

Campins H., Rieke G.H., Lebofsky M.J., 1985, AJ 90, 896

Castelli, F. and Kurucz, R., 1993, Models for Vega. In: Dworetsky M.M., Castelli F., Faraggiana R. (eds) Proc IAU Coll. 138, PASPC 44, p. 496

Dreiling D.E., Bell R.A., 1980, ApJ 241, 736

Hayes D.S.,1979, A new absolute calibration of infrared photometry. In: Davis Philip (eds) Problems of Calibration of Multicolor Photometric Systems, p. 297

Knyazeva L.N., Kharitonov A.V., 1990, SvA 34, 626

Mégessier C., 1995, A&AS 296, 771

Mégessier C., 1997, A&AS (submitted)

Mountain C.M., Leggett S.K., Selby M.J., Blackwell D.E., Petford A.D., 1985, A&A 151, 399

Neckel H., Labs D., 1981, Solar Phys., 74, 231

Oke J.B., Schild R.E., 1970, ApJ 161, 1015

Petford A.D., Leggett S.K., Blackwell D.E., Booth A.J., Mountain C.M., Selby M.J., 1985, A&A 146, 195

Selby M.J., Mountain C.M., Blackwell D.E., Petford A.D., Leggett S.K., 1983, MNRAS 203, 795

Terez E.I. 1985, Absolute calibration of energy distribution in the Vega spectrum. In: Morozhenko (Ed) Photometric and polarimetric investigations of celestial bodies, Akademia Nauk Ukrainskaya SSR, Kiev 1985, p.55

Tüg H., White N.M., Lockwood G.W., 1977, A&A 61, 679

Wamsteker W., 1981, A&A 97, 329

DISCUSSION

PIERRE MAXTED: Has any attempt been made to model the IR excess of Vega in the 1-4 μm region?

CLAUDE MEGESSIER: Attempts exist to find an explanation for the near-IR excess of Vega. At wavelengths longer than 10 μm, it can be accounted for. In the near-IR, 1 to 5 μm, no explanation has been found and then it is not possible to model it.

REVIEW OF THE ULTRAVIOLET FLUX CALIBRATION

JEFFREY W. KRUK
Johns Hopkins University, Baltimore, MD, 21218, USA

1. Introduction

The practical difficulties in performing laboratory calibrations of instruments at ultraviolet wavelengths are considerably greater than at visible wavelengths, and the concommitant uncertainties are greater as well. In recent years theoretical models of white dwarf atmospheres have been adopted as UV flux standards, with impressive results. In this review, I will discuss the methodology of laboratory flux calibrations in the UV, the internal consistency and potential shortcomings of calibrations based on white dwarf model atmospheres, recent laboratory results, and future prospects.

2. Methodology

Absolute calibrations of the sensitivity of ground-based telescopes are performed by direct comparison of the signal observed from a standard star with that of a laboratory standard source (such as a platinum or copper blackbody) located on a nearby mountain top. The major inherent uncertainty is in correcting for the differences in absorption by the atmosphere between the horizontal line of sight to the laboratory blackbody and the vertical line of sight to the star. Different groups have been able to obtain results that are consistent to within about 1.5%. Reviews of these measurements can be found in Megessier (1997) and Hayes (1985).

At UV wavelengths, observations are performed by sounding rockets or by orbiting spacecraft. There is no atmospheric absorption to be corrected, but there is also no means of observing a laboratory flux standard while the instrument is in space. There are no laboratory blackbody sources hot enough to provide a usable flux in the FUV, so electron synchrotrons are the only practical sources whose flux can be calculated from first principles. At present, the only suitable facility for calibration purposes in the US

T.R. Bedding et al. (eds.),
Fundamental Stellar Properties: The Interaction between Observation and Theory, 67–72.

is the Synchrotron Ultraviolet Radiation Facility (SURF) at the National Institute for Standards and Technology (NIST). Direct calibration of instruments at SURF is usually impractical, however, and is rarely done. In lieu of direct calibration, instruments are usually calibrated by comparison with secondary (or tertiary) quantum efficiency standards such as photodiodes. Secondary standard photodiodes are typically calibrated by NIST at fixed wavelengths using the continuous spectrum produced by SURF and a double monochromator (the details vary with wavelength; see Canfield & Swanson 1987). A user's photodiode is then calibrated as a tertiary standard at fixed wavelengths by comparison with the secondary standard photodiodes. The quoted uncertainty in the calibration at this point is typically 8% in the FUV, and 10 – 15% in the EUV. These photodiodes require far more intense photon beams to produce a usable signal than the photon-counting detectors typically used in space astronomy, so the user ordinarily calibrates intermediate reference detectors by comparison with the tertiary standard photodiodes. At this point, the user's instrument can be calibrated by comparison with the intermediate reference detectors. The cumulative uncertainties in the calibration at this point are typically 15%.

3. Results Prior to Adoption of the White Dwarf Standard

Prior to 1994, the UV calibration of the International Ultraviolet Explorer (IUE) and the Hubble Space Telescope Faint Object Spectrograph (HST FOS) were based on measurements of the star η UMa obtained by a series of rocket flights and the OAO-2 satellite (which in turn was calibrated by rocket flights). These sounding rocket instruments were calibrated using NIST photodiodes as described above, and the measurements of η UMa were consistent to within about 10% (see Bohlin *et al.* (1980) and references therein). When spectra obtained with IUE of numerous hot DA white dwarf stars were compared with theoretical predictions, similar wavelength-dependent discrepancies of 10–15% were observed in each case. Similar discrepancies between observation and expectation were also obtained for other hot stars and such dissimilar sources as BL-Lac objects and quasars. The uncertainties in the model atmosphere predictions for hot DA's were expected to be about 1% relative to the observed fluxes at visible wavelengths, so these discrepancies ultimately led to the abandonment of the laboratory-based calibration and the adoption of white dwarf models as calibration standards in the UV.

Prior to the Hopkins Ultraviolet Telescope (HUT) and Orbiting Retreivable Far and Extreme Ultraviolet Spectrograph (ORFEUS) space shuttle flights, the only spectral data available below 1150Å with absolute fluxes were obtained with the Voyager UVS or by sounding rocket instruments.

Spectra obtained by the rocket instruments were usually consistent with the corresponding Voyager spectra to within their measurement uncertainties (typically 15%) longward of 1200 Å, but discrepancies sometimes grew dramatically at shorter wavelengths, ranging from factors of two to as much as seven over 950–1000 Å. The Voyager calibration is described in Holberg *et al.* (1982), and a summary of the controversies at short wavelengths can be found in Holberg *et al.* (1991).

4. White Dwarf Models

In 1994 the HST FOS switched to defining the UV sensitivity based on theoretical models of DA white dwarf atmospheres. Initially the white dwarf G191–B2B was used for this purpose; subsequently a combination of stars of widely varying effective temperatures was employed: G191-B2B (61,300 K), HZ 43 (50,000 K), GD 153 (38,500 K), and GD 71 (32,300 K). Fluxes for these stars were calculated by D.Finley using the model code of D. Koester. Effective temperatures and gravities for each star were determined from fits to Balmer line profiles, and the models were normalized to match Landolt V photometry (Finley *et al.*, 1997). These four independent models provide internally consistent results: when using this calibration the spectrum of each of these four stars agreed with its theoretical model flux to within 2% over the full FOS wavelength range of 1150 - 8000 Å(Bohlin *et al.*, 1995).

The final archive of IUE low dispersion spectra was reprocessed with a flux calibration defined by a model for G191-B2B (Nichols & Linsky, 1996). The model code used was the same as for the FOS, but the effective temperature was slightly lower, and the normalization was chosen so that fluxes for the brightest IUE standard stars matched the corresponding OAO-2 fluxes over 2100-2300 Å. The difference in T_{eff} (58,000 K *vs.* 61,300 K) causes the IUE fluxes at Ly α to be about 2% lower than the FOS relative to the visible, and the different normalization causes IUE fluxes to be 6% lower than the FOS for the same star.

The internal consistency of white dwarf models down to 912 Å can be tested by HUT spectra obtained during the Astro-2 mission. HUT provides spectrophotometry over 820 – 1840 Å at approximately 3 Å resolution (Davidsen *et al.*, 1992). In addition to the four FOS standard DA stars (which have surface gravities in the range of 7.7–8.0), HUT also observed GD 50 (41,000 K, $\log g = 9.0$), RE 0512-004 (32,000 K, $\log g = 7.4$), and Wolf 1346 (20,000 K, $\log g = 8.0$). The final calibration was defined by first computing the effective area from comparisons of the observed spectrum with model atmosphere predictions separately for each of the stars HZ 43, GD 153, and GD 71, and then averaging the results. The models are the same as used by the FOS. For all 7 stars, the resulting observed flux agrees with the predicted flux to within 5% (usually much better) over the entire wavelength range of 912 Å– 1840 Å, despite the very wide range of effective temperatures and surface gravities. This is particularly noteworthy, since the flux at the Lyman edge is a very sensitive function of the effective temperature. The star Wolf 1346 is cool enough that opacity arising from quasi-molecular states of Ly β had to be added to the model code (Koester *et al.*, 1996). Further information on the Astro-2 HUT calibration can be found in Kruk *et al.* 1995.

5. EUV Calibration

Models of white dwarfs cannot be used as flux standards at EUV wavelengths because there is no independent means of determining the opacity of interstellar H I. The Extreme Ultraviolet Explorer (EUVE) is therefore using its preflight laboratory calibration, which has uncertainties of 15% - 20%. The instrument sensitivity has been stable since launch, and the its observed fluxes have been consistent with the few other measured fluxes that exist. The biggest systematic error in EUVE fluxes arises from overlapping diffraction orders longward of 560 Å(Dupuis *et al.*, 1995).

6. New Laboratory Calibration

Following the first flight of HUT on the space shuttle *Columbia* in December 1990, we were able to perform a thorough post-flight calibration. This calibration differed from previous UV radiometric calibrations in that we were able to calibrate the spectrograph throughput directly at SURF, thereby avoiding buildup of uncertainties through the long sequence of transfer standards. The primary mirror reflectivity was measured in the calibration facilities at Johns Hopkins, since the instrument as a whole was too large to install in the vacuum chamber at SURF. The ratio of the white dwarf derived effective area to the laboratory effective area had a mean value of 1.0044 over the wavelength range 912-1840 Å, a slope of 7.61×10^{-6}Å^{-1} (which corresponds to a relative difference of only 0.35% at either end of the spectrum), and fluctuations about the mean with an RMS amplitude of 3%. The uncertainties in the laboratory calibration are: 4% for the overall normalization, 5% relative to the mean on scales of 50–100 Å, and an uncertainty in the overall slope that corresponds to 1% at 1840 Å and which increases smoothly to 3% at 912 Å. This calibration is described in detail by Kruk *et al.* (1997).

Astro-1 HUT spectra of 5 hot stars were compared with the corresponding *Voyager* fluxes. When averaged over broad bands, the spectra agreed to within 10%. Given the absolute laboratory calibration of HUT, this should help resolve the controversies over the *Voyager* flux measurements.

The uncertainties in this laboratory calibration are not that much larger than the intrinsic uncertainties in the white dwarf calibration at wavelengths shortward of Ly α. The combined uncertainties in the flux of Vega and the V-band photometry give an absolute normalization uncertainty of about 2%. The internal uncertainties in the effective temperature for a well-studied star such as G191-B2B are typically about 300 K, leading to negligible uncertainties in the FUV flux. External uncertainties, such as different treatments in pressure broadening when fitting the line profiles, do lead to significant uncertainties in the FUV flux. For G191-B2B, these uncertainties in the predicted flux are estimated to be 1% at 1840 Å, 2.3% at 1000 Å, and almost 7% at the Lyman edge. The uncertainties in the surface gravity have little effect on the continuum flux, but do affect the Lyman line profiles.

An additional intrinsic source of uncertainty in the theoretical predicted flux arises from the treatment of Stark broadening of the Lyman lines. At present, the method of Schöning & Butler (1989) is used to calculate the Stark-broadened Lyman and Balmer line profiles. This is based on the theory of Vidal *et al.* (1970), which makes the assumption that the individual lines do not overlap. This assumption is badly violated in the high gravity atmosphere of a white dwarf. An empirical correction for this effect is incorporated into the models for calculating Balmer line profiles (Bergeron, 1993). The same correction is also used for the Lyman lines, but the only justification for doing so is that it permits an internally consistent fit to the HUT spectra. The lack of a proper theory of Stark broadening is the major shortcoming in models of hot DA white dwarfs at present.

7. Modelling of metals in G191-B2B

The star G191-B2B is a potentially poor choice as a calibration standard, because of the presence of trace quantities of heavy elements in its photosphere. Despite the low abundances, these elements cause strong line blanketing shortward of 250 Å(Dupuis *et al.*, 1995), which in turn causes backwarming of the atmosphere and an apparent increase in the flux longward of 240 A. The stars HZ 43, GD 71, and GD 153 were also measured by Dupuis *et al.* and were found to have atmospheres that were truly pure hydrogen. The internal consistency of the HST FOS and Astro-2 HUT calibrations, and the laboratory calibration of the Astro-1 HUT sensitivity all indicate that using a pure hydrogen model to fit the Balmer lines of G191-B2B *does* result in a model that accurately predicts the flux at FUV wavelengths. However, it was only very recently that a theoretical model was able to predict both the EUV spectrum of G191-B2B and the UV–visible spectrum simultaneously (Lanz *et al.*, 1996). While there is still some difficulty with detailed fits to the EUV He lines, the model of Lanz *et al.* is an important advance, not only in the state of the art in modelling techniques, but also for validating the previously adopted flux of G191-B2B at FUV wavelengths for use as a flux standard.

8. Prospects for the Future

The techniques used to calibrate HUT at SURF could certainly be improved upon. A further factor of two reduction in the ultimate laboratory incertainties seems feasible, especially in light of planned upgrades to the SURF accelerator. However, I am not aware of any plans for a suitable sounding rocket or other retrievable spectroscopic instrument that could be so calibrated and used to test white dwarf models. The most likely opportunity will be with a sounding rocket. However, given the limited possible aperture and the short duration of a sounding rocket flight, it might be prefereable to observe a hot sdO star rather than a white dwarf in order to obtain a high signal to noise spectrum. Such a star could then serve as a standard, much as Vega is used in the visible.

References

Bergeron, P. 1993, in White Dwarfs: Advances in Observation and Theory, NATO ASI Series, ed. M.A. Barstow (Kluwer: Dordrecht), p. 267

Bohlin, R., Holm, A., Savage, B., Snijders, M., & Sparks, W. 1980, A&A 85, 1

Bohlin, R.C., Colina, L., & Finley, D.S. 1995, AJ 110, 1316

Canfield, L.R., & Swanson, N. 1987, J. Res. NBS 92, 97

Davidsen, A.F., Long, K.S., Durrance, S.T., Blair, W.P., Bowers, C.W., Conard, S.J., Feldman, P.D., Ferguson, H.C., Fountain, G.H., Kimble, R.A., Kriss, G.A., Moos, H.W., Potocki, K.A., 1992, ApJ 392 264

Dupuis, J., Vennes, S., Bowyer, S., Pradhan, A.K., & Thejll, P. 1995, ApJ 455, 574

Finley, D.S., Koester, D., & Basri, G. 1997, ApJ, submitted

Hayes, D.S. 1985, in Calibration of Fundamental Stellar Quantities, Proc. IAU Symp. No. 111, eds. D. Hayes, L. Pasinetti, & A. Phillip (Reidel, Dordrecht), p. 225

Holberg, J.B., Forrester, W.T., Shemansky, D.E., & Barry, D.C. 1982, ApJ 257, 656

Holberg, J.B., Ali, B., Carone, T.E., & Polidan, R.S. 1991, ApJ 375, 716

Koester, D., Finley, D.S., Allard, N.F., Kruk, J.W., & Kimble, R.A. 1996, ApJ 463, L93

Kruk, J.W., Durrance, S.T., Kriss, G.A., Davidsen, A.F., Blair, W.P., Espey, B.R., & Finley, D.S. 1995, ApJ 454, L1

Kruk, J.W., Kimble, R.A., Buss, R.H., Davidsen, A.F., Durrance, S.T., Finley, D.S., Holberg, J.B., & Kriss, G.A. 1997, ApJ 482, in press

Lanz, T., Barstow, M.A., Hubeny, I., & Holberg, J.B. 1996, ApJ 473, 1089

Megessier, C., these proceedings

Nichols, J.S., & Linsky, J.L. 1996, AJ 111, 517

Schoning, T., & Butler, K. 1989, A&AS 78, 51

Vidal, C.R., Cooper, J., & Smith, E.W. 1970, ApJS 25, 37

Discussion of this paper appears at the end of these Proceedings.

THE BOLOMETRIC LUMINOSITIES OF STARS

R. C. BLESS AND J. W. PERCIVAL
University of Wisconsin - Madison, 53706, USA

Abstract. The total energy emitted by a star is one of its basic parameters. It is also one of the most difficult to determine, requiring space-based as well as ground-based measurements of the absolute flux from stars. We review the current status of these observations in the ultraviolet, visible, and near infrared, and estimate the uncertainties of the absolute flux measurements in these spectral regions. We compare recent determinations of the bolometric luminosity of early-type stars with the first (mostly) empirical results of 20 years ago. We close with a few comments on possible improvements of these measurements in the future.

1. Introduction

The link between stellar atmospheric models (theory) and real stars (observations) has been through their effective temperatures and bolometric luminosities. It is only in the last 20 years or so, with observations made from space along with techniques for measuring angular diameters, that these latter two quantities could be directly determined by observations for stars other than the sun. Previous speakers have given the basic relationships. Let us just remind you that the effective temperature of a star is proportional to the square root of its angular diameter and the fourth root of the total flux measured at the Earth. If necessary, this flux must be corrected for interstellar extinction, which of course introduces uncertainty. Except for the use of models in finding the limb-darkened angular diameter, the star's effective temperature can be found completely empirically (at least in principle), subject to the errors in the measured flux and angular diameter. Neither of these, however, enters strongly into its determination.

If, in addition, the parallax of the star is well-determined, then its radius follows directly from its angular diameter. With the effective temperature

T.R. Bedding et al. (eds.),
***Fundamental Stellar Properties: The Interaction between Observation and Theory*, 73–82.**

and radius (assuming, of course, that the star has a well-defined radius), the luminosity of the star can be calculated and a fundamental HR diagram (luminosity vs. effective temperature) can be plotted. Thus the physics involved is quite straightforward; getting the relevant data is much less so. Before describing the latter, let us quickly remind you in which wavelength regions most of the flux emerges for the various spectral types.

We integrated the emergent flux from several Kurucz (1992) models at a variety of effective temperatures over several wavelength bands, corresponding roughly to the extreme ultraviolet, the 300Å between Ly-α and the Lyman continuum (the FUV), the conventional UV, the visual and two bands in the infrared. The fractional fluxes in these bands, plotted as a function of effective temperature, are shown in Figure 1. Note that the EUV region becomes significant only for stars having effective temperatures greater than about 30,000 K, while even the coolest stars radiate little in the 4-10μ region. The 1-4μ region, however, is about as important as the visual region for the sun and cooler stars. We must keep this in mind when assessing the flux accuracy possible for a given spectral type.

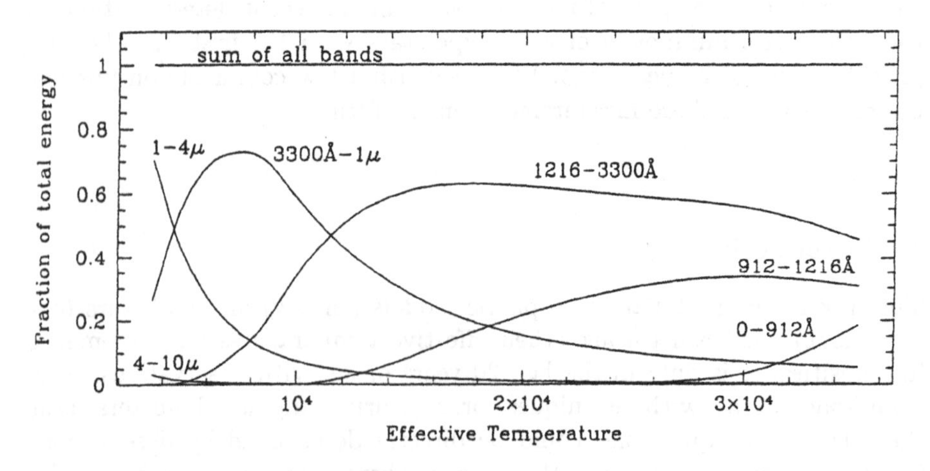

Figure 1. Kurucz models by wavelength region

Time does not allow a discussion of all absolute calibration work and for the most part in this review we will concentrate on the most recent efforts.

2. The Ultraviolet Spectral Region

We will begin this discussion with the conventional ultraviolet because it was in this spectral region that the white dwarf model calibration system, now the standard, was first developed. With the capability of observing stars from above the earth's atmosphere about 40 years ago came the challenge of absolute calibration in the ultraviolet. Kruk (1997) briefly describes some of the earlier calibration efforts and their difficulties that led to the use of white dwarf model atmospheres for calibration purposes.

Though the use of hot DA white dwarf model atmospheres for absolute flux calibrations was apparently first suggested by Greenstein and Oke in 1979, it has been only in the last several years that this idea has been put into wide practice. One chooses the appropriate model effective temperature and surface gravity from the best fit to the profiles of several hydrogen lines in each of the white dwarf calibration stars. The model fluxes are then normalized to the visual magnitude of the white dwarf being modeled. An instrument and telescope, for example, the Faint Object Spectrograph on the Hubble Space Telescope, can then be calibrated by comparing its response to the predicted white dwarf model flux.

White dwarf models are attractive for several reasons. The hot (> 20,000 K) DA white dwarfs have nearly pure hydrogen, plane-parallel, non-convective, essentially LTE atmospheres and have no winds. Apart from the hydrogen lines, their spectra are very nearly featureless. Above temperatures of about 30,000 K, relative fluxes are nearly independent of gravity. At 50,000 K, an effective temperature error of 1000 K leads to errors of only 0.5% in the long wavelength portion of the Balmer continuum, increasing to about 1% at Ly-α. Internal accuracies of 1.5% or less should be attainable over the optical and UV spectral regions. The models do have some deficiencies of course, e.g. the lack of a rigorous Stark broadening theory. Nevertheless, model atmospheres of hot DA white dwarfs represent an attractive alternative to laboratory calibrations, so attractive, in fact, that today the calibrations of both the IUE archive and HST data are based on DA white dwarfs.

The fundamental standards used by Bohlin (1996) to define the HST white dwarf scale are an average of the pure hydrogen LTE models calculated by D. Finley using D. Koester's code for HZ 43, GD 71, GD 153, and G191-B2B of which the last is by far the brightest. Hubeny (1996) computed non-LTE atmospheres for the same temperatures and surface gravities as Koester's models. The agreement with all but G191-B2B is very good, but for the latter, Hubeny's fluxes are about 0.5% lower shortward of 3500Å and up to 3.5% higher at 2.5μ. This latter difference may be a reflection of the poorer model accuracy in spectral regions where little flux is emitted. Lanz et al. (1996) computed a non-LTE model of G191-B2B blanketed by about 9,000,000 lines. With an effective temperature of $55,200 \pm 1000$ K it agrees with the Koester model within 1% from 20μ to about 1700Å. From 1700-1000Å many absorption lines depress the spectrum by a few per cent.

As attractive as this approach may be, to have such an important stellar datum based on theory is *not* an acceptable state of affairs. Consequently, the calibration of the Hopkins Ultraviolet Telescope on ASTRO just described by Jeff Kruk is very important (Kruk et al. 1997). He and his colleagues are to be congratulated for carrying out this program. Their work confirms the white dwarf scale over the spectral region from 912-1840Å to within the error of their laboratory calibration, roughly 5%. Bohlin feels that the white dwarf models give calibrations good to 3% from 1150Å through the optical, a number in line with the Hopkins calibration work.

Now, as just noted, the atmosphere of G191-B2B contains some metals. However, the Hopkins group showed that within their error limit, the spectrum of this white dwarf is indistinguishable from that of a pure hydrogen atmosphere at HUT resolution in the FUV. It appears, therefore, that despite its metal content, G191-B2B is a satisfactory calibration star in the FUV and UV, at least at the 5% level, and that the calibration does not depend strongly on the white dwarf model employed.

These numbers suggest that if you are careful and lucky, it should be possible to produce spectrophotometry in the UV to the Lyman limit that is accurate to about 5%. Part of the luck just mentioned is the correction for interstellar extinction. If the (B-V) color excess is in error by just 0.01 magnitudes, the extinction correction at 1500Å and 1000Å could be in error by 5 and 10%, respectively. For stars that radiate much of their energy in the UV, this could be the dominant source of error. Buss et al. (1995) in analyzing observations of about a dozen OB stars obtained by HUT from 912-1840Å, estimate their calibration uncertainty to be about 5%, but extinction uncertainties to be as large as 15%

Both the final IUE archive and HST spectra are being calibrated on the white dwarf scale. Though the spectra of the same star observed by IUE and HST will have about the same shape, they will not have the same absolute flux level. This results from two factors: first, the IUE calibration takes the effective temperature of its standard, the white dwarf, G191-B2B, to be 58,000 K, which is 3,000 K lower than the temperature taken for this star in the HST calibration. This results in a maximum difference of about 1.3%. Secondly, the HST zero point is based on the measured V magnitudes of Landolt (1992). The IUE calibration, however, takes as its zero point OAO-2 fluxes of the brightest IUE standards in the spectral region of 2100-2300Å, chosen because there the OAO-2 and TD-1 observations agree best. The argument for normalization in the UV rather than in the visual is that it avoids an extrapolation from the visual to the UV and thereby minimizes the effects of errors in both the model and the observations which otherwise might be introduced. This difference in zero points amounts to about 5%. Thus, the HST fluxes are up to 6% brighter than the new IUE fluxes, independent of wavelength. The Hopkins laboratory calibration gives a normalization that is about 4% brighter than that of IUE spectra on the white dwarf scale. This difference is in the same sense and nearly the same amount as found by Bohlin and suggests that the HST normalization is to be preferred over the IUE scale.

3. The Visual and Infrared Spectral Regions

We turn next to the calibration of the visual region of the spectrum on which work has been done for the longest period of time and which is probably in reasonably good shape. Vega has traditionally been taken as the standard star for this region of the spectrum. Hayes (1985) has reviewed the absolute calibration of Vega and recommends a weighted mean of six determinations from 3300-10,500Å. He feels that the internal error of this average is good to 1-1.5%, that is, about as accurate spectrophotometry as can be done. The measurements of Hayes and Latham (1975) do not differ significantly from this mean curve and so some authors still use these observations. In the optical, the white dwarf scale agrees with the Hayes Vega calibration to about 3% and so gives a somewhat poorer calibration.

As we have just heard, Dr. Megessier has reviewed the various determinations of the absolute flux from Vega and recommends that the value of $F_{5556\mathring{A}} = 3.46 \times 10^{-9}$ ergs cm^{-2} s^{-1} Å^{-1}. This differs by about 0.5% from the value recommended by Hayes, well within the error of about 1.5% he quotes. Realistically, the absolute visual flux of Vega is probably known about as well as can be at the present time.

An observer using these data must compare spectrophotometry of the program star with that of Vega or some other well-observed standard star, as well as determine accurately the apparent magnitude of the star in a well-defined bandpass. With great care, spectrophotometry can be done to 1-2% and the apparent magnitude to perhaps 0.5%. Adding all these errors (those of the standard star as well as the program star) in quadrature gives a mean error for program stars of about 2.5% on the Vega scale or about 3.5% on the white dwarf scale. This is probably the best that can be done now and many observations will not achieve this accuracy. Hall (1996) gives a compilation of 9 major sources of absolute visual spectrophotometry for which the average accuracy is 4%, probably a realistic figure.

Ground-based absolute IR observations, for example in the broad J, H, K, and L bands, are difficult to make accurately. In this spectral region, absorption by the earth's atmosphere can change rapidly during a night. To minimize these effects, some observers use much narrower filters than those of the Johnson system. Care is required in relating such observations to those in the Johnson system, however, where the shape of the bandpass, at the temperature at which they are actually used, must be determined accurately.

Just as in the visible, Vega has been used as a standard star in the IR. A model atmosphere fit to the visual is assumed to represent Vega in the infrared and provide the absolute calibration. Some argue, however, that in the near IR Vega is systematically brighter than the model by a small amount. That is, the model atmosphere does not represent the real Vega. Megessier (1995) emphasizes this point and derives a near IR flux calibration based on solar analog stars or on blackbody furnaces, but not on model atmospheres. In doing this she finds consistent results among various authors and estimates errors of about 2% in the absolute calibration and 2-3% in the relative photometry. Thus, the IR calibration from 1-4μ is thought to be about as accurate as the absolute calibration in the visual.

4. The Extreme UV Region

We will turn finally to the data produced by the EUV Explorer. Three spectrometers, each with its own detector, cover the spectral range from 70-760Å with resolution of about 200. The Berkeley group, in preparing an EUV spectral atlas of many stellar objects, is establishing a final instrumental calibration. The contamination of the long wavelength spectra by overlapping orders from the shorter wavelengths is just one of several problems that must be addressed. Before launch, a sensitivity model for the instrument that was no better than about 20% was produced. After launch, observations of white dwarfs, especially HZ 43, were combined with the lab calibrations to reduce the errors to about 10%, except over the 600-700Å region where the calibration is poorer. This is thought to be representative of the current calibration state of the instrument.

Figure 2 shows the pure hydrogen, LTE model spectra of the four fundamental HST calibration white dwarfs and their spectra observed by EUVE, HUT and HST from 100-10,000Å. That they fit together nicely should be no surprise since they are all based on the same calibration. Probably the most striking feature of the observations is that of G191-B2B which shows the effect of a large amount of interstellar absorption by hydrogen and helium in the FUV. Though the shape of this absorption is well known from theory, the total amount is less certain. Unless this absorption can be accurately predicted independently, it is hard to see how this star can be a reliable calibration object in the EUV. This is particularly true since Lanz et al. (1996) point out that the He I edge in G191-B2B may be partly stellar or circumstellar rather than entirely interstellar. From that point of view, GD 71 is much to be preferred, but unfortunately it is faint in the EUV region. Until a detailed account of the EUVE calibration procedures and results are published, it will be difficult to assess the accuracy of its calibration.

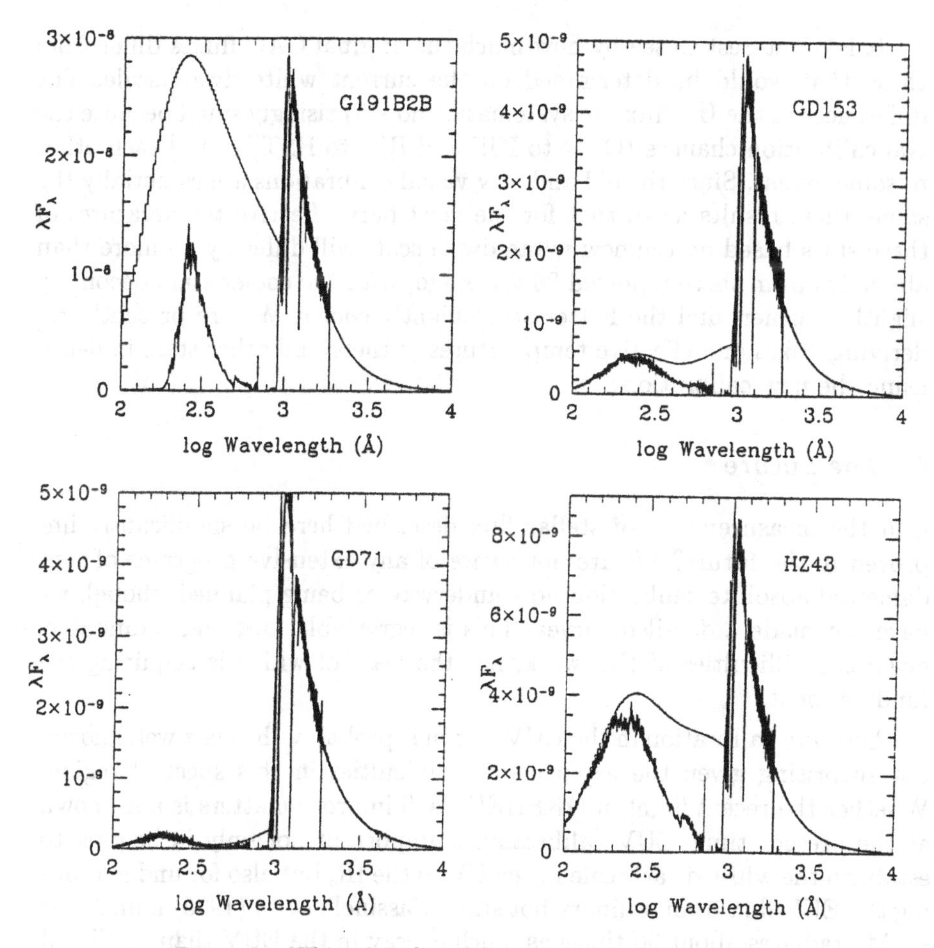

Figure 2. White dwarf models compared to observations

In summary, it appears that with great care and accurate extinction corrections, program stars can be absolutely calibrated to about 5% from the Lyman limit to the visible and to about 3-4% from the visible to 3μ.

5. Comparison With Previous Data

About 20 years ago Hanbury Brown and colleagues (1974) completed measurements of the angular diameters of 32 mostly early-type stars with the Narrabri interferometer. At the same time, the Wisconsin group provided the ultraviolet fluxes from these stars measured with the OAO-2 satellite observatory. Sonce these stars radiate most of their energy in the UV and visual, the derived effective temperatures and bolometric corrections were, in large part, empirically determined.

It is of interest to see by how much the original OAO fluxes differ from those that would be determined on the current white dwarf scale. The differences in the UV flux are systematic and surprisingly small because the two calibration changes (OAO to IUE and IUE to HST) cancel each other to some extent. Since the old and new visual calibrations are essentially the same, these results mean that for the most part, effective temperatures of these stars based on the new white dwarf scale will differ by no more than about 1% from those reported 20 years ago, with the cooler stars becoming slightly warmer, and the hotter stars slightly cooler. We are presently re-deriving fluxes and effective temperatures of these and other stars in detail using the new calibrations.

6. The Future

Can the measurements of stellar flux described here be significantly improved in the future? We are not aware of any extensive programs of fundamental absolute calibration now underway or being planned, though we have not made a detailed survey. This is regrettable, but understandable, given the difficulties of this work, not the least of which is acquiring the funding for it.

Absolute calibration in the EUV region is probably the least well known, not surprising given the experimental difficulties in this spectral region. Whether the recent flight of ORPHEUS will improve matters is not known at the present time. EUV calibration is important not only in helping to establish the white dwarf scale in the UV to the IR, but also for understanding the EUV flux from ordinary hot stars. Cassinelli et al. (1995) found that ϵ CMa radiates about 30 times as much energy in the EUV than predicted by an appropriate B1 giant model. This result indicates that ϵ CMa is the dominant source of ionizing flux for interstellar gas within a few parsecs of the sun, a completely unexpected result. This suggest that not only our knowledge of stellar models for hot stars, but also of the ionization equilibrium in the interstellar medium, is incomplete. DA white dwarfs used for absolute calibration may well have simpler atmospheres than B-type giants. Nevertheless, it is a bit worrisome that this discrepancy between model and star exists in the spectral region, the EUV, that is largely calibrated by a model.

A relatively simple but quite useful near IR program would be absolute photometry of the white dwarf calibration stars. Since this could be done from the ground, the opportunity exists to establish by observation the validity of the white dwarf calibration in this spectral region rather than relying on the model atmospheres, which, as noted earlier, may not be as accurate in the IR as in the UV or visible.

The 2.5-45μ spectrometer ($R = 1000$) on ISO should be useful for absolute calibration purposes. Its calibration is based primarily on observations of mostly late-type standard stars, not white dwarfs. In turn these depend on the careful absolute calibration work given by Cohen et al. (1996 and references therein) in several papers. They estimate that from 1.2-35μ the uncertainty in their calibration is about 3%. They feel, however, that the case for a small near IR excess in Vega is not convincing. This should be kept in mind when using ISO data.

A light source in the sky, but calibrated on the ground, would be desirable. It might be possible to mount such a calibration source on the space station, assuming it ever gets built. In principle, lamps could be calibrated on the ground and gotten into use on relatively short time scales, and similarly returned to ground for recalibration. Many technical problems would have to be overcome, for example, tracking the source from a moving satellite. However, it would also provide some useful science for the space station to do.

ESTEC is developing a promising new detector based on superconducting tunnel junctions. Photoabsorption in a niobium junction, cooled to 0.3 K, produces hundreds of free charge carriers for each incident event with a resolution of 100Å or less. These are detected when they quantum mechanically tunnel through an insulating barrier. Such a detector is a complete low resolution spectrometer having a small dark noise and no read-out noise, with high sensitivity from the UV to the near IR. Eventually, this sort of detector might be accurately calibrated in the laboratory and provide useful checks on the assumed calibrations.

Finally, a brief comment concerning fluxes for stars with measured angular diameters. As of now, angular diameters have been measured only for bright stars. Mid- to late-spectral type stars radiate most of their flux in the visual and infrared which can be measured from the ground. For early-type stars the UV is important, but HST cannot measure the bright stars for which there are angular diameters. One must hope that IUE observations made in the fast trail mode exist for the stars of interest. It may well turn out that some of the old observations made by OAO-2 and other satellites might have a rebirth of usefulness in this regard. We are looking into this possibility.

In summary, a fundamental change in the absolute calibration of stellar fluxes has been taking place over the last few years. It is less and less of an experimental enterprise and increasingly becoming one in which a computer is the primary piece of equipment. Model makers have become the calibrators. White dwarf model atmospheres do indeed seem to be reasonably successful as calibrating objects over much of their spectra and can be computed to seductively high internal precisions. It is important to remember, however, that the models are deemed successful in the visual and ultraviolet BECAUSE they are constrained by laboratory calibrations and observations. There appears to be less such constraint in the EUV region, so results from that spectral regions should be carefully evaluated. In general, caution should be exercised in accepting model results uncritically. In particular, we hope that independent calibration work does not disappear entirely, but will be continued, especially at the two ends of the spectrum-the IR and the EUV.

Acknowledgements

We are grateful to M. Abbott, R. Bohlin, J. Cassinelli, and J. Kruk for many helpful discussions, and to the Berkeley EUVE group (Craig, et al., 1996) for providing EUVE spectra prior to publication. We also thank I. Hubeny for the emergent flux over the whole spectrum of the Lanz et al. model of G191-B2B.

References

Bohlin, R.C. 1996, AJ, 111, 1743.

Brown, R.H., Davis, J., & Allen, L.R. 1974, MNRAS, 167, 121.

Buss, R.H., Kruk, J.W. & Ferguson, H.C. 1995, ApJ, 454, L55.

Cassinelli, J.P. et al. 1995, ApJ 438, 932.

Cohen, M., Witteborn, F.C., Bregman, J.D., Wooden, D.H., Salama, A., & Metcalfe, L. 1996, AJ, 112, 241.

Craig, N., Abbott, M., Finley, D., Jessop, H., Howell, S., Mathioudakis, M., Sommers, J. Vallerga, J., & Malina, R. 1996, private communication.

Hall, J.C. 1996, PASP, 108, 313.

Hayes, D.S. 1985, in Calibration of Fundamental Stellar Quantities, Proc. IAU Symposium No. 111, eds. D.S. Hayes, L.E. Pasinetti, & A.G. Davis Phillip (Reidel: Dordrecht), p. 225.

Hayes, D.S. & Latham, D.W. 1975, ApJ, 197, 593.

Hubeny, I., see Bohlin, 1996.

Greenstein, J.L. & Oke, J.B. 1979, ApJ, 229, L141.

Kruk, J.W. 1997, these proceedings.

Kruk, J.W., Kimble, R.A., Buss, R.H., Davidsen, A.F., Durrance, S.T., Finley, D.S., Holberg, J.B., & Kriss, G.A. 1997, ApJ 482, in press.

Kurucz, R.L. 1992, Smithsonian Astrophysical Observatory, CD-ROM No.

Landolt, A. 1992, AJ, 104, 340.

Lanz, T., Barstow, M.A., Hubeny, I., & Holberg, J.B. 1996, ApJ, 473, 1089.

Megessier, C. 1995, A&A, 296, 771.

Discussion of this paper appears at the end of these Proceedings.

THE COLOURS OF THE SUN

B. J. TAYLOR
Brigham Young University, Provo, UT 84602, USA

ABSTRACT. In this paper, arguments are developed for treating $(R-I)_C$ as the most important colour to be derived for the Sun. The solar value of $(R-I)_C$ is then found to be 0.335 ± 0.002 mag. This result updates a counterpart given by Taylor in 1992.

1. Introduction

Which colour index should be derived for the Sun?

Papers on this subject rarely consider this question. Authors sometimes explain that one wants to know the solar colours to learn how the Sun compares to similar stars. In addition, authors cite a number of previous studies—likely including those which piqued their interest in the first place. However, one seldom sees an explanation for the fact that (almost always) $B-V$ is being determined. Was $B-V$ the best choice historically? Is it still the best choice now? Let us see what answers these questions may have.

2. $B-V$: Blanketing Effect And Accidental Errors

When the solar-colours problem "came of age" in 1964, at least some people knew that field stars should be measured to the red of the V passband to minimize blanketing effects (see Sandage & Smith 1963). Putting this idea into practice for the Sun, however, seems to have been ruled out by a shortage of high-precision red photometry for solar stars and a tacit "1P21 limit" in general thinking. The Strömgren system was just getting started, so $b-y$ was not yet a realistic option. Only $B-V$ and $U-B$ were serious contenders, and since $B-V$ is less blanketed than $U-B$, one can see why $B-V$ would have been an obvious choice historically.

Let us re-examine that choice, using resources that were not available in 1964. The blanketing effect on $B-V$ is an obvious concern, so one should probably derive a numerical estimate of its size before doing almost

T.R. Bedding et al. (eds.),
Fundamental Stellar Properties: The Interaction between Observation and Theory, 83–88.

anything else. This may be done by using the calculations of VandenBerg & Bell (1985). Data for stars with the solar temperature and surface gravity are used. The blanketing estimate may be expressed as the value of $|\Delta(B-V)|$ as [M/H] increases from -0.5 dex to 0.0 dex. $|\Delta(B-V)|$ is increased by a factor of 1.5 to allow for the difference between theoretical and empirical calculations (see Table 3 of Taylor 1994).

Since $(B-V)_{\odot}$ is often determined from field-star values of $B-V$, one would also like to know something about the rms errors of the field-star data. The best rms error found commonly for $B-V$ has been calculated by Nicolet (1978), who obtains a value of 0.009 mag. Since $|\Delta(B-V)|$ turns out to be 0.053 mag, one sees at once that the blanketing effect is inescapable if one cannot restrict attention to field stars with [Fe/H] ~ 0. This is an especially serious matter if $B-V$ is used as a temperature proxy.

3. Tactics For Deriving $(B-V)_{\odot}$

The next issue that might be considered is the best choice of tactics for determining a solar colour index. For $B-V$, one procedure which has attracted much attention is the search for solar twins. High-dispersion analysis is used to identify dwarfs whose metallicity resembles that of the Sun closely (see, for example, Cayrel de Strobel & Bentolila 1989). With a "short list" of such stars in hand, one can then correct their values of $B-V$ for residual temperature, gravity and metallicity differences between the stars and the Sun. (See Edvardsson et al. 1993, who use this procedure for $b-y$).

The obvious advantage in this procedure is that if the corrections are small, the uncertainties they introduce into the final value of $(B-V)_{\odot}$ will be small as well. There is also a disadvantage, though: can one obtain a reliable rms error for $(B-V)_{\odot}$ by using only a small number of stars? This question will be considered again below.

A second common tactic is to measure an index (usually a spectroscopic index) which can readily be secured for both field stars and the Sun. By comparing field-star indices and values of $B-V$ to the solar index, one can then determine $(B-V)_{\odot}$. The chief question here is what to do about the $B-V$ blanketing problem. If the adopted index has no metallicity sensitivity, the blanketing problem has its full scope, and metallicities for the field stars will be required in order to mitigate it (Cayrel de Strobel 1996). One might cancel out the blanketing problem by choosing an index with compensating metallicity sensitivity, and an assumption is often (if tacitly) made that such cancellation takes place. This assumption is not always tested, however (see, for example, Croft et al. 1972).

4. Is The Traditional Approach The Best Approach?

As of 1994, determinations of $(B - V)_\odot$ fell in two groups: a "short-wavelength" group with $(B - V)_\odot = 0.665 \pm 0.003$ mag, and a "long-wavelength" group with $(B - V)_\odot = 0.633 \pm 0.009$ mag. These two means differ at better than 99.5% confidence. (See Taylor 1994.)

Is this dichotomy the real problem here, or could the choice of colour index be more significant? For some time now, the real reason for this choice has been tradition, as Griffin & Holweger (1989) note. In a discipline which uses the traditional magnitude scale and the traditional name "planetary nebulae," the force of tradition is not surprising. Nonetheless, it is fair to ask whether one can improve on tradition.

5. The Solar-Colours Problem in 1997

An inducement for determining a different solar colour index in 1997 is the fact that the Cousins *VRI* system is now available. Measurements in this system are widely available, and they have high precision and coherency (see, for example, Cousins 1974 and Taylor & Joner 1996). This is in clear contrast to the state of *VRI* photometry in 1964.

Theoretical work shows that $(V - I)_C$ is the most blanketing-free Cousins index for G dwarfs (VandenBerg & Bell 1985, Buser & Kurucz 1992). $(R - I)_C$ is the next-best choice where blanketing is concerned, but one must also realize that especially in the northern hemisphere, $(R - I)_C$ is available for many more stars than $(V - I)_C$ (compare Tables 7 and 8 of Taylor 1986). Since astronomy, like politics, is the art of the possible, it seems better to determine $(R - I)_C$ for the Sun at the moment than it would be to determine $(V - I)_C$. (Note, though, that this judgment could easily change when the Hipparcos photometry becomes available.)

6. $(R - I)_C$: Blanketing Effect And Accidental Errors

What are the relative blanketing sensitivities of $(R - I)_C$ and $B - V$? In addition, what is the ratio of their best common rms errors? To answer these questions meaningfully, one must allow for the fact that $(R - I)_C$ and $B - V$ respond differently to temperature changes. To allow for this difference, one may use the Hyades relation between the two indices given by Taylor (1994, Eq. 1). The derivative of this relation is evaluated at the solar value of $(R - I)_C$ (see below).

Taylor (1996, Appendix B) gives pertinent information about rms errors for $(R - I)_C$. With this information and the dynamic-range allowance in hand, the basic procedure used above for $B - V$ may be employed.

The best commonly-found rms error for $(R-I)_C$ turns out to be 0.003 mag before the dynamic-range correction and 0.008 mag afterwards. Recalling that the counterpart for $B-V$ is 0.009 mag, one sees that there is no great difference between the two colour indices—at least where accidental error is concerned. Blanketing changes are a different story, however: $|\Delta(R-I)_C|$ is 0.008 mag before the dynamic-range correction and 0.021 mag afterwards. Recalling that $|\Delta(B-V)| = 0.053$ mag, the blanketing problem is reduced by a factor of 2.5 by replacing $B-V$ with $(R-I)_C$. (If the calculations of VandenBerg and Bell 1985 are replaced by those of Buser and Kurucz 1992, this estimate favors $(R-I)_C$ even more decisively.)

7. Deriving $(R-I)_C$ For The Sun

If $(R-I)_C$ replaces $B-V$, one has taken a step toward an analysis which allows minimum scope for blanketing effects. To be sure, this aim is also satisfied by using solar twins, as noted above. However, the change in colour index allows the use of many more data than are available if one restricts the analysis to solar twins. To secure this advantage, one collects published indices for the Sun and field stars which are insensitive to metallicity. If necessary, one also imposes the condition that $[Fe/H](*) \sim [Fe/H](\odot)$.

There are a number of indices that can be used for this problem. Photometry yields measurements of Hα and Hβ. From spectroscopy, one can use temperatures from Balmer-line wings, Gray's (1995) line-strength ratios, and excitation temperatures. As a precaution, the latter are adopted only if equivalent widths have been measured by the same observer for both the Sun and the field stars. [Taylor (1992) has done this kind of calculation, but without using Gray's ratios, excitation temperatures, or Balmer-line results published by Chmielewski et al. (1992) and Friel et al. (1993).]

Results from analyses based on these indices are given in Table 1. The analyses turn out to be very insensitive to the exact value one assumes for the blanketing derivative of $(R-I)_C$, but corrections based on that derivative have been made nonetheless. No allowance has been made for reddening, since the stars used for the analyses should be too close to the Sun for even small values of reddening to interfere (see, for example, Leroy 1993).

The results from Fuhrmann et al. (1994) attract first attention because they stand off from all the others. According to the Dixon (1951) statistics, the result from the Fuhrmann et al. Hα measurements may be rejected at 99% confidence. For this reason and because the Fuhrmann et al. data are on two different zero points, it seems fair to set all of those data aside. The next issue deserving notice is the scatter in the remaining results. The rms error per entry turns out to be 0.0068 mag (0.018 mag when rescaled

TABLE 1. Results for the solar value of $(R-I)_C$

$(R-I)_C$	Original (secondary) source	Index
(0.300)	Fuhrmann et al. 1994	Balmer-line wings (Hα)
(0.307)	Fuhrmann et al. 1994	Balmer-line wings (other lines)
0.325	Herbig 1965 group	Excitation temperatures
0.328	(Taylor 1992)	Balmer-line wings, group
0.331	(Taylor 1992)	Balmer-line wings, group
0.332	Clegg et al. 1981	Excitation temperatures
0.332	(Taylor 1992)	Balmer-line wings, group 2
0.334	Wallerstein 1962	Excitation temperatures
0.334	Price 1966 (Taylor 1992)	Photometric Hα
0.337	Olsen 1976 (Taylor 1992)	Photometric Hβ
0.338	Gray 1995	Gray's ratios
0.343	(Taylor 1992)	Balmer-line wings, group 8
0.349	(Taylor 1992)	Balmer-line wings, group 4

to the $B-V$ dynamic range). These "external" errors are larger than the "internal" errors one can derive for the various entries. The external errors cannot be sampled adequately by using only a small number of stars, so one expects an analysis using solar twins to yield underestimated errors (and perhaps a systematic bias as well). This is an argument for performing the analysis done here instead.

By averaging the tabular data (except those of Fuhrmann et al.) and using the methods of Taylor (1992), one finds that $(R-I)_C = 0.335 \pm 0.002$ mag, $(V-K)_J = 1.474 \pm 0.012$ mag, and $B-V = 0.628 \pm 0.008$ mag. The quoted value of $(R-I)_C$ differs from that of Taylor (1992) by only 1σ. The quoted value of $B-V$ is an updated version of $B-V$ for the "long-wavelength group" mentioned above. This updated version appears to agree with Cayrel de Strobel's (1996) most recent result (0.642 ± 0.004 mag) to within the errors. However, it must not be forgotten that the redder value of $B-V$ from the "short-wavelength group" remains unexplained.

8. Summing Up: A Question Of Strategy

Given the agreement between Taylor's (1992) value of $(R-I)_C$ and the updated value, it seems fair to conclude that the $(R-I)_C$ analysis is mature in some sense. However, it is certainly not definitive, since one must expect it to be repeated yet again in the future. Note, in addition, that nothing has been said here about $(R-I)_C$ from the solar irradiance curve. For the present, it seems reasonable to say that if one uses $(R-I)_C$ in a minimum-metallicity analysis, the results look promising. Now the question to be settled is whether $B-V$ or $(R-I)_C$ is to be the colour index of choice. Astronomers are invited to consider the evidence and then make this choice.

References

Buser, R., and Kurucz, R. L. 1992, A&A 264, 557
Cayrel de Strobel, G. 1996, A&AR 7, 243
Cayrel de Strobel, G., and Bentolila, C. 1989, A&A 211, 324
Chmielewski, Y., Friel, E., Cayrel de Strobel, G., and Bentolila, C. 1992, A&A 263, 219
Clegg, R. E. S., Lambert, D. L., and Tomkin, J. 1981, ApJ 250, 262
Cousins, A. W. J. 1974, M. N. A. S. So. Africa 33, 149
Croft, S. K., McNamara, D. H., and Feltz, K. A. Jr. 1972, PASP 84, 515
Dixon, W. J. 1951, Ann. Math. Stat. 22, 68
Edvardsson, B., Andersen, J., Gustafsson, B., Lambert, D. L., Nissen, P. E., and Tomkin, J. 1993, A&A 275, 101
Friel, E., Cayrel de Strobel, G., Chmielewski, Y., Spite, M., Lèbre, A., and Bentolila, C. 1993, A&A 274, 825
Fuhrmann, K., Axer, M., and Gehren, T. 1994, A&A 285, 585
Gray, D. F. 1995, PASP 107, 120
Griffin, R. E. M., and Holweger, H. 1989, A&A 214, 249
Hardorp, J. 1978, A&A 63, 383
Herbig, G. H. 1965, ApJ 141, 588
Leroy, J.L. 1993, A&A 274, 203
Nicolet, B. 1978, A&AS 34, 1
Olsen, E. H. 1976, A&A 50, 117
Price, M. J. 1966, MNRAS 133, 449
Sandage, A., and Smith, L. L. 1963, ApJ 137, 1057
Taylor, B. J. 1986, ApJS 60, 577
Taylor, B. J. 1992, PASP 104, 500
Taylor, B. J. 1994, PASP 106, 444
Taylor, B. J. 1996, ApJS 102, 105
Taylor, B. J., and Joner, M. D. 1996, AJ 111, 1338
VandenBerg, D. A., and Bell, R. A. 1985, ApJS 58, 561
Wallerstein, G. 1962, ApJS 6, 407

DISCUSSION

ROGER BELL: Have you considered determining the solar $V-R$, since this showed to be even less sensitive to abundance (as shown by the Barnes Evans effect) and would also avoid the extinction prob with the O_2 and H_2O bands in the I filter passband?

BENJAMIN TAYOR: Extinction problems in the I filter don't appear to compromise either the accuracy of the precision of $R-I$. Morec - if I remember rightly - the theoretical colours show that $V-R$ is actually more metallicity-sensitive than $R-I$. Since $V-R$ has also been measured much less often in the northern hemisphere the choice in favor of $R-I$ seems to be clearcut for the moment.

THE ISO-SWS FLUX STANDARD STARS: SYNTHETIC SPECTRA AND OBSERVATIONS

N.S. VAN DER BLIEK
Sterrewacht Leiden, The Netherlands

P.W. MORRIS
SRON Utrecht, The Netherlands

B. VANDENBUSSCHE
Instituut voor Sterrenkunde, K.U. Leuven, Belgium

L.B.F.M. WATERS AND P. ZAAL
Sterrenkundig Instituut Amsterdam, The Netherlands

R.A. BELL
Astronomy Department, University of Maryland, USA

B. GUSTAFSSON AND K. ERIKSSON
Uppsala Observatory, Sweden

AND

TH. DE GRAAUW
SRON Groningen, The Netherlands

1. Introduction

We present flux-calibrated, synthetic spectra for the calibration stars of the Short Wavelength Spectrometer of the Infrared Space Observatory ISO-SWS (cf. Kessler et al. 1996). ISO-SWS covers a wavelength range of 2.4 to 45 μm (cf. de Graauw et al. 1996), and although in the NIR the flux calibration of Vega can be used, at longer wavelengths the flux calibration has to be extrapolated by other means, because (i) Vega has at $\lambda \geq 25\,\mu$m an IR excess due to circumstellar dust and (ii) there is an, as yet, unresolved controversy about Vega's flux at 10 μm (cf. Rieke et al. 1985). The majority of the standard stars of SWS are of MK class G and K III, as cool giants are amongst the brightest objects in the IR, and model atmospheres for cool giants are available and well-studied (cf. Jørgensen and Gustafsson 1994).

T.R. Bedding et al. (eds.),
Fundamental Stellar Properties: The Interaction between Observation and Theory, 89–92.

To obtain flux-calibrated synthetic spectra for these stars we take the following steps: (i) we adopt the absolute calibration of Vega by Mégessier (1995), $F_{555.6\mathrm{nm}} = 3.56 \times 10^{-23}\mathrm{Wm}^{-2}\mathrm{Hz}^{-1}$; (ii) we extrapolate this flux calibration into the NIR, with the spectrum of Vega by Dreiling and Bell (1980); (iii) we convert NIR narrowband K photometry of the standard stars into flux densities; and (iv) we extrapolate the flux calibration over the complete wavelength region of SWS by means of synthetic spectra of the standard stars. The *Kn* band is located in a relative clean spectral region of the spectra of G and K giants as well as in a clean region of the atmospheric transmission and, moreover, late G and K stars emit most of their energy in the NIR.

2. Synthetic spectra

Synthetic spectra were generated using the model atmosphere code MARCS (Gustafsson et al. 1975, updated version) and the spectrum generating program by Bell (Bell and Gustafsson 1989; Bell 1993). For Vega we use the model atmosphere by Dreiling and Bell (1980). Stellar effective temperatures T_{eff} are determined by fitting synthetic colours of a small grid of stellar atmospheres to the observed colours of the stars. Stellar surface gravities and metallicities are taken from the literature. For HR 5755, HR 5981 and HR 7341 no literature data is available, and their parameters are based on their spectral classification.

For each star we have calculated optical and IR colours by folding the synthetic spectra with filter passbands. Zero points were established using the synthetic spectrum of Vega and for the NIR colours the spectra were also folded with the earth's atmospheric transmission. The comparison between observed and synthetic colours is very satisfactory: the average differences are of the same order of magnitude as the observational uncertainties.

We have also calculated the IRAS PSC fluxes and find that the synthetic 12 and 25 μm fluxes are systematically too low, although the differences in the 12 μm fluxes are well within the observational uncertainties. Bell (1993) calculated IRAS fluxes for G and K giants, whereby he corrected for the geometrical dilution using stellar diameters. His sample includes three of the SWS flux-calibration stars and the comparison with our results is reassuring: the calculated fluxes are, within the observational uncertainties, equal for both studies and the calculated fluxes agree with the observed fluxes.

3. Uncertainties in the flux-calibrated spectra

The uncertainties in the continuum fluxes of the calibration stars are of the order *2.5 to 4.5 %*. This includes: (a) the accuracy of the absolute flux calibration of Vega (0.7 % Mégessier 1995); (b) uncertainties in the extrapolation to the NIR. We estimate on the basis of the results of van der Bliek et al. (1996) for their hottest model, that uncertainties in Vega's synthetic spectrum due to uncertainties in the temperature structure of the model atmosphere are of the order of 1.5 %; (c) uncertainties in the observed narrowband photometry which are of the order of 0.02 mag.; and (d) the accuracy with which the continuum far-IR fluxes are predicted, which for cool stars is of the order of 2 to 4 % (van der Bliek et al. 1996).

Uncertainties in the detailed spectra, especially for $\lambda \geq 5\mu$m, are likely to be larger than the uncertainties quoted here. The discussion by van der Bliek et al. was confined to continuum fluxes; errors in the detailed spectra have not been studied so far. Moreover, IR line lists have, as yet, not been check in detail against high-resolution spectra, and might be found to be far from complete, and need much improvement.

4. Comparing with the SWS observations

The spectra of the flux-calibration standard stars are used to derive the photometric responses of the SWS grating detectors in dedicated calibration observations: the measured flux at so-called "key" wavelengths, within a narrow bandpass, is compared with the synthetic spectra (cf. Schaeidt et al. 1996). The instrument calibration and synthetic spectra will be improved iteratively, by comparing the observed and synthetic spectra. In addition, a comparison between observed and synthetic spectra can be used to asses the reliability of the *photometric* calibration.

Fig. 1 shows such a comparison for γ Dra for the CO fundamental (Δv = 1). The observed SWS spectrum was obtained in the slowest AOT1 observing mode, with the full wavelength range and $\lambda/\Delta\lambda \simeq$ 1500-2000. The signals from each array of 12 detectors covering the 1st overtone and fundamental bandheads were flatfielded to the median response, sigma-clipped, and rebinned to instrumental resolution. The uncertainty in the relative spectral response is about 2-3%; the uncertainty in the absolute flux levels is of the order of 10-12 %. The synthetic spectra presented in Fig. 1 have been rebinned to the instrumental resolution and offseted for clarity.

The overall agreement between the observed and synthetic $^{12}C^{16}O$ bandhead profiles is quite excellent. Some mismatch in the weaker $^{13}C^{16}O$ $\Delta v =$ 2 bandhead strengths are present, possibly due $T(\tau)$ or $P(\tau)$ of the model atmospheres, or to the $^{12}CO/^{13}CO$ ratio. However, this has little affect on SWS calibration over the 0.08μm bandpass, which is centered at 2.48μm.

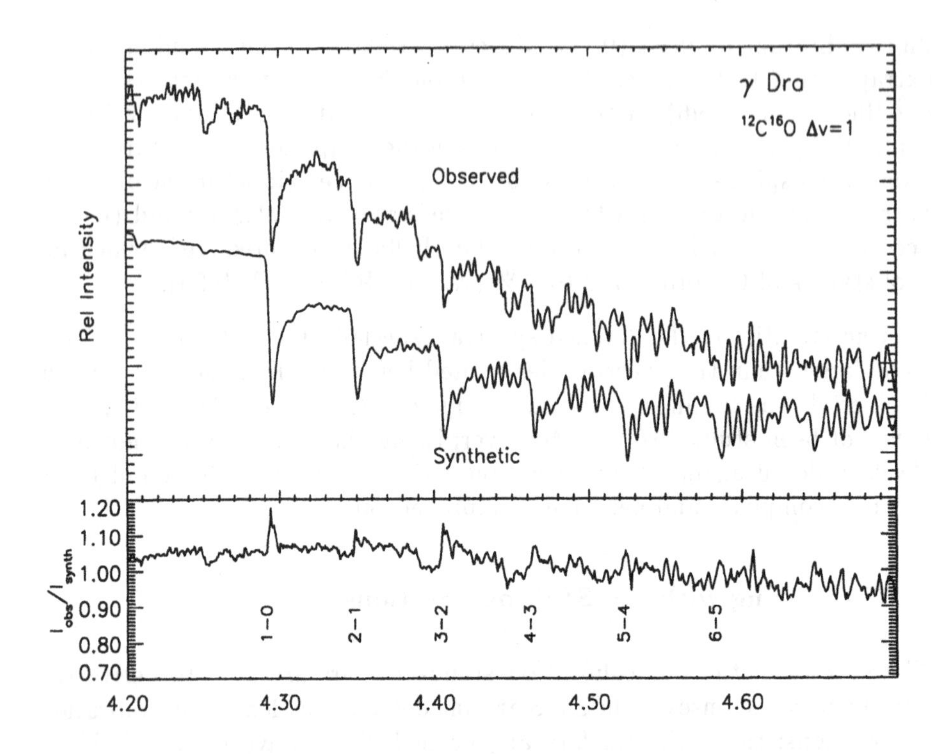

Figure 1. A comparison between observed and synthetic spectra for the fundamental ($\Delta v = 1$) CO bandheads in the spectrum of γ Dra. The top panel shows both the observed (top) and synthetic (bottom) spectrum. The bottom panel shows the observed spectrum relative to the synthetic spectrum, in order to show in more detail the differences between observations and calculations.

Acknowledgements LBFMW acknowledges financial support from the Royal Netherlands Academy of Art & Sciences KNAW and through a "Pionier" grant from NWO.

References

Bell R.A. 1993, MNRAS 264, 345
Bell R.A., Gustafsson B. 1989, MNRAS 236, 653
de Graauw T., et al. 1996, A&A 315, L49
Dreiling L.A., Bell R.A. 1980, ApJ 241, 736
Gustafsson B., Bell R.A., Eriksson K., Nordlund Å. 1975, A&A 42, 407
Kessler M.F., et al. 1996, A&A 315, L27
Mégessier C. 1995, A&A 296, 771 (M95)
Rieke G.H., Lebofsky M.J., Low F.J. 1985, AJ 90, 900
Schaeidt S.G., et al. 1996, A&A 315, L55
van der Bliek N.S., Gustafsson B., Eriksson K. 1996, A&A 309, 849

UV ENERGY DISTRIBUTIONS OF ACTIVE COOL STARS.

C. MOROSSI AND M. FRANCHINI
Osservatorio Astronomico di Trieste, Via Tiepolo 11, I-34131, Trieste, Italy

AND

M.L. MALAGNINI
Dipartimento di Astronomia, Università degli Studi di Trieste, Via Tiepolo 11, I-34131, Trieste, Italy

Abstract. We present semi-empirical models based on the "minimum-temperature-concept" in order to provide a description of the UV spectra of solar type stars consistent with the observational data.

1. Introduction

The analysis of spectral energy distributions (SED's) of cool stars received a strong incentive in the last years from the availability of improved theoretical flux distributions (Kurucz 1993, hereafter K93; Edvardsson et al. 1993, hereafter EAGLNT). Recently, the predicted theoretical fluxes have been compared with observations of K (Malagnini et al. 1992, Morossi et al. 1993), F, and G stars (EAGLNT). The main result of the analysis of the K stars is the failure of K93 models in predicting the near UV (2200 $< \lambda <$ 3000 Å) SED's of stars showing chromospheric features. Morossi et al. (1993) presented semi-empirical models (hereafter Tmin models), based on a modified temperature structure with respect to the K93 one, obtained by taking into account the presence of a minimum temperature region ($T = T_{\mathrm{min}}$) in the upper photosphere. The fluxes computed starting from this new set of models provide a fairly good description of the observations not only in the visible and IR regions but also in the UV. In this paper we will extend the "minimum-temperature-concept" approach to effective temperatures higher than those studied in Morossi et al. (1993). In

T.R. Bedding et al. (eds.),
***Fundamental Stellar Properties: The Interaction between Observation and Theory*, 93–96.**

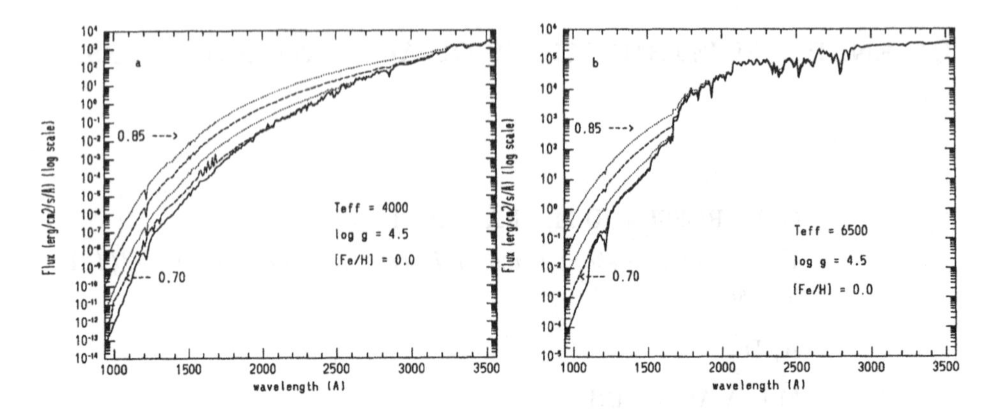

Figure 1. Ultraviolet fluxes computed from K93 "standard" models (solid lines), and from Tmin models with four different values of $R = \frac{T_{\rm min}}{T_{\rm eff}}$ (i.e. 0.70, 0.75, 0.80, 0.85). The cases $T_{\rm eff}$=4000 K and $T_{\rm eff}$=6500 K (log g=4.5 dex) are shown in a) and b) respectively

particular, we will analyse a set of stars which are proposed as "candidate solar analogs" (see Cayrel de Strobel 1996 for further details).

2. Tmin models

The nine K93 models characterized by the $T_{\rm eff}$, log g and $[M/H]$ values given in Table 1 have been modified according to the following formula:

$$T_{\rm emp}(\tau) = max(T(\tau), R \times T_{\rm eff})$$

where R is the ratio $\frac{T_{\rm min}}{T_{\rm eff}}$. The ATLAS9 code was then used to compute the energy distributions. Figure 1 shows the UV SED's for two $T_{\rm eff}$'s for illustrating the differences introduced by the presence of a Tmin region in the $T(\tau)$ atmosphere structure. It results that the use of Tmin models produces an enhancement of the ultraviolet fluxes due to a lower blanketing. The higher the $\frac{T_{\rm min}}{T_{\rm eff}}$ ratios, the larger the effect. The wavelength shorter of which the increase in the UV fluxes becomes significant depends mainly on the $T_{\rm eff}$ of the original K93 model. The higher the $T_{\rm eff}$, the shorter the limiting wavelength.

3. Observations versus predictions

In this paper we pay attention to a narrow temperature range around the solar value. EAGLNT introduced six UV bands (see their Table 5): at the shortest wavelength, band (A), they found that "the model flux is typically

TABLE 1. Program stars and results

HD	B-V	T_{eff}	log g	[Fe/H]	$\frac{T_{min}}{T_{eff}}$
1835	0.66	5781.	4.50	0.18	0.83
10307	0.62	5538.	4.40	0.07	0.85
20630	0.68	5631.	4.45	0.02	0.85
25680	0.62	5794.	4.30	-0.03	0.84
30495	0.63	5915.	4.40	-0.02	0.82
44594	0.66	5777.	4.40	0.15	0.80
72905	0.62	5843.	4.40	-0.11	0.85
95128	0.61	5860.	4.31	-0.02	0.81
186427	0.66	5765.	4.30	0.05	0.80

TABLE 2. Comparison of Tmin and K93 model predictions

T_{eff}	$\frac{T_{min}}{T_{eff}}$	A	B	C	D	E	F
6500	0.85	-0.76	-0.22	-0.11	-0.06	-0.04	-0.02
	0.80	-0.37	-0.06	-0.02	-0.02	-0.01	-0.00
	0.75	-0.10	-0.00	-0.00	-0.00	-0.00	-0.00
	0.70	-0.00	0.00	0.00	0.00	0.00	0.00
6000	0.85	-1.34	-0.72	-0.43	-0.12	-0.06	-0.04
	0.80	-0.88	-0.35	-0.16	-0.04	-0.02	-0.01
	0.75	-0.42	-0.11	-0.04	-0.00	-0.00	-0.00
	0.70	-0.14	-0.03	-0.01	-0.00	-0.00	-0.00
5500	0.85	-1.75	-1.12	-0.79	-0.26	-0.08	-0.06
	0.80	-1.23	-0.65	-0.38	-0.08	-0.02	-0.01
	0.75	-0.67	-0.24	-0.10	-0.01	-0.00	0.00
	0.70	-0.18	-0.03	-0.00	0.00	0.00	0.00

five times lower than the observed fluxes". Analogously, we computed in the same bands the logarithms of the ratios between K93 and Tmin model fluxes for $T_{eff} = 5750$ K (see Table 2).

The ratios in Table 2 show the same trend with wavelength as those reported in EAGLNT Table 5. Actually, a good agreement between observations and theoretical fluxes can be obtained by using Tmin models characterized by the R values given in Table 1 (see for example fig. 2).

Our analyses suggest that, for active cool stars, Tmin models are well suited for interpreting the observed ultraviolet fluxes. Our results may also explain why, for the Sun, EAGLNT do not find deviations in the A passband between observations and their solar model. In fact, assuming that the Tmin

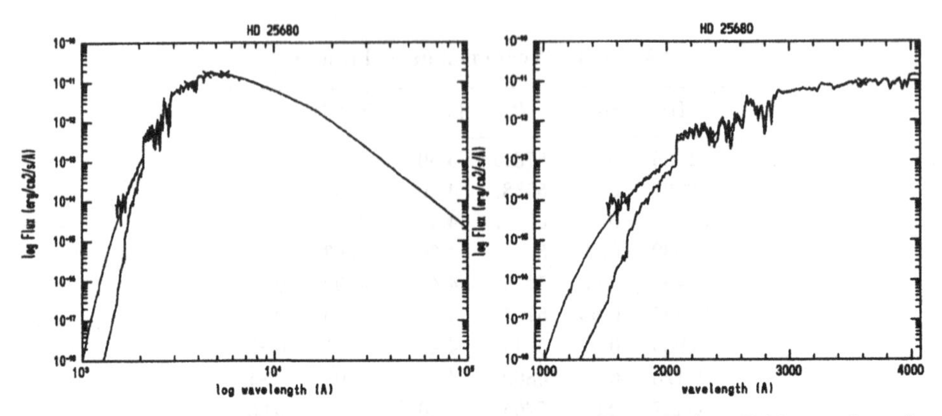

Figure 2. Observed and predicted fluxes for the star HD 25680 (see Table 1 and text)

models are required to take into account stellar activity, the agreement found by EAGLNT in the case of the sun can be understood since the far UV solar observations they used "correspond to regions of the quiet Sun, where the inhomogeneities have been smoothed out and all active regions avoided" (Samain 1979).

Acknowledgements

Partial support from CRA 60% (Osservatorio Astronomico di Trieste) and from MURST 40% and 60% (Università degli Studi di Trieste) grants is acknowledged.

References

Cayrel de Strobel, G. 1996, A&AR 7,243

Edvardsson B., Andersen J., Gustafsson B., Lambert D.L., Nissen P.E., Tomkin J. 1993, A&A 275, 101 (EAGLNT)

Kurucz R.L. 1993, CD-ROM 13, ATLAS9 Stellar Atmosphere Programs and 2 km/s Grid (Cambridge: Smithsonian Astrophys. Obs.) (K93)

Malagnini M.L., Morossi C., Buser R., Parthasarathy M. 1992, A&A 261, 558

Morossi C., Franchini M., Malagnini M.L., Kurucz R.L., Buser R. 1993, A&A 277, 173

Samain D. 1979, A&A 74, 225

DISCUSSION

ANDREAS KELZ: Do you expect that the 'minimum-temperature concept' would be a good description for main sequence stars hotter than K and G as well? Alternatively, for which stars wouldn't the Tmin layer make any difference in the UV?

CARLO MOROSSI: I still have to look at F type stars, but I suppose that for these kind of stars you'll have to check EUV fluxes, probably shortward of 912Å, which is very difficult from an observational point of view.

5. STELLAR MASSES AND SURFACE GRAVITIES

PRECISE STELLAR MASS AND MASS-LUMINOSITY DATA

J. ANDERSEN

Astronomical Observatory; Niels Bohr Institute for Astronomy, Physics, and Geophysics; Juliane Maries Vej 30, DK-2100 Copenhagen, Denmark

Abstract. Recent progress in observing and data reduction methods for precise mass and mass-luminosity determinations in binary systems are briefly reviewed. The foundations appear to have been laid for a new burst of accurate data. Detailed model simulations of the individual systems are the best way to use these data to critically test the theoretical models and advance our understanding of the evolution of single and binary stars.

1. Introduction

Second in importance only to its actual existence (Maeder, this meeting!), the *mass* of a star is its most fundamental physical property, and its apparent luminosity is usually the parameter first observed. Similarly, stellar models are usually specified primarily by their initial mass, and their luminosity is computed so as to allow comparison with the real stars. Because luminosity, like most other observable stellar properties, depends very sensitively on mass, great accuracy is required of empirical data that are to be used in tests of theoretical models.

Precise stellar mass and radius determinations, including the techniques, available data, and their interpretation, were reviewed in detail a few years ago (Andersen, 1991). Here, we recall some of the uses of these data and the accuracies required for various purposes, review recent improvements in the techniques for precise mass and mass-luminosity determinations, and illustrate some recent applications. To avoid duplication, this review will be limited to stars more massive than ~0.75 $M_{\odot}$, while stars below this limit are reviewed in the following paper by M. Mayor.

T.R. Bedding et al. (eds.),
***Fundamental Stellar Properties: The Interaction between Observation and Theory*, 99–108.**

2. Needs for Precise Stellar Mass and Mass-Luminosity Data

Accurate, fundamental stellar mass and mass-luminosity data can be derived only from precise studies of stars in binary systems. Suitable systems must be selected so as to yield accurate results for the individual stars, and only data from non-interacting systems can be expected to be valid for normal, single stars.

The data are used in a variety of studies in galactic and stellar astronomy. In galactic astronomy, applications include:

- Binary stars with well-determined radii and effective temperatures, hence luminosities, can be used as distance indicators within the Milky Way and to nearby galaxies (see, e.g., Bell et al. 1991, 1993), mass being essentially a by-product in this context.
- Data for well-defined stellar types can be used to estimate the total mass of the components of our own and other galaxies (halo, bulge, disk).
- Similarly, accurate mass-luminosity data are needed when comparing the mass density in the disk as determined from an inventory of the local stars with dynamical estimates, from which the amount of any remaining dark matter follows.
- Mass-luminosity data are used to infer total masses of (open and globular) star clusters, estimating the IMF in such clusters, and assessing the degree and time scale of their dynamical evolution.
- Finally, on the border between galactic and stellar astronomy, mass and mass-luminosity data are essential in establishing ages for subgroups of stars that can define an age scale in our galaxy and help clarify the sequence of formation of its major components.

Examples of the use of precise mass and mass-luminosity data in stellar astrophysics include:

- Calibration of the fundamental properties of single stars.
- Validation of models of single-star structure and evolution.
- Study of the action of tidal forces on stellar interiors (internal rotation, rotational synchronization, orbital circularization).
- Observational constraints on models of close binary evolution.

In view of sampling and other basic uncertainties, mean errors of 5% or so in mass and/or luminosity are satisfactory for most applications in galactic astronomy. In stellar astrophysics, however, demands are stricter: Mass differences due to evolution within the main sequence for a given spectral type or colour can easily amount to ~30%, while abundance variations in disk stars of the same colour and $\log g$ may correspond to mass differences of only ~5%. Accordingly, stellar masses to be used in critical tests of stellar evolution models must be accurate to ~1% or better in order to yield truly useful results (Andersen, 1991).

Fortunately, these demands can now be routinely met in suitable eclipsing or spectroscopic-interferometric binary systems using modern techniques, which will be outlined briefly in the following. It is essential to remember, however, that even the most accurate mass has no useful astrophysical application unless the nature of the star itself is specified to matching precision: Giving a number of, say, 1.500 $M_{\odot}$ for the mass of a given star conveys no useful information whatever if all that is otherwise known is that it is an F5 main-sequence star – a class within which actual masses vary by 30% and luminosities by factors of several. To allow meaningful interpretation, accurate masses and luminosities must be accompanied by precise indicators of age (radius or $\log g$ being the most direct and sensitive observational diagnostics) as well as chemical composition in order to derive conclusions beyond those obtainable from standard handbooks.

3. Advances in Observational Techniques

Optical interferometers may eventually yield accurate absolute rather than relative astrometric orbits for stars in binary systems. Until then, mass ratios and, thus, *individual* binary masses will continue to require accurate spectroscopic orbits for both components. These must be combined with a determination of the inclination of the orbital plane, which in eclipsing systems is done from a light-curve analysis and in non-eclipsing systems from a visual/interferometric orbit. Because the derived masses are proportional to the third power of the radial-velocity amplitudes, these are always critical for the accuracy of the result. In low-inclination systems the inclination may become equally important, however, a point worth recalling as improved interferometers bring more spectroscopic/visual binaries within reach.

3.1. PROGRESS IN SPECTROSCOPIC ANALYSIS METHODS

The accuracy of the derived spectroscopic orbital elements depends, first, on the resolution and S/N ratio of the spectra from which the radial velocities are measured. The great strides made in recent years in instrument and detector technology have also brought great advances in the quality of the spectra available for mass determinations in spectroscopic binary stars, and hence in both their accuracy and efficiency (Popper, 1993). Equally importantly, however, new data analysis techniques have been developed which considerably improve the reliability of radial velocity determinations from the blended spectra of double-lined binaries, in which no single spectral feature necessarily lends itself to a "clean" measurement of any one component. These methods fall in three basic classes:

Popper & Jeong (1994) studied the accuracy of radial velocities derived by cross-correlation of individual diffraction orders in cross-dispersed echelle spectra, and the consistency of results from different orders of the same spectrum. In continuation of earlier work by Popper & Hill (1991), they constructed synthetic double-lined binary spectra from broadened and co-added spectra of single stars to test the effect of their reduction procedures, and to derive corrections to initial radial velocities determined from simple one-dimensional cross-correlations. This basic, but powerful technique for validating the results is now generally used by all authors.

Simon & Sturm (1993) developed the so-called "disentangling" technique for double-lined spectra, somewhat analogous to the Doppler tomographic algorithm developed by Bagnuolo & Gies (1991). In the "disentangling" method one determines, by a single-value decomposition technique, the set of spectroscopic orbital elements plus two best-estimate component spectra which together yield the best overall fit to an ensemble of double-lined spectra of the system, obtained over a wide range of orbital phases. In addition to its potential for accurate orbit and mass determinations, the method has the considerable merit of yielding optimum mean, high S/N spectra of both stars. Thus, conventional model atmosphere analyses can be performed to determine individual effective temperatures and chemical compositions for the two components – key information which has previously been buried in the blended spectra. While final documentation of the performance of the method as regards the ultimate precision obtainable for the masses is still in progress, preliminary indications are very encouraging (Maxted, 1996).

A different approach is taken by Zucker & Mazeh (1994), who explicitly model the blended spectrum of a double-lined binary by a two-dimensional cross-correlation technique and develop an elegant, efficient algorithm (TODCOR) for the corresponding computations. The technique appears to work very well, even on spectra covering a short spectral range and correspondingly few lines, but care is needed with end masking of the spectra and other fine details of the cross-correlation procedure in order to avoid small residual velocity errors. Latham et al. (1996) find no significant systematic errors when applying TODCOR with optimised synthetic template spectra to the equal-component eclipsing binary DM Vir, but later experience indicates that the precaution of checking the results on a set of synthetic binary spectra constructed for an initial set of velocities – simple with today's computing power – should always be taken.

These techniques carry the promise that stellar masses can be determined with errors as low as perhaps 0.3%. At this level of accuracy, the time-honoured value 1.0385×10^{-7} for the numerical constant in the classical formula for the masses of a spectroscopic binary is actually noticeably wrong: As pointed out by Torres (1995, priv. comm.), modern values for the solar mass and astronomical unit give a value of 1.036055×10^{-7} - almost 0.2% lower than the canonical number(!) and a non-negligible difference in front-line work (Hummel et al., 1994).

Finally, in close systems, tidal deformation and mutual irradiation of the components will affect the measured radial velocities and masses to some degree. If a physical model of the binary is constructed, as routinely done for eclipsing systems, these effects can either be approximated as luminosity-weighted mean corrections to the measured radial velocities (Wilson, 1990) or directly modelled in the line profiles of a synthetic binary spectrum subjected to the same analysis as the real binary (Hill, 1993).

3.2. ADVANCES IN INTERFEROMETRY

Given accurate double-lined spectroscopic orbital elements, information on the orbital inclination is needed in order to compute absolute masses. In eclipsing binary systems, this is derived from an analysis of the light curves (see, e.g., Andersen 1991 for further references on the subject). In non-eclipsing systems, the inclination is derived from the apparent orbit on the sky as determined, with increasing accuracy, from visual, speckle, or long baseline interferometric observations. Because, again, masses are proportional to the cube of sin i and the major axis of the visual orbit, demands on accuracy are very high, and data from non-eclipsing systems have traditionally not met the selection criterion of 2% individual accuracy required for critical tests of stellar evolution models (Andersen, 1991).

However, the coming-of-age of interferometric techniques is now changing the picture. Capella (α Aur) provides a good illustration of the progress: From new spectroscopic and speckle data, Barlow et al. (1993) redetermined the masses of the components to an accuracy of ~2%, with about equal contributions to the error from the two types of data. Shortly after, however, Hummel et al. (1994) determined a long baseline orbit from the Mark III interferometer which improved the accuracy of the apparent orbit by an order of magnitude (to some 0.1%) and sent the ball right back in the spectroscopists' court. Because both stars are in quite rapid evolutionary phases, knowing the masses to better than 1% does have significant impact on the precision of the astrophysical interpretation.

In spectroscopic/interferometric binaries, the orbital parallax and hence luminosity of the system result directly from the combination of absolute and angular orbital dimensions. If the luminosity ratio can be accurately established from the data, such systems yield direct anchor points on the mass-luminosity relation(s), see below. In eclipsing systems, the absolute stellar radii are first determined by combining spectroscopic and photometric orbital elements, and luminosities follow when effective temperatures can be assigned from observed colours and a suitable temperature scale; the distance of the system is a by-product, but usually only of secondary interest. Conversely, one may use effective temperatures in a spectroscopic/interferometric binary to determine absolute stellar radii from accurate observed luminosities if the angular diameters cannot, as in Capella (Hummel et al., 1994), be determined directly from the interferometric data.

Finally, we note that spectroscopic/interferometric data, even when of less than ultimate accuracy, remain valuable in special cases: While binary masses with errors of the order of 15% are not particularly useful in themselves at the present stage, an absolute distance determination to the systems themselves with an error of 5% certainly is when these systems happen, e.g., to be members of the Hyades (Torres et al., 1997).

4. The Data and Their Interpretation

Since the earlier review of accurate masses and radii (Andersen, 1991), not many new determinations in that category have been added[1]. A particular need is for more metal abundance data, a crucial parameter for the interpretation. Fig. 1 shows the basic distribution of the data in a way that highlights the fine structure of evolution (mostly) within the main-sequence band. No pretence to completeness is made, and in view of ongoing work to provide more data for cool stars (Clausen et al. 1997 and unpublished work by others), any such list would be rapidly superseded anyway. New data are included, however, for V539 Ara (Clausen, 1996), Capella (Hummel et al., 1994), DM Vir (Latham et al., 1996), RT And and GG Cyg (Popper, 1994), and CM Dra (Metcalfe et al., 1996), supplementing the previous list, primarily at lower masses.

4.1. OVERALL RELATIONS

Figs. 2 and 5 of the earlier review (Andersen, 1991) showed overall colour-mass and mass-luminosity relations for the sample; not much would be gained by repeating them here for the slightly larger sample. At first sight, these relations are deceptively tight and give the impression of a unique re-

[1]Dr. D.M. Popper points out an unfortunate error in the entry for EW Ori B in Table 1 of Andersen (1991): The correct data are $R = 1.090 \pm 0.011 R_\odot$, $\log g = 4.426 \pm 0.010$, $\log L = 0.08 \pm 0.03$, and $M_v = 4.64 \pm 0.07$.

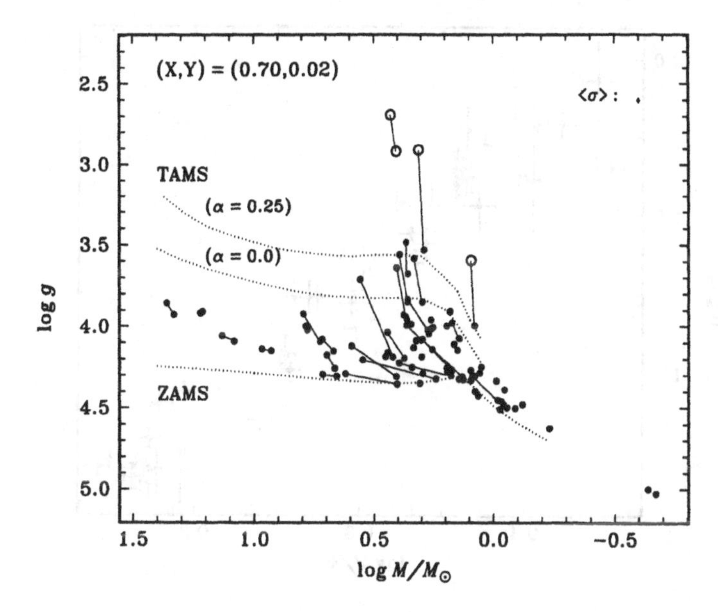

Figure 1. log M – log g diagram for binary components with individual masses and radii known to within ±2% (see text). Note that typical errors are smaller than the plotted symbols. Circles denote giant stars; lines connect members of the same system. Model boundaries are from Claret (1995).

lation with a bit of observational scatter added. And the model boundaries in Fig. 1 do generally contain the observed points. Apparently, observation and theory agree and harmony reigns.

However, difficulties set in the moment one begins to fully exploit the completeness and accuracy of the data. Stothers & Chin (1991) used model boundaries and observed binary components in the mass-radius and mass-luminosity diagrams to show that, assuming a high enough metal abundance, standard models (i.e. no overshooting) could accommodate the data. Yet, Andersen et al. (1990) showed that such overall agreement can be achieved without the models fitting individual systems satisfactorily at all. And the near-ZAMS system GG Lup is found (Andersen, 1991) to actually lie *below* the ZAMS unless a sub-solar metallicity is assumed.

A recent comparison of mass estimates from modern solar-abundance models with precise binary data (Schönberner & Harmanec, 1995) finds "remarkably good" agreement from 1.3 to 25 $M_\odot$ – again reassuring at first sight. But surely the conclusion cannot be that all stars have the same abundance? The deviations from the 45° line in their Fig. 4 are small indeed, but are they explained by the observational errors? In other words, does the diagram tell us whether or not current stellar evolution models are adequate to account for the best existing observations?

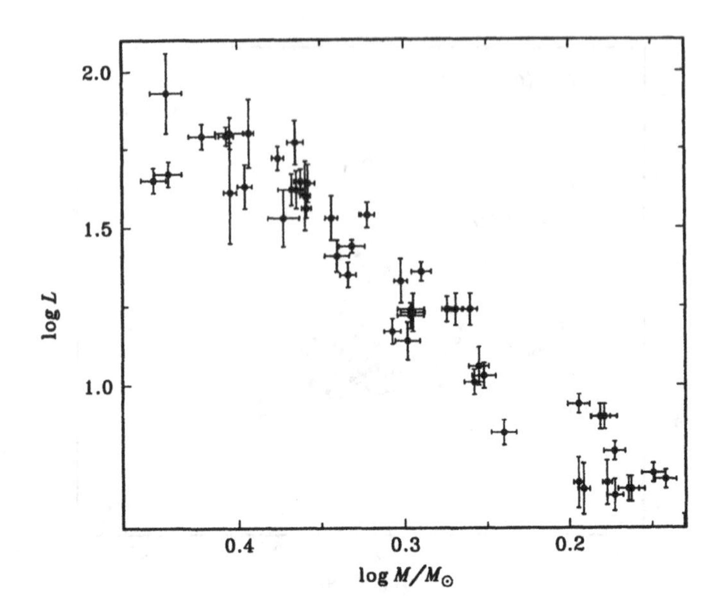

Figure 2. $\log M - \log L$ diagram for a subset of the binaries in Fig. 1.

Fig. 2 illustrates the same point by zooming in on a small section of the mass-luminosity diagram. The apparently smooth overall relation obviously breaks up, and the deviation of individual stars from a mean relation are clearly significant. This is fundamentally no surprise, because the luminosity of a stellar model depends on both age and chemical composition. What Fig. 2 tells us is that, when assessing the performance of contemporary stellar models, studying average trends must be replaced by accurate modelling of individual systems, respecting all possible observational constraints.

Although published almost a decade ago, the study of the solar-type stars in AI Phe by Andersen et al. (1988) probably remains the best example of just how far one can go in the detailed modelling of a well-observed binary; but the similar studies of TZ For (Andersen et al., 1991) and Capella (Barlow et al., 1993) also yielded much useful insight into the stellar and tidal evolution of somewhat more massive stars. Much progress in the models has been made since those studies, but the number of binary systems with accurate determinations of masses, radii, luminosities *and* chemical abundances has not yet increased materially. Still, one of these few systems did provide a valuable, refined constraint on models used to fit a precise new colour-magnitude diagram of an open cluster (Nordström et al., 1997) and study the dynamical evolution of the cluster.

Most of the above examples pertain to evolutionary models for single stars, in part because accurate mass and mass-luminosity data are best determined in non-interacting binaries, and in part because of their interest for galactic and extragalactic applications. But another important reason is that the starting parameters for such models (basically mass and composition) are sufficiently few and preserved during most of a stellar lifetime that the model predictions are strongly constrained by the observations.

In contrast, processes of mass loss and mass transfer in interacting binaries may change both the masses, sizes, chemical compositions, separation, and even the apparent ages of the (refuelled) stars. As a result, starting conditions can no longer be uniquely specified, and model predictions are correspondingly poorly constrained. Nonetheless, accurate mass-luminosity data for Algol binaries have recently been used very effectively by Maxted & Hilditch (1996) to show that fundamental weaknesses still remain in current models for the evolution of Algol systems. Yet, the theory of mass exchange was thought to have solved "the Algol paradox" already some 30 years ago!

5. Conclusions

Our subject seems poised for a renaissance: New spectroscopic and interferometric tools have laid the foundations for a new burst of very accurate data on stellar masses, radii, luminosities *and* abundances, which will enhance our ability to not only test our theoretical models of stellar evolution, but also to apply the results to problems of the evolution of our own and other galaxies. Confronting the best theoretical models with all these new data, in every conceivable way, will help us ensure that we advance our understanding, not just our complacency!

Acknowledgment. Financial support from the Danish Natural Science Research Council and the Carlsberg Foundation is gratefully acknowledged.

References

Andersen J. 1991, A&AR 3, 91
Andersen J., Clausen J.V., Giménez A.: 1993, A&A 277, 439
Andersen J., Clausen J.V., Gustafsson B., Nordström B., VandenBerg D.A. 1988, A&A 196, 128
Andersen J., Clausen J.V., Nordström B., Tomkin J., Mayor M. 1991, A&A 246, 99
Andersen J., Nordström B., Clausen J.V. 1990, ApJ 363, L33
Bagnuolo W.G., Jr., Gies, D.R. 1991, ApJ 376, 266
Barlow D.J., Fekel F.C., Scarfe C.D. 1993, PASP 105, 476
Bell S.A., Hill G., Hilditch R.W., Clausen J.V., Reynolds A.P., Giménez A. 1991, MNRAS 250, 119
Bell S.A., Hill G., Hilditch R.W., Clausen J.V., Reynolds A.P. 1993, MNRAS 265, 1047
Claret A. 1995, A&AS 109, 441
Clausen J.V. 1996, A&A 308, 151
Clausen J.V., Helt B.E., Olsen E.H., Garcia J.M. 1997, this meeting; poster no. 17

Edvardsson B., Andersen J., Gustafsson B., Lambert D.L., Nissen P.E., Tomkin J. 1993, A&A 275, 101
Hill G. 1993, in New Frontiers in Binary Star Research, eds. K.C. Leung and I.-S. Nha, ASPC 38, 127
Hummel C.A., Armstrong J.T., Quirrenbach A., Buscher D.F., Mozurkevich D., Elias N.M., Wilson R.E. 1994, AJ 107, 1859
Latham D.W., Nordström B., Andersen J., Bester M., Torres G., Stefanik R.P., Thaller M.: 1996, A&A 314, 864
Metcalfe T.S., Mathieu R.D., Latham D.W., Torres, G. 1996, ApJ 456, 356
Maxted P.F.L. 1996, A&A, submitted
Maxted P.F.L., Hilditch R.W. 1996, A&A 311, 567
Nordström B., Andersen J., Andersen M.I. 1997, this meeting; poster paper no. 64
Popper D.M. 1970, in Mass Loss and Evolution in Close Binaries (IAU Colloquium No. 6), eds. K. Gyldenkerne and R.M. West, Univ. of Copenhagen, Copenhagen, p. 13
Popper D.M., Hill G. 1991, AJ 101, 600
Popper D.M. 1993, PASP 105, 721
Popper D.M. 1994, AJ 108, 1091
Popper D.M., Jeong Y.-C. 1994, PASP 106, 189
Schönberner D., Harmanec P. 1995, A&A 294, 509
Simon K.P., Sturm E. 1993, A&A 281, 286
Stothers R.B., Chin C.-w. 1991, ApJ 381, L67
Torres G., Stefanik R.P., Latham D.W. 1997, ApJ, in press
Wilson R.E. 1990, ApJ 356, 613
Zucker S., Mazeh T., 1994, ApJ 420, 806

DISCUSSION

DAVID ARNETT: In your last graph, were the mass loss models done with angular momentum loss as well? Has that parameter been explored?

JOHANNES ANDERSEN: Those models did assume loss of 50% of the mass and angular momentum, but a vast parameter space remains to be explored.

TIM BEDDING: How well is mixing length constrained in these stars? Is it the same as the solar value?

JOHANNES ANDERSEN: All the single-star model series assume a mixing length around 1.5-1.6 H_p, consistent with that of the solar model in each series.

INTERFEROMETRIC MEASUREMENTS OF BINARIES

H.A. MCALISTER
CHARA, Georgia State University, Atlanta, Georgia, USA

1. Introduction

This paper is a brief overview of the past performance and future promise of interferometry as applied to the study of binary stars. For the sake of brevity, the important results from infrared techniques in recent years are regretfully omitted here. It is probably unnecessary to remind the reader that the analysis of binary star orbital motions is the only direct means for the determination of stellar masses. Historically, stellar masses have resulted primarily from orbits that are sufficiently wide in angular separation to permit the astrometric determination of the motions of the individual components about the center of mass as well as the parallax of the system or from short-period spectroscopic/eclipsing binaries. The advent of long baseline optical interferometers holds the promise of a wealth of new stellar mass data through the direct resolution of spectroscopic binaries.

Interferometry in the context of binaries is a very narrow-field, differential astrometric method. It is time for formal reconsideration of such terms as "visual" and "astrometric" binaries, but we must resist the temptation to compound the problem by introducing the term "interferometric" binary.

2. Techniques and Milestones

The application of Michelson's then newly proposed technique of astronomical interferometry to the study of double stars was first proposed more than a century ago by Karl Schwarzschild (1895). He described an adjustable objective grating method for obtaining and varying the spacing of interferometric fringes whose variable visibility would tell the observer the orientation and separation of the components of a binary. The first important application of Michelson interferometry in the field was made by Anderson (1920) and Merrill (1922) who used an adjustable slit device on

T.R. Bedding et al. (eds.),
Fundamental Stellar Properties: The Interaction between Observation and Theory, **109–114.**

the 100-inch Hooker telescope on Mt. Wilson to measure Capella and other close binaries and inspected a number of spectroscopic binaries for possible resolution. Their measures are noteworthy for extreme accuracy, and the combination of the Mt. Wilson measures of Capella with subsequent speckle data obtained many revolutions later was instrumental in providing an orbit of extreme accuracy for that classic "friend of the interferometrist."

By far, the most significant interferometric contribution prior to the discovery of speckle interferometry was the work of William S. Finsen at the Johannesburg Observatory. Finsen designed and fabricated with his own hands an "eyepiece interferometer." For more than two decades, Finsen proceeded to discover numerous new, close binaries and to measure then with high accuracy. At the end of his career, he gave a sober evaluation of the method describing its virtues but warning against its susceptibility to misuse by incautious observers (Finsen 1971). The work of Finsen in South Africa and Anderson and Merrill on Mt. Wilson provide fine examples of binary star measures whose time dependent quality cannot be entirely superseded by observations at subsequent epochs.

Even if this meeting were not being held in honor of Hanbury Brown, it would be appropriate to emphasize the significance of the observations of the Spica system in 1966 and 1970 from the stellar intensity interferometer at Narrabri, New South Wales (Herbison-Evans et al. 1971). Their elegant combination of interferometric, spectroscopic and photometric data not only yielded a rare determination of the masses of early spectral type stars, but also determined the parallax and luminosities of the components as well as the linear radius of the primary. This landmark achievement will not be duplicated until the present generation of long-baseline interferometers become fully operational during the next few years.

Labeyrie's invention of speckle interferometry in 1970 provided a wonderful new tool for the measurement of binary stars with angular separations limited by the Airy criteria of whatever large telescopes could be brought to bear. A number of speckle programs which emphasized double stars were originated in France, Britain, Germany, Russia, and the U.S., with the majority of the results generated by the author's own program, which is now under the direction of William Hartkopf, at Georgia State University. Rather than give what would be a very incomplete bibliography, the reader is referred to the regularly updated Catalogue of Interferometric Measurements of Binary Stars accessible at http://www.chara.gsu.edu. While great efforts were expended over the years to make speckle interferometry a widely applicable method for high resolution astronomical imaging, the technique is ideally suited to binary star measurement. The CHARA program is now the longest, continuously pursued program of interferometric studies of binaries. In 1990, the U.S. Naval Observatory initiated a binary star speckle interferometry program using the 26-inch refractor in Washington, and the first results of this highly productive effort should appear in the literature in 1997.

Speckle interferometry is presently being surpassed in its ability to resolve close binaries by long-baseline interferometers. These instruments employ the technique of amplitude interferometry over baselines of tens to hundreds of meters to achieve resolutions ranging from a few milliarcseconds to perhaps 100 microarcseconds. The greatest body of binary star data to date using such interferometers comes from the Mark III interferometer built jointly by the Naval Research Laboratory and the USNO and operated on Mt. Wilson from the mid-1980's until the early 1990's. The results from the Mark III have been summarized at this meeting in the poster presentation by Christian Hummel. The Mark III provided considerable overlap with speckle interferometry, and the comparison of the two data types on the same system shows the clear advantage in obtaining higher accuracy by going to longer baselines. This is particularly demonstrated in the Mark III orbit of Capella (Hummel et al. 1994)

3. Advantages and Performance Limits

In comparison with classical visual methods for studying binary stars, interferometry offers the dual advantages of increased resolution and increased accuracy. While skilled visual micrometrists (currently comprising a single-digit sub-population of astronomers) can measure binaries with separations perhaps as small as 0.15 to 0.12 arcseconds, these measures tend to be qualitative rather than quantitative. Using 4-m class telescopes, speckle observers can achieve limiting resolutions of 0.030 arcsecond with an accuracy of perhaps 10% in a single measurement. Visual observers have occasionally complained of unfair comparison, pointing out that they are using aperture five or six times smaller than the typical telescope used by speckle teams. However, the much larger number of speckles encountered at these large apertures leads this writer to believe that micrometer measures at 4-m telescopes would be very difficult to undertake. Regrettably, this assertion has never been tested. For binaries which are well resolved by both techniques, speckle clearly has the advantage in accuracy as can be immediately seen by perusal of the orbits calculated by Hartkopf et al. (1989).

Interferometric measurements also have the potential for providing accurate measurements of the intensity ratios of the components. In terms of placing points on the empirical mass-luminosity relation, Δm is as important as the astrometric orbit. It has proven challenging to extract this photometric information reliably from speckle data, however. It is most certainly not feasible to measure intensity ratios from Labeyrie's classical approach to analyzing speckle data, and although Δm's have been published which have resulted from image reconstruction algorithms, aperture masking or triple correlation analysis, their reliability has not been verified through external comparisons. It therefore remains a challenge to single aperture interferometrists to prove and exploit the potential in this important area. The extraction of intensity ratios from amplitude interferometry is dependent upon the accurate calibration of visibilities, but may ultimately be more straightforward than the efforts exerted on speckle data.

The Δm detection limits of various techniques are also an important point for comparison. An approximate but reasonable expectation from speckle methods is 3-4 magnitudes while one might hope that long-baseline interferometers can perhaps go one magnitude greater in limiting Δm. These limits are dependent on the combined magnitude of the target and rapidly approach zero as the object brightness approaches the limiting magnitude of the technique. Speckle interferometry with photon counting detectors is perhaps limited to objects brighter than V=+12 while interferometric arrays are not likely to reach such a limit without very sophisticated adaptive optics and moderately large apertures.

4. Resolution of Spectroscopic Binaries

The resolution of spectroscopic binaries by interferometric techniques provides a powerful combination of complementary capabilities. Traditionally, few spectroscopic binaries have been directly resolved because classical astrometric techniques do not possess the resolution required for objects in the period domain to which spectroscopy is normally sensitive. When the happy circumstance of having a resolved spectroscopic binary does arrive, the interferometrically determined orbital semi-major axis and inclination combine with the spectroscopically determined $a\sin i$ and mass ratio to yield masses, distance and luminosities. The distance thus determined is referred to as the "orbital parallax."

A conservative estimate of the productivity of interferometer arrays in this regard is seen by noting that 180 double-lined systems from the "Eighth Catalogue of the Orbital Elements of Spectroscopic Binary Systems" (Batten et al. 1989) have predicted angular separations exceeding 1 milliarcsec. Recognizing that several interferometers will have limiting resolutions 10 times small than this value, literally hundreds of double-lined pairs will be resolved. This will completely close the gap between the "visual" and "spectroscopic" binaries and provide a wealth of new data for the mass/luminosity and mass/radius relations.

For single-lined systems, two directions of progress are noteworthy. First, techniques like tomographic separation of spectra (Bagnuolo et al. 1994) and the two-dimensional, cross-correlation method of Zucker and Mazeh (1994) are turning single-lined systems into double-lined pairs with known mass ratios. The latter technique is currently being applied in a series of papers in which Hyades spectroscopic binaries resolved by lunar occultation observations, speckle interferometry and the Mark III interferometer are finally giving up their masses *and* new geometrically determined distances to the cluster. Secondly, at the time of this writing, we anticipate the release of new parallax data from the HIPPARCOS satellite. If these parallaxes are of sufficient accuracy and free from the effects of orbital submotions, they will provide the missing information needed to make resolved single-lined binaries give up their secrets.

5. The Third CHARA Catalogue

The continuously updated "Catalogue of Interferometric Measurements of Binary Stars," maintained on-line at CHARA by William Hartkopf, is perhaps the best summary of the current status of the field. This catalogue includes speckle data as well as data from the modern generation of optical arrays which at present is limited to the Mark III. Also included are occultation measurements (because of their strong overlap with interferometry) and measurements from other amplitude interferometers except for the large body of work from Finsen's eyepiece interferometer. Relevant statistics from the catalogue are shown in Table 1 below. The column head "current" indicates the present status while the subsequent two columns include new data anticipated from the USNO and CHARA programs during 1997. The final column indicates the anticipated status of the catalogue after those results are included. Hopefully, additional material will be provided by other practitioners of interferometry.

6. Conclusion

During the past 20 years, speckle interferometry has essentially replaced visual micrometry as the preferred technique for measuring the angular motions in resolved binary systems. This transition is perhaps best represented by Charles Worley's decision in the late 1980's to set aside his micrometer and develop a speckle program at the USNO on the same telescope where he had for years applied his "double star eye."

Binary star observations will be a major part of the observing program at Georgia State University's CHARA Array now under construction on Mt. Wilson, California. This facility will employ five 1-meter aperture telescopes in a Y-shape configuration to achieve limiting resolutions of 200 microarcseconds at the V spectral bandpass and 1 milliarcsecond at the K

TABLE 1. CHARA's Catalogue of Interferometric Measurements of Binary Stars

Category	Current	USNO-97	CHARA-97	New
No. of Resolved Stars	3,455	467	1,146	4,000
No. of Unresolved Stars	2,843	0	197	3,012
No. of New Resolved Binaries	375	0	5	380
No. of Measurements	17,500	2,329	2,981	22,810
CHARA	13,408	-	2,981	16,389
Occultation	695	-	-	629
Other	3,397	2,329	-	5,726
Median Separation (arcsec)	0.25	1.14	0.45	0.31
No. of Negative Results	6,610	0	316	6,926
CHARA	4,816	-	316	5,132
Other	1,794	0	-	1,794
Contributing Papers	201	1	3	205

band. It can be stated with confidence that binaries will also be emphasized at the Navy's Prototype Optical Interferometer (NPOI) on Anderson Mesa, Arizona, as well as by our host's facility near Narrabri, the Sydney University Stellar Interferometer (SUSI). These powerful new instruments will yield masses, distances, luminosities, effective temperatures and physical radii in unprecedented numbers for stars of almost all MK spectral types. The interferometric revolution in the field of binary star astrometry will be complete by the end of the century.

7. References

References

Anderson, J.A. (1920) *Astrophys. Journ.*, **51**, p. 263.
Bagnuolo, W.G., Gies, D.R., Hahula, M.J. and R. Wiemker, R. (1994) *Astrophys. Journ.*, **423**, p. 446.
Finsen, W.S. (1971) *Astrophysics and Space Science*, **11**, p. 13.
Hartkopf, W.I., McAlister, H.A. and Franz, O.G. (1989) *Astron. Journ.*, **98**, p. 1014.
Herbison-Evans, D., Hanbury Brown, R., Davis, J., and Allen, L.R. (1971) *Mon. Not. Roy. Astron. Soc.*, **151**, p. 161.
Hummell, C.A., Armstrong, J.T., Quirrenbach, A., Buscher, D.F., Mozurkewich, D. and Elias, N.M. (1994) *Astron. Journ.*, **107**, p. 1859.
Merrill, P.W. (1922) *Astrophys. Journ.*, **56**, p. 43.
Schwarzschild, K. (1895) *Astronomische Nachrichten*, **139**, p. 23.
Zucker, S. and Mazeh. T. (1994) *Astrophys. Journ.*, **420**, p. 806.

DETERMINING THE MASSES OF WHITE DWARFS IN MAGNETIC CATACLYSMIC VARIABLES FROM X-RAY OBSERVATIONS

KINWAH WU
RCfTA, School of Physics
University of Sydney, NSW 2006, Australia
AND
MARK CROPPER AND GAVIN RAMSAY
MSSL, University College London
Holmbury St. Mary, Surrey, RH5 6NT, United Kingdom

Abstract.
A method to determine the masses of white dwarfs in magnetic cataclysmic variables from X-ray observations is presented. The method is appropriate for both eclipsing and non-eclipsing systems, for it is insensitive to the orbital inclination of the systems.

1. Introduction

Magnetic cataclysmic variables (mCVs) are close binaries consisting of a magnetic white dwarf which accretes matter from a Roche-lobe-filling M dwarf companion star. There are two types of mCVs: (i) AM Herculis binaries (AM Hers), in which the white dwarf has a very strong magnetic field that locks all components in the binaries into synchronously rotation and prevents the formation of an accretion disk, and (ii) Intermediate polars (IPs) and DQ Herculis binaries (DQ Hers), which are asynchronously rotating systems with a weaker white dwarf magnetic field only partially disrupting the accretion disk (see Warner 1995). The white dwarf mass is one of the fundamental parameters of mCVs and has implications for our understanding of the origins, evolution and emission of these binaries.

Most mCVs are strong X-ray sources, with a keV X-ray spectrum typical of an optically thin bremsstrahlung spectrum. The X-ray emission regions are located near the white dwarf magnetic poles, where the accreting matter

T.R. Bedding et al. (eds.),
Fundamental Stellar Properties: The Interaction between Observation and Theory, 115–118.

is heated up by an accretion shock. The accretion shock is formed when the speed of the accreting matter decelerates from supersonic to subsonic near the white dwarf surface. The shock temperature is $T_s = 3\,G\mu m_H M_w/8R_w k$, a function of the mass (M_w) and radius (R_w) of the white dwarf only. (Here, G is the gravitational constant, k the Boltzmann constant, m_H the mass of hydrogen atom, and μ the mean molecular mass.) Thus, provided that the shock temperature is measured, the white dwarf mass can be deduced.

By fitting the observed X-ray spectrum with a single-temperature optically thin bremsstrahlung spectrum, it is possible to obtain an effective temperature T_e which characterises the emission region. Ishida (1991) assumed that $T_s \approx T_e$ and, by fitting the GINGA X-ray data, estimated the white dwarf masses of 5 AM Hers and 8 IPs. However, the X-ray emission regions are vertically structured in temperature and density, so that these factors must be taken into account when determining M_w.

2. Spectra of X-rays from Structured Emission Regions

The thickness of the post-shock region ($\sim 10^7$cm) is generally much smaller than the white dwarf radius ($\sim 10^9$cm), so the post-shock accretion flow can be considered as 1-dimensional. The density and temperature structures are therefore determined by a set of 1-D hydrodynamic equations when the appropriate cooling functions are specified. For mCVs, the dominant processes are bremsstrahlung and cyclotron cooling. The treatment of bremsstrahlung cooling is relatively straight-forward because it is optically thin, but the treatment of the optically thick, angle-dependent cyclotron cooling is more complicated. However, for typical parameters of mCVs, the effects of cyclotron cooling can be approximated by a cooling function of power-laws of density and temperature (see Wu, Chanmugam & Shaviv 1994).

In calculating the X-ray spectra from structured emission regions, we used the MEKAL optically thin plasma code (Mewe, Kaastra & Liedahl 1995). The absorption of the cooler material in the pre-shock accretion flow was modelled using a modified warm absorber model based on the ABSORI code (Zdziarski & Magdziarz 1996). The reflection of X-rays from the white dwarf surface was modelled following van Teeseling, Kaastra and Heise (1996). (For the details of the calculations of the model X-ray spectra and the fitting to the observed data, see Cropper, Ramsay & Wu 1997.) The fit to the GINGA X-ray spectra of the system AM Her at phases of maximum brightness is shown in Figure 1. The white dwarf mass for the best fit is $M_w = 0.75 M_\odot$ ($\chi^2 = 0.74$, 100 dof). The white dwarf masses of 4 other AM Hers and 8 IPs/DQ Hers obtained by fitting the GINGA data are shown in Table 1.

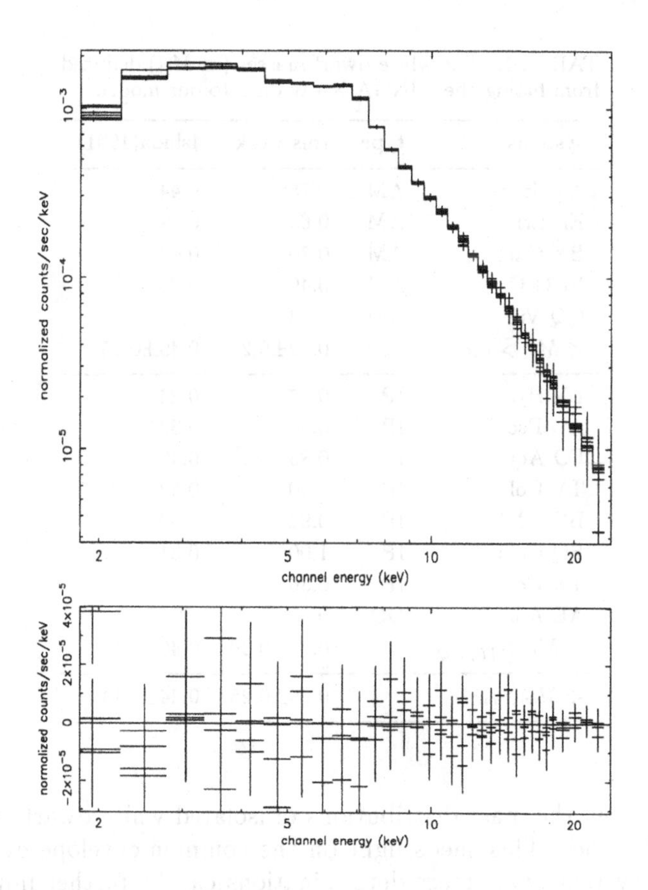

Figure 1. The GINGA data of the system AM Her along with the model fits and residuals.

3. Discussions

As shown in Table 1, the mass deduced by Ishida (1991) is systematically lower than the masses obtained in this work. The mass underestimation due to single-temperature fitting can be understood because the local temperatures everywhere in the post-shock region is always lower than T_s. Thus, the mass estimates of Ishida (1991) are lower limits to the actual values.

Generally, the mass of white dwarfs in mCVs are determined by orbital variations of optical spectral lines emanating from the secondary star. Uncertainties in the orbital inclination of the binary system strongly affects the accuracy with which the mass can be measured using this technique. The method presented here, however, depends only very weakly on the orbital inclination.

The sample we considered is homogeneous, thus it allows a direct com-

TABLE 1. The white dwarf masses (in $M_\odot$) deduced from fitting the GINGA X-ray data to our model.

systems	type	this work	Ishida(1991)
AM Her	AM	0.75	0.44
EF Eri	AM	0.63	0.39
BY Cam	AM	0.70	0.63
V834 Cen	AM	0.49	0.33
QQ Vul	AM	1.03	0.66
$< M_w >_{AM}$		0.72±0.20	0.49±0.15
EX Hya	IP	0.37	0.24
AO Psc	IP	0.48	0.32
FO Aqr	IP	0.85	0.49
TV Col	IP	1.00	0.52
BG CMi	IP	0.92	0.50
PQ Gem	IP	1.00	0.31
TX Col	IP	0.36	
AE Aqr	DQ	0.43	
$< M_w >_{IP/DQ}$		0.68±0.29	0.40±0.12
$< M_w >_{mCV}$		0.69±0.25	0.44±0.14

parison between the mass distributions of isolated white dwarfs and white dwarfs in binaries. This sheds light on the common-envelope evolution of the binary systems. Our mass determinations can be further improved by refinement of the physical model (e.g. better treatment of cyclotron cooling, two-fluid effects) and by the availability of the X-ray data with higher spectral resolution.

References

Cropper, M., Ramsay, G. and Wu, K. (1997), *MNRAS*, in press.
Ishida, M. (1991), *PhD Thesis*, University of Tokyo, (ISAS RN 505).
Mewe, R., Kaastra, D.S. and Liedahl, D.A. (1995), *Legacy (Journal of HEASARC)*, **6**, 16.
van Teeseling, A., Kaastra, J.S. and Heise, J. (1996), *A&A*, **312**, 186.
Warner, B. (1996), *Cataclysmic Variable Stars*, Cambridge University Press.
Wu, K. Chanmugam, G. and Shaviv, G. (1994), *ApJ*, **426**, 664.
Zdziarski, A.A. and Magdziarz, P. (1996), *MNRAS*, **279**, 21.

EMPIRICAL SURFACE GRAVITIES

(from spectra and from binaries)

P. F. L. MAXTED
University of Southampton, UK

Abstract. The surface gravity of a star (log g) is a fundamental parameter in models of stellar atmospheres. Given suitable spectra, log g can be determined from such models with an accuracy of 0.1dex, at best. Detached eclipsing binary stars can provide values of log g an order of magnitude more accurate than this, though for a more limited range of stars. Naturally, less accurate surface gravities can be obtained for a wider range of eclipsing binary stars.

These facts are well known, so in this short review I will outline the types of stars to which the two methods have be usefully applied and might be applied in the near future. This naturally leads to the question of where the two ranges overlap and the comparison of results from the two methods. Techniques for allowing this comparison to made directly will be described. Surface gravities derived from winds in hot stars and (indirectly) from gravitational redshifts in white dwarf stars will also be covered briefly.

1. Introduction

The surface gravity of a star (log g) is a fundamental parameter in models of stellar atmospheres and arises naturally from the masses and radii determined for stars in eclipsing binary systems. This short review will summarise the reliability and accuracy of the two methods and describe methods for directly comparing the results.

2. Overview

The determination of log g for a star from its spectrum has been applied to stars of almost every type with quoted accuracies $\sim$ 0.1dex being typical. Since this method relies on fitting a model to the observed data further

T.R. Bedding et al. (eds.),
***Fundamental Stellar Properties: The Interaction between Observation and Theory*, 119–124.**

uncertainty is introduced i.e. values of log g determined from spectra are prone to systematic errors.

Eclipsing binary stars (EBS) yield masses and radii for the component stars from the analysis of the lightcurve and the spectroscopic orbit. With high quality data for stars in systems uncomplicated by proximity effects, log g can be reliably determined to $\sim$ 0.01dex. Provided care is taken to do the analysis properly, these values are free from systematic errors. This method is, of course, limited to those stars found to be EBS, although the method can be applied to a wider range of stars at the expense of lower accuracy.

The direct comparison of values of log g determined by these two methods is now feasible and desirable.

3. Surface gravities from spectra

A complete catalogue of log g determinations is not feasible in a short review and so I will simply summarise two recent reviews of the modeling of spectra which capture the flavour of the work that has been done.

Figure 1. Surface gravities of 5 cool giants

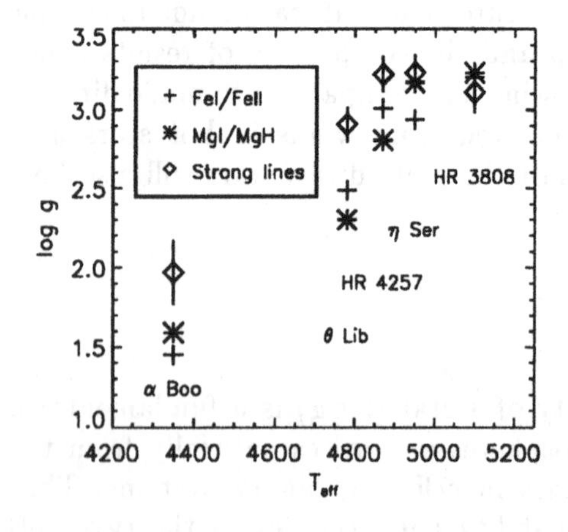

3.1. SPECTROSCOPY OF COOL STARS

Gustafsson(1989) has reviewed the chemical analysis of cool stars. The determination of log g is fundamental to this process and so Gustafsson has covered this topic in some detail. Three techniques are described: ionisation equilibria, dissociation equilibria and damping wings of strong lines. Since

Gustafsson's review all three methods have been applied to the 5 cool giant stars as shown in Fig. 1. The error bars shown for the data of Edvardsson are the result of a thorough discussion of the uncertainties involved and similar errors ($\sim$ 0.1dex) are quoted for the other log g determinations. Although the methods tend to agree, differences of up to 0.5dex are evident. There is further uncertainty in these log g determinations due to correlations between log g and $T_{\rm eff}$ (though this is less of a problem for the "strong line" method). These differences are presumably due to deficiencies in the modelling of cool stellar atmospheres e.g. convection stubbornly defies a simple treatment. The result is that log g for cool giant stars is often determined from an adopted absolute magnitude and a mass predicted from evolutionary tracks, which yields accuracies $\sim$ 0.5dex. Systematic errors in these determinations due to composition and mass variations are difficult to quantify.

3.2. SPECTROSCOPY OF HOT STARS

The atmospheres of hot stars are relatively simple when compared to those of cool stars, or at least, they are sufficiently uncomplicated that our understanding of the relevant physics is adequate for *quantitative spectroscopy of hot star* (QSHS) to be feasible. The concept of QSHS is reviewed by Kudritzki & Hummer and refers to "the systematic acquisition and analysis of accurate spectroscopic data ... in order to determine accurate values of the stellar parameters ... ". Stellar winds are commonplace among hot stars and the modelling of spectral features due to the wind provides valuable data that complements the data derived from the photospheric features. In particular, log g can be determined independently using both methods with comparable accuracy. This is precisely what has been done by Herrero *et al.*(1992) for 25 luminous galactic OB stars. Fig. 2. shows the values of log g derived from their data for the 21 stars to which the method is applicable. The mean difference between the methods is 0.07 ± 0.10, a very encouraging result.

4. Eclipsing binary stars

Very accurate values of log g($\pm \lesssim 0.02$) can be determined in the case of detached EBS. Andersen's review (1991) of radii and masses for normal stars lists 45 binaries that fall into this category. They cover the majority of the main sequence although there are very few systems at the extremity of the mass range. Away from the main sequence, only two normal giant stars are listed (AI Phe and TZ For). Work continues to extend the range of normal stars for which accurate parameters are available. These and other notable EBS are:

Figure 2. Surface gravities for Hot stars.

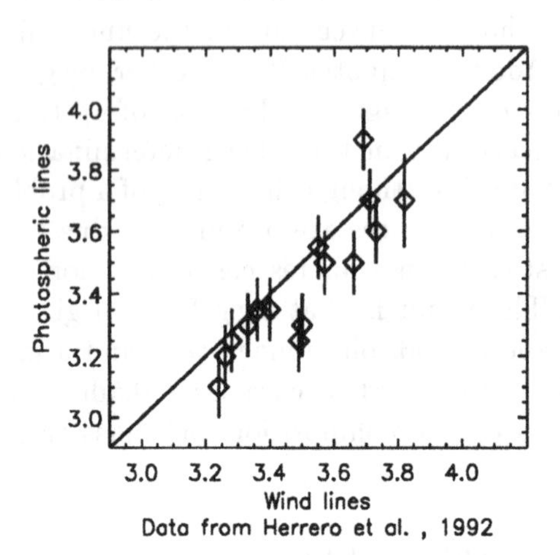

TABLE 1. Surface gravities for giants in ζ Aurigae binaries

Name	Sp. Type	log g
ζ Aur	K4Ib + B5V	0.86 ± 0.02
22 Vul	G2Ib + B8V	1.40 ± 0.17
τ Per	G8IIIa + AV	2.41 ± 0.07
HR 6902	G9IIb + B8-9V	1.99 ± 0.04
31 Cyg	K4Ib + B3-4V	0.92 ± 0.10

CM Dra A remarkably useful EBS comprising two old M dwarfs (Metcalfe *et al.*, 1995). The weak chromospheric activity should enable more accurate parameters to be determined than for YY Gem, the only other known M dwarf EBS. This star is currently under intensive study.

V643 Ori This pair of eclipsing giant stars cannot be the product of single star evolution but is nevertheless an interesting system for which accurate parameters will be published soon(ish)

LZ Cen, V346 Cen, GL Car Several new, massive ($> 10 M_{\odot}$) eclipsing binaries have been identified in recent years. Lightcurves have been secured for the three systems listed and the analysis of spectra to determine spectroscopic orbits continues. The Magellanic clouds are a fruitful source of high mass binaries although more effort is required to derive accurate parameters and so results may be slow to appear.

HR 7940 Clayton(1996) has observed primary and secondary eclipses in this reddened B2III star. Accurate parameters of this rare system will be interesting but may take some time given that the period is 99.76days

ζ Aurigae systems Several of these G/K bright giant + BV eclipsing binaries have been analysed by Griffin and others. Their results for log g summarised in Table 1.

Detached white dwarf binaries The work of Marsh *et al.*(1995) in which he finds 7 out of 9 low-mass DA white dwarfs to be short-period binaries has led to a concerted effort to find more of these systems. Until we are lucky enough to find an eclipsing system, the gravitational redshift can aid in the determination of log g (since it determines the ratio M/R). E.g. Reid(1996) has measured gravitational redshifts in 53 WD systems in clusters and in wide binaries. In systems with a measured parallax, $T_{\rm eff}$ leads to an estimate of the radius and so log g can be determined to $\sim$ 0.1dex

5. Direct Comparison of the methods

The reasons for making a direct comparison between values of log g derived from EBS and spectra are nicely illustrated by Fig. 3 which is based on the work of Moon & Dworetsky(1985). The diamonds show the log g values predicted by the $uvby\beta$–log g calibration of Balona(1984) compared to the observed values in EBSs with similar components. The mean difference in log g is 0.22 ± 0.15dex, which again shows that log g values from "spectra" show an intrinsic accuracy $\sim$ 0.1dex but that this may hide systematic errors. Moon & Dworetsky have used the observed log g of these EBSs to apply corrections to the $uvby\beta$–(log g,$T_{\rm eff}$) calibration of Relyea & Kurucz(1978). The success of this approach is evident from the agreement between the observed and predicted values (crosses) for which the mean difference is -0.01 ± 0.10dex. In general, for EBS with dissimilar components, some method is required to separate the combined spectra of the components so that they can be analysed as though they were singled stars. Three such methods exist, all of which take advantage of the varying Doppler shift to distinguish the two components. Bagnuolo & Gies(1991) have used their "Doppler tomography" technique to establish spectral types for the binary O-star AO Cas and others. Simon *et al.*(1994) have applied their "Disentangling" method to the massive EBS Y Cyg. Their analysis of the resulting spectra yields surface gravities for the two components of 4.16 ± 0.10 and 4.18 ± 0.10, in excellent agreement with the actual values of 4.140 ± 0.012 and 4.149 ± 0.012. Finally, Hadrava(1995) has developed a method based on Fourier transforms, although no results with this method

Figure 3. Surface gravities of EBS from *uvbyβ* photometry.

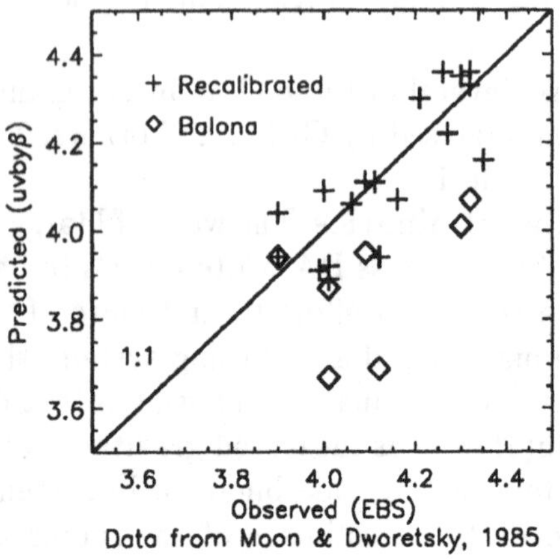

have been published yet.

6. Conclusion

Now that techniques for the direct comparison of log g values of EBS with those predicted from their spectra are available, the comparison should be made for the increasing range of EBS for which accurate log g values are available. This has already been shown to be an effective technique for revealing any systematic errors that may be present.

References

Bagnuolo W.G. and Gies D.R., (1991) *Astroph. J.* **376** pp. 266-271
Balona L.A. (1985) *Mon. Not. R. Astron. Soc.* **211**, pp. 973-980
Bonnell J.T. and Bell R.A., (1993) *Mon. Not. R. Astron. Soc.* **264**, pp. 334-344
Clayton G.C., (1996) *Pub. Astron. Soc. Pacific* **108**, pp. 401-403
Edvardsson B. (1988) *Astron. & Astrophys.* **190**, pp. 148-166
Gustafsson B. (1989) *Ann. Rev. Astron. Astrophys.* **27**, pp. 701–756
Hadrava P., (1995) *Astron. Astrophys. Suppl. Ser.* **114**, pp. 393-396
Herrero A., Kudritzki R.P., Vilchez D. Kunze K., Butler K. and Haser S. (1992) *Ann. Rev. Astron. Astrophys.* **261**, pp. 209-234
Kudritzki R.P. and Hummer D.G.,(1990) *Ann. Rev. Astron. Astrophys.* **28**, pp. 303-345
Marsh T.R., Dillon V.S. and Duck S.R., (1995) *Mon. Not. R. Astron. Soc.* **275**, pp. 89-99
Metcalfe T.S., Mathieu R.D., Latham D.W. and Torres G. (1996) *Astroph. J.* **456**, pp. 356-364
Moon T.T. and Dworetsky M.M., (1985) *Mon. Not. R. Astron. Soc.* **217**, pp. 305-316
Reid I.N. (1996) *Astron. J* **111**, pp. 2000-2016
Relyea L.J.and Kurucz R.L., (1978) *Astroph. J. Suppl.* **37**, pp. 45
Simon K.P., Sturm E. and Fiedler A. (1994) *Astron. & Astrophys.* **292**, pp. 507-518

6. THE STELLAR EFFECTIVE TEMPERATURE SCALE

COOL STAR EMPIRICAL TEMPERATURE SCALES

MICHAEL S. BESSELL
Mt. Stromlo & Siding Spring Observatories,
Private Bag, Weston Creek PO, ACT 2611, AUSTRALIA

1. Introduction

The empirical temperatures scale for cool stars is generally well established. Temperatures are now known with reasonable precision for stars covering the range of spectral types from A to M. In the historical paper by Code, Davis, Bless and Hanbury Brown (Code et al. 1976), six stars between 10000K and 6500K had radii measured by the intensity interferometer and these six, together with the sun formed the basis of the empirical temperature calibration at the time. Since then, many temperatures have been derived for A-K stars (Blackwell & Lynas-Gray 1994; Alonso et al. 1996a) using the Infra-Red Flux Method (see Megessier 1994,5 and this volume), while lunar occultations (Ridgway et al. 1980) and more recently Michelson interferometry (Di Benedetto & Rabbia 1987; Dyck et al. 1996), have been used to measure the radii of K and M giants. It is a tribute to Hanbury Brown's Intensity Interferometer that temperature scales based on its measurements are essentially unchanged by the new data.

The empirical temperatures of metal-deficient and metal-rich stars had been virtually non-existent, but very recently, the IRFM has been applied by Alonso et al. (1996a,b) to a sample of such stars with excellent results.

There are however, some kinds of stars where the empirical temperature scale is not as well established. For M dwarfs, eclipsing binaries provide temperature for only 2 stars although the IR flux method has been used for a few additional M dwarfs (Tsuji et al. 1995, 96a).

Significant advances have also been made in atmospheric modelling for cool stars incorporating improved metal-line and molecular line opacities. Synthetic spectra and synthetic photometry generated from these models show good agreement with the empirical temperature scales and now allow us to confidently extend the temperature calibrations to stars over the full range of parameter space. We will discuss and illustrate some of those comparisons here.

T.R. Bedding et al. (eds.),
***Fundamental Stellar Properties: The Interaction between Observation and Theory*, 127–136.**

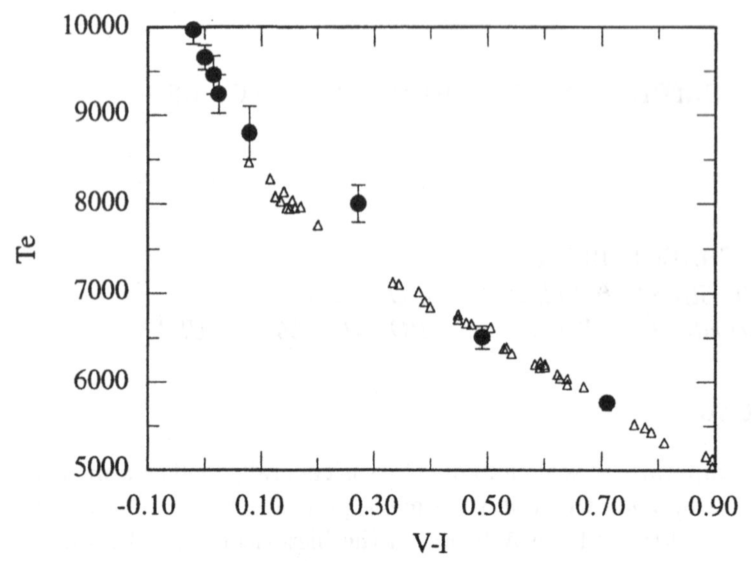

Figure 1. Closed circles with error bars are data from Code et al. (1976); without error bar, the sun. Open triangles are dwarfs from Blackwell & Lynas-Gray (1994).

2. Empirical temperatures

2.1. A–G DWARFS

In Fig. 1 is shown the comparison between the Code et al temperatures for A-F stars and those for the hotter stars measured with the IRFM by Blackwell & Lynas-Gray (1994). The colour index V-I in the Cape-Cousins system is used for the abscissa. These data are in excellent agreement except for Alpha Aql. In Fig. 2 all the Blackwell & Lynas-Gray (1994) temperatures are plotted against V-K. There is impressively little scatter about the mean locus.

2.2. K–M GIANTS

A few years after Code et al. published their main sequence stellar radii, another important paper was published by Ridgway, Joyce, White & Wing (1980) on the radii of red giants measured by lunar occultations. This paper has essentially defined the temperature scale for red giants over the past 17 years. However, Michelson interferometer observations of red giants have recently produced more precise angular diameters and in the future we look forward to this technique delivering very precise radii for most kinds of stars. In Fig. 3 are shown the best data from Ridgway et al. (1980) together with the Michelson interferometer data from DiBenedetto & Rabbia (1987). Although the precision of the occultation data was much lower than that of the Michelson data, the mean Te versus colour relations defined by the two data sets is almost identical. A comparison between the cool end of the Blackwell & Lynas-Gray (1994) IRFM based temperature calibration

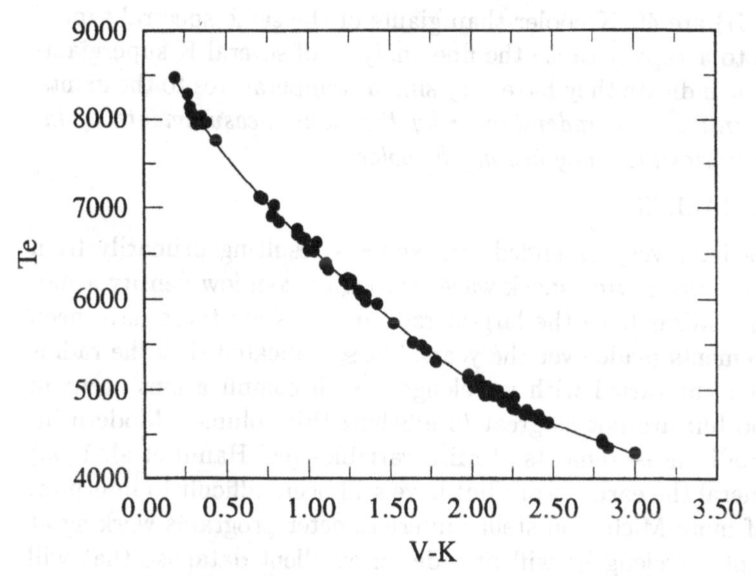

Figure 2. **Closed circles are IRFM temperatures from Blackwell & Lynas-Gray (1994). The continuous line is a polynomial fit to the data.**

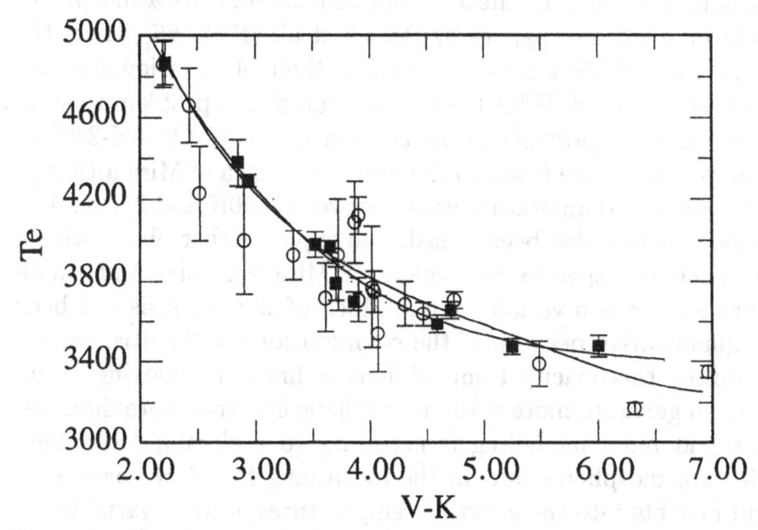

Figure 3. **The Te versus V-K diagram for the best occultation data of Ridgway et al. (1980) and the Michelson interferometer data of Di Benedetto and Rabbia (1987) . Error bars in the measured temperatures are shown.**

and the hot end of the occultation based temperature calibration from Ridgway et al. (1980) shows that the BLG94 scale is too hot below 4400K. Dyck et al. (1996) summarize the current temperature scale for the KM giants and supergiants and for the giants their result is essentially identical to that shown here. However, they also conclude that the K4.5 supergiants (luminosity I, II) are 400K cooler than giants of the same spectral type. I find this hard to accept because the fine analyses of several K supergiants that I have done indicate they have very similar temperatures to the giants. *It is of great interest to understand why the radii measurements of the supergiants indicates that they are much cooler.*

2.3. MIRA VARIABLES

Mira variables have very extended atmospheres resulting primarily from the radial pulsations driving shock waves through a cool low density atmosphere. Because Miras have the largest radii of any stars there have been direct measurements made over the years. These indicated that the radius was not unique but varied with wavelength. Such complications occur in other stars too but are not as great (see Scholz this volume). Modern interferometer radii measurements of Mira variables (eg. Haniff et al. 1995) support in general the earlier work but have still been difficult to interpret. The advent of more Michelson stellar interferometer programs working at several different wavelengths will provide an excellent database that will enable the remaining longstanding problems of Mira atmospheres to be better diagnosed.

A start to understanding the effective temperatures of Mira atmospheres has been made in a series of papers by Bessell et al. (1989, 96) where the theoretical structure of Mira atmospheres and their observational consequences have been explored. Whilst this is too complex a problem to summarise here, it has been possible to understand how cool (2200K-2400K) black-body temperatures are fitted to the near-IR spectra of Mira although the underlying effective temperature varies between 3050K and 2700K during the pulsation. It has also been possible to explain other observational anomalies between the spectra and colors of Mira variables and small-amplitude variables or non-variable stars. Much of this analysis has been only crudely quantitative because of the computational difficulties associated with handling the opacity from millions of lines in a moving atmosphere in order to generate more realistic synthetic spectra. More diameter measurements and more modelling is necessary to understand the complexities of Mira atmospheres but, in the meantime, the above papers do provide useful insights into the effective temperatures of Mira variables.

2.4. K–M DWARFS

The K and M dwarfs are such small objects that it is unlikely that ground-based stellar interferometers will ever be able to directly measure diameters. However, there is another way. There are two M dwarfs with measured diameters. These stars are members of eclipsing binaries where the occultation of one of the components by the other enables the diameters to be measured. Although the details of the orbits and the timings can be

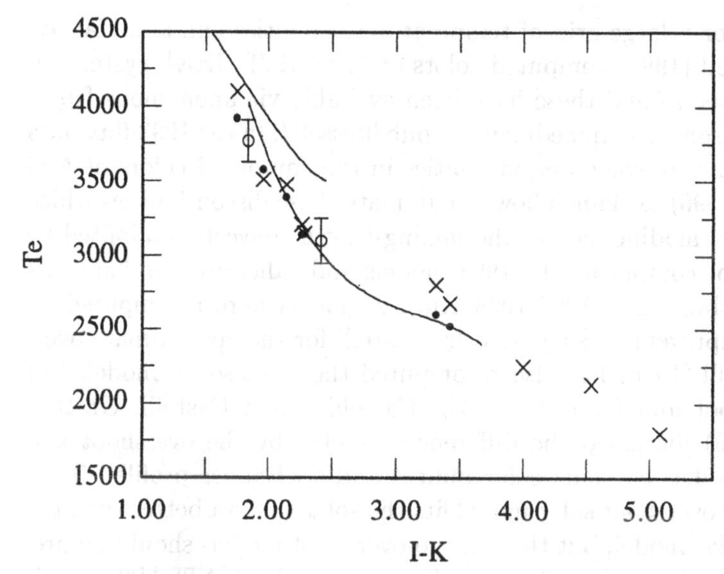

Figure 4. The comparison between model Te versus I-K color and the empirical temperatures for M dwarfs. The open circles with error bars are YY Gem and CM Dra; the crosses are IRFM data from Tsuji et al. (1995, 1996); the filled circles are Brett(1995a,b) fits to far-red spectra. The upper line represents the no-overshoot ATLAS9 models; the lower line represents the NMARCS models

very precise, the factor limiting the precision of the effective temperature is usually the parallax. New parallax measurements are badly needed for these stars. The two stars are YY Gem (3770± 200K) (Kron 1951; Habets & Heintze 1981) and CM Dra (3120± 150K) (Lacy 1977; Metcalfe et al. 1996). CM Dra, rather unfortunately has halo kinematics but its spectrum suggests that its metallicity is not more than a factor of 3 less than solar. Its halo nature should be remembered, however, when using it as a temperature standard. Popper (1993, 96) has reported several other possible F-K eclipsing binaries whose orbits should be worked on. None of these are cooler than CM Dra.

Tsuji et al. (1993, 96) has used the IRFM to deduce the temperature of some M dwarfs. These temperatures, which extend from 4000K to 1600K, are in excellent agreement with the two eclipsing binary star temperatures. Brett (1995a,b) derived temperatures for some M dwarfs from fitting the spectra to his model spectra. In Fig. 4 are plotted the empirical temperatures together with some theoretical colour temperature relations from model atmospheres and Brett's spectral fitting. The model atmospheres will be discussed below.

3. Model atmosphere color-temperature relations

a) The Kurucz-Castelli ATLAS9 models

Kurucz (1993) has made available models, hydrogen line profiles, spectra and colors for a large grid of temperatures, gravities and abundances. Wood and Bessell (1994) computed colors on the UBVRIJHKL system for these Kurucz spectra and these have been available via anonymous ftp as ubvrijhkl.dat.z from mso.anu.edu.au at /pub/bessell/. In the 1993 flux data there was evidence of some discontinuities in the computed colors of A-G stars. Castelli (1996) explained how to eliminate these discontinuities which were related to a modification of the mixing-length convection adopted by Kurucz(1993) for computing the 1993 models and called by him "approximate overshooting". The 1995-1996 Kurucz models were recomputed by adopting the improvement suggested by Castelli for the approximate overshooting. Castelli (1996) has also recomputed the same set of models but with no-overshoot and for $l/H = 1.25$ (Castelli 1995). Castelli, Gratton & Kurucz (1996) discussed the differences yielded by the overshoot and no overshoot models on some color indices and on Balmer profiles. They showed that the overshoot solar model fits the solar spectra better than the no-overshoot solar model, but that the no-overshoot models should be preferred for stars different from the sun. The no-overshoot ATLAS9 models are available from castelli@astrts.oat.ts.astro.it.

b) The NMARCS models.

The new revised MARCS program incorporating statistical line opacities has been used by Plez, Brett and Nordland to model M giants and dwarfs. Plez, Brett & Nordland (1992) and Plez (1996) modelled giants while Brett and Plez (1993) and Brett (1995a,b) have modelled the M dwarfs. These models are available from plez@nbivax.nbi.dk. The colors and bolometric corrections of these models are discussed by Bessell et al. (1970). A complete grid (A–M stars) is currently being computed by Gustafsson et al. (1997).

3.1. A–K DWARFS

In Fig. 5 are shown the no-overshoot ATLAS9 model Te versus V-I relation in comparison with the BLG94 IRFM data. In Fig. 6 is shown the comparison between the model temperature V-I color relations and the empirical relations. Another model comparison in I-K for the M stars was shown in Fig. 4. There is quite good agreement between the model temperatures and the empirical temperatures. The ATLAS9 models are seen to be less reliable for temperatures below 4250K. The published NMARCS M dwarf models are suspect below 2800K due to inadequate H_2O opacity. Brett (1995a,b) used a smoothed empirical treatment of H_2O opacity for these preliminary models but in the NMARCS grid currently being computed, H_2O opacity will be incorporated on the basis of the more than 20 million lines that have recently been computed by Jorgensen (1996) and we anticipate much better agreement. Models by Allard et al. (1994) had similar problems with water but new models being computed will also incorporate the new opacities.

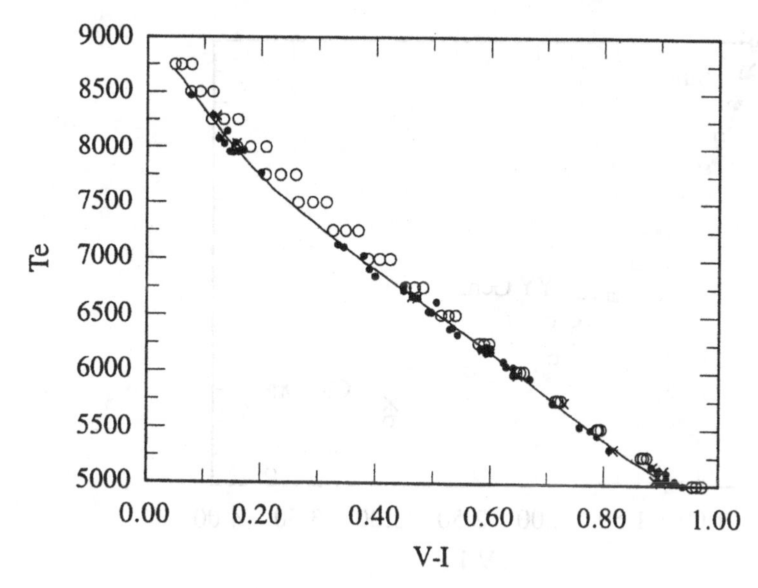

Figure 5. Comparison between model temperatures derived from V-I colors and the IRFM temperatures for A-G stars. The crosses indicate measured V-I values; the filled circles are transformed V-I values from V-K. There is no obvious separation of the data points. Model colors from ATLAS9 no-overshoot models are plotted for log g = 4.5, 4.0, 3.5; higher gravity gives redder color.

However modelling of the atmospheres of the coolest M dwarfs requires in addition the inclusion of grain opacities and an understanding of how grains form and segregate (see Tsuji et al. 1996b).

3.2. K–M GIANTS AND SUPERGIANTS

In Fig. 7 are compared the observed and model Te versus V-K color relation for the KM giants. The plotted observational data are from Blackwell & Lynas-Gray (1994) and Di Benetto & Rabbia (1987). The continuous line is the mean empirical relation discussed above. The model colors are in extremely good agreement with the observations even given the complication of a range in model color depending on gravity and extension. The models predict temperatures to better than 100K between 5000K and 3200K. Below 3000K all giants are variable, many of them Mira or long-period variables. The colors of cool models are affected not only by temperature, gravity and abundance but also by mass (or sphericity). For temperatures between 4000K and 3400K the effect of mass is comparable or greater than the effect of gravity on the V-K color but below 3200K the effect of gravity is very great. Accurate Michelson interferometer radii for late-M stars with known abundances and a range of masses will enable such predicted variation in the color-temperature relations to be tested. More discussion of the model colors are given in Bessell et al. (1997).

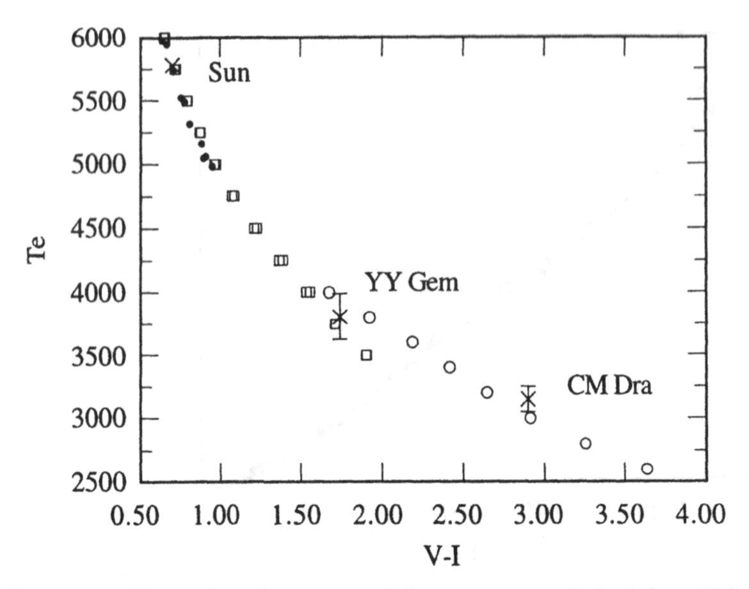

Figure 6. Comparison between model temperatures derived from V-I colors and the empirical temperatures for G-M dwarfs. The dwarfs with IRFM temperatures are plotted as solid circles. The dwarfs with directly measured radii (large crosses with error bars) are the sun and the eclipsing binaries YY Gem and CM Dra. The ATLAS9 no-overshoot models for log g =4.5 and 4.0 are indicated by open squares; the open circles are the NMARCS models for log g = 4.5.

4. Distances and Bolometric Corrections

Parallaxes provide the fundamental distances to the nearest stars, in particular to the low mass K-M dwarfs and white dwarfs. Lists of parallaxes are available in the Yale Parallax Catalog (Van Altena et al. 1995) and in other catalogs such as the 3rd Gliese & Jahreiss (1996) Catalog. The US Naval Observatory (eg. Monet et al. 1992) and the University of Virginia (eg. Ianna et al. 1996) continue strong ground based CCD programs to provide parallaxes for intrinsically faint stars. However, the HIPPARCOS satellite (Perryman et al. 1995) is producing extremely accurate parallaxes for bright stars which will revolutionize our knowledge of the upper main sequence. The catalog is expected to be available in 1997.

Bolometric corrections are required to enable comparison between theoretical HR diagrams and colour-magnitude diagrams. Bolometric corrections have historically been used to correct the visual magnitude to the bolometric magnitude (ie the luminosity) but the term nowadays refers to the magnitude correction to apply to any passband. The bolometric correction is defined as $M_{bol} = M + BC_M$ where M refers to a particular passband magnitude. The bolometric correction zeropoints are arbitrary but have usually been defined so that no star has a positive visual bolometric correction or else have adopted some particular visual bolometric

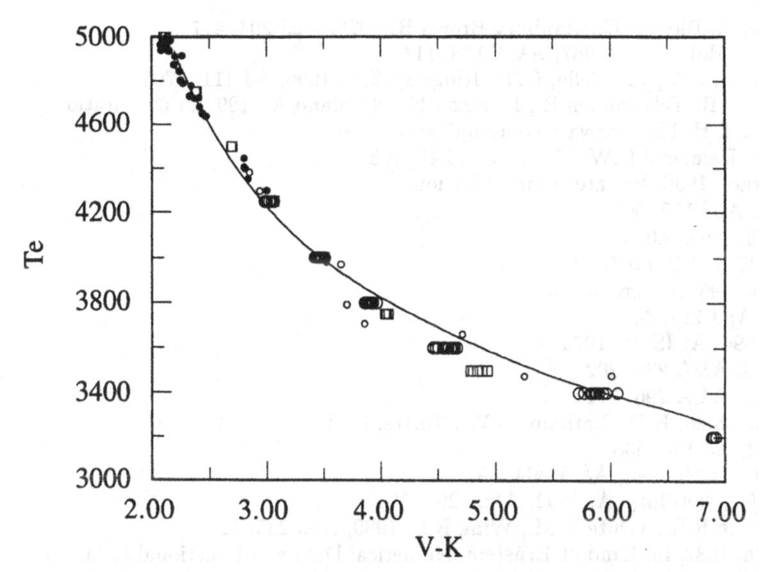

Figure 7. Comparison between model temperatures derived from V-K colors and the IRFM and interferometer temperatures for GK stars. The closed circles are IRFM data; the small open circles are Michelson interferometer data. The open squares are ATLAS9 no-overshoot model colors for log g = 0.0, 0.5, 1.0, 1.5; the large open circles are NMARCS models for the same gravities.

correction for the sun. This leads to lots of confusion in the literature. We support the proposal to adopt a bolometric correction (BC_V) of -0.07 mag for a solar model. But note that this zeropoint is different to that of -0.194 adopted by Kurucz (1979) and followed by Schmidt-Kaler (1982). We recommend that the theoretical bolometric corrections from the ATLAS9 and NMARCS models be used. These are in good agreement with empirical corrections and are available for a wide range of temperatures, gravities and abundances. They are discussed in Bessell et al. (1997) and available by anonymous ftp from the authors.

References

Allard F., Hauschildt P.H., Miller S., Tennyson J. 1994, ApJ 426, L39
Alonso, A., Arribas, S., Martinez-Roger,C. 1996a A&AS 117, 227
Alonso, A., Arribas, S., Martinez-Roger,C. 1996b A&A 313,873
Bessell M.S., Brett J.M., Scholz M., Wood P.R. 1989, A&A 213, 209
Bessell M.S., Scholz M., Wood P.R. 1996, A&A 307, 481
Blackwell D.E., Lynas-Gray A.E. 1994, A&A 282, 899
Brett J.M. 1995a, A&A 295, 736
Brett J.M. 1995b, A&AS 109, 263
Castelli F. 1995 Private communication.
Castelli F. 1996 Proceedings of the Workshop on Model Atmospheres and Spectrum Synthesis, Vienna July 1995, ASP Conference Series, ed. S. Adelman, F. Kupka & W. Weiss In press
Castelli F., Gratton R., Kurucz R.L. 1997 A&A 318, 841

Code A.M., Davis J., Bless R.C., Hanbury Brown R. 1976, ApJ 203, 417
DiBenedetto, G.P., Rabbia, Y. 1987, A&A 188, 114
Dyck, H.M., Benson, J.A., van Belle, G.T., Ridgway, S.T. 1996, AJ 111, 1705
Gustafsson B., Plez B., Edvardsson B., Eriksson K., Nordland A., 1997 in preparation
Gliese, W., Jahreiss, H. 1996, private communication
Habets G.M.H.J., Heintze J.R.W. 1981, A&AS 46, 193
Jorgensen, U. Grae-, 1996, Private communication
Kron, G.E. 1952, ApJ 115, 301
Kurucz R.L. 1979, ApJS 40, 1
Kurucz R. L. 1993, CD-ROM No 13.
Kurucz R.L 1995-1996 in preparation
Lacy C.H. 1977, ApJ 218, 444
McWilliam, A. 1990, ApJS 74, 1075
Megessier C. 1994, A&A 289, 202
Megessier C. 1995, A&A 296, 771
Metcalfe, T.S., Mathieu, R.D., Latham, D.W., Torres, G. 1996, ApJ 456, 356
Monet et al. 1992, AJ 103, 638
Perryman, M.A.C. et al. 1995, A&A 304, 69
Plez B., Brett J.M., Nordlund A. 1992, A&A 256, 551
Ridgway S.T., Joyce R.R., White N.M., Wing R.F. 1980, ApJ 235, 126
Schmidt-Kaler Th. 1982, in: Landolt-Brnstein, Numerical Data and Functional Relationships in Science and Technology, Vol. 2. (eds.) K. Schaifers & H.H. Voigt Springer-Verlag, Berlin
Tsuji T. 1981b A&A 99, 48
Tsuji T., Ohnaka K., Aoki W. 1996a, A&A 305, L1
Tsuji T., Ohnaka K., Aoki W. 1996b, A&A 308, L29
Wood P.R., Bessell M.S. 1993 Private communication by anonymous ftp

DISCUSSION

DAVID ARNETT: What about the M supergiants?

MIKE BESSELL: Dyck et al. (1997) claim that the M supergiants are also cooler than M giants of the same spectral types by about 150K. However, I find that like the K supergiants, the early M supergiants seem to follow the same Te versus V-K as the giants. But the models do indicate that there should be differences in the temperature color relation due to gravity in the mid to late M giants.

GIUSEPPE BONO: The discrepancy between the theoretical atmosphere models and observational data in the temperature color relations you showed could be due to the fact that in this temperature range the stars should be located inside the instability strip.

NICHOLAS ELIAS: Are the effects of starspots taken into account for M star calibration?

MIKE BESSELL: No. All the models are homogeneous, plane parallel or spherically symmetrical atmospheres and convection was handled using the mixing length technique. It is true that star spots will affect the colors of M dwarfs but I think mainly in the UV and blue. Colors obtained in the far-red and near-IR should be good effective temperature indicators. In our Mira models we have also neglected the the non-LTE aspects of the shock front, such as the emission lines in the computation of synthetic spectra or colors.

THE EFFECTIVE TEMPERATURES OF HOT STARS

P.A. CROWTHER
Dept. of Physics & Astronomy, University College London, Gower Street, London, WC1E 6BT, U.K.

Abstract. We review the effective temperature scale of hot ($\geq$ 10 kK) stars, including results from direct, continuum and ionization equilibrium techniques. We discuss the impact of recent developments in theoretical model atmospheres for OB subdwarfs, dwarfs and supergiants, white dwarfs and Wolf Rayet stars and present a revised $T_{\rm eff}$ scale for OB stars. Direct techniques coupled with Kurucz model atmospheres allow Strömgren photometry to be used as a sensitive $T_{\rm eff}$ indicator for normal stars with $T_{\rm eff} \leq 25$ kK. Reliable $T_{\rm eff}$ determinations for hotter, low surface gravity and H-deficient stars require sophisticated ionization equilibrium techniques, generally considering non-LTE and line-blanketing effects.

Key words: stars:temperatures – early-type – fundamental parameters

1. Continuum techniques

1.1. DIRECT METHODS

The effective temperature of a star – providing the essential link between observed properties and results of theoretical stellar structure and atmosphere models – can be directly obtained from measurements of its angular diameter together with the total absolute flux at the Earth, integrated over the entire spectrum The prime source in this field remains the seminal work of Code et al. (1976), who combined angular diameters measured by the Narrabri stellar intensity interferometer with UV, visual and IR observations for 32 hot stars, corrected for interstellar extinction. Smalley & Dworetsky (1995) repeated their analysis using more recent observational and theoretical results, and obtained values on average only one percent lower for stars with $T_{\rm eff} \leq 25$ kK.

T.R. Bedding et al. (eds.),
Fundamental Stellar Properties: The Interaction between Observation and Theory, 137–146.

While this method is termed 'direct', the allowance for fluxes in the unobservable Lyman continuum are taken from model calculations, and so becomes increasingly reliant on suitable models for very hot stars. For O and early B stars a significant fraction of the total energy is emitted shortward of $\lambda = 912$Å and obtaining $T_{\rm eff}$ from observed energy distributions requires accurate knowledge of surface gravities (Abbott & Hummer 1985; Hummer et al. 1988). Therefore, more sophisticated techniques need to be employed for such stars. We now turn to indirect techniques used for the temperature determination in the vast majority of hot stars.

1.2. CONTINUUM FLUX METHODS

The infrared flux method is a semi-direct technique developed by Blackwell & Shallis (1977) for cool stars, which uses the ratio of the integrated flux relative to a monochromatic IR flux to obtain a star's angular diameter and $T_{\rm eff}$ using an iterative procedure based on model atmospheres. Napiwotzki et al. (1993) found that the V-band flux provided an better calibration for B, A and F stars. Fitting model spectra to observations over the *entire* spectral range provides improved results for B stars (e.g. Malagnini et al. 1986; Drilling et al. 1984). Temperatures for DB white dwarfs can be obtained from UV continuum fits (e.g. Thejll et al. 1991), while hotter stars tend to be poorly constrained (e.g. 'R-index' technique of Schönberner & Drilling 1984), often suffering the added complication of varying UV interstellar extinctions from star to star.

1.3. PHOTOMETRIC METHODS

Strömgren $uvby\beta$ photometry, coupled with Kurucz (1979, 1991) model atmospheres, provides a powerful technique for the determination of $T_{\rm eff}$ in B, A and F stars. Napiwotzki et al. (1993) have provided a discussion of the various calibrations developed over the past decade, with the reddening free Strömgren colour-indices $[c_1]$ and $[u-b]$ widely employed. Fig. 1 demonstrates the temperature sensitivity of the Balmer jump for B stars based on Kurucz model atmospheres, and includes the Strömgren filter bandpasses. Temperatures of hot stars are poorly constrained using the Johnson reddening free index (e.g. Moehler et al. 1990).

Photometric calibrations applied to supergiants and stars showing non-solar He-contents can lead to significant discrepancies since their colours typically lie either at the edge of, or beyond current calibrations and the strength of their Balmer jumps are affected by composition effects (Kudritzki et al. 1989). For example, Venn (1995) obtained $T_{\rm eff} = 9.7$ kK from a detailed spectral analysis of HD 46300 (A0 Ib) using Mg I-II lines, which compares with 8.4kK from the $(b-y)_0$ calibration of Napiwotzki et al.

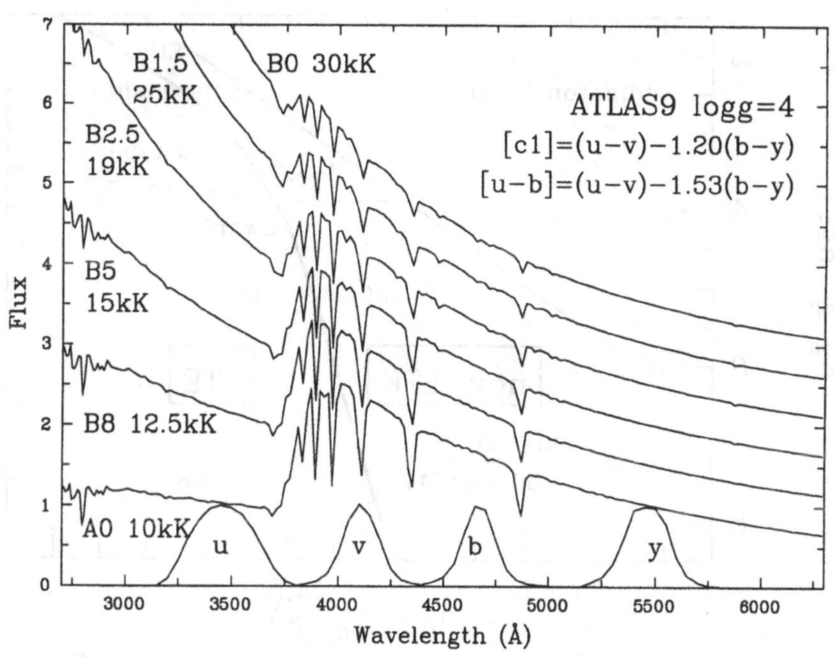

Figure 1. Kurucz (ATLAS9) theoretical energy distributions for B0–A0 dwarfs showing the strong temperature sensitivity of the Balmer jump, measured by the Strömgren colour-indices $[c_1]$ (favoured below ~20kK) and $[u-b]$ (above ~20kK)

(1993), and 10.0kK from the β versus $[c_1]$ relation of Lester et al. (1986). Saffer et al. (1994) have discussed the systematic errors obtained using standard photometric calibrations (e.g. Lester et al. 1986) for temperatures of (He-poor) sdB stars.

2. Spectroscopic techniques

While line blanketed, LTE Kurucz model atmospheres are widely employed for early-type stars, the assumption of LTE breaks down for hot, low gravity stars (see Fig. 2). Considerable effort has gone into developing realistic non-LTE model atmospheres for early-type stars in recent years. Ideally, temperatures are obtained using ionization equilibrium techniques, i.e. from suitable diagnostic lines of adjacent ionization stages showing a strong temperature dependence. Naturally, determinations of $T_{\rm eff}$ from model atmospheres are critically dependent on assumptions going into those models and the accuracy with which surface gravities and elemental abundance ratios are determined. Unfortunately, due to the complexity of consistently treating metal line blanketing in extended atmospheres, i.e. considering the influence of thousands to millions of spectral lines on the atmospheric struc-

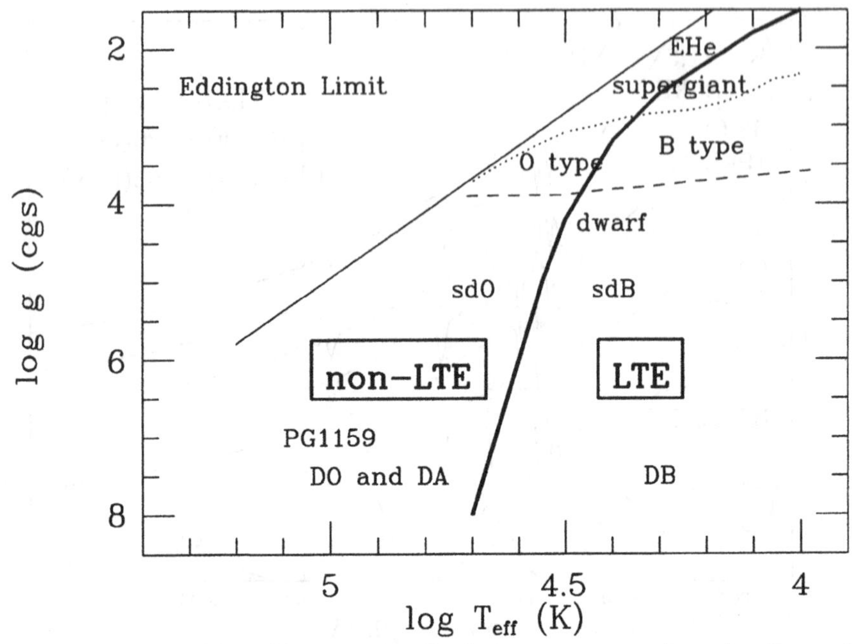

Figure 2. The position of hot stars in effective temperature and gravity discussed in this review. The Eddington limit for radiative stability is indicated (thin solid), as are the LTE and non-LTE domains (thick solid) and OB dwarf/supergiant (dashed/dotted) scales

ture (via backwarming) and emergent spectrum (via line blocking), various statistical simplifications are employed (Hillier, these proc.)

2.1. OB DWARFS AND SUPERGIANTS

The majority of detailed analyses of OB stars have been subject to the assumption of hydrostatic equilibrium in a plane-parallel atmosphere (e.g. Herrero et al. 1992; Puls et al. 1996). However, should stellar winds be strong, absorption lines will be contaminated by wind effects (Schaerer & Schmutz 1994; Crowther & Bohannan 1997). Optical He I-II lines generally provide temperature diagnostics for O stars (e.g. Herrero et al. 1992) while Si II-IV lines are used for B stars (e.g. Becker & Butler 1990), though Kudritzki (1992) discussed the complicating role of velocity fields for investigating supergiants. Kilian et al. (1991) employed both He and Si diagnostics for early B dwarfs and giants which agreed relatively well, though with differences of up to 1 800K. Kilian et al. (1991) and Grigsby et al. (1992) used line blanketed atmospheres to obtain effective temperatures systematically higher (2 000K) than those from c_0–$T_{\rm eff}$ photometric calibrations (e.g. Lester et al. 1986). The $T_{\rm eff}$ scale for B supergiants has been improved upon recently by McErlean et al. (1997) who used Si III-IV diagnostics for early-types, with non-LTE corrections to results from the Lester et al. (1986) calibration applied to mid/late B supergiants.

Recent ionization equilibrium techniques for O stars represent considerable advances relative to earlier (insensitive) continuum methods. For example, Bohannan et al. (1986) obtained $T_{\rm eff} = 42$ kK from a non-LTE wind blanketed analysis of ζ Pup (O4 I(n)f), with $T_{\rm eff} = 32$–50 kK obtained from continuum techniques. However, widely varying results have also been found in recent non-LTE spectroscopic analyses of late O/early B dwarfs. For instance, for HD 214680 (10Lac, O9 V) Grigsby et al. (1992) derived $T_{\rm eff} = 30.0$ kK using a line-blanketed model atmosphere, in sharp contrast to $T_{\rm eff} = 37.5$ kK from the unblanketed analysis of Herrero et al. (1992). Despite these conflicting results, HD 214680 remains widely used as a standard in $T_{\rm eff}$–photometric calibrations following Malagnini et al. (1986) who used LTE Kurucz continuum fits to obtain $T_{\rm eff} = 31.75$ kK. Studies based on similar non-LTE models can result in discrepant temperatures when based on optical or UV diagnostics (e.g. Melnick 42, Heap et al. 1991, Pauldrach et al. 1994). Indeed, within a single analysis different diagnostic line ratios often yield considerably different temperatures – for HD 46150 (O5 V(f)), Herrero et al. (1992) obtained $T_{\rm eff} = 42.5$ kK from He I λ4922/He II λ4542 compared with $T_{\rm eff}$=47 kK from He I λ4471/He II λ4542.

Detached OB binaries provide useful tests of spectroscopic analyses since their masses, radii, gravities can be independently and accurately determined, though temperature determinations generally rely on Strömgren photometric calibrations (Schönberner & Harmanec 1995; Hilditch et al. 1996) which are imprecise for O and early B stars.

Various MK spectral type–$T_{\rm eff}$ correlations have been attempted for OB stars over the past couple of decades, including Böhm-Vitense (1981), Schmidt-Kaler (1982) and Humphreys & McElroy (1984). While these remain widely used, in ionizing flux calibrations for instance (necessary for photoionization studies of H II regions and starbursts), more recent scales have been obtained by Howarth & Prinja (1989), Theodossiou & Danezis (1991) and Vacca et al. (1996) using more recent observational and theoretical results. Due to the plethora of new results in this field we present a revised $T_{\rm eff}$ scale of OB dwarfs and supergiants in Fig. 3, based on results from various direct, continuum and ionization equilibrium techniques. Previous standard scales are also included – note that we find considerable differences relative to the recent Vacca et al. (1996) calibration for O dwarfs since we incorporate results from blanketed studies (e.g. Voels et al. 1989; Grigsby et al. 1992) and do not impose a linear $T_{\rm eff}$ scale.

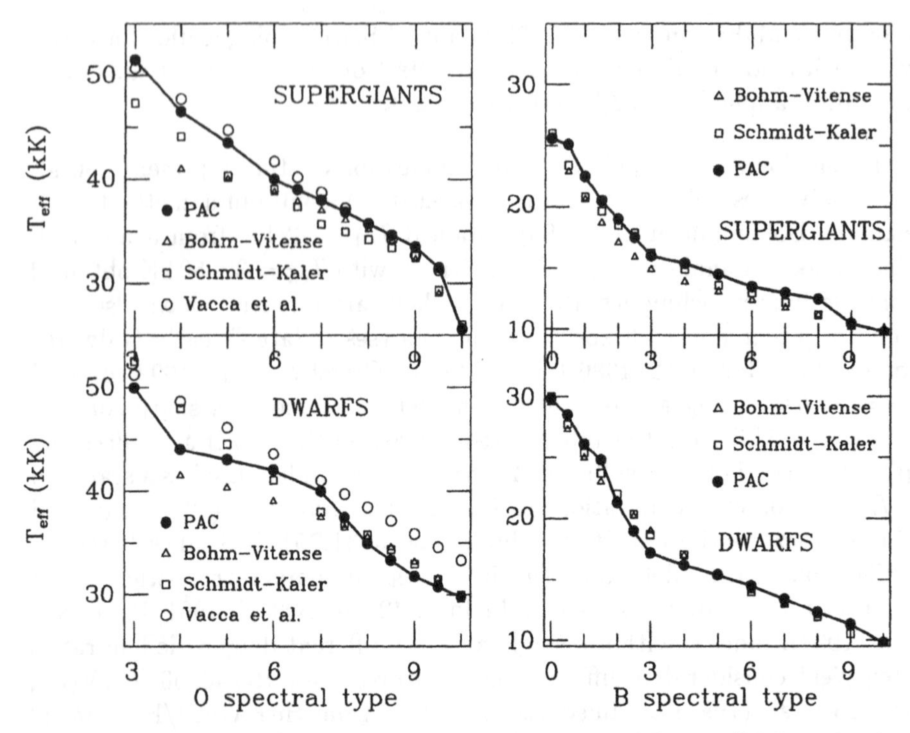

Figure 3. Comparison of our revised temperature scale for OB stars (PAC, thick solid line), available in tabular form on request, with the standard calibrations of Böhm-Vitense (1981) and Schmidt-Kaler (1982), plus the recent *linear* O star scale of Vacca et al. (1996)

2.2. OB SUBDWARFS

Recent theoretical progress in modelling H-deficient hot subdwarfs has had a greater influence on temperature determinations than normal OB stars since the continuum opacity, dominated by hydrogen in normal stars, is severely reduced in helium stars. Therefore the line opacity is much more effective in blocking and redistributing radiation, so the treatment of line-blanketing is critical (e.g. Haas et al. 1996). In the case of He-rich sdO stars, Dreizler (1993) has used ionization equilibrium techniques with line-blanketed non-LTE models, revealing 44.5kK for LS IV+10° 9, substantially lower than 65kK obtained from UV continuum methods (Schönberner & Drilling 1984). Temperatures of sdO stars using blanketed models (Dreizler et al. 1990) are systematically 5−10kK lower than unblanketed results (Thejll et al. 1994).

Temperature determinations in He-poor sdB stars use Balmer line profile fits to blanketed LTE model atmospheres (e.g. Saffer et al. 1994) and are in good agreement with results from suitable Strömgren photometric calibrations (e.g. Bergeron et al. 1984). As with sdO stars, temperatures of H-deficient extreme helium (EHe) stars obtained from line-blanketed analyses are systematically lower (~2 000 K) than unblanketed studies (e.g. Jeffery & Heber 1992).

2.3. WHITE DWARFS AND PG1159 STARS

Recent temperature determinations of hot H-rich (DA) and H-poor (DO, DB) white dwarfs and PG1159 stars are based on model Balmer line profile fits, supplemented by far/extreme UV continuum fits or ideally ionization equilibrium techniques, where observations permit.

Pure H-atmosphere models are generally used to analyse DA white dwarfs, with temperatures obtained from optical and EUV analyses showing good consistency (see Barstow et al. 1993), and show improved temperature sensitive relative to earlier UV continuum/Lyα fits (e.g. Holberg et al. 1986). However, there is a well known 'Balmer line problem' for some high gravity H-rich central stars of Planetary Nebulae (CSPNe), well illustrated by Napiwotzki (1995b) who obtained discrepant temperatures using different Balmer series members for the CSPNe of the Helix nebula NGC 7293, ranging from 48 kK (using Hα) to 107 kK (Hδ)! While unsolved, the incorporation of metal line blanketing improves the situation, and temperatures obtained from higher Balmer lines are in better agreement with alternative spectroscopic ($\geq$70 kK since He I absent) and photoionization (120 kK from Clegg & Walsh 1989) results.

Napiwotzki (1995a) discussed the importance of non-LTE effects even for cool DO white dwarfs, despite their high surface gravities, while Dreizler & Werner (1996) have compared results based on blanketed non-LTE models with previous LTE analyses finding the greatest difference between 60–70 kK. Werner et al. (1991) found that the non-LTE analyses are critical for the higher temperature PG1159 stars. In the absence of superior spectroscopic diagnostics, temperature determinations in DB white dwarfs rely on He I profile fits using LTE models (e.g. Beauchamp et al. 1995).

Stellar temperatures of CSPNe (and OB stars in H II regions) can also be obtained by means of their nebulae, most frequently via Zanstra analyses (Zanstra 1931; Harman & Seaton 1966). However, results from He II are frequently inconsistent with those from H I (the 'Zanstra discrepancy') due to the use of inappropriate (often blackbody) flux distributions (Gabler et al. 1991), or an optically thin PNe. Use of realistic O-type model fluxes by Sellmaier et al. (1996) provides a possible solution to the [Ne III] problem in H II-regions, though model atmospheres are not yet fully capable of explaining both the stellar *and* nebular spectrum simultaneously.

2.4. WOLF-RAYET STARS

Obtaining a reliable temperature scale for Wolf-Rayet (WR) stars, the chemically evolved descendents of OB stars or evolved H-deficient CSPNe, is a formidable challenge. Continuum techniques were used to estimate $T_{\rm eff}$ in WR stars (e.g. Underhill 1983) though were shown to be unrealistic by Garmany et al. (1984). Due to extreme mass-loss, their winds are so optically thick that their photospheres are invisible, with the usual plane-parallel, LTE approximations completely invalid, with radii highly wavelength-dependent (Cherepachchuk et al. 1984). As for O stars, much progress has been made in recent years using extended non-LTE model atmospheres (e.g. Hillier 1987; Schmutz et al. 1989). Unfortunately, an added complication of their dense winds is that the usual radius at $\tau_{\rm Ross} = 2/3$ becomes a poor indicator of excitation in WR stars, with $\tau_{\rm Ross} = 20$ usually defining the stellar temperature, though this is difficult to relate with interior evolutionary models (Hamann 1994). Unfortunately, stellar temperatures of WR stars are also critically dependent on *assumed* velocity distributions (Hillier 1991), so that observational determinations of WR velocity distributions are keenly sought.

Initially, ionization balance studies restricted to lines of helium from model grids where used to obtain temperatures (e.g. Schmutz et al. 1989). These have subsequently been extended to tailored models including nitrogen in WN stars (e.g. Crowther et al. 1995a) and carbon in WC stars (e.g. Koesterke & Hamann 1995), with heavy element line blanketed models now starting to appear. In general, results from pure helium analyses tend to underestimate stellar temperatures, with the discrepancy greater at earlier spectral type. For HD 151932 (WN7), Underhill (1983) obtained a temperature of 25 kK from the integrated flux method, increasing to 32 kK from non-LTE model grids by Schmutz et al. (1989) and 34 kK from line profile fits by Crowther et al. (1995a). However, results obtained for different species in non-blanketed models generally lead to discrepant temperatures: For HD 211564 (WN3), Crowther et al. (1995b) obtained $T_{\rm eff} = 49$ kK (69 kK) from helium (nitrogen) diagnostics! Observations of WR nebulae provide excellent tests of current theoretical flux distributions below the Lyman limit (see Esteban et al. 1993).

To summarise, huge advances have been made over the past decade towards a reliable temperature scale of hot stars. However, significant uncertainties remain, especially for O and WR stars, which may be resolved using nebulae as tests of theoretical energy distributions. Since real stars rotate and pulsate it is unclear what scale of temperature fluctuations are occurring across their stellar disks.

References

Abbott D.C., Hummer D.G., 1985, ApJ 294, 286
Barstow M.A., Fleming T.A., Diamond C.J., et al. 1993, MNRAS 264, 16
Beauchamp A., Wesemael F., Bergeron P. et al. 1995, in: European Workshop on White Dwarfs, Koester D., Werner K. (eds.), Lecture Notes in Physics 433, Springer, p.108
Becker S.R., Butler K., 1990, A&A 235, 326
Bergeron P., Fontaine G., Lacombe P. et al., 1984, AJ 89, 374
Blackwell D.E., Shallis M.J., 1977, MNRAS 180, 177
Bohannan B., Abbott D.C., Voels S.A., Hummer D.G., 1986, ApJ 308. 728
Böhm-Vitense E., 1981, ARA&A 19, 295
Cherepachchuk A.M., Eaton J.A., Khaliullin Kh.F., 1984, ApJ 281, 774
Clegg R.E.S., Walsh J.R., 1989, in: Planetary Nebulae, Torres-Peimbert S. (ed.), Proc IAU Symp. 131, Kluwer, p.223
Code A.D., Davis J., Bless R.C., Hanbury-Brown R., 1976, ApJ 203, 417
Crowther P.A., Bohannan B., 1997, A&A 317, 532
Crowther P.A., Smith L.J., Hillier D.J., 1995a, A&A 293, 403
Crowther P.A., Smith L.J., Hillier D.J., 1995b, A&A 302, 457
Dreizler S., 1993, A&A 273, 212
Dreizler S., Werner K., 1996, A&A 314, 217
Dreizler S., Heber U., Werner K., et al., 1990, A&A 235, 234
Drilling J.S., Schönberner D., Heber U., Lynas-Gray A.E., 1984, ApJ 278, 224
Esteban C., Smith L.J., Vílchez J.M., Clegg R.E.S., 1993, A&A 272, 299
Gabler R., Kudritzki R.-P., Méndez R.H., 1991, A&A 245, 587
Garmany C.D., Massey P., Conti P.S., 1984, ApJ 278, 233
Grigsby J.A., Morrison N.D., Anderson L.S., 1992, ApJS 78, 205
Haas S., Dreizler S., Heber U., et al. 1996, A&A 311, 669
Hamann W.-R., 1994, Space Sci. Rev. 66, 237
Harman R.J., Seaton M.J., 1966, MNRAS 132, 15
Heap S.R., Altner B., Ebbets D., et al. 1991, ApJ 377, L29
Herrero A., Kudritzki R.-P., Vilchez, J.M. et al. 1992, A&A 261, 209
Hilditch R.W., Harries T.J., Bell S.A., 1996, A&A 314, 165
Hillier D.J., 1987, ApJS 63, 947
Hillier D.J., 1991, in: Wolf-Rayet stars and Interrelations with Other Massive Stars in Galaxies, Proc. IAU Symp. 143, van der Hucht K.A., Hidayat B. (eds.), Kluwer, p.59
Holberg J.B., Vasile J., Wesemael, F., 1986, ApJ 306, 629
Howarth I.D., Prinja R.K., 1989, ApJS 69, 527
Hummer D.G., Abbott D.C., Voels D.A., Bohannan B., 1988, ApJ 328, 704
Humphreys R.M., McElroy D.B., 1984, ApJ 284, 565
Jeffery C.S., Heber U., 1992, A&A 260, 133
Kilian J., Becker S.R., Gehren T., Nissen P.E., 1991, A&A 244, 419
Koesterke L., Hamann W.-R., 1995 A&A 299, 503
Kudritzki R.-P., 1992, A&A 266, 395
Kudritzki R.-P., Gabler A., Gabler R., et al. 1989, in: Physics of Luminous Blue Variables, Davidson K., et al. (eds.), Kluwer, p.67
Kurucz R.L., 1979 ApJS 40, 1
Kurucz R.L., 1991, in: Precision Photometry: Astrophysics of the Galaxy, Philip A.G.D., et al. (eds.), L. Davis Press, p.27
Lester J.B., Gray R.O., Kurucz R.L., 1986, ApJS 61, 509
Malagnini M.L., Morossi C., Rossi L., Kurucz R.L., 1986, A&A 162, 140
McErlean N., et al. 1997, in prep.
Moehler S., Heber U., 1990, De Boer K.S., A&A 239, 265
Napiwotzki R., 1995a, in: European Workshop on White Dwarfs, Koester D., Werner K. (eds.), Lecture Notes in Physics 433, Springer, p.132
Napiwotzki R., 1995b, in: European Workshop on White Dwarfs, Koester D., Werner K.

(eds.), Lecture Notes in Physics 433, Springer, p.176
Napiwotzki R., Schönberner D., Wenske V., 1993, A&A 268, 653
Pauldrach A.W.A., Kudritzki R.-P., Puls J., et al. 1994 A&A 283, 525
Puls J., Kudritzki R.-P., Herrero A. et al. 1996, A&A 305, 171
Saffer R.A., Bergeron P., Koester D., Liebert J., 1994, ApJ 432, 351
Schaerer D., Schmutz W., 1994, A&A 288, 231
Schmidt-Kaler Th., 1982, in: Landolt-Börnstein, Group VI, Vol 2b, Schaifers K., Voigt H.H. (eds.), Springer-Verlag
Schmutz W., Hamann W.-R., Wessolowski U., 1989, A&A 210, 236
Schönberner D., Drilling J.S., 1984, ApJ 278, 702
Schönberner D., Harmanec P., 1995, A&A 294, 509
Sellmaier F.H., Yamamoto T., Pauldrach A.W.A., Rubin R.H., 1996, A&A 305, L37
Smalley B., Dworetsky M.M., 1995, A&A 293. 446
Thejll P., Vennes S., Shipman H.L., 1991, ApJ 370, 355
Thejll P., Bauer F., Saffer R., et al. 1994, ApJ 433, 819
Theodossiou E., Danezis E., 1991, Ap&SS 183, 91
Underhill A.B., 1983, ApJ 266, 718
Vacca W.D., Garmany C.D., Shull J.M., 1996, ApJ 460, 914
Venn K.A., 1995, ApJS 99, 659
Voels S.A., Bohannan B., Abbott D.C., Hummer D.G., 1989, ApJ 340, 107
Werner K., Heber U., Hunger K., 1991, A&A 244, 437
Zanstra H., 1931, Publ. Don. Astrophys. Obs 4, 209

DISCUSSION

ANDRE MAEDER: Can you comment on the uncertainty in the $T_{\rm eff}$ calibration of O-type stars, in view of the fact that many of them show sizeable He and N enrichment?

PAUL CROWTHER: While helium enrichment in O stars certainly plays a role in affecting their temperature scale, uncertainties are dominated by the limited treatment of wind/line blanketing to date. More concerning is the current $T_{\rm eff}$ scale for B supergiants, which rely on observations of the Balmer Jump (via Strömgren photometric calibrations), which is very sensitive to He-enrichment.

THE STELLAR TEMPERATURE SCALE FROM ANGULAR DIAMETERS AND FLUX DISTRIBUTIONS

A.J.BOOTH
University of Sydney,
School of Physics A28, NSW 2006, Australia

Abstract. The current situation for the measurement of effective temperatures from interferometrically determined angular diameters and spectrophotometry in the UV, visible and IR is considered. Accuracies and reliabilities of the resulting temperatures are assessed and coverage of stellar types is discussed.

1. Introduction

The effective temperature, T_e, of a star (or equivalently the total emergent flux, F) can be determined from two quantities that are in principle directly observable: the angular diameter θ, and the total flux received at the Earth F_E. The definition of effective temperature is

$$\sigma T_e = F, \tag{1}$$

and, in the absence of interstellar absorption,

$$F = \frac{4}{\theta^2} F_E; \tag{2}$$

hence

$$T_e = \frac{1}{\sigma} \frac{4}{\theta^2} F_E. \tag{3}$$

Angular sizes can be determined from speckle interferometry (for very large θ stars), lunar occultation, or long baseline optical interferometry. This paper will be concerned with the last (see Richichi, this volume, for effective temperatures determined by lunar occultation).

Total fluxes must be found by integrating the observed monochromatic stellar flux from $0 \leq \lambda \leq \infty$. This requires spectrophotometric and calibrating photometric observations. A distinction should be noted with the

T.R. Bedding et al. (eds.),
Fundamental Stellar Properties: The Interaction between Observation and Theory, 147–152.

Infra-Red Flux Method, IRFM, (Megessier, this volume), which requires only a *ratio* of fluxes (total to infra-red) whereas we require an absolute flux.

This paper will describe the present situation with particular regard to coverage, and accuracy and reliability. It will also discuss only "normal", and single stars. As yet very few "special" stars (eg. pulsating stars, shell stars), or stars in binaries have measured angular sizes.

2. Coverage

At present the number of stars with accurate, interferometrically determined temperatures are rather few, and this number is limited by the available angular size measurements. In principle the flux is more easily measurable than θ in that it does not require specialised instruments or observing techniques.

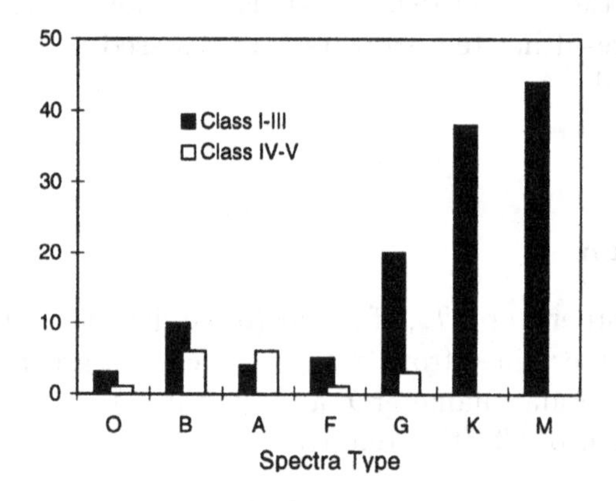

Figure 1. Histogram of measured angular sizes as a function of spectral type and luminosity class.

Davis (this volume) gives a review of the available angular diameters and their accuracies. Fig. 1 summarises these. Although Fig. 1 might at first suggest quite extensive coverage a few points should be noted. The figure includes all measurements with uncertainties $\lesssim 10\%$. if only high accuracy values are accepted ($\lesssim 3$–4%) the numbers drop by about a factor of two. In the O, B, A, range of type the measurements are mostly from the Narrabri Intensity Interferometer (Code *et al.*, 1976). It is a tribute to Hanbury Brown and his co-workers that these measurements made some 20 years ago still form the basis of the temperature scale for hot stars. However, both the angular size and flux measurement accuracies can now be improved for these stars. Of the other stars in Fig. 1 most are from the Mk.III interferometer (Hutter *et al.*, 1989), with some from the IOTA (Dyck *et al.*, 1996) or I2T interferometers (Di Benedetto & Rabbia, 1987). Of the Mk.III angular diameters few have been turned into effective temperatures (Mozurkewich, private communication), though the flux measurements to do so are often available. Further, there are in general very few measurements for dwarfs, and in particular none for types K and M. Indeed the only direct T_e measurements for type MV comes from two eclipsing binary systems (Habets & Heintze, 1981). So the main sequence in the K, M region has a very poorly defined temperature scale. Finally, there is a particular lack of any measurements for types A and F.

3. Accuracy and Reliability

Since $T_e \propto \theta^{-2}$ and $T_e \propto {F_E}^4$, the error in T_e is insensitive to errors in θ or F_E. Roughly speaking, the formal error in T_e, $\sigma(T_e)$, is given by

$$\sigma(T_e) \propto \frac{1}{2}\sigma(\theta) \quad \text{and} \quad \sigma(T_e) \propto \frac{1}{4}\sigma(F_E) \tag{4}$$

Too much can be read into this result, however, as T_e is an many ways an unphysical quantity, and should really be thought of as a label for a particular set of atmospheric conditions. The physical quantity is F ($\propto T_e^4$) which does not have these desirable dependencies of its accuracy. Having said this, I shall continue to use T_e here. A good target accuracy for T_e determinations is 1%, which for example matches the best atomic data available for abundance determinations, and the best determinations of $\log(g)$ (Maxted, this volume). Hence, the target accuracy of θ should be $\lesssim 2\%$ and of $F_E \lesssim 4\%$.

As indicated above, the situation for the accuracy of θ measurements is quite good. Angular sizes with formal accuracy $\lesssim 2\%$ (Davis, this volume) can be found quite readily, at least for a limited number of stars. However, the quoted accuracies are usually the internal consistency errors. Interferometric measurements are very sensitive to calibration uncertainty and there is some evidence of systematic errors of up to 10%, for example between IOTA and I2T values (Dyck *et al.*, 1996). Both the coverage problem and this calibration uncertainty are due to the limited number of currently operating interferometers and their restricted baselines. When the next generation of interferometers (eg. CHARA, NPOI, SUSI, VLTI) come into operation this situation is likely to be greatly improved (eg. Booth *et al.*, 1997 for improvement in coverage).

The situation for the accuracy of the flux measurements is less satisfactory and it turns out that F_E measurements limit the accuracy of the present T_e determinations. It is useful to break up the F_E measurements into 4 or 5 different wavelength regimes:

3.1. FAR UV FLUXES

In the region $\lambda \lesssim 100$nm interstellar absorption blocks most of the flux, so direct measurements are largely impossible. Models must be used, and their accuracy is uncertain, one may guess at 20%. This region is very important for the hottest stars and makes the measurement of their T_e particularly difficult.

3.2. UV FLUXES

The region 100nm $\lesssim \lambda \lesssim$ 350nm can be sampled by IUE or HST measurements. Recently the calibration in this region has been improved using white dwarf model atmospheres, but a good direct calibration would be more satisfactory to avoid modeling errors (Kruk, this volume). Internal consistency errors are typically 2–7% on well observed IUE stars, so, including a contribution for calibration uncertainty, a reasonable flux uncertainty in this region might be taken as 5%.

3.3. VISIBLE FLUXES

Megessier (this volume) reviews the current calibration situation, with the absolute calibration of Vega being secure at about 0.7%. Careful spectrophotometry can give internal consistency at about the 1% level (eg. Petford *et al.*, 1988), though there are systematic differences at the 1–2% level between observers. A reasonable level of uncertainty in the range 350nm $\lesssim \lambda \lesssim 1\mu$m might thus be 1–2%.

3.4. INFRA-RED AND FAR INFRA-RED FLUXES

The situation in the infra-red is less clear (Megessier, this volume). The direct calibration of Vega for 1μm $\lesssim \lambda \lesssim 5\mu$m has internal consistency at the 3% level, but differs from models at the 7% level. The implication from the application of the IRFM is that the models are correct (Blackwell *et al.*, 1991). A good test of this would be to compare IRFM deduced angular sizes with those found from interferometry. At longer wavelengths IRAS flux measurements can be used, but their calibration rests on model atmosphere colours (Cohen *et al.*, 1992) and consequently are prone to unknown levels of systematic uncertainty. Fortunately, this region makes only a small contribution for most stars, and can often be well approximated by a black body function. Until these questions are resolved, a level for the flux accuracy in the infra-red could be about 5%, but this may be optimistic.

Fig. 2 shows how these errors combine to produce a total error for T_e. This graph is based on black body curves not atmosphere models, but shows

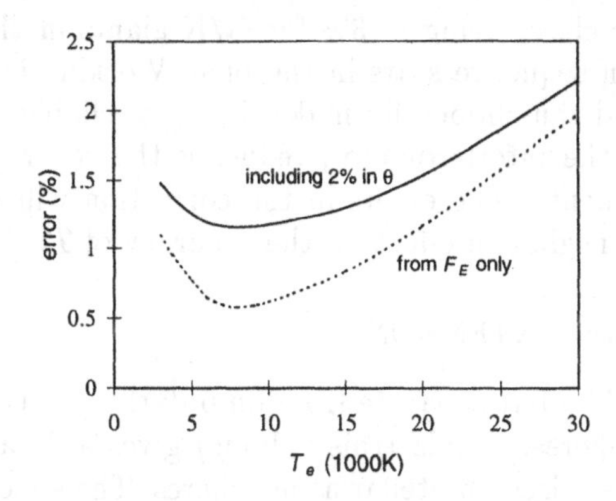

Figure 2. Total uncertainty in T_e due to flux and angular size errors.

the general trends. Clearly we are well placed with regard to the types A to G, but for hotter and cooler stars the calibrations of the infra-red and UV fluxes, respectively, need to be better defined. Some caution is called for, however, as many of the calibration errors are likely to be systematic rather than random. As such they are not improved by averaging, and further I may be underestimating their effect here by taking as a "typical" error the middle of the estimated range.

4. Potential Problems

There are also several caveats that must be placed on the above analysis, where systematic errors may make Fig. 2 overly optimistic.

4.1. INTERSTELLAR ABSORPTION

It is particularly difficult to determine an accurate value for A_V for a given star. Different methods tend to give quite different results when stars are more than about 100pc away. As an example, an error of only 0.05mag in A_V when $T_e = 10000$K gives a 3% error in T_e; more for hotter stars. Obviously this effect is greatest for hot stars due to their greater blue flux and tendency to be further away, and it may place the ultimate limit on how well we can calibrate their T_e values. It is also a problem even for cool giants, however, as they also tend to be at large distances.

4.2. LIMB DARKENING

The θ measured by interferometry is not a true size, but that of an *equivalent uniform disk* (Davis, this volume). Model atmospheres are needed to provide a correction for limb darkening. The correction varies with wavelength and stellar class, being $\sim$ 3% for G/K giants in the infra-red, and $\sim$ 10% for B main sequence stars in the blue. Working in the red reduces the correction, and thus hopefully makes it more reliable, but also reduces the resolution of the interferometer, reducing the coverage and accuracy of the θ measurements. The errors in the correction will usually be small enough to have a negligible effect on the accuracy of T_e.

4.3. ATMOSPHERIC EXTENSION

A problem related to but separate from limb darkening is the extension of giant star atmospheres. Scholz (this volume) gives a detailed appraisal of the difficulties this raises for stellar atmospheres. The effect is that angular sizes change with measurement wavelength (Quirrenbach *et al.*, 1993), so care must be taken to ensure that enough information is gathered to characterise these changes and account for them in the analysis to obtain T_e. Measurements at one wavelength, even a continuum one are not enough.

5. Summary

At present the best interferometrically determined angular diameters combined with the best flux measurements can give effective temperatures to 1–2% accuracy. The available angular sizes limit the coverage for types of star, but this is due to the limitations of currently operating interferometers. The next generation of interferometers will greatly improve this situation. There is still room for improvement in the accuracy of flux calibrations and spectrophotometry, particularly in the UV and IR. Problems with these will continue to limit the accuracy of T_e determinations for stars hotter than A and cooler than G in the near future.

References

Blackwell, D.E, Lynas-Gray, A.E. & Petford, A.D. (1991), *Astron. Astrophys.*, **245**, 567.

Booth, A.J., Davis, J. & Bessell, M. (1997), *Proceedings of Science with the VLTI, Ed. Paresce, F.*, in press.

Code, A.D., Davis, J., Bless, R.C. & Hanbury Brown, R. (1976), *Astrophys. J.*, **203**, 417.

Cohen, M., Walker, R.G., Barlow, M.J. & Deacon, J.R. (1992), *Astron. J.*, **104**, 1650.

Di Benedetto, G.P. & Rabbia, Y. (1987), *Astron. Astrophys.*, **188**, 114.

Dyck, H.M., Benson, J.A., van Belle, G.T. & Ridgway, S.T. (1996), *Astron. J.*, **111**, 1705.

Habets, G.M.H.J. & Heintze, J.R.W. (1981), *Astron. Astrophys.*, **46**, 193.

Hutter, D.J., Mozurkewich, D., Simon, R.S., Colavita, M.M. Pan, X.P., Shao, M., Hines, B.E., Staelin, D.H., Hershey, J.L., Hughes, J.A. & Kaplan, G.H. (1989), *Astrophys. J.*, **340**, 1103.

Petford, A.D., Blackwell, D.E., Booth, A.J., Haddock, D.J., Leggett, S.K., Mountain, C.M., Selby, M.J. & Arribas, S. (1988), *Astron. Astrophys.*, **203**, 341.

Quirrenbach, A., Mozurkewich, D., Armstrong, J.T., Buscher, D.F. & Hummel, C.A. (1993), *Astrophys. J.*, **406**, 215.

STELLAR EFFECTIVE TEMPERATURES THROUGH THE INFRARED FLUX METHOD

C. MEGESSIER
Observatoire de Paris-Meudon
F-92195 Meudon Cedex France

Abstract.
The principle of the method and its advantages are recalled, as well as the works giving such T_{eff}. Although the method is almost model independent, the monochromatic red fluxes play an important part. The resulting T_{eff} depends on the blanketing in the models. The stellar metallicity and gravity have to be known to avoid uncertainties of about 1%. In principle an accuracy around 1% on T_{eff} could be obtained, but the uncertainty on the absolute red flux calibration, due to the unsolved problem of the Vega red flux, prevents this from being achieved.

1. Introduction and principle of the method

More and more accurate stellar effective temperatures are required, either for atmosphere modelling or to study the internal structure, to better scale the HR diagram, specially in connection with the high accuracy of Hipparcos parallaxes, and so to fix better the evolutionary track for a star. Then we ask: what accuracy is achievable on T_{eff}, is it possible to obtain effective temperatures within 1% or better, as asked for and claimed in some works? Why the infrared flux method (IRFM)?

The most direct method to derive T_{eff} relies on the Stefan law. It requires the stellar angular diameter θ and the total flux received from the star, integrated over the whole wavelength range f. Indeed :

$$F = \sigma T^4 \qquad \text{and} \qquad \theta = 2R/d \qquad \text{so that} \qquad f/F = \theta^2/4 \quad (1)$$

where R is the stellar radius and d the stellar distance.

However the angular diameter is known for the brightest stars only and it has a poor accuracy compared to the aim of T_{eff} within 1%. As reported

T.R. Bedding et al. (eds.),
Fundamental Stellar Properties: The Interaction between Observation and Theory, 153–158.

by Davis (these Proceedings), for a magnitude 3 star the uncertainty is at least 5%, leading to 2.5% on T_{eff}.

In their method, Blackwell and Shallis (1977, 1980) assume that in the infrared spectrum, the relation (1) holds also for monochromatic wavelengths :

$$f_\lambda/F_\lambda = \theta^2/4 \quad (2) \qquad R = F/F_\lambda = \sigma T^4/F_\lambda = f/f_\lambda = R_{obs} \quad (3)$$

so that the ratio of (1) over (2) leads to (3) where the angular diameter enters no more. The ratio R_{obs} deduced from the observations is compared to the ratio R computed from the atmosphere models. Blackwell et al. justify their assumption by several arguments: the small sensitivity of the infra-red flux to T_{eff}, the smaller blanketing in the infrared, the more certain stellar opacity sources.

First we list the existing works on the IRFM, then we discuss in turn the required observations and models, and the uncertainties they induce on T_{eff}.

2. Stars with effective temperatures through the IRFM

The groups who derived T_{eff} through the IRFM are listed in Table 1 for the normal stars, with the type of the stars and their number. Table 2 lists the works on Ap and Am stars.

3. Influence of the models and of the stellar metallicity and gravity

In the works earlier than 1990, it was assumed that the ratio F/F_λ, where F is given directly by the Stefan law and F_λ by atmosphere models, does not contribute to the uncertainty on T_{eff}. Meǵessier (1994) showed qualitatively and quantitatively that, in fact, the model itself has an influence on T_{eff} derived from a given observed ratio R_{obs}. That is logically explained by the increase of blanketing in the more recent codes. The numerous UV spectral lines absorb flux that is redistributed in the Paschen continuum, which is then enhanced. The near infrared monochromatic fluxes being larger from the more recent codes, the resulting $R = F/F_\lambda$ will be smaller. The relation R vs T_{eff} is lowered and then it leads to larger T_{eff} for a given R_{obs}. This is illustrated in Table 3, which compares, for few stars, IRFM T_{eff} obtained from the same R_{obs} and from R (T_{eff}) for different models. The change in T_{eff} is larger than 1% between the oldest codes considered and ATLAS9, and it can be as large as 350K, i.e., 2.4%.

Mégessier (1994) showed that the relations R (T_{eff}) computed for different metallicities and gravities differ, so that using a relation with inappropriate [M/H] or log g for a star will induce systematic errors in T_{eff}. This

TABLE 1. Stars T_{eff} from the IRFM

Authors	N	Type	Class
Blackwell and Shallis (1977)	7	A0 to F8	V, III, II, I
Blackwell et al. (1980)	28	A, F, G	V, IV, III
Saxner et al. 1985	31	F, G	V, IV, III
Leggett et al. (1986)	22	O to M	V, IV, III
Mégessier (1988)	9	B3 to A2	V, IV, III
Blackwell et al. (1990)	114	F to M	V, IV, III
Blackwell et al. (1994)	114	F to M	V, IV, III
Alonso et al. (1994)	31	B to M	V, IV, III, II
Alonso et al. (1996)	475	F to M	V, IV

TABLE 2. Stars T_{eff} from the IRFM for Ap, Am stars

Ap stars	N	Am stars	N
Shallis and Blackwell (1979)	6	Lane and Lester 1984	10
Lanz (1984)	12	Mégessier et al. (1990)	1
Mégessier (1988)	12	Mégessier et al. (1993)	2
		Smalley (1993)	25

is illustrated in Table 4 for few stars. The influence of [M/H] is larger than that of log g. It is effective in the same way as the more or less blanketed models. The error on T_{eff} is around 1% and it depends on T_{eff}.

TABLE 3. IRFM T_{eff} from different models

		$T_{a,b}$	T_{atlas8}	T_{atlas9}	ΔCG-9	%	Δ9-8	%
β Sex	B6V	14505 [a]	14746	14860	355	2.4	114	0.80
α Leo	B7V	12025 [a]	12075	12145	120	1.0	70	0.58
γ Lyr	B9III	9950 [a]	10062	10102	152	1.5	40	0.40
α CMa	A1V	9750 [a]	9870	9900	150	1.5	30	0.30
μ And	A5V		8000	7985			-15	-0.18
α Oph	A5III		7890	7882			-0.8	-0.10
45 Boo	F5V	6663 [b]	6720	6765	102	1.52	45	0.67
π Peg	F5III	6170 [b]	6220	6280	110	1.77	60	0.96

[a] T_{eff} from Carbon and Gingerich (1969) models
[b] T_{eff} from old MARC models (Gustafsson et al. 1975)

TABLE 4. Influence of the model metallicity and gravity on IRFM T_{eff} for 8 bright stars

←		[M/H]			→	←	log g	→
	0.0	0.3	0.5	$\Delta^{.3}_{.0}$	$\Delta^{.5}_{.0}$	%	0.4	0.3	$\Delta^{.3}_{.4}$
β Sex	14860	14957	15037	97	177	1.18	14860	14905	45
α Leo	12145	12207	12260	62	115	0.94	12145	12187	42
γ Lyr	10102	10139	10172	37	72	0.72	10102	10080	-22
α CMa	9900	9940	9967	40	67	0.67	9900	9875	-25
μ And	7985	8028	8057	43	72	0.90	7985	7911	-74
α Oph	7882	7930	7957	48	75	0.95	7882	7828	-54
45 Boo	6845	6900	6935	55	90	1.30	6845	6845	0
π Peg	6280	6336	6378	56	98	1.55	6280	6280	0

4. The required observations and and their accuracy

The total integrated flux received at the earth f is obtained from the flux distribution, calibrated in the visible at $\lambda5556$ Å by f^*_{5556} given in Eq. 4

$$f^*_{5556} = f^{Vega}_{5556} 10^{-0.4(V_* - V_{Vega})} \quad (4) \qquad f_\lambda = C_\lambda 10^{-0.4 m_\lambda} \quad (5)$$

where $f^{Vega}_{5556} = 3.46\ 10^{-11} \mathrm{Wm^{-2}nm^{-1}}$ is the reference visual flux calibration (Mégessier 1995) as recalled in my previous talk. The construction of the energy distribution has been presented by Bless (this proceedings). The observed red monochromatic flux is given by Eq. 5, where C_λ is the near infrared flux calibration at the wavelength λ.

The uncertainty on f, which is mainly due to the spectrophotometry accuracy, is different in the three wavelength ranges: UV, visible and red. As a whole, it is generally less than 1% for the brightest stars. (see Mégessier 1997 for a complete discussion of IRFM Teff accuracy). The uncertainty on f^{Vega}_{5556} is 0.7% as given in Mégessier (1995). The visual magnitudes V can be measured to better than 0.01mag. The problem of the accuracy on the red calibration C_λ has been discussed earlier (Mégessier in this proceedings). If one excludes the problem of the difference between Vega observed flux and model's, C_λ is known within about 2% or 3% and the red magnitudes are measured to 0.02 magnitudes or better. Combining all these uncertainties leads to a relative error on $T_{\rm eff}$ less than or equal to 1%. Thus the aim of an accuracy better than 1% seems to be achieved.

But the values given here are valid for well observed bright stars, specially with good spectrophotometry. Moreover, the error due to the problem of the red calibration has to be added i.e. 2% or more, depending on the red monochromatic wavelength, and if the metallicity and gravity of the star are not certain, one more error has to be taken into account, which can be around 1% or more.

5. Conclusion

The infrared flux method is the closest to the Teff definition, it relies almost not on atmosphere models. Then it should give Teff on an absolute scale. The IRFM Teff accuracy can be as good as 1%, may be somewhat better. Then the method is very attractive.

However several limitations exist. The energy distribution is not available for a large amount of stars, and its measurement is not as easy as that of photometric colours. If it is derived from photometric indices or models, the accuracy is lowered. The method is not totally independent of the models and one should be really confident in the model red fluxes. If the visible flux calibration is well constrained, it is not the case of the red ones and the additional error induced is likely the largest among those mentioned here. Also, the metallicity and the gravity of the stars have to be known.

Then we can retain the advantages of the infrared flux method, but also remember its present limitations. It would provide Teff of stars bright enough, used themselves as references. The problem of the red flux calibration is the most important to solve in oder to get an accuracy around 1% on Teff.

References

Alonso A., Arribas S., Martnez-Roger C., 1994, AA 282, 684
Alonso A., Arribas S., Martnez-Roger C., 1996, AAS 117, 227
Blackwell, D.E., Shallis, M.J.,: 1977, MNRAS 180, 177
Blackwell, D.E., Lynas-Gray, A.E., Petford, A.D.:1991, A&A 245, 567
Blackwell, D.E., Lynas-Gray, A.E., 1994, A&A 282, 899
Blackwell, D.E., Petford A.D., Shallis, M.J.,1980, A&A 82, 249
Blackwell, D.E., Petford, A.D., Arribas, S., Haddock, D.J., Selby, M.J.,: 1990, A&A 232, 410
Bless R.C.,1997, this proceedings
Carbon, D.F., Gingerich, O.: 1969, in: Third Smithsonian Conference on Stellar Atmospheres, MIT press, p.377 (CG69)
Davis, J.,1997, this proceedings
Gustafsson, B., Bell, R.A.,Eriksson, K.E., Nordlund, A. : 1975, A&A 42, 407
Kurucz, R., 1979, ApJS 40, 1
Kurucz, R.L., 1991a, in Stellar Atmospheres: Beyond Classical Models, eds. L. Crivllari, I. Hubeny, D.G. Hummer, NATO ASI-Series C, vol. 341, p. 441
Kurucz, R.L.,1991, in: Stellar Atmospheres : Beyond Classical Models, NATO ASI Series C, vol. 341, p.441
Lane, M.C. and Lester, J.B., 1984, ApJ 281, 723
Lanz, T., 1984, A&A 139, 161
Leggett, S.K., Mountain, C.M., Selby, M.J., Blackwell, D.E., Booth, A.J., Haddock, D.J. and Petford, A.D., 1986, A&A 159, 217

Mégessier C., 1988, A&AS 72, 551
Mégessier C., 1994, A&AS 289, 202
Mégessier C., 1995, A&AS 296, 771
Mégessier C., 1997, A&AS (submitted)
Mégessier C, van 't Veer C., 1990, in Evolution of stars: the photospheric abundance connection. IAU Symposium 145 (poster paper edition) Eds G. Michaud and A. Tutukov, p.99
Mégessier C, van 't Veer C., 1993, in Peculiar versus normal phenomena in A-type and related stars. IAU Colloquium 138, ASP Conf. Ser. 44, 208
Saxner M., Hammarbck G., 1985, A&A, 151, 372
Shallis M.J., Blackwell D.E., 1979, A&A 79, 48
Smalley, B. 1993, A&A, 274, 391

TESTS OF EFFECTIVE TEMPERATURE - COLOUR RELATIONS

R.A.BELL
Department of Astronomy, University of Maryland, College Park, MD 20742, USA

1. Abstract

This paper discusses some tests of synthetic spectra used in the calculation of synthetic colours, as well as some of the problems which arise in these calculations. Synthetic colours are tested in two ways. Firstly, observed angular diameters of stars are compared with ones predicted from stellar temperatures derived from synthetic colours and the infrared flux method. The average difference is less than 2%. Secondly, a very metal-poor isochrone is transformed to the M_V, B–V and M_V, V–I colour planes and compared with observations of stars in the globular cluster M92. The agreement is good in both colour–magnitude diagrams.

2. Introduction

The problem of finding the effective temperatures of stars from their colours or, conversely, finding the colours of stellar models has been studied extensively. A general approach to this problem has been to mimic the process by which the colours of stars are measured, i.e., to synthesize the spectrum of a stellar model, to multiply this spectrum by the sensitivity functions of the passbands of different photometric systems and then to normalize the results to the observational systems.

The quality of these calculations can be checked in various ways. Such checks include:

1) Comparisons of synthetic line spectra with the observed spectra of one or more "standard" stars;

2) Comparisons of synthetic fluxes with the observed fluxes of "standard" stars;

3) Comparisons of the angular diameters derived from models for various stars with their observed angular diameters;

T.R. Bedding et al. (eds.),
Fundamental Stellar Properties: The Interaction between Observation and Theory, 159–168.

4) Comparisons of temperatures derived from different colours.

Furthermore, there are a number of points which must be considered in the calculation of the synthetic spectra. It is often necessary to allow for abundance changes caused by evolution, either stellar or galactic, as well as for variations of microturbulent velocity with surface gravity.

3. Comparisons of Line Spectra

The Sun is an obvious choice for tests of computed line spectra. Its effective temperature, surface gravity and chemical composition are well known. Its brightness allows it to be observed with high signal/noise and high resolution from the UV to the IR. Spectral atlases have been published by Kurucz et al. (1984) and Delbouille et al. (1973) for the 3000 Å – 12000 Å region and by the NSO for the IR. These data, obtained with ground-based telescopes, often contain both solar and telluric lines. Other infrared data have been obtained by the ATMOS experiment on the space shuttle (Farmer & Norton, 1989).

While very accurate oscillator strengths are available for some lines, e.g. O'Brian et al. (1991) and from the Oxford group (e.g. Blackwell et al. 1979), the quantity of such data is insufficient to allow us to calculate detailed line spectra which match the observations. It is necessary to alter the oscillator strengths of known atomic lines to make a good fit.

This problem is clearly seen when considering lines in the H band. Johansson & Learner (1990) reported about 360 new Fe I lines in the infrared, identifying them as transitions between $3d^64s(^6D)4d$ and $3d^64s(^6D)4f$. More than 200 of these lines coincide in wavelength with lines in the solar spectrum. These identifications have been checked by comparing the line intensities from the laboratory source with those in the solar spectrum. Only 4 of these lines deviate from this relationship by being stronger than implied from the laboratory data, thereby indicating a high probability that the identification of the remaining lines as Fe I. Few lines in the IR have measured gf values, and so the gfs of the majority of the solar Fe I and other lines must be derived from the solar spectrum.

Lines of Na I, Mg I, Al I, Si I, S I and Ca I with gf values from the Opacity Project were used in IR synthetic solar spectrum calculations. Several of these lines are much stronger in the synthetic spectrum than the observed one.

Since examples of comparisons of observed and computed solar and stellar spectra in the optical have been given elsewhere (Bell et al. 1994, Bell & Tripicco 1995), an example of the fit obtained to the solar spectrum in the IR is given in Fig. 1. It is hoped to obtain this level of fit over the J, H and K bands. Many of the solar lines were not included in the synthetic spectra calculated by Bell & Gustafsson (1989) and so the J, H and K magnitudes presented there will be too bright, probably by a few hundredths of a magnitude. The effect is probably greatest in the H band, owing to the larger number of Fe I lines there.

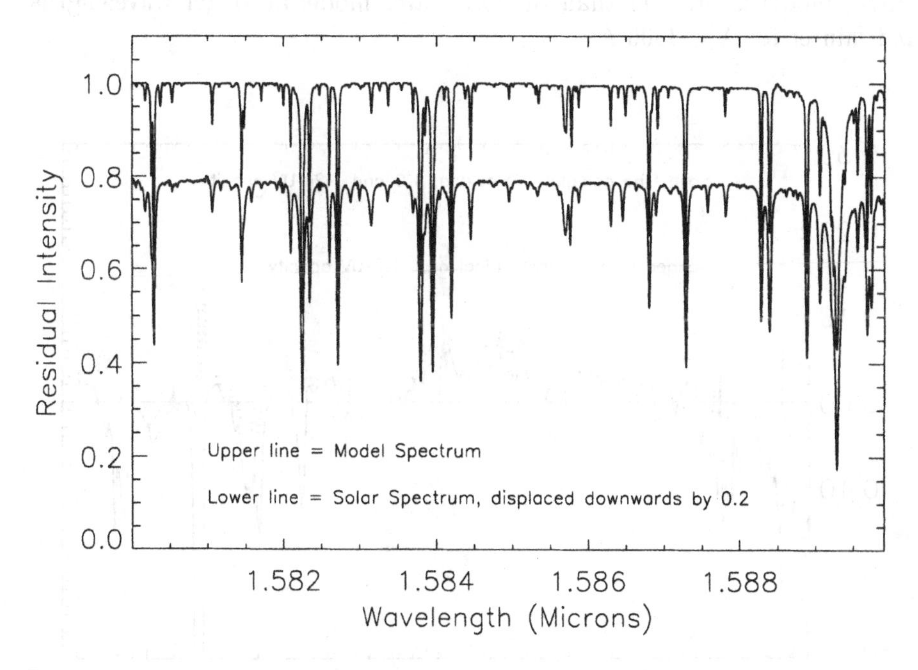

Figure 1. A comparison of observed and calculated solar spectra in the H band region. Some solar lines, e.g. that at 1.5875 μm, have not yet been identified.

4. Comparison of Fluxes

The brightness of the Sun, a virtue for getting line spectra, is a problem for measuring fluxes. The data of Neckel & Labs (1984), Lockwood et al, (1992, LTW) and Arvesen et al. (1969) disagree with one another. For this talk, I will compare calculated values with the results of LTW.

The continuous opacity used in the models includes contributions from H^-, H, electron scattering, Rayleigh scattering by H and H_2, He^-, OH, CH, AlI, SiI, MgI and Fe I. The SiI and MgI are Opacity Project data, the Fe I is that of Dragon & Mutschlecher (1980).

As a test of numerical accuracy, the OS Marcs solar model (Edvardsson, private communication) has been used to compare continuous fluxes calculated at Uppsala and at Maryland, The programs used for both opacity and flux calculations are quite independent and yet the results at 10 wavelengths between 3800 and 6000Å agree to better than 1%, with one exception. The inclusion of metal opacities from Al I, Mg I, Si I and Fe I reduces the fluxes shorward of 3800 Å by about 3%.

A Maryland Marcs solar model is fainter than the OS Marcs model at all wavelengths between 3000 and 6000 Å whereas the Holweger–Muller (1974) model is brighter than the OS Marcs model at longer wavelengths and fainter for $\lambda < 3800$ Å.

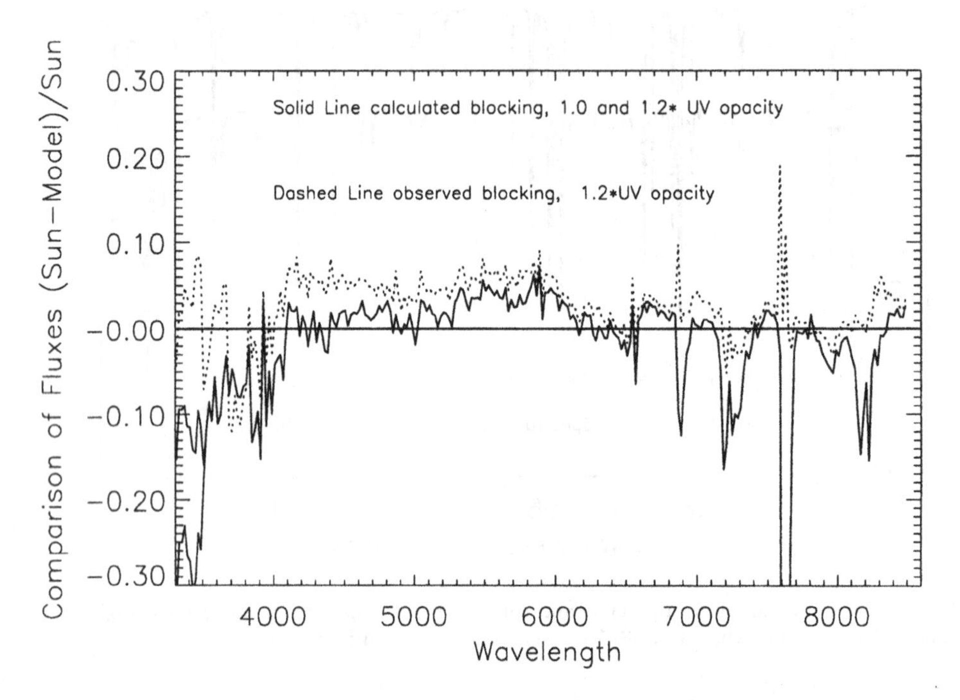

Figure 2. A comparison of the Maryland solar model and the LTW observed solar fluxes. The model fluxes (solid lines) include calculated line spectra, with the continuous opacity for $\lambda < 3500$ Å having the standard value and also being increased by 20%. The dips in the comparisons beyond 6800 Å are due to telluric lines in the observational data. The dashed line results from the model fluxes when the calculated line blocking is replaced by that derived from the Kurucz et al. solar atlas.

A direct comparison of fluxes of the Maryland Marcs solar model (T_{eff} = 5780K) and the LTW observational data shows that the model is too bright in the UV (Figure 2). One way of obtaining better agreement is increasing the model's continuous opacity shortward of 3500 Å by about 30%. If we measure the observed line blocking from the Kurucz et al. (1984) solar atlas and combine these data with the continuous fluxes of the solar model, the resulting fluxes are still brighter than the Sun in the UV. An increase in the continuous opacity of about 20% is needed to get agreement.

However, if the current Dragon & Mutschlecner opacity of Fe I is too small, an increase by over a factor of ten would be needed to solve the problem. It would be very helpful to have Opacity Project data for Fe I.

Balachandran and I are studying this problem further by making fits to OH lines in the 3100 Å region. Since the solar oxygen abundance is known accurately and the oscillator strengths of the OH lines are believed to be accurate to 10–20%, they allow us to check the continuous opacity to that accuracy.

Short & Lester (1996) have also found that an additional UV opacity source is needed in models of both the Sun and Arcturus. They suggest diatomic molecules may be the source of this opacity.

5. Details of the calculations

Since errors of 1% in fluxes correspond to errors of 0.01 in magnitudes, it is necessary to treat the hydrogen lines, at least in the hotter stars, with care. In work on the Balmer lines, I include lines up to H_{20} and, when appropriate, add the line absorption coefficients of the two nearest lines. I do not include the blueward component of H_{20} and alter the Balmer limit from 3646 Å to 3676.36 Å. Similar approximations are made for other hydrogen series.

Replacement of the Van der Waals pressure broadening treatment by the theory of Anstee & O'Mara (1995) will be a welcome improvement, in view of the work of Anstee, O'Mara & Ross (1997) in resolving a conundrum in the solar abundance of iron – see Blackwell et al. (1995a, b) and Holweger et al. (1995).

6. Abundance effects

In some circumstances, it is necessary to include abundance changes resulting from stellar evolution or to use non-solar abundance ratios in order to best match stellar spectra.

For example, observational work (e.g. Kjaergaard et al. 1982) shows that C is depleted, N enhanced and $^{12}C/^{13}C$ decreased in Population I giant stars. The same effect is seen in globular cluster giants, the C depletion increasing with evolution on the RGB and with decreasing [Fe/H]. This has an effect on the DDO colour index C(42–45), since CH lines occur in the 42 passband. Tripicco & Bell (1995) showed that this could cause an error of up to 250K in T_{eff} if the C depletion was not included in Population I giant star models. The strength of the CO bands will also be affected by this evolutionary change.

Recent work on isochrones for metal poor stars (VandenBerg et al. 1997) have used abundances [α/Fe] of 0.0, 0.3, and 0.6, where the "α" elements include O, Ne, Mg, Si, S, Ar, Ca and Ti, while Al and Mn are depleted, with [Al/Fe] = – [α/Fe] and [Mn/Fe] = – 0.5 [α/Fe]. While evolutionary abundance changes in globular cluster giants make the results somewhat uncertain, a value of [α/Fe] = 0.3 seems most appropriate for the metal poor globular clusters. Such a value should be used in color calculations, since Stetson et al. (1997) have found that cool giant models with [Fe/H] = –2.14 and [α/Fe] = 0.6 are somewhat bluer than those with [α/Fe] = 0.0.

7. Comparison of Angular Diameters

Bell & Gustafsson (1989, BG89) presented temperature scales for G and K dwarf and giant stars. They used their stellar temperatures and relative absolute fluxes to deduce angular diameters for a sample of 95 stars. The new data on stellar diameters presented by Mozurkewich (1997) at this meeting gives an excellent opportunity to compare observed and predicted diameters.

The BG89 temperatures were derived from tables of infrared colours as well as from the infrared-flux-ratio-method (IRFM, Blackwell & Shallis 1977). The IRFM finds stellar temperatures from the ratio of the apparent integrated flux of a star to its apparent K-band flux, which is, in effect, a K-band bolometric correction. The ratio is also, of course, the ratio of absolute integrated flux to absolute K-band flux, which can be found from model atmospheres.

The point which is quite critical for the application of the IRFM is the accuracy of the K-band flux. Two factors affecting this quantity are the filter profile of the K passband, which will affect the derivation of the K-band flux from stellar models, and the conversion of the K-band apparent magnitudes of stars to fluxes. This latter conversion is based either upon the K-band flux radiated by a Vega model and the angular diameter of Vega or upon absolute measurements of the flux from Vega or the use of solar analogues. BG89 used the fluxes from the Dreiling & Bell (1980) model of Vega.

Megessier (1995) has argued that the calibration of K band fluxes should be based upon measurements using either solar analogue stars or terrestrial light sources, which give larger values than do models. However, as discussed below, it is unclear if the use of solar analogues gives reliable results. No specific criticism can be made of the absolute measurements of Vega fluxes that would make them in better agreement with the Vega model.

The measurements of the infrared fluxes of solar analogues (Wamsteker 1981, Campins et al. 1986) are based upon the flux data of Labs & Neckel (1968). Vernazza et al. (1976) have questioned the way in which Labs & Neckel combined the relative observations of solar central intensity of Peyturaux (1952) and Pierce (1954) between 1.0 and 2.5 μm with their own absolute measurements for $\lambda < 1.25$ μm. Vernazza et al. quote Labs & Neckel – this fitting was done "not by fitting Pierce's data at about 1.0 μ to our data but rather by using the model–distribution between 1.0 and 1.5 μ as a linking medium between both observational data sets. By this procedure, Pierce's values at around 1.0 μ become about 6% higher than our values." (The model distribution referred to is from Gingerich & de Jager 1968). Vernazza et al. present an alternative way of fitting the two data sets together by studying the region of overlap. This uncertainty in fitting the different data sets leads directly to uncertainty in the absolute infrared fluxes from solar analogs.

The ground–based measurements, where Vega is compared with terrestrial light sources, have to be corrected for extinction. This is complicated by water vapour lines in the terrestrial spectrum. Examination of the telluric line spectrum in the NSO solar atlas shows a rich H_2O spectrum, with some lines being as much as 40% deep in the 3.8 μm region observed by Selby et al. (1983) while the line at 4.9163 μm, which is 80% deep, may have affected the 4.92 μm observations of Mountain et al. (1985). However, an underestimate of the extinction, which is possible since Selby et al. and Mountain et al. assumed that the extinction between 1 and 2 airmasses could be extrapolated to give that between 0 and 1 airmasses, would make their IR fluxes even brighter than those predicted by the Dreiling & Bell (1980) Vega model.

In order to compare angular diameters from Mozurkewich (1997, M97) with those of BG89, we need to convert the M97 uniform-disc diameters to limb-darkened ones. The U.S. Navy Prototype Optical Interferometer project has derived limb-darkened diameters of 5.58 and 6.87 milli-arc seconds (mas) for α Cas and α Ari, respectively (Hindsley, private communication). The ratios of these diameters to the M97 uniform-disc diameters measured at 0.8 μm are 1.073 and 1.037, respectively, for an average ratio of 1.055. Hanbury Brown et al. (1974) found the very similar ratio of 1.044 for their data. The ratio of 1.055 has been applied to the M97 data for the 17 stars in common with BG89. The details of the comparison are given in Table 1. The mean ratio of angular diameters then becomes(BG89)/(M97) = 1.017. While the uncertainty in the limb-darkening correction makes it unprofitable to explore the consequences of the ratio departing from unity, the agreement is very gratifying.

Using the same limb-darkening correction for Vega as for the K giants, the M97 limb-darkened diameter for Vega is about 1.4% larger than that of Hanbury Brown et al. (1974), being 3.285 versus 3.24 mas, with an error of 0.015 versus 0.07.

TABLE 1. Comparison of Angular Diameters

Star		Sp T	ϕ(Obs, 0.8 μm)	err	ϕ(BG89)	ϕ(BG89)/ϕ(Obs)	T
617	α Ari	K2 III	6.754	0.013	6.91	1.02	45
1409	ϵ Tau	G9.5 III	2.648	0.016	2.64	0.985	49
1457	α Tau	K5 III	20.776	0.036	20.62	1.01	39
3249	β Cnc	K4 III	5.108	0.037	5.17	1.025	40
3748	α Hya	K3 II-II	9.536	0.043	9.44	0.99	41
3873	ϵ Leo	G1 II	2.612	0.036	2.72	1.04	53
4301	α UMa	K0 IIIa	6.962	0.038	6.79	0.975	46
4335	ψ UMa	K1 III	4.077	0.025	4.18	1.025	45
4932	ϵ Vir	G8 III	3.252	0.018	3.30	1.015	50
5340	α Boo	K2 IIIp	21.061	0.067	21.07	1.00	43
5602	β Boo	G8 IIIa	2.455	0.026	2.61	1.06	49
5681	δ Boo	G8 III	2.731	0.018	2.80	1.025	48
5854	α Ser	K2 III	4.717	0.014	4.96	1.05	45
6220	η Her	G8 IIIb	2.627	0.034	2.61	0.99	49
6418	π Her	K3 IIab	5.177	0.042	5.52	1.075	41
6623	μ Her	G5 IV-V	1.994	0.020	1.99	1.00	55
8684	μ Peg	G8 III	2.469	0.021	2.47	1.00	50

8. Comparison of isochrones and colour magnitude diagrams

One test of synthetic photometry is to ask whether different colours give the same T_{eff} for a star. An alternative test is to ask if observations of the same cluster in different colours give colour magnitude diagrams which match the same isochrone.

This comparison has been carried out for M92, using an M_V, B–V diagram consisting of mean points for the fainter stars from Stetson & Harris (1988) and the brighter stars from Sandage (1970) and an M_V, V–I diagram consisting of mean points from Johnson & Bolte (1997). These are compared with isochrones calculated by VandenBerg for [Fe/H] = −2.14 and [α/Fe] = 0.3. (These isochrones are intended for comparison with HST observations of globular clusters – see Stetson et al. 1997). VandenBerg (private communication) has found that these isochrones gives a very good fit to the Frogel et al. (1981) red giant branch for M92 in the M_V, V–K plane.

The stars and the isochrone are compared in the M_V, B–V and M_V, V–I colour magnitude diagrams in Figure 3. The zero points of the colours have been found by using the Vega fluxes of Hayes (1985) to predict colours for Vega and the bolometric corrections found using a solar model and a

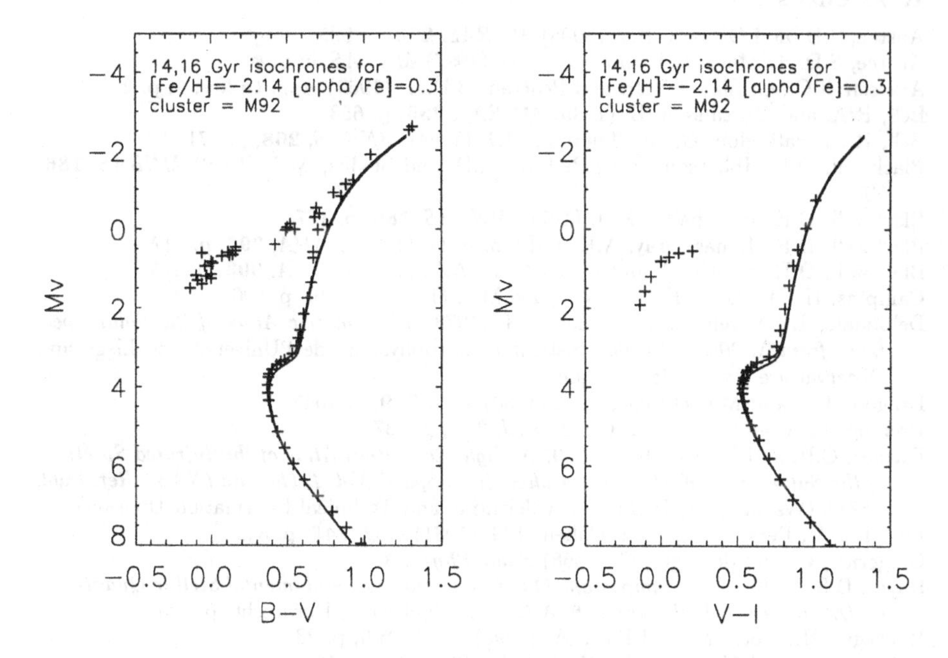

Figure 3. A comparison of the colour magnitude diagrams of M92 in M_V, B–V and M_V, V–I with transformed VandenBerg 14 and 16 Gyr isochrones, computed for [Fe/H] = -2.14, $[\alpha/Fe] = 0.3$

bolometric correction of -0.12 for the Sun. The distance modulus used for M92 is $V-M_V = 14.65$. Reddening corrections of E(B-V) = 0.02 and E(V-I) = 0.026 have been applied.

In the M_V, B–V diagram, the fit to the 14 Gyr isochrone is very good from the tip of the RGB down to $M_V = 6.0$, after which the stars are redder and/or brighter than the isochrone. In the M_V, V–I diagram, the 14 Gyr isochrone is slightly bluer than the stars at $M_V = 6.0$ as well as at the turn off, while the reverse is true on part of the subgiant branch and part of the giant branch.

9. Acknowledgments

This work was supported by NSF under grant AST93–14931 and by STScI under grant GO–06106.03–94A. The NSO/Kitt Peak FTS data used here were produced by NSF/NOAO. Part of this work was carried out with P. B. Stetson, W.E. Harris, M. Bolte, H.E. Bond, G.G. Fahlman, J.E. Hesser, H.B. Richer, D.A. Vandenberg, and S. van den Bergh, in a collaboration to interpret HST observations of globular clusters.

References

Anstee, S.D. and O'Mara, B.J. (1995) *MNRAS*, **276**, p. 895
Anstee, S.D., O'Mara, B.J. and Ross, J.E. (1997) *MNRAS*, in press
Arvesen, J.C., Griffin, R.N.Jr. and Pearson, B.D.Jr. (1969) *Appl Optics*, **8**, p. 2215
Bell, R.A. and Gustafsson, B. (1989) *MNRAS*, **236**, p. 653
Bell, R.A., Paltoglou, G. and Tripicco, M.J. (1994) *MNRAS*, **268**, p. 771
Blackwell, D.E., Ibbetson, P.A., Petford, A.D. and Shallis, M.J. (1979) *MNRAS*, **186**, 633
Blackwell, D.E. and Shallis, M.J. (1977) *MNRAS*, **180**, p. 177
Blackwell, D.E., Lynas-Gray, A.E. and Smith, G. (1995a) *A&A*, **296**, p. 217
Blackwell, D.E., Smith, G. and Lynas-Gray, A.E. (1995b) *A&A*, **303**, p. 575
Campins, H. , Rieke, G.H. and Lebofsky, M.J. (1985) *AJ*, **90**, p. 896
Delbouille, L., Neven, L. and Roland, G. (1973) *Photometric Atlas of the Solar Spectrum from λ3000 to λ10000* Institut d'Astrophysique de l'Universite de Liege and Observatoire Royale de Belgique
Dragon, J.N. and Mutschlecner, J.P. (1980) *ApJ*, **239**, p. 1045
Dreiling, L. A. and Bell, R.A. (1980) *ApJ*, **241**, p. 737
Farmer, C.B. and Norton, R.H. 1989, *A High-Resolution Atlas of the Infrared Spectrum of the Sun and the Earth's Atmosphere from Space, Vol. 1:The Sun (NASA Ref. Publ. 1224)*. (Washington, D.C.: NASA Scientific and Technical Information Division)
Frogel, J.A., Persson, S.E. and Cohen, J.G. (1981) *ApJ*, **246**, p. 842
Gingerich, O. and de Jager, C. (1968) *Solar Phys.* , **3**, p. 5
Hayes, D.S. (1985) *IAU Symposium 111, Calibration of Fundamental Stellar Quantities, ed D.S.Hayes, L.E. Pasinetti & A.G.D. Philip* Reidel, Dordrecht, p. 225
Holweger, H., Kock, M. and Bard, A. (1995) *A&A*, **296**, p. 233
Holweger, H. and Muller, E.A. (1974) *Solar Phys.*, **39**, p. 19
Johansson, S.E. and Learner, R.C.M. (1992) *ApJ*, **354**, p. 755
Johnson, J. and Bolte, M. (1997) *AJ* submitted.
Kjaergaard, P., Gustafsson, B., Walker, G.A.H. and Hultquist, L. (1982) *A&A*, **115**, 145
Kurucz, R.L., Furenlid, I., Brault, J. and Testerman, J. (1984) *Solar Flux Atlas from 296 to 1300 nm* (National Solar Observatory, Sunspot, NM.)
Labs, D. and Neckel, H. (1968) *Z.Astrophys*, **69**, p. 1
Lockwood, G.W., Tug, H. and White, N.M. (1992) *ApJ*, **390**, p. 668 (LTW)
Megessier, C. (1995) *A&A*, **296**, p. 771
Mountain, C.M., Leggett, S.K., Blackwell, D.E., Selby, M.J. and Petford, A.D. (1983), *A&A*, **151**, p. 399
Mozurkewich, D. (1997), poster paper presented at IAU Symposium 189.
Neckel, H. and Labs, D. (1984) *Solar Phys*, **90**, p. 245
O'Brian, T.R., Wickliffe, M.E., Lawler, J.E., Whaling, W. and Brault, J.W. (1991) *J Opt Soc Am B*, **8**, p. 1185
Peyturaux, R. (1952) *Ann d'Ap*, **15**, p. 302
Pierce, A.K. (1954) *ApJ*, **119**, p. 312
Sandage, A.R. (1970) *ApJ*, **162**, p. 848
Selby, M.J., Mountain, C.M., Blackwell, D.E., Petford, A.D. and Leggett, S.K., (1983), *MNRAS*, **203**, p. 795
Short, C.I. and Lester, J.B. (1996) *ApJ*, **469**, p. 898
Stetson, P.B. and Harris, W.E. (1988) *AJ*, **96**, p. 909
Stetson, P.B., Harris, W.E., Bell, R.A., Bolte, M., Bond, H.E., Fahlman, G.G., Hesser, J.E., Richer, H.B., Vandenberg, D.A. and van den Bergh, S. (1997) in preparation
Tripicco, M.J. and Bell, R.A. (1991) *AJ*, **102**, p. 744
Tripicco, M.J. and Bell, R.A. (1995) *AJ*, **110**, p. 3035
Vernazza, J.E., Avrett, E.H. and Loeser, R. (1976) *ApJS*, **30**, p. 1
Wamsteker, W. (1981), *A&A*, *97*, p. 329

Discussion of this paper appears at the end of these Proceedings.

7. STELLAR ABUNDANCES

REVIEW OF ABUNDANCES FOR MID-TEFF RANGE STARS

P.E. NISSEN
Institute of Physics and Astronomy
University of Aarhus, DK-8000 Aarhus C, Denmark

1. Introduction

Among the mid-$T_{\rm eff}$ (5000 - 7000 K) range stars, those on or near the main sequence are particularly important, because they provide much information on Galactic evolution and nucleosynthesis of the elements. This is due to the fact that these stars represent an age range from 0 to 15 Gyr, and a metallicity range from [Fe/H] $\simeq -4.0$ to $+0.3$. Furthermore, it is likely that the present composition of elements in their atmospheres is similar to the composition of the interstellar gas from which they were formed, and that the present kinematics of the stars give information about their birthplaces. Hence, by observing selected samples of nearby F and G main sequence and subgiant stars one has the possibility to study the chemical and dynamical history of our Galaxy. Due to space limitation the present review is limited to these F and G stars. Supergiants are not discussed, but it should be emphasized, that their abundances are important for studies of advanced stellar evolutionary phases and determinations of the chemical composition of nearby galaxies. e.g. the Magellanic Clouds (Hill et al. 1995).

Important problems in connection with studies of the element abundances in F and G stars are: *i*) Li, Be and B abundances vs. [Fe/H] and $T_{\rm eff}$ in order to gain insight into the complex problems of stellar depletion and cosmic ray production of these elements with the ultimate goal of determining an accurate value of the primordial abundance of ^{7}Li. This problem has gained new interest with the possible detection of deuterium in distant quasar absorption systems at a level of D/H $\simeq 2 \times 10^{-4}$ (Songaila et al. 1994), which requires a ^{7}Li abundance of Li/H $\simeq 2 \times 10^{-10}$ according to standard Big Bang nucleosynthesis theory. *ii*) Oxygen and other α-element abundances as a function of [Fe/H] in order to study the production of these elements in supernovae of type II and Ia as a function of time. *iii*) Ba and Eu vs. [Fe/H] in order to study the sites and details of the *s*- and *r*-processes.

T.R. Bedding et al. (eds.),
***Fundamental Stellar Properties: The Interaction between Observation and Theory*, 171–178.**

iv) The metallicity distribution of G dwarfs, age-metallicity relations, and abundance ratios vs. kinematics in order to test Galactic evolution models. In all of these cases we need abundances of the elements to a rather high accuracy of say 25%. Similarly high accuracy is needed in connection with tests of stellar models from observations of stellar oscillation frequencies or in connection with the determination of stellar ages.

In the following, we shall try to to assess the accuracy of abundances that can be determined from high resolution spectra of F and G main sequence and subgiant stars.

2. The Sun

Element abundances in the Sun have recently been reviewed by Holweger (1996). Abundances derived from the photospheric spectrum show an impressive agreement with abundances in carbonaceous chondrites of type I, when Si is adopted as the reference element. For most elements the agreement is better than 10% showing that accurate abundances can indeed be derived from a stellar spectrum. Furthermore, the agreement strongly suggests that both the solar photosphere and the meteorites have the original elemental composition of the solar nebulae. A few deviations are, however, present. Li and perhaps Be have been depleted in the Sun, and C, N and O have probably escaped the meteorite formation by forming volatile gases like N_2 and CO. Hence, one cannot get any confirmation of the solar photospheric CNO abundances from the meteorites. This leaves us with some uncertainty, because the abundances of these elements are difficult to determine from the solar spectrum. In the case of oxygen, Lambert (1978) derived $A_O \equiv \log(n_O/n_H) + 12 = 8.92$ from the forbidden [O I] lines at 6300 and 6363 Å, and Sauval et al. (1984) derived $A_O = 8.91$ from infrared OH lines. Both analyses were based on the empirical, plane-parallel solar model of Holweger & Müller (1974) and assumed LTE. A 3D, non-LTE study of Kiselman & Nordlund (1995) of the same lines suggests, however, that the solar oxygen abundance is around 8.80. Such a change, amounting to 25% , may have quite a significant effect when oscillation frequencies of the Sun are used to test solar models, because oxygen contributes about half of the total heavy element abundance in the Sun.

The iron abundance in the solar atmosphere has been another longstanding problem. On the basis of accurate experimental oscillator strengths of Fe I lines, Blackwell et al. (1984) derived $A_{Fe} = 7.67 \pm 0.03$, about 50 % higher than the meteoritic value of $A_{Fe} = 7.51$. A number of recent works based on Fe II lines with oscillator strengths from lifetime measurements and experimental branching ratios (Holweger et al. 1990, Biemont et al. 1991, Hannaford et al. 1992) resulted, however, in a photospheric iron abundance that agrees well with the meteoritic value. Milford et al. (1994) and Holweger et al. (1995) obtained also a near-meteoritic value from Fe I lines. Blackwell et al. (1995), on the other hand, maintain the high photospheric Fe abundance, although they stress the sensitivity of the result to the photospheric model adopted, and their high value is supported by Kostik et al. (1996). The differences between these results seem to arise from a combination of errors in the gf-values and equivalent widths as well as uncertainties in the values of the microturbulence and the damping constants. In addition, small non-LTE effects and inhomogeneities in the solar atmosphere may affect the results. The lesson seems to be, that abundances in stellar atmospheres should as far as possible be derived from weak lines belonging to the major ionization stage of the element (Fe II in the case of iron for F and G stars), although it should be mentioned that the new broadening theory of Fe I lines by Anstee & O'Mara (1995) leads to the meteoritic iron abundance, when strong lines are applied (Anstee et al. 1997).

The excellent agreement between abundances of elements heavier than O in the solar atmosphere and in meteorites does not exclude that there has been a general diffusion of all heavy elements relative to hydrogen in the upper layers of the Sun. Proffitt (1994) and Bahcall & Pinsonneault (1995) have calculated that settling of heavier elements has reduced the present surface value of Z by about 10% relative to the original Z of the Sun. This result gets support from low-degree l oscillation frequencies, which do not agree with a model having the present solar atmospheric abundance, $Z = 0.018$, in the interior, but require $Z = 0.020$ (Basu et al. 1996). The effect is small, but may be larger in the older, metal-poor stars and it may also depend on the depth of the convection zone, i.e. T_{eff}.

3. Solar-type stars

If the basic parameters of a star are similar to those of the Sun, then a highly accurate study of element abundances relative to the Sun can be carried out. One should apply the same set of theoretical model atmospheres to the star and the Sun, because errors in the models are then expected to cancel out. Furthermore, gf-values of the lines can be determined from the solar spectrum and applied to the stellar spectrum. As an example of such differential studies we briefly review recent work on abundances in the Hyades stars. Cayrel et al. (1985) determined [Fe/H] for late F and G dwarfs in the Hyades from weak iron lines with T_{eff} derived from Hα wing profiles to an accuracy of ± 30 K. Boesgaard & Friel (1990) have continued this work for the earlier F stars with T_{eff} derived from colours. Altogether [Fe/H] is practically independent of T_{eff} in the range 5200 – 7000 K with a mean value of 0.12 and a dispersion of ± 0.04 dex only.

Carbon abundances for the Hyades F dwarfs have been determined by Friel & Boesgaard (1990) from weak, high excitation C I lines. Again there is no trend with $T_{\rm eff}$. The mean value of [C/H] is 0.05 with a scatter of ±0.08 dex. Finally, oxygen abundances in 25 F dwarfs have been derived by Garcia López et al. (1993) from a non-LTE analysis of the O I triplet at 7774 Å. Apart from an apparent increase of the oxygen abundances of stars with $T_{\rm eff}$ around 7000 K, [O/H] is nearly constant as a function of $T_{\rm eff}$ with a dispersion of ±0.08 dex. The mean value of [O/H] is as low as −0.08, which points to a surprisingly large O/Fe deficiency for the Hyades. As discussed by King & Hiltgen (1996) part of this apparent deficiency seems to arise from a systematic error of the oxygen triplet equivalent widths measured in the stellar spectra relative to the equivalent widths measured from the solar spectrum. From a careful analysis of the [O I] 6300 Å line in two Hyades dwarfs, King & Hiltgen determine $[O/H] = 0.15 \pm 0.10$ in good agreement with the Hyades [Fe/H] value.

We conclude that there is no sign of systematic abundance errors or different degrees of settling of heavy elements in the Hyades as a function of $T_{\rm eff}$.

4. Metal-poor stars

In the case of more metal-poor stars, especially those with $[Fe/H] < -1.0$, differential abundance studies with respect to the Sun are rather uncertain. The structure of the atmospheres are quite different from that of the Sun due to the decrease of P_e and the increase of P_g, which changes the depth at which convection sets in. The ultraviolet flux is higher due to the decrease of line blocking causing a higher degree of over-ionization of many elements. Furthermore, gf-values are difficult to determine from the solar spectrum, because the weak lines in the metal-poor stars are strong in the Sun. Finally, there has recently been much debate about the $T_{\rm eff}$ scale for metal-poor F and G stars.

The uncertain situation is illustrated in Table 1, that lists some recent results for the bright, metal-poor star, HD 140283, based on LTE model atmosphere analyses of weak Fe lines with accurately measured equivalent widths. As seen, the $T_{\rm eff}$ values differ by up to 300 K and [Fe/H] by as much as 0.4 dex. $T_{\rm eff}$ was determined from colours except by Axer et al. (1994), who used the Balmer lines. The gravity was determined by requesting that the iron abundance derived from Fe I and Fe II lines should be the same except by Nissen et al. (1994), who adopted $\log g$ from the isochrones of Bergbusch & VandenBerg (1992). In this case there is a difference in [Fe/H] derived from Fe I and Fe II lines indicating a non-LTE problem. The true $\log g$ value of HD 140283 may be derived from the recent accurate parallax

TABLE 1. Atmospheric parameters for HD 140283. The [Fe/H] values given refer to $A_{Fe,\odot} = 7.51$

Reference	$T_{\rm eff}$	log$g$	[Fe/H]
Magain (1989)	5640 K	3.10	−2.58
Tomkin et al. (1992)	5640	3.28	−2.60
Nissen et al. (1994)	5540	3.50	−2.75 (from Fe I lines)
–	–	–	−2.56 (from Fe II lines)
Axer et al. (1994)	5814	3.27	−2.36
Ryan et al. (1996)	5750	3.40	−2.54

given by Dahn (1994), $\pi = 0.0145 \pm 0.0017$. Using the relations $g \propto M/R^2$, $L \propto R^2 T_{\rm eff}{}^4$ and adopting a mass of $M/M_\odot = 0.8 \pm 0.1$ from the 14 Gyr isochrone of Bergbusch & VandenBerg (1992), one derives log$g = 3.58 \pm 0.12$. Hence, the high gravity adopted by Nissen et al. (1994) is supported indicating that [Fe/H] derived from Fe I lines is too low due to a non-LTE (over-ionization) effect. A further problem with using the Fe I lines as the primary indicator of the iron abundance is their high sensitivity to temperature and hence to inhomogeneities in the stellar atmosphere. In the future it seems more promising to base [Fe/H] on Fe II lines using Hipparcos parallaxes to determine log$g$. Better gravities will also allow us to determine more accurate Be, B and and CNO abundances.

As seen from Table 2, recent determinations of the lithium abundance of HD 140283 also disagree, although the equivalent widths measured for the Li I line at 6708 Å are in good agreement. The Li abundance determined is very sensitive to both $T_{\rm eff}$ and the temperature gradient in the line forming region. Thus, the high value derived by Thorburn (1994) is due to the application of the ATLAS9 models of Kurucz (1993) with convective overshoot and a corresponding shallow temperature gradient. The very high Li abundance estimated by Kurucz (1995) is based on an inhomogeneous model, where Li in the cold regions is strongly over-ionized by the UV flux from the hot regions. It would be a surprise if this effect is indeed as large as postulated, because then one would also have expected to see a very large difference in the iron abundance derived from Fe I and Fe II lines, but Kurucz is certainly right in pointing out that the coupling between non-LTE effects and inhomogeneities in metal-poor stellar atmospheres should be studied in detail.

TABLE 2. Li abundances derived for HD 140283

Reference	$T_{\rm eff}$	W(Li)	A_{Li}
Pilachowski et al. (1993)	5650 K	42 mÅ	2.05
Thorburn (1994)	5742	46	2.30
Norris et al. (1994)	5750	48	2.18
Kurucz (1995)	5750	45	$\simeq$ 3.0
Spite et al. (1996)	5540	46	1.96

5. Abundance ratios and stellar kinematics

Some abundance ratios can be determined more accurately than the Fe and Li abundances discussed in Sect. 4. Despite the large differences of [Fe/H] quoted in Table 1, the [Mg/Fe] ratios derived by the same authors agree within ± 0.1 dex. In general, the abundance ratio between two elements with about the same ionization potential, will be rather insensitive to errors in $T_{\rm eff}$ and model atmospheres, if weak lines from the same ionization stage are used. Hence very accurate abundance ratios can be obtained especially when stars having similar atmospheric parameters are compared. In this way some interesting correlations between abundance ratios and stellar kinematics have recently been discovered.

Edvardsson et al. (1993) determined the ratio between the abundances of the α-elements (O, Mg, Si, Ca, and Ti) and iron in 189 disk stars as a function of [Fe/H]. In the metallicity range, $-0.8 <$ [Fe/H] < -0.4, corresponding to the transition from the thick to the thin disk, [α/Fe] depends on the kinematics. Stars with large Galactic orbits, $R_{mean} > 9$ kpc, have smaller [α/Fe] values than stars with $R_{mean} < 7$ kpc. The correlation can be interpreted as due to a more rapid chemical evolution in the inner regions of the Galaxy, i.e. SNe of type Ia start to contribute iron at a lower metallicity in the outer parts than in the inner parts.

More recently, Nissen & Schuster (1997) have studied two groups of stars with $-1.3 <$ [Fe/H] < -0.4. The stars were selected from the $V_{\rm rot}$-[Fe/H] diagram of Schuster et al. (1993), where $V_{\rm rot}$ is the stellar velocity component in the direction of Galactic rotation. Stars in the first group (the halo stars) have $V_{\rm rot} \simeq 0$ km s^{-1}, whereas stars in the second group (the disk stars) have $V_{\rm rot} \simeq 200$ km s^{-1}. The stars are confined to rather narrow ranges in $T_{\rm eff}$ and gravity, $5400 < T_{\rm eff} < 6500$ K and $4.0 < \log g < 4.6$ respectively, so that very accurate differential abundance ratios could be derived from high resolution, high S/N spectra obtained with the ESO NTT. As seen in Fig. 1, the majority of the halo stars (shown by open circles)

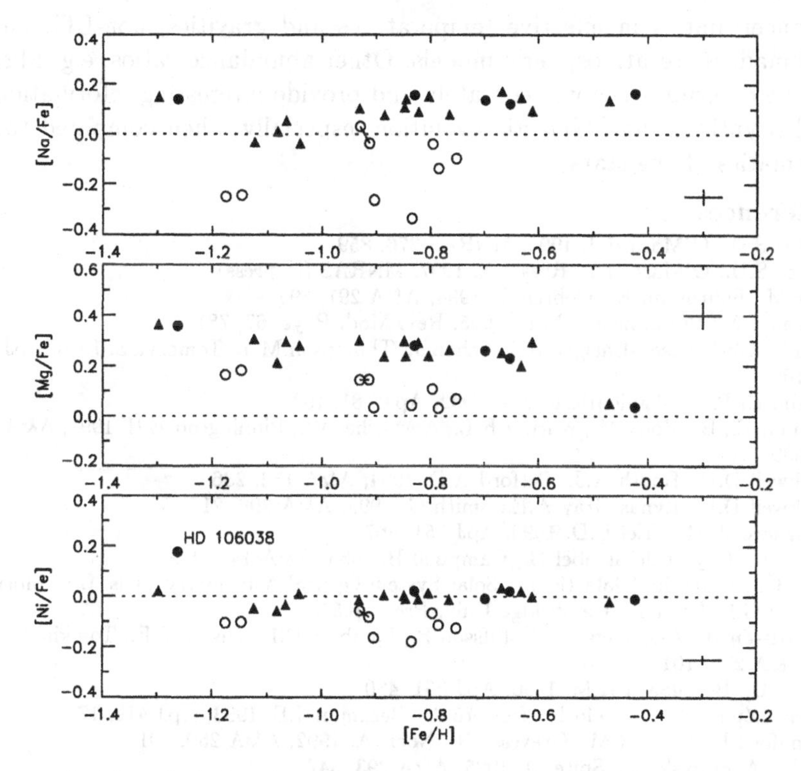

Figure 1. Na, Mg and Ni abundances relative to Fe as a function of [Fe/H]. Disk stars are shown with filled triangles and halo stars with filled or open circles. The two components of a binary star are connected with a straight line. One sigma error bars, valid for the relative abundance ratios, are shown to the right

have Na/Fe, Mg/Fe and Ni/Fe ratios that deviate very significantly from the corresponding ratios for the disk stars. Furthermore, the deficiencies in Na and Ni are strongly correlated. These deviating halo stars tend to have larger Galactic orbits than the other stars and may therefore have been accreted from dwarf galaxies with a chemical evolution history different from that of the Galactic disk and the inner halo.

6. Conclusion

For the Sun and solar-type stars with an overall metal-abundance similar to the solar value, relative abundances of the most important elements can be determined to an accuracy of about 10% with the possible exception of oxygen. For more metal-deficient stars, in particular those with $[Fe/H] < -1.0$, there may, however, be severe errors (up to 0.3 dex) in such important abundance ratios as Li/H, Be/H, B/H, Fe/H and O/Fe due to uncertainties in effective temperatures and gravities, non-LTE effects and inadequate atmospheric models. Other abundance ratios, e.g. Mg/Fe, can be determined more accurately and provide interesting information on nucleosynthesis and Galactic evolution, especially when compared to the kinematics of the stars.

References

Anstee S.D., O'Mara B.J. 1995, MNRAS 276, 859
Anstee S.D., O'Mara B.J., Ross J.E. 1997, MNRAS (in press)
Axer M., Fuhrmann K., Gehren T. 1994, A&A 291, 895
Bahcall J.N., Pinsonneault M.H. 1995, Rev. Mod. Phys. 67, 781
Basu S., Christensen-Dalsgaard J., Schou J., Thompson M.J., Tomczyk S. 1996, ApJ 460, 1064
Bergbusch P.A., VandenBerg D.A. 1992, ApJS 81, 163
Biémont E., Baudoux M., Kurucz R.L., Ansbacher W., Pinnington E.H. 1991, A&A 249, 539
Blackwell D.E., Booth A.J., Petford A.D. 1984, A&A 132, 236
Blackwell D.E., Lynas-Gray A.E., Smith G. 1995, A&A 296, 217
Boesgaard A.M., Friel E.D. 1990, ApJ 351, 467
Cayrel R., Cayrel de Strobel G., Campbell B. 1985, A&A 146, 249
Dahn C.C. 1994, in "Galactic and Solar System Optical Astrometry", eds. L.V. Morrison and G.F. Gilmore, Cambridge Univ. Press, p.55
Edvardsson B., Andersen J., Gustafsson B., Lambert D.L., Nissen P.E., Tomkin J. 1993, A&A 275, 101
Friel E.D., Boesgaard A.M. 1990, ApJ 351, 480
Garcia López R.J., Rebolo R., Herrero A., Beckman J.E. 1993, ApJ 412, 173
Hannaford P., Lowe R.M., Grevesse N., Noels A. 1992, A&A 259, 301
Hill V., Andrievsky S., Spite M. 1995, A&A 293, 347
Holweger H. 1996, Physica Scripta 65, 151
Holweger H., Müller E.M. 1974, Solar Phys. 39, 19
Holweger H., Heise C., Kock M. 1990, A&A 232, 510
Holweger H., Kock M., Bard A. 1995, A&A 296, 233
King J.R., Hiltgen D.D. 1996, AJ 112, 2650
Kiselman D., Nordlund Å. 1995, A&A 302, 578
Kostik R.I., Shchukina N.G., Rutten R.J. 1996, A&A 305, 325
Kurucz R.L. 1993, CD-ROM 13, Smithsonian Astrophys. Obs.
Kurucz R.L. 1995, ApJ 452, 102
Lambert D.L. 1978, MNRAS 182, 249
Magain P. 1989, A&A 209, 211
Milford P.N., O'Mara B.J., Ross J.E. 1994, A&A 292, 276
Nissen P.E., Schuster W.J. 1997, A&A (submitted)
Nissen P.E., Gustafsson B., Edvardsson B., Gilmore G. 1994, A&A 285, 440
Norris J.E., Ryan S.G., Stringfellow G.S. 1994, ApJ 423, 386
Pilachowski C.A., Sneden C., Booth J. 1993, ApJ 407, 699
Proffitt C.R. 1994, ApJ 425, 849
Ryan S.G., Norris J.E., Beers T.C. 1996, ApJ 471, 254
Tomkin J., Lemke M., Lambert D.L., Sneden C. 1992, AJ 104, 1568
Thorburn J.A. 1994, ApJ 421, 318
Sauval A.J., Grevesse N., Brault J.W., Stokes G.M., Zander R. 1984, ApJ 282, 330
Schuster W.J., Parrao L., Contreras Martinez M.E. 1993, A&AS 97, 951
Songaila A., Cowie L.L., Hogan C.J., Rugers M. 1994, Nature 368, 599
Spite M., François P., Nissen P.E., Spite F. 1996, A&A 307, 172

Discussion of this paper appears at the end of these Proceedings.

THE BROADENING OF METALLIC LINES IN COOL STARS

S. D. ANSTEE, B. J. O'MARA AND J. E. ROSS
The University of Queensland,
St. Lucia Queensland 4072, Australia

Abstract. A theory for the broadening of spectral lines by collisions with atomic hydrogen developed by Anstee and O'Mara [1] is described and when applied to strong lines of elements in the sun leads to abundances which are independent of non-thermal motions in the photosphere and in excellent agreement with meteoritic values. In the case of the solar abundance of iron the results lead to a resolution of recent conflict concerning possible differences between the abundance of iron in the sun and meteorites. Excellent results obtained with the Mg b-lines permit these lines to be used in abundance determinations in other stars and also makes them particularly useful in determining the temperature structure in cool stars.

1. Introduction

In cool stars neutral hydrogen atoms outnumber electrons by four orders of magnitude, consequently the broadening of most spectral lines is dominated by collisions with neutral hydrogen atoms. Conventional van der Waals' theory for this broadening process is known to underestimate the broadening of spectral lines in the sun by about a factor of two. Development of a satisfactory theory is important as it would allow strong lines with well determined f-values to be used to determine abundances in cool stars in a manner which is independent of photospheric motions. Also such lines could also be used to determine surface gravities in cool stars.

2. Theory Outline

Collisions with neutral hydrogen atoms are sufficiently fast for the impact approximation of spectral line broadening theory to be valid. In this ap-

T.R. Bedding et al. (eds.),
Fundamental Stellar Properties: The Interaction between Observation and Theory, 179–183.

proximation the line has a Lorentz profile with a half half-width which is given by:

$$w = N \int_0^\infty v f(v) \sigma(v) dv,$$

where N is the hydrogen atom number density, v is the relative collision speed, $f(v)$ is the speed distribution and the line broadening cross section

$$\sigma(v) = 2\pi \int_0^\infty < \Pi(b, v) >_{av} bdb,$$

where the integrand contains the product of the geometrical cross section $2\pi bdb$ and a line broadening efficiency factor $< \Pi(b, v) >_{av}$ for collisions with impact parameter b and relative speed v.

$< \cdots >_{av}$ indicates that the efficiency factor has to be averaged over all orientations of the perturbed atom. The efficiency factor can be expressed in terms of the S-matrix elements for the collision which are functionally dependent on the interaction energy between a hydrogen atom in the ground state and the perturbed atom in its upper and lower states. The only essential difference between various theoretical treatments is in the method employed to determine this interaction energy.

In the theory developed by Anstee and O'Mara [1] the interaction energy is calculated using Rayleigh-Schrödinger perturbation theory. If exchange effects are neglected, the shift in energy of the two-atom system as a result of the electrostatic interaction V between them is given by

$$\Delta E_i =< i|V|i > + \sum_{j \neq i} \frac{< i|V|j >< j|V|i >}{E_i - E_j},$$

where the unperturbed eigenstates of the two-atom system $|i >$ are products of the unperturbed eigenstates of the two atoms. As first pointed out by Unsöld the above expression can be greatly simplified if $E_i - E_j$ can be replaced by a constant value E_p. Closure can then be used to complete the sum over j to obtain

$$\Delta E_i =< i|V|i > + \frac{1}{E_p}(< i|V^2|i > - < i|V|i >^2)$$

The second term accounts for the inter-atomic interaction resulting from fluctuations of both atoms simultaneously (dispersion)and the fluctuation of each atom in the static field of the other(induction). The second term dominates the interaction. To develop the theory further the perturbed atom is described by an optical electron outside a positively charged core so that product states of the two-atom system have the form $|100 > |n^*lm >$.

With reference to the accompanying diagram, the electrostatic interaction energy V, in atomic units, is

$$V = \frac{1}{R} + \frac{1}{r_{12}} - \frac{1}{r_2} - \frac{1}{p_1}.$$

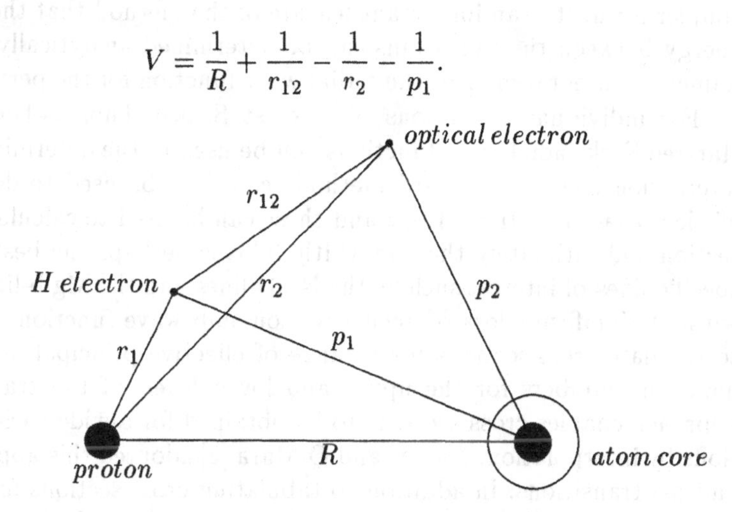

For the state $|i>=|100>|n^*lm>$, the interaction energy can be expressed in the form

$$\Delta E_{n^*l|m|} = <i|V|i> + \frac{1}{E_p}\int_0^\infty R^2_{n^*l}(p_2)I_{l|m|}(p_2,R)p_2^2dp_2 - \frac{1}{E_p}<i|V|i>^2,$$

where $R_{n^*l}(p_2)$ is the radial wave function for the optical electron in the perturbed atom and $I_{l|m|}(p_2,R)$ are lengthy complicated analytic functions of p_2 and R, which have a logarithmic singularity at $p_2 = R$ and which can be can expressed as an asymptotic expansion in powers of $\frac{1}{R^2}$ when R is large. It can be shown that the leading term in this expansion leads to

$$<\Delta E_{nl|m|}>_{av} \sim \frac{1}{E_p}\frac{<p_2^2>}{R^6},$$

and if E_p is chosen to be 4/9 atomic units $1/E_p = 9/4$ which is the polarizability of hydrogen in atomic units. Thus

$$<\Delta E_{nl|m|}>_{av} \sim \alpha_H \frac{<p_2^2>}{R^6},$$

the standard expression for the van der Waals interaction between the two atoms. However the impact parameters important in the determination of the cross section are always too small for this asymptotic form of the interaction to be valid.The terms in $< i|V|i >$ can be expressed in a similar but simpler form. It is an important feature of the method that the interaction energy between the two atoms can be determined analytically to within a numerical integration over the radial wave function for the perturbed atom.

For individual transitions of interest Scaled Thomas-Fermi-Dirac or Hartree-Fock radial wave functions can be used in the determination of the interaction energy. Standard methods can then be used to determine the efficiency factor $< \Pi(b, v) >_{av}$ and these can be used to calculate the cross-section and ultimately the line width. This is perhaps the best method for specific lines of interest such as the Na D-lines and the Mg b-lines. However without significant loss of accuracy Coulomb wave functions can be used to tabulate cross-sections for a range of effective principal and azimuthal quantum numbers for the upper and lower levels of the transition. This approach enables cross-sections to be obtained for a wide variety of transitions by interpolation. Anstee and O'Mara [2] adopted this approach for s-p and p-s transitions. In addition to tabulating cross-sections for a collisions speed of $v_0 = 10^4 ms^{-1}$ for a range of effective principal quantum numbers for the upper and lower level, they also determined by direct computation, velocity exponents α, on the assumption that $\sigma(v) \sim v^{-\alpha}$.

With this dependence of the cross-section on collision speed the integration over the speed distribution can be performed to obtain the line width per unit H-atom density which is given by

$$\frac{w}{N} = \left(\frac{4}{\pi}\right)^{\alpha/2} \Gamma\left(\frac{4-\alpha}{2}\right) v_0 \sigma(v_0) (\bar{v}/v_0)^{1-\alpha},$$

where $\bar{v} = \left(\frac{8kT}{\pi\mu}\right)^{1/2}$, and μ is the reduced mass of the two atoms. Typically α is about 0.25, which leads to a temperature dependence of $T^{0.38}$ for the line width. At present tabulated values of σ and α are only available for s-p and p-s transitions but work is in progress to extend the results to p-d, d-p and d-f, f-d transitions.

3. Application to the solar spectrum

The theory has been used by Anstee and O'Mara to synthesize strong lines in the solar spectrum using the spectrum synthesis program of Ross [4] and the solar model of Holweger and Müller [5]. Good fits were obtained to a selection of solar lines using abundances consistent with meteoritic values. More recently the theory has been applied to the solar Mg b-lines with excellent results. The theory has also been used by Anstee, O'Mara and Ross [3] to determine the solar abundance of iron from strong lines of Fe I.

The resulting abundance is in agreement in all respects with the meteoritic value, thus ending recent controversy concerning the solar abundance of iron. They also show that the theory accounts both qualitatively and quantitatively for the bifurcation in the empirical curve of growth for iron first observed by Carter [6].

4. Conclusions

Strong lines can now be used with some confidence to determine the chemical composition of cool stars, with the derived abundance being independent of non-thermal motions in the atmospheres of these stars. The abundance derived from strong lines can be used to determine the microturbulent velocity from medium-strong lines. Also selected strong lines can be used to obtain surface gravities and to test model atmospheres for cool stars.

References

1. Anstee S. D. and O'Mara B. J. *MNRAS*, 253(4A):549, 1991.
2. Anstee S. D. and O'Mara B. J. *MNRAS*, 276:859, 1995.
3. Anstee S. D., O'Mara B. J., and Ross J. E. *MNRAS*, 284:202, 1997.
4. Ross J. E. Spectrum synthesis program. http://www.uq.edu.au/~phjross/syn.htm.
5. Holweger H. and Müller E. A. *Solar Phys.*, 39:19, 1974.
6. Carter W. W. *Phys. Rev.*, 76:962, 1949.

DISCUSSION

ROBERT KURUCZ: What about eight equivalent electrons?

JIM O'MARA: The question refers to the broadening of Fe I lines. I do not believe that there is a problem for lines involving single-electron excitations where the binding energy and hence the effective principal quantum of the optical electron is well defined, as the line-broadening cross-sections are largely determined by the exponential tail of the wave function. There is a definite problem when two-electron excitations are involved. In such cases the binding energy is somewhat ambiguously determined by using the energy of the appropriate excited state of Fe II. In future, better line-broadening cross-sections will be calculated for Fe I lines by using more appropriate wave functions.

ABUNDANCES OF THE ELEMENTS IN THE HALO STARS, INTERACTION BETWEEN OBSERVATION AND THEORY

M.SPITE
Observatoire de Paris-Meudon
F-92195 Meudon principal Cedex, FRANCE

1. Introduction

The halo stars are very old stars formed at the beginning of the evolution of the Galaxy. Their main characteristic is that their atmosphere is very metal-poor since at the time of their birth, the matter of the Galaxy had been enriched by only a few number of supernovae. The first very metal deficient stars have been studied in the sixties: Baschek (1959), Aller and Greenstein(1960) Wallerstein et al. (1963). The stellar models used in these first studies were rather crude: the atmosphere of the stars was represented by only one layer with a mean temperature and a mean electronic pressure, the continuum opacity was supposed to be due to H^- and included in addition only the contribution by Rayleigh scattering. Later rescaled solar models were used and finally, at the end of the seventies, grids of theoretical models more and more sophisticated became available.

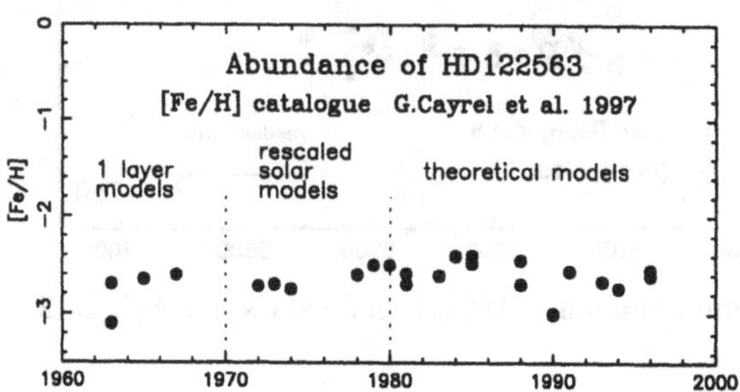

Figure 1: Variation of the metallicity of HD122563 with time

In figure 1 are compared the different determinations of the metallicity of HD122563 versus time. It can be seen that he abundance determinations, as far as an accuracy of 0.2 dex is satisfactory, are robust since even with rather crude models, a good estimation of the metallicity has been obtained.

T.R. Bedding et al. (eds.),
***Fundamental Stellar Properties: The Interaction between Observation and Theory*, 185–192.**

In fact, the variation of the iron abundance [Fe/H] from author to author is mainly due to the variation of the temperature assumed for the star. When the temperature of the star increases by 250K the metallicity of the star increases by about 0.2dex. This effect is in general almost the same for all the elements and thus the ratios like [Mg/Fe], [Ba/Fe] etc... are more stable.

Let us remark that the important parameter in an abundance determination is not the effective temperature of the star but the temperature in the layers where the metallic lines are formed ($\log \tau \approx -1$). If two models have the same effective temperature (they predict the same flux) but have different structures the abundances will be different.

Recently difficulties appear with two elements which have in particular an importance for cosmology: lithium and oxygen.

2. The determination of the lithium abundance.

The abundance of lithium in the atmosphere of the old halo stars is a test of the Big-bang theory since lithium is essentially built in the primordial nucleosynthesis. The lithium abundance in the halo dwarfs is independent of the metallicity of the star (as far as [Fe/H]< -1.4dex) and independant of the temperature (and thus of the mass) in the interval 5600K$< T_{eff} <$6300K (Fig. 2). Two important parameters are: the level of this plateau because if lithium is not depleted in these stars this level represents the Big-bang production, and the spread of the lithium abundance around the plateau since this spread could be the witness of a lithium depletion.

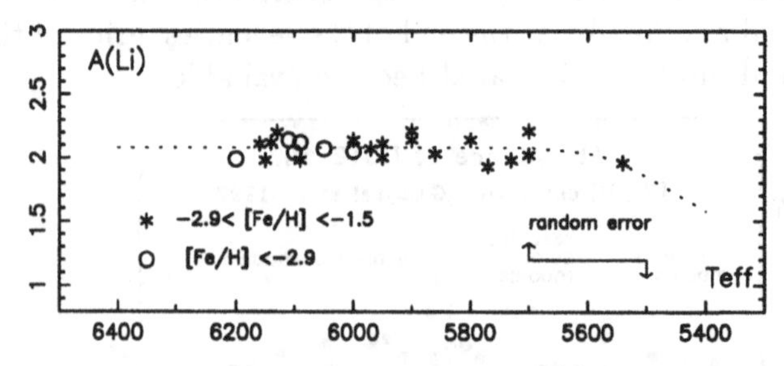

Figure 2: Lithium abundance A(Li) in halo stars, for A(H)=12.

2.1. LEVEL OF THE "PLATEAU"

Since the lithium line is a resonance line of the neutral species of a highly ionised atom, it is very sensitive to the temperature assumed for the star. From author to author, there is a variation of the iron abundance in a given star mainly because they have chosen different effective temperature, on the same way there will be a variation of the lithium abundance. In Spite et al. (1996) we have used the new NMARCS models to study a sample of 25 turn-off stars. Three different temperature scales wer used:

-1- The temperature ($Teff_1$) was deduced from the excitation equilibrium of the FeI lines and then the level of the lithium plateau was found to be: A(Li) = 2.08dex (Fig.2).

-2- The temperature ($Teff_2$) was deduced from the (b-y) color and the relation Teff (b-y) computed by Nissen from the NMARCS models, then the level of the plateau was found to be: A(Li) = 2.15dex.

-3- The temperature ($Teff_3$) was deduced from the profile of the wings of the $H_\alpha$lines), and we then found that the level of the lithium plateau was: A(Li) = 2.22dex.

(For this sample of stars, mainly turn-off stars, we have found that approximately $Teff_3 = Teff_2 + 100K = Teff_1 + 200K$)

These different values of the level of the "plateau" must be compared to the values found by Thornburn 1994: (A(Li) = 2.3 using the Carney's scale), by Molaro et al. (1995): A(Li)= 2.21, using a temperature deduced from the hydrogen wings, by Ryan et al. (1996): A(Li) $\approx$ 2.18, using the Carney's scale, and by Bonifacio & Molaro (1997): A(Li)= 2.24 using the IRFM scale.

The problem of the exact level of the plateau of the lithium abundance in the halo stars can be solved only when the temperature scale of the halo stars will be fixed...

2.2. SPREAD OF THE LITHIUM ABUNDANCE

Is the spread of the lithium abundance around the plateau explained by the measurement errors, only ?

Some sophisticated models predict an almost uniform depletion in the halo stars with $T_{eff} > 5500K$ (Pinsonneault et al. 1992, Charbonnel & Vauclair 1995 etc..) but in this case, a dispersion of the lithium abundance around the plateau is expected: (in the rotational models this dispersion would reflect the distribution of the initial stellar rotation velocities).

In our recent analysis we found an rms scatter of 0.08dex, a value expected from the measurements errors only, contrary to the findings of Thornburn (1994) who found in similar conditions a rms scatter of 0.11 0.13dex.

• The duplicity of some stars can partly explained this discrepancy. Duplicity affects the colors of the star and the equivalent width of the lithium line. In our sample of stars, (Spite et al. 1996) if we only remove the two stars which have been found spectroscopic binaries from our spectra, the rms scatter around the plateau drops by 0.015dex. Since it is estimated that

TABLE 1. Comparison of the measurements of the equivalent widths of the lithium line

	Spite et al. spectra	Thornburn spectra	Norris et al. spectra	New meas. of Norris et al. spectra
HD140283	48	50	48	48
LTT 815-43	27	22	15	20
CD -33 1173	17	12	10	14

more than 20% of the halo stars are binaries, duplicity may significantly increase the expected spread.

• Moreover, the accuracy of the equivalent widths is often overestimated. The error on the equivalent widths due to photon statistic is given by Cayrel (1988):

$$\sigma_W = \frac{1.5}{S/N} \times \sqrt{FWHM \times \delta x}$$

where S/N is the signal to noise ratio, $FWHM$ the full width of the line at half maximum, and δx the width of one pixel (in Å).

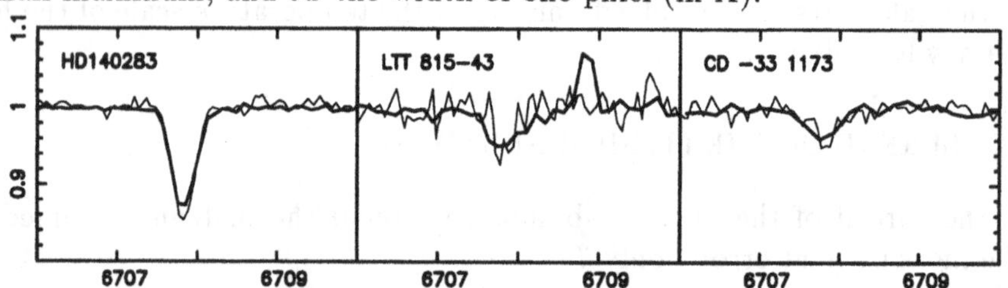

Figure 3: Spectra of three halo stars obtained at ESO by Spite et al.,1996 (thick lines) and by Norris et al. 1994, (thin lines).

In Spite et al. (1996) we observed three stars in common with Norris et al.(1994): HD140283, LTT815-43 and CD-33°1173). The graphs of the observed spectra are given in Norris et al., and it is possible to compare directly the spectra obtained (the resolution is almost the same). The agreement is rather good (Fig. 1). However for the two stars which in the paper of Norris et al. have a lower S/N ratio per pixel the difference in the equivalent widths is larger than expected (Table 1).

But, if we repeat the measurements, assuming that the lithium line is a doublet and that the $FWHM$ is the same in the three stars (as it has been observed on our spectra) then the agreement of the equivalent widths is much better. A precise determination of the $FWHM$ is an important information when an accurate measurement of the equivalent widths is required. Let us remark that if, in the paper of Thornburn (1994), we select only the stars which have a S/N ratio higher than 120, (and where the *FWHM* is well defined the observed scatter drops to $\sigma A(Li) = 0.083$ a value compatible with the expected errors on the temperature and the equivalent widths.

3. The determination of the oxygen abundance in the halo stars.

Oxygen is an abundant element and thus is crucial for the evolution of a star: an oxygen rich star evolves more rapidly than an oxygen poor star with the same metallicity. The oxygen abundance is thus an important parameter to compute the age of the Galaxy.

On the other hand, oxygen is formed in the massive SNII which have a very short life time. On the contrary iron is mainly formed in the SNI which have a much longer life time. The ratio [O/Fe] at different ages of the Galaxy gives the ratio SNII/SNI which is a key parameter of the models of the chemical evolution of our Galaxy.

Unfortunately few lines of oxygen can be measured in a spectrum. The forbidden lines at $\lambda = 630nm$ (which are almost freee of non LTE effect) are used to determine the oxygen abundance in the giant stars. In the dwarfs they are weak and can be used only if [Fe/H] > -1.5. In dwarfs with a lower metallicity -3 $<$ [Fe/H] $<$ -1.5 the permitted oxygen lines (a triplet at $\lambda \approx 777nm$) with a very high excitation potential ($\chi = 9.14$), or the OH molecular bands at $\lambda \approx 360$ *or* $620nm$ are used.

3.1. TREND OF OXYGEN ABUNDANCE IN FIELD GALACTIC GIANTS

Following Gratton & Ortolani(1986), Barbuy & Erdelyi-Mendes(1989), Sneden et al.(1991), and Shetrone (1996) the mean value of the ratio [O/Fe] in the field halo giants is ≈ 0.4 dex (Fig. 4). The decline of [O/Fe] toward solar value is interpreted as the growing importance of type I supernovae

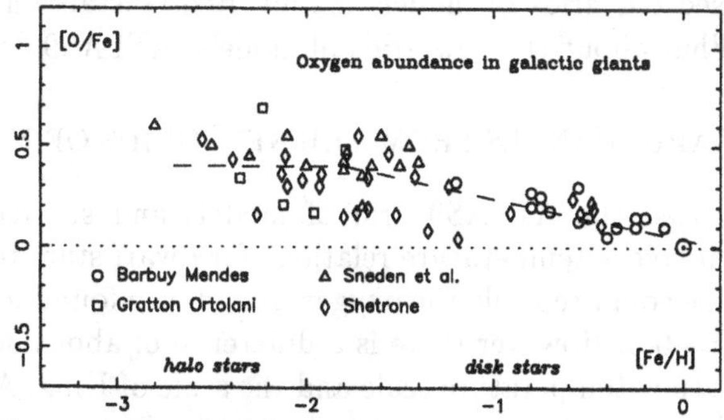

Figure 4: Trend of the oxygen abundance in giant stars

(where iron is mainly formed) relative to type II supernovae (where oxygen is formed.

3.2. TREND OF OXYGEN ABUNDANCE IN HALO DWARFS

Tomkin et al. (1992) and Cavallo et al. (1997) have recently measured the oxygen abundance in halo dwarfs or halo subgiants and found that [O/Fe] $\approx$ 0.8 dex, a value significantly higher than the value found in the giant stars. Some years ago Spite & Spite (1991) and Bessell et al. (1991) have shown that there is indeed a systematic difference between the abundances found from the permitted lines generally used to determine the oxygen abundance in the dwarf stars, and abundances found from the forbidden lines or the OH molecular bands.

3.3. WHAT IS THE CAUSE OF THE DISCREPANCY ?

The non-LTE corrections have been calculated by Kiselman (1991) and Tomkin et al.(1992) and are too modest (0.05dex) to explain the observations. Let us remark that the ratio [O/Fe] when determined from the permitted oxygen lines is very sensitive to the determination of the temperature of the star. The excitation potential of the oxygen lines (9.14eV) is higher than the ionisation potential of iron. As a consequence when the *temperature of the star is increased*, the iron abundance , the oxygen abundance deduced from the forbidden lines or the OH molecular band, increase but *the oxygen abundance deduced from the permitted lines decreases.* Moreover the high excitation permitted lines of oxygen are formed deeper in the atmosphere than the FeI lines and thus the ratio $[O/Fe]_{permitted}$ is sensitive to the temperature structure of the model.

It has been also suggested that since the inhomogeneities are larger in the deep layers the neglect of granulation in the computations could explain part of the discrepancy.

Up to now, all the (even recent) papers which have been quoted hereabove, have used old grids of models mainly Bell-Gustafsson (Gustafsson et al. 1975). What about the new grids of models, ATLAS9 or NMARCS ?

3.4. OXYGEN ABUNDANCES FROM THE NEW GRIDS OF MODELS

In 1993 King used the ATLAS9 grid of models and showed that shifting the classical colors temperature relations for dwarf stars toward higher temperatures, he could reoncile the oxygen abundance found for dwarfs and giants: [O/Fe]=+0.5. However there is a difference of about 300K between the classical Carney temperature scale and the scale of King. And for many astronomers this large shift appeared to be "ad hoc".

Recently, Nissen computed from the NMARCS models the relation T_{eff} versus $(b-y)$. This relation is close to the relation of Carney, at least for the turn-off stars, and the subgiants. With this relation Nissen et al. (1994) deduced the oxygen abundance from the OH molecular band in a sample of 9 very metal deficient dwarfs ($-3 < [Fe/H] < -2$) and found that in these stars $[O/Fe] \approx 0.58$dex. But the oxygen permitted lines have been observed for three of these stars by Tomkin et al.(1992) it is thus possible to compute the oxygen abundance from these lines using the same NMARCS model as Nissen et al.(1994). Then, a discrepancy of ≈ 0.3dex appears between the abundances deduced from the permitted lines and the OH band.

However in 1996, Nissen (in Spite et al. 1996) taking into account the reddening, gives for the same stars different temperatures. It is possible to estimate the corresponding new value of [Fe/H] and of $[O/Fe]_{OH}$ with this new temperature and also to compute the new value of $[O/Fe]_{permitted}$. And finally now a good agreement is obtained between the permitted lines and the molecular bands: the mean values are $[Fe/H] = -2.2$, $[O/Fe]_{OH} = 0.66$ and $[O/Fe]_{permitted} = 0.68$. Let us remark however that we could only "estimate" for these three stars the new values of the iron abundance and of the oxygen abundance deduced from the OH band. It would be worth computing directly these values.

But recently Balachandran and Carney (1996), using ATLAS9 models without overshooting to study HD103095 ([Fe/H] = -1.2), found also a good agreeement between the [O/Fe] ratios computed from the forbidden lines (+0.33dex) the permitted triplet (+0.41dex) and the near infra-red OH and CO lines (+0.29dex).

When I repeated the computation of the atomic lines with the NMARCS models, the good agreement remained: $[O/Fe]_{permitted} = 0.31$dex, and $[O/Fe]_{forbidden} = 0.36$.

Finally if the values obtained are confirmed, the overabundance of oxygen could be ≈ 0.40dex at $[Fe/H] = -1$ and ≈ 0.65dex at $[Fe/H] = -2$.

4. Conclusion

It has been shown that the determinations of critical abundances are reliable only with careful measurements of the equivalent widths, with reliable models, providing a unique determination of T_{eff} , whatever criterium is used (excitation equilbrium, $H_\alpha$wings, color ...). The improvement of the models (especially for very metal deficient stars) is now an urgent task.

References

Aller L.H., Greenstein J.L. (1960) *ApJS* 5, 139
Balachandran S., Carney B.W. (1996) *AJ* 111, 946

Barbuy B., Erdelyi-Mendes M. (1989) *A&A* 214, 239
Baschek B. (1959) *Zs. f. Ap* 48, 95
Bessell M.S., Sutherland R.S., Ruan K. (1991) *ApJ* 383, L71
Bonifacio P., Molaro P. (1997) *MNRAS* in press
Cavallo R.M., Pilachowski C.A., Rebolo R. (1997) *PASP* submitted
Cayrel de Strobel G., Soubiran C., Friel E.B., Ralite N., François (1997) *A&AS* in press
Cayrel R., (1988) in IAU Symp 132 "The impact of Very High S/N Spectroscopy on Stellar Physics", G. Cayrel de Strobel and M. Spite eds., Kluwer, p.345
Charbonnel C., Vauclair S. (1995) *A&A* 295, 715
Edvardsson B., Anderssen J., Gustafsson B., Lambert D.L., Nissen P.E., Tomkin J. (1993) *A&A* 275, 101
Gustafsson B., Bell R.A., Eriksson K., Nordlund Å (1975) *A&A* 42, 407
Gratton R., Ortolani S. (1986), *A&A* 169, 201
King J.R. (1993) *AJ* 106, 1206
Kiselman D. (1991) *A&A* 245, L9
Molaro P., Primas F., Bonifacio P. (1995), *A&A* 295, L47
Nissen P.E., Edvardsson B. (1992) *A&A* 261, 255
Nissen P.E., Gustafsson B., Edvardsson B., Gilmore G. (1994) *A&A* 285, 440
Norris J. E., Ryan S. G., Stringfellow G. S. (1994) *ApJ* 423, 386
Pinsonneault M.H., Deliyannis C.P., Demarque P. (1992) *ApJS* 78, 179
Ryan S.G., Beers T.C., Deliyannis C., Thorburn J.A. (1996) *ApJ* 458, 543
Shetrone M.D. (1996) *AJ* 112, 1517
Sneden C., Kraft R.P., Prosser C.F., Langer G.E. (1991) *AJ* 102, 2001
Spite M., Spite F.: (1991) *A&A* 252, 689
Spite M., François P., Nissen P.E., Spite F. (1996) *A&A* 307, 172
Tomkin J., Lemke M., Lambert D.L., Sneden C. (1992) *AJ* 104, 1568
Thorburn J.A., (1994) *ApJ* 421, 318
Wallerstein G., Greenstein J.L., Parker R., Helfer H.L., Aller L.H. (1963) *ApJ* 137, 280

Discussion of this paper appears at the end of these Proceedings.

ABUNDANCE ANOMALIES IN GLOBULAR CLUSTERS

G. S. DA COSTA
Mt. Stromlo & Siding Spring Observatories,
Private Bag, Weston Creek PO, ACT 2611, AUSTRALIA

1. Introduction

The galactic globular clusters presumably formed rapidly as high density concentrations at the centres of extensive star forming regions and the natural expectation is that they will be chemically homogeneous. In general, this is confirmed by observation — most globular clusters are extremely homogeneous with regard to most elements. (There are two exceptions: ω Cen and M22 both have intrinsic internal abundance ranges. While of considerable interest, e.g. Norris et al. 1996, this type of "abundance anomaly" will not be discussed here). But we have known for more than twenty years[1] that the surface abundances, which are what we observe, of elements such as C, N and O can vary substantially from red giant to red giant within an individual globular cluster. Indeed it has become clear that "abundance anomalies" of this type are common in the galactic globular cluster population. Briefly, the observed anomalies can be summarized as: *(i)* the "anomalous" stars are depleted in C and enhanced in N. Depletions of O also often accompany the depletions in C. *(ii)* The C, N and O variations are usually accompanied by enhancements of Na and Al and when the O depletion and the Al enhancement are both large, Mg is depleted. No other elements, including r- and s-process indicators, vary.

There are two basic competing explanations for these observations, though it is quite possible that both processes occur to varying extents in most clusters. The first is the *Evolutionary Hypothesis* in which the abundance anomalies are the result of internal processes. An example would be the mixing of nuclear processed material from the interior into the surface layers. The second explanation is the *Primordial Hypothesis* in which the

[1]The origin of this subject is often ascribed to the DDO photometry of Osborn (1971) and Hesser et al. (1976). However, Harding (1962) had discovered a CH-star in ω Cen, and even earlier, Popper (1947) noted that the M13 red giant L199 had strong CN bands when compared to other giants in the cluster. Modern work has confirmed L199 as a CN-strong star.

T.R. Bedding et al. (eds.),
Fundamental Stellar Properties: The Interaction between Observation and Theory, 193–202.

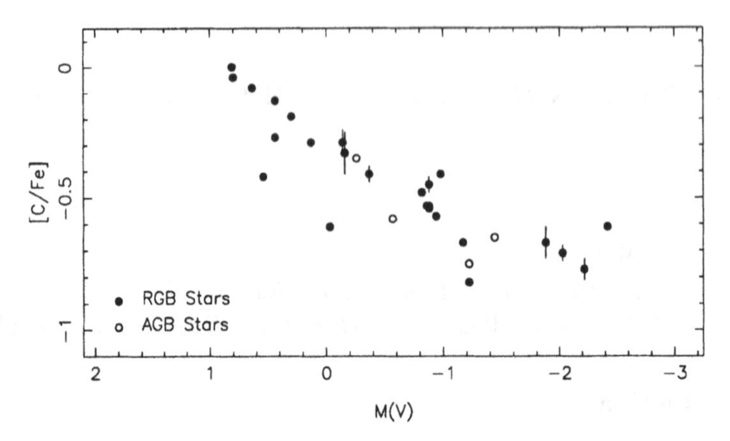

Figure 1. Carbon abundances on the giant branch as a function of luminosity in the metal-poor globular cluster NGC 6397. Data from Briley et al. (1990).

abundance anomalies result from external processes, most probably at the time the cluster formed. For example, if the proto-cluster gas cloud was not well-mixed chemically, then the stars that form from it would not have identical compositions.

2. Metal-Poor Clusters

2.1. THE OBSERVATIONS

There are a number of galactic globular clusters with [Fe/H] ≤ -1.0 for which spectroscopic observations of large samples of cluster members exist. From these datasets a number of general results for the C, N and O abundances can be inferred. *(i)* In those clusters where there are data over a sufficiently large luminosity interval, the C abundance increases, or more correctly, is less depleted, with decreasing luminosity. This is illustrated in Fig. 1 where the results of Briley et al. (1990) for the cluster NGC 6397 are shown. Similar trends of smaller C depletions at lower luminosities are also seen in the clusters M4 and NGC 6752 (Suntzeff & Smith 1991) and M92 (Langer et al. 1986). In M92 in particular, the lowest luminosity stars with C abundance determinations are almost at the main sequence turnoff, and at this luminosity, the abundance ratio [C/Fe] is comparable to that for field subdwarfs of similar [Fe/H]. Further, in all these clusters, at a fixed luminosity there is a range in the observed C abundance that is not due to observational error only; an intrinsic range in the C depletions is present. *(ii)* The depletion of C is accompanied by an enhancement in N. *(iii)* A range of O depletions of up to 1 dex is seen in many clusters (e.g. Kraft et al. 1995 and references therein). More specifically, where such data exist, the O and C depletions are tightly coupled (e.g. Dickens et al. 1991, Smith et al. 1996). *(iv)* As demonstrated by, for example, Smith et al. (1996) for

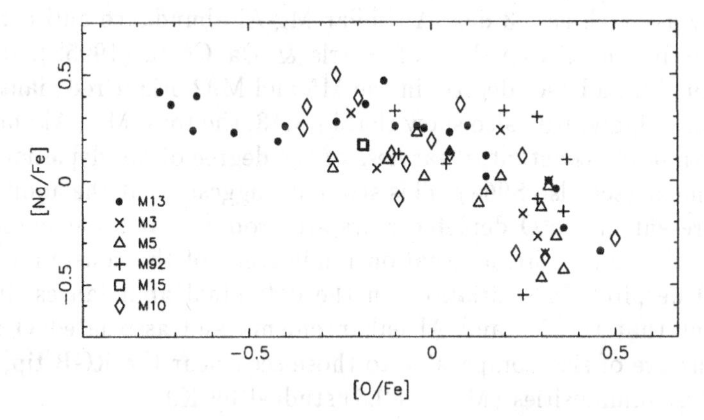

Figure 2. The relationship between Na enhancement and O depletion for bright red giants in six globular clusters (see Kraft et al. 1995 and references therein).

M13 and M3, and Dickens et al. (1991) for NGC 362 and NGC 288, the total C+N+O abundance for the red giants in a cluster is generally constant, despite the large ranges seen in the individual C, N and O abundances.

This well-documented change in abundance with evolutionary state is incompatible with any primordial process and so can only be explained as an evolutionary process. The constancy of the C+N+O abundance strongly suggests that we are seeing the mixing of material that has been processed through the CN-cycle (in which C is converted to N) and the ON-cycle (in which O is converted to N) from the interior of the star into its outer layers.

Are the C, N and O abundance variations correlated with abundance anomalies in any other elements? Cottrell & Da Costa (1981) and Norris et al. (1981) were the first to notice that in the cluster NGC 6752, the C-depleted, N-enhanced stars also possess enhanced Na and Al abundances relative to the C- and N-normal red giants, while Peterson (1980) noted an apparent connection between Na enhancements and O depletions for red giants in M13. These results have now been extended to many clusters and there is a well established anti-correlation between the size of the Na and Al enhancements and the degree of O depletion. The relation for Na is shown in Fig. 2 while the equivalent relation for Al is shown in Norris & Da Costa (1995a) or Shetrone (1996a, hereafter S96a).

Most of these results are derived from spectra of the brightest giants in each cluster. Pilachowski et al. (1996) and Kraft et al. (1997, hereafter K97), however, have extended the observations in M13 to luminosities almost as faint as the horizontal branch. These observations reveal that the relations seen among the brighter red giants are still prevalent at fainter luminosities. These studies also provide information on Mg abundance variations in M13 (see also S96a). In particular, K97 show that in those M13 stars that have large Al enhancements and large O depletions, the Mg abundance is depleted by as much as 0.3 dex. A similar Mg-Al abundance anti-correlation is present in the ω Cen data of Norris & Da Costa (1995b). It is also seen, though to a lesser degree, in the M5 and M92 bright red giant data of S96a. Intriguingly, K97 also show that in M13, the total Mg+Al abundance is approximately constant, regardless of the degree of Mg depletion and Al enhancement (see also S96a). This strongly suggests that the additional Al atoms present in the O depleted stars are produced at the expense of Mg atoms in the same star, a situation reminiscent of the constancy of total C+N+O despite large variations in the individual abundances. It is also significant that the Na and Al enhancements and associated O and Mg depletions are of size comparable to those seen near the RGB tip, even at the faintest luminosities ($M_V \approx +0.3$) studied by K97.

2.2. DOES STELLAR EVOLUTION THEORY OFFER AN EXPLANATION?

The simple answer to the question posed in the section title is "No" — standard stellar evolution sequences for red giants do not predict *any* of the observations described in the previous section. At most, they predict a modest C depletion and N enhancement as a result of "first dredge-up" at relatively bright luminosities on the red giant branch. Nevertheless, we can use stellar evolution to provide guidance in interpreting the observations. For example, as first noted by Denisenkov & Denisenkova (1990), in the region of the H-burning shell in the interior of a red giant, conditions are such that proton capture reactions on Ne and Mg seeds can generate Na and Al abundance enhancements. This possibility has been explored in more detail recently — see Cavallo et al. (1996) and the references therein. Of course the region where these reactions occur is also that where O and C are depleted and N is enhanced. Consequently, at least in principle, the observations can be explained by mixing the processing products from the vicinity of the H-burning shell into the surface layers.

There are, however, some quantitative problems with this scenario. First, observationally, in the O-depleted stars the Al enhancement is as large as, if not larger, than the Na enhancement. But in the nucleosynthesis calculations, generally more Na than Al is produced, except perhaps near the RGB tip (Cavallo et al. 1996). Second, in the theoretical calculations, there is little change in the total Mg abundance (which is dominated by the abundance of ^{24}Mg) and the Al enhancement is produced principally from ^{25,26}Mg. Yet, at least in M13, the observations show that in many of the stars with large O depletions and Al enhancements, the (total) Mg abundance is depleted by a factor of $\sim$2 (K97), in disagreement with the theoretical calculations.

It is possible that some of these discrepancies could be removed by changes in the reaction rates used in the calculations, since, for example, Arnould et al. (1995) have shown that the rates of some of the reactions in the NaNe and AlMg cycles are uncertain by orders of magnitude. Indeed this may be a situation where observations will serve as a guide for theory. However, it is also possible that the discrepancy with the observations may be revealing a more significant problem in the standard evolutionary models. For example, Arnould et al. (1995) show that significant depletions in ^{24}Mg occur in the AlMg cycle only for temperatures above ~70 million degrees. But such temperatures do not occur in the H-burning shells of standard red giant models. The observation of Mg (and O) depletion and Al enhancement at relatively low luminosities on the red giant branch in M13 (K97) is particularly worrisome in this regard since, at least in standard models, the H-burning shell in such stars has a temperature well below 70 million degrees. Further, the work of Shetrone (1996b) has demonstrated that we cannot appeal to unusual Mg isotopic ratios as the explanation. His observations of a small sample of bright red giants in M13 demonstrate unequivocally that the Mg depletions in the Al enhanced stars are due to depletions in ^{24}Mg. The ^{25}Mg and ^{26}Mg abundances do not vary in any significant way with the degree of Al enhancement (Shetrone 1996b).

2.3. OTHER CONSTRAINTS

There are at least three other issues that should be kept in mind when considering the origin of these abundance anomalies. First, it is extremely unlikely that any large scale mixing process operates on the main sequence in these clusters, for the following reason. Any mixing out of processed material to the surface layers of the star inevitably requires mixing in of fresh hydrogen, and this modifies the star's evolution significantly (e.g. VandenBerg & Smith 1988). Thus a hypothesized stochastical distribution of mixing extents on the main sequence in a cluster would not generate the tight narrow cluster main sequence turnoff loci that are, in fact, observed. The observations of the C abundance decline with evolutionary state in clusters such as M92 are not incompatible with this requirement, as long as the mixing begins after the turnoff (VandenBerg & Smith 1988) as appears to be the case (Langer et al. 1986). However, Suntzeff & Smith (1991) claim the existence of anti-correlated CN and CH band strength variations on the main sequence (at $M_V \approx +6$) in NGC 6752. If (when) verified by larger samples and spectra with higher S/N, these observations will be very difficult to interpret in terms of any evolutionary mixing hypothesis.

Second, a "bump" in the luminosity function for the red giant branch is seen in many clusters, including clusters such as M5, M3 and NGC 6752 which show significant abundance anomalies. This feature, which results from the slow down in RGB evolution when the H-burning shell reaches the composition discontinuity left behind by the earlier deepest penetration of the convective envelope, occurs at approximately the luminosity predicted by standard evolutionary models. The existence of this bump indicates that not all stars in a cluster can have mixed the region between the H-burning shell and the surface at relatively low luminosities on the red giant branch, else the luminosity function feature would not be detected.

Finally, it is vital to remember that the abundance anomalies discussed here are confined primarily to globular cluster stars. Field stars show these anomalies only to a much smaller extent (e.g. Fig. 8 of Pilachowski et al. 1996). Hence any explanation of the globular cluster star abundance anomalies must also be capable of accommodating this striking cluster star/field star difference.

3. 47 Tucanae - A Special Case?

The comparatively metal-rich ([Fe/H] = −0.7) globular cluster 47 Tuc has been the subject of a number of intensive studies. Norris & Freeman (1979) were the first to show that the distribution of strengths of the λ4216Å CN-band in the spectra of cluster red giants was bimodal — stars could be classified as "CN-strong" or "CN-weak". Further, this bimodality is also present on the horizontal branch (Norris & Freeman 1982) and on the asymptotic giant branch (Norris & Freeman 1979). In these data there are no indications that the CN-strong to CN-weak number ratio is any different between the red giant, horizontal and asymptotic giant branches, i.e. there is no support for any change in this ratio with evolutionary state, as might have been expected in an evolutionary mixing scenario. However, as demonstrated in Norris et al. (1984), there is a general anti-correlation between the CN band strengths and the strength of the G band (CH) — CN-strong stars have weaker CH features (and vice versa) consistent with the involvement of the CN-cycle.

Norris & Freeman (1979) and later Paltoglou (1989) have also shown that among the red giants, there is a radial gradient in the CN-strong to CN-weak number ratio in the sense that CN-strong stars are considerably more common in the inner regions of the cluster. Figure 3 shows data from Paltoglou (1989) which reveals a decline in the number ratio with radius out to ~10 arcmin, beyond which the ratio appears to remain approximately constant. Aside from ω Cen where there is a distinct radial abundance gradient (e.g. Norris et al. 1996), no other globular cluster shows evidence for a similar radial gradient.

As regards other elements, Cottrell & Da Costa (1981) first showed that there are Na abundance variations which correlate positively with CN strength in 47 Tuc red giants. Subsequent work (e.g. Brown & Wallerstein 1992) has confirmed a Na abundance difference of ~0.3 dex between the two groups of giants, though the sample studied at high dispersion remains small. In contrast to the more metal-poor globular clusters, however, there is no indication within this small sample of any Al, Mg or O abundance differences between the CN-strong and CN-weak stars (Cottrell & Da Costa

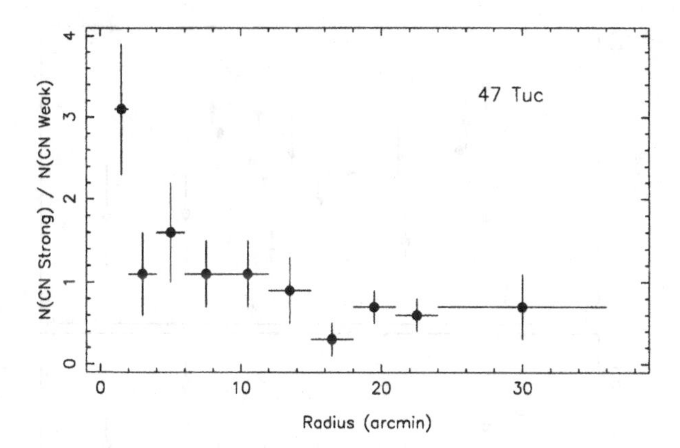

Figure 3. **Ratio of the number of CN-strong red giants to the number of CN-weak red giants in 47 Tuc as a function of distance from the cluster centre. There are ~30 to 40 stars per radial bin; data from Paltoglou (1989).**

1981, Brown et al. 1990, Brown & Wallerstein 1992). It appears that in this cluster the abundance anomalies are restricted to C, N and Na only.

There is one further characteristic that distinguishes 47 Tuc from other globular clusters and that is the availability of reasonable quality spectra for a modest sample of stars at the cluster main sequence turnoff (Briley et al. 1994). The results for this sample of stars are shown in Fig. 4. It is clear from the upper panel of Fig. 4 that there is also a bimodal distribution of CN-band strengths among these main sequence turnoff stars — they can be classified "CN-strong" or "CN-weak" in the same way as the more luminous red giants. Indeed to within the uncertainties inherent in this limited sample, the CN-strong to CN-weak number ratio for these main sequence turnoff stars is the same as it is at higher luminosities. In other words, again there is no evidence for a change in the CN-strong to CN-weak number ratio with evolutionary state. In the lower panel of Fig. 4 it is also apparent that there is a general anti-correlation between the CN- and CH-band strengths, just as is found at higher luminosities. The existence of a CH/CN anti-correlation is usually regarded as a signature of a mixing process but for these main sequence turnoff stars, as noted in the previous section, it is difficult to invoke a mixing mechanism to explain the abundance differences.

In a further paper, Briley et al. (1996) present observations at the Na D lines of a subset of the 47 Tuc main sequence turnoff stars shown in Fig. 4. Significantly, the three CN-strong stars observed have, on average, Na D lines that are ~30% stronger than for three CN-weak stars. In other words, there is evidence for a *positive sodium-cyanogen correlation at the main sequence turnoff in 47 Tuc.* In addition, the Na abundance difference between the two groups of stars is comparable, at ~0.3 dex, to the Na

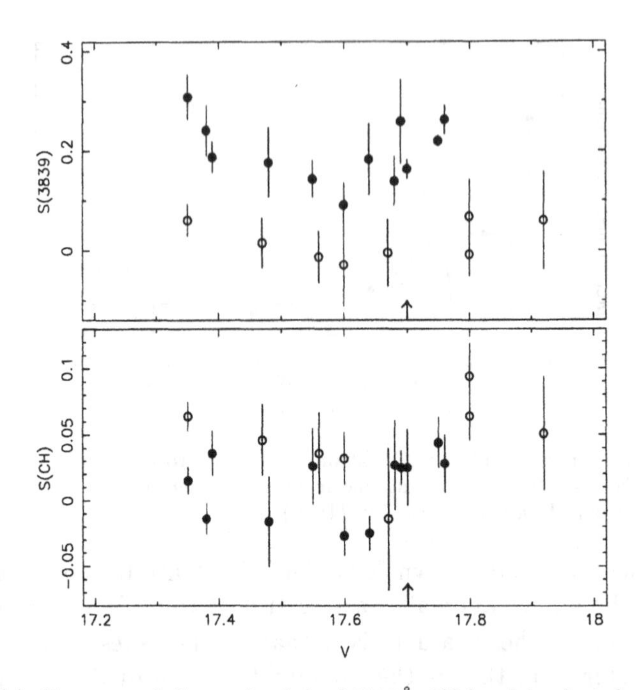

Figure 4. (a) *Upper panel.* Strength of the $\lambda3883$Å CN band is plotted against V for stars near the main sequence turnoff in 47 Tuc. The main sequence turnoff is indicated by the arrow on the x-axis. Stars can be unambiguously classified as CN-strong (filled symbols) or CN-weak (open symbols). (b) *Lower panel.* Strength of the G band (CH) at $\lambda \sim 4300$Å is plotted against V for the same stars. In both panels the band strength increases upwards. Note the general tendency for CN-strong stars to be CH-weak, and vice versa. All data from Briley et al. (1994).

abundance difference seen at much higher luminosities. This observation of Na/CN line strength correlation at the main sequence turnoff presents two difficulties for the mixing hypothesis. First, it is extremely unlikely that the NaNe cycle operates to any significant extent in stars of this luminosity — the interior temperatures are simply not high enough (even supposing the existence of a mixing mechanism to bring Na up to the surface layers which, as noted above, is unlikely). Second, the apparent agreement between the Na abundance difference at the main sequence turnoff and that near the red giant branch tip, strongly argues that no mixing driven processing can have occurred during the post-turnoff evolution. This contrasts markedly with the situation in the more metal-poor clusters where abundances are seen to change with evolutionary state.

Hence it seems likely that in 47 Tuc a primordial process is responsible for the abundance anomalies. If this is the case then we are presented with a dilemma: we can either claim that 47 Tuc is a rare or possibly unique cluster in that it possesses primordial abundance anomalies, or we must face the possibility that primordial processes play a significant role in generating the Na abundance anomalies (at least) in other clusters. However, as noted originally by Norris et al. (1981) and Smith & Norris (1982), the fundamental problem with the operation of a primordial process is not only to make it specific to the particular elements, here C, N and Na, but also to be able to generate the C depletions. In particular, in 47 Tuc the number of CN-weak and CN-strong stars are approximately equal, and they differ in C abundance by about a factor of two (e.g. Brown et al. 1990, Norris et al. 1984) with the CN-strong (Na-strong) stars being C depleted. One is then forced to the requirement that if the CN-strong stars form from a mixture of roughly equal amounts of original and processed/enriched material, then the processed/enriched material must contain only very small amounts of carbon. It is not at all clear that such a requirement can be met.

4. Conclusions

Despite intensive efforts on both the theoretical and observational fronts it is clear from the discussion above that we are far from a complete understanding of the origin(s) of abundance anomalies in globular cluster stars. So I will end this contribution with a list of goals that, if achieved, will allow considerable progress towards solving this very complex problem.

On the *observational side*, there are two questions (at least) that can be tackled. First, the observations of M13 show that Mg and Al abundance variations are still occurring at relatively faint luminosities ($M_V \approx +0.3$). Nevertheless, in the context of the mixing hypothesis, there should be a luminosity on the RGB below which the Mg variations are no longer seen. This luminosity would correspond to the point in the evolution where the temperature in the vicinity of the H-burning shell first becomes hot enough to sustain significant ^{24}Mg to Al conversion. Second, it is very important to establish whether or not Na abundance variations are commonly seen on the main sequence in other clusters besides 47 Tuc. If such main sequence Na variations are common, then it will be very difficult to explain them within the evolutionary mixing hypothesis.

On the *theoretical side* there are also a number of issues that can be tackled. In terms of the mixing hypothesis, we need to resolve if *any* "simple" artificial mixing scheme can reproduce all the existing observational detail on the red giant branch. In other words, what modifications are required to reaction rates and internal temperature structure to reproduce the observations, and are such modifications plausible? Are we instead begining to see that "perturbations" of spherically symmetric 1D stellar models with mixing length convection are no longer adequate for this problem — a full 3D hydrodynamic calculation of giant branch evolution incorporating rotation and time dependent convection may be required instead! More fundamentally, we need further work on the physical mechanisms that could drive the mixing. But here it is especially important to keep in mind that the mechanism must be almost exclusively a globular cluster star phenomenon; field stars do not show these anomalies to anything like the same extent. As regards the primordial hypothesis, the fundamental requirement is to identify a process that is specific to the elements that are observed to vary, and to explain how such an enrichment (and depletion!) process can work in the context solely of globular cluster formation. Once again it is important to reproduce the observations quantitatively, not just qualitatively.

5. References

Arnould, M., Mowlavi, N., & Champagne, A. 1995, in Stellar Evolution: What Should Be Done, edited by A. Noels et al. (Univ. Liège, Liège), p. 17
Briley, M.M., Bell, R.A., Hoban, S., & Dickens, R.J. 1990, ApJ, 359, 307
Briley, M.M., et al. 1994, AJ, 108, 2183
Briley, M.M., et al. 1996, Nature, 383, 604
Brown, J.A., & Wallerstein, G. 1992, AJ, 104, 1818
Brown, J.A., Wallerstein, G., & Oke, J.B. 1990, AJ, 100, 1561
Cavallo, R.M., Sweigart, A.V., & Bell, R.A. 1996, ApJ, 464, L79
Cottrell, P.L., & Da Costa, G.S. 1981, ApJ, 245, L79
Denisenkov, P., & Denisenkova, S.N. 1990, Sov. Astron. Lett. 16, 275
Dickens, R.J., Croke, B.F.W., Cannon, R.D., & Bell, R.A. 1991, Nature, 351, 212
Harding, G.A. 1962, Observatory, 82, 205
Hesser, J.E., Hartwick, F.D.A., & McClure, R.D. 1976, ApJ, 207, L113
Kraft, R.P., et al. 1995, AJ, 109, 2586
Kraft, R.P., et al. 1997, AJ, 113, 279 (K97)
Langer, G.E., et al. 1986, PASP, 98, 473
Norris, J.E. & Da Costa, G.S. 1995a, ApJ, 441, L81
Norris, J.E. & Da Costa, G.S. 1995b, ApJ, 447, 680
Norris, J., & Freeman, K.C. 1979, ApJ, 230, L179
Norris, J., & Freeman, K.C. 1982, ApJ, 254, 143
Norris, J., Freeman, K.C., & Da Costa, G.S. 1984, ApJ, 277, 615
Norris, J.E., Freeman, K.C., & Mighell, K.J. 1996, ApJ, 462, 241
Norris, J., Freeman, K.C., Cottrell, P.L., & Da Costa, G.S. 1981, ApJ, 244, 205
Osborn, W. 1971, Observatory, 91, 223
Paltoglou, G. 1989, PhD Thesis, Australian National University
Peterson, R.C. 1980, ApJ, 237, L87
Pilachowski, C.A., Sneden, C., Kraft, R.P., & Langer, G.E. 1996, AJ, 112, 545
Popper, D.M. 1947, ApJ, 105, 204
Shetrone, M.D. 1996a, AJ, 112, 1517 (S96a)
Shetrone, M.D. 1996b, AJ, 112, 2639
Smith, G.H., & Norris, J.E. 1982, ApJ, 254, 594
Smith, G.H., et al. 1996, AJ, 112, 1511
Suntzeff, N.B., & Smith, V.V. 1991, ApJ, 381, 160
Vandenberg, D.A. & Smith, G.H. 1988, PASP, 100, 314

STELLAR PARAMETERS IN THE BULGE CLUSTER NGC 6553

B. BARBUY
University of São Paulo, CP 9638, São Paulo 01065-970, Brazil
S. ORTOLANI
University of Padova, Vicolo dell'Osservatorio 5, I-35122 Padova, Italy
E. BICA
University of Rio Grande do Sul, CP 15051, Porto Alegre 91500-970, Brazil
AND
A. RENZINI AND M.D. GUARNIERI
European Southern Observatory, Karl-Schwarzschild Str. 2, D-85748 Garching bei München, Germany

1. Introduction

Globular clusters in the Galactic bulge form a flattened system, extending from the Galactic center to about 4.5 kpc from the Sun (Barbuy et al. 1997). A study of abundance ratios in these clusters is very important for a more complete understanding of the bulge formation. In this work we present a spectroscopic analysis of individual stars in NGC 6553. This cluster is a key one because it is located at $d_{\odot} \approx 5.1$ kpc, therefore relatively close to us, and at the same time it is representative of the Galactic bulge stellar population: (a) Ortolani et al. (1995) showed that NGC 6553 and NGC 6528 show very similar Colour-Magnitude Diagrams (CMDs), and NGC 6528 is located at $d_{\odot} \approx 7.83$ kpc, very close to the Galactic center; (b) the stellar populations of the Baade Window is also very similar to that of NGC 6553 and NGC 6528 as Ortolani et al. (1995) have shown by comparing their luminosity functions.

The temperature calibration of metal-rich cool giants needs high quality photometric data. For giant stars in the metal-rich bulge globular cluster NGC6553 we have obtained JK photometry at ESO and VI photometry

T.R. Bedding et al. (eds.),
Fundamental Stellar Properties: The Interaction between Observation and Theory, 203–206.

with the Hubble Space Telescope. This allowed us to derive effective temperatures. A detailed analysis of 3 stars was then carried out using échelle spectra obtained at the ESO 3.6m telescope.

TABLE 1. Magnitudes

star	V	I	J	K
III-3	15.825	13.408	11.536	10.328
IV-13	15.586	13.114	11.282	10.028
II-85	15.515	13.010	11.040	9.789

TABLE 2. Derived temperatures

star	T(V-I)	T(V-K)	T(J-K)	M(bol)	log g
III-3	4021	3968	4220	-2.26	0.95
IV-13	3980	3944	4125	-2.50	0.85
II-85	3930	3880	4130	-2.57	0.80

2. Stellar Parameters

2.1. TEMPERATURES

The crucial issue in the detailed analysis of these cool and metal-rich giants is the determination of their effective temperatures T_{eff}. We succeeded to obtain very high quality V and I colours using the Hubble Space Telescope (HST) (Ortolani et al. 1995), and J and K colours using the detector IRAC2 at the 2.2m telescope of ESO (Guarnieri et al. 1997). In Table 1 are given the magnitudes. The calibration of our old ESO Danish telescope data (Ortolani et al. 1990) showed an offset of about 0.28 mag, very probably due to crowding.

A further step in the derivation of colours is the reddening value adopted. In Guarnieri et al. (1997) we derived a colour excess of $E(V-I) = 0.95$ for NGC 6553. Assuming a ratio $E(V-I)/E(B-V) = 1.35$ (Terndrup 1988; Grebel & Roberts 1995) the resulting $E(B-V)$ is 0.7. Assuming E(V-K)/E(B-V) = 2.744 and E(J-K)/E(B-V) = 0.527 (Rieke & Lebofsky 1983), and $E(V\text{-}I)_C/E(B\text{-}V) = 1.33$ (Dean et al. 1978) the available colours

were de-reddened.
In order to derive temperatures we used several methods:
(i) Infrared flux method: Based on absolute measurements of stellar monochromatic fluxes in the infrared region for a sample of 80 solar metallicity stars, Blackwell & Lynas-Gray (1994) established the relation T(V-K) = 8862 - 2583(V-K) - 353.1$(V-K)^2$.
(ii) Relations by McWilliam (1990): Based on a sample of 671 giants of about solar metallicity McWilliam (1990) derived the following relations: T(B-V) = 8351 - 4936(B-V) + 1456$(B-V)^2$ - 78$(B-V)^2$; T(V-K) = 8595 - 2349(V-K) + 321$(V-K)^2$ - 8$(V-K)^2$; T(V-I) = 9853 - 6733(V-I) + 2564$(V-I)^2$ - 335$(V-I)^2$. T(J-K) = 8176 - 8715(J-K) + 6850$(J-K)^2$ - 2489$(J-K)^2$.
(iii) Tables by Bessell et al. (1996): Using Bessell et al. (1996) calibrations based on NMARCS models (Plez et al. 1992), give the temperatures reported in Table 2, which correspond essentially to T_{eff} = 3950 K for the three sample stars.
The final adopted temperatures for our stars are those corresponding to the Bessell et al. calibration of (V-I) and (V-K) colours.

2.2. GRAVITIES

The classical relation $\log g_* = 4.44 + 4 \log T_*/T_\odot + 0.4(M_{\rm bol} - 4.67) + \log M_*/M_\odot$ was used, where $T_\odot$ = 5770 K, M_* = 0.8 $M_\odot$, $M_{bol\odot}$ = 4.67 were adopted; for deriving M_{bol*} we adopted M_V(HB) = 1.06 (Buonanno et al. 1989), V(HB) = 15.8 (Ortolani et al. 1995) and bolometric magnitude corrections by Persson et al. (1980), and the resulting M_{bol} are given in Table 2.

2.3. METALLICITY

The metallicities were derived by plotting curves of growth of FeI, where the equivalent widths where a selected list of lines were measured using IRAF, and the code RENOIR by M. Spite was employed for plotting the curves of growth. The resulting metallicity for the cluster is [Fe/H] ≈ -0.35.

3. Average final abundance ratios

We find the preliminary abundance ratios: [Mg/Fe] ≈ +0.15, [Ti/Fe] ≈ +0.3, [Si/Fe] ≈ +0.6, [Ca/Fe] ≈ 0.0 and [Eu/Fe] ≈ +0.3.

It is interesting to compare these results with abundance ratios for the bulge from the literature. McWilliam & Rich (1994) analyzed 11 Baade Window K giants, for which they obtained: [Mg/Fe] ≈ [Ti/Fe] ≈ +0.3 and [Ca/Fe] ≈ [Si/Fe] ≈ 0.0. Idiart, Freitas Pacheco & Costa (1996) derived from and integrated spectrum of the Baade Window [Mg/Fe] ≈ +0.45.

Terndrup et al. (1995), Sadler et al. (1996) used medium-resolution spectra of 400 giants in the Baade Window for which they obtained [Mg/Fe] $\approx$ 0.2 - 0.3. It is also interesting to point out that all these results are also in agreement with Mg-to-Fe ratios in bulges of ellipticals: Worthey et al. (1992), Bender (1996), Barbuy et al. (1996) presented evidences pointing to enhanced [Mg/Fe] in bulges of ellipticals.

References:
Barbuy, B., Bica, E., Ortolani, S.: 1997, A&A, submitted
Barbuy, B., Castro, S., Ortolani, S., Bica, E.: 1992, A&A, 259, 607
Barbuy, B., Freitas Pacheco, J.A., Idiart, T.: 1996, in IAU S. 171, New Light on Galaxy Evolution, Kluwer, eds. R. Bender, R. Davies, p. 340
Bender, R.: 1996, 1996, in IAU S. 171, New Light on Galaxy Evolution, Kluwer, eds. R. Bender, R. Davies, p. 181
Bessell, M.S., Castelli, F., Plez, B.: 1996, preprint
Blackwell, D.E., Lynas-Gray, A.E.: 1994, A&A, 282, 899
Buonanno, R., Corsi, C.E., Fusi-Pecci, F.: 1989, A&A, 216, 80
Dean, J., Warpen, P., Cousins, A.: 1978, MNRAS, 183, 569
Grebel, A.K., Roberts, W.: 1995, A&AS, 109, 293
Guarnieri, M.D., Ortolani, S., Montegriffo, P., Renzini, A., Barbuy, B., Bica, E., Moneti, A.: 1997, A&A, in press
Idiart, T., Freitas Pacheco, J.A., Costa, R.D.D.: 1996, AJ, 111, 1169
McWilliam, A.: 1990, ApJS, 74, 1075
McWilliam, A., Rich, R.M.: 1994, ApJS, 91, 749
Ortolani, S., Barbuy, B., Bica, E.: 1990, A&A, 236, 362
Ortolani, S., Renzini, A., Gilmozzi, R., Marconi, G., Barbuy, B., Bica, E., Rich, R.M.: 1995, Nature, 377, 701
Persson, S.E., Frogel, J.A., Cohen, J.G., Aaronson, M., Matthews, K.: 1980, ApJ, 235, 452
Plez, B., Brett, J.M., Nordlund, A.: 1992, A&A, 256, 551
Rieke, G.H., Lebofsky, M.J.: 1985, ApJ, 288, 618
Sadler. E.M., Terndrup, D.M., Rich, R.M.: 1996, AJ, 112, 117
Terndrup, D.M.: 1988, AJ, 96, 884
Terndrup, D.M., Sadler, E.M., Rich, R.M.: 1995, AJ, 110, 1774
Worthey, G., Faber, S.M., Gonzalez, J.J.: 1992, ApJ, 398, 69

DISCUSSION

SHETRONE: Are your results not similar to the Edvardsson et al. work for disk stars? That is, when Mg is slightly enhanced, Ti is significantly enhanced, regardless of Ca. The same pattern is repeated in this cluster.

BEATRIZ BARBUY: I believe these clusters are not a typical disk population like the stars studied by Edvardsson et al., but a typ bulge population. As a matter of fact, the bulge shows a mean met of [Fe/H] $\simeq -0.2$ (McWilliam & Rich 1994, ApJS, 91, 749).

8. STELLAR ATMOSPHERES

NON-LTE LINE BLANKETED ATMOSPHERES FOR HOT STARS

D. J. HILLIER
University of Pittsburgh, USA

Abstract.

The modeling of hot star atmospheres falls into two broad classes: those where the plane parallel approximation can be used, and those where the effects of spherical extension and stellar winds are important. In both cases non-LTE modeling is a necessity for reliable spectroscopic analyses.

While simple ions (e.g., H, He I, and He II) have been treated routinely in non-LTE for many years it is only recently that advances in computing power, computational techniques, and the availability of atomic data have made it feasible to perform non-LTE line blanketing calculations. Present models, with varying degrees of approximation and sophistication, are now capable of treating the effects of tens of thousands of lines. We review the latest efforts in incorporating non-LTE line blanketing, highlighting recent advances in the modeling of O stars, hot sub-dwarfs, Wolf-Rayet stars, novae, and supernovae.

1. Introduction

An essential tool for understanding stars is spectroscopic analysis. Through spectroscopic analysis we can determine fundamental parameters such as the star's effective temperature, surface gravity, and composition and, assuming the star's distance is known, its luminosity and mass. Other parameters such as mass-loss rate, outflow velocity, rotational speed, and magnetic field strength can also be determined. With these determinations we can place constraints on the evolutionary state of the star. Chemical abundance analyses, for example, can provide information on nuclear processes that

T.R. Bedding et al. (eds.),
***Fundamental Stellar Properties: The Interaction between Observation and Theory*, 209–216.**

have occurred deep within the core of the star, providing fundamental tests of stellar evolution.

Hot stars provide special difficulties when it comes to performing spectroscopic analyses. First, due to the combination of low atmospheric densities and high temperatures, the usual assumption of local thermodynamic equilibrium (LTE) is invalid. The non-LTE effect is not subtle, and is readily evident in low resolution spectra of O stars. LTE models of O stars produce H lines that are much weaker than observed. Methods for handling non-LTE calculations in O stars date back to the late 1960's (e.g., Auer and Mihalas 1969).

Second, hot stars emit a significant fraction (in some cases over 90%) of their energy in the extreme UV where it cannot be directly observed. Thus accurate spectroscopic analysis is essential and provides the only method for determining fundamental data.

Third, many hot stars (particularly Pop. I) suffer significant interstellar extinction. In conjunction with the hot temperatures this renders continuum fluxes useless for providing information on stellar temperatures (e.g., Hummer et al. 1988).

Until recently a major deficiency of most codes used to analyse hot stars was the neglect of line blanketing. Only within the last decade, and primarily within the last several years, has progress been made to incorporate non-LTE line blanketing in stellar atmosphere codes. Non-LTE line blanketing is a formidable problem. Ideally tens of thousands of lines should be included and, in addition, the populations of the levels (which number in the hundreds, and perhaps thousands) involved in producing the line transitions must be explicitly solved for. It is the latter requirement, and the lack of the necessary atomic data, which has hindered the incorporation of non-LTE line blanketing into existing codes.

Below we discuss recent advances in the construction of non-LTE blanketed model atmospheres for hot stars. Due to similarities with hot stars, we will also discuss recent work on non-LTE line blanketing in novae and supernovae. For a more general review on the status of non-LTE stellar atmospheres for hot stars the reader is referred to Kudritzki and Hummer (1990).

2. The Need for Line Blanketing in Hot Star Analyses

Recently a large spectroscopic study of O stars was performed by Herrero et al. (1992) using plane-parallel non-LTE non-blanketed model atmospheres. In that study they found that the derived spectroscopic O star masses were systematically lower than evolutionary masses. Since the spectroscopic masses were similar to those obtained using the wind-momentum relationship of Kudritzki et al. (1992), they concluded that this discrepancy was not due to their analysis. The discrepancy, however, is largest for the most evolved stars, which is of concern since the most evolved stars are expected to show the largest effects due to extension and blanketing by the stellar wind. Recently Lanz et al. (1996) argued that metal line blanketing can partly alleviate the mass discrepancy problem. The mass discrepancy is of fundamental importance; if the spectroscopic masses are valid the inferences for stellar evolution are profound.

Wolf-Rayet stars provide another example of where blanketed models are urgently required. Current non-blanketed models are incapable of matching the line strengths of all ionization stages (e.g., N III, N IV and N V in WN stars) simultaneously. This is of concern since it limits the accuracy with which stellar parameters and abundances can be discerned, and consequently also limits the constraints on stellar evolution. Further, the mass-loss rates of W-R stars cannot be derived from first principals. While radiation pressure is generally believed to be responsible (e.g., Lucy and Abbott 1993) this has yet to be demonstrated.

3. Iron

Iron is believed to be the dominant blanketing agent in hot stars, and is also believed to be the primary species responsible for driving mass loss via radiation pressure from hot stars. It is also an important opacity source in stellar evolution models. The importance of Fe is in part due to its relatively high abundance, and is in part due to its atomic structure with its partially filled 3d subshell. As a consequence of the latter, Fe ions give rise to a large number of terms and an enormous number of transitions.

Consider, for example, Fe VI whose three lowest configurations are $3d^3$ (8 terms), $3d^2\,4s$ (7 terms) and $3d^2\,4p$ (19 terms). Transitions between $3d^3$ and the $3d^2\,4p$ configurations lie in the extreme UV (EUV) around 300Å (note that transitions between terms of the 2 lower configurations are forbidden since they both have even parity). The upper configurations are also connected via permitted transitions to the $3d^2\,4s$ terms ($\lambda \sim 1300\mathring{A}$) and consequently there is a strong coupling between the radiation field in the EUV, and that in the IUE UV. In W-R stars this coupling is reflected via the presence of significant Fe emission shortward of 1500 Å — emission that is the direct result of the processing of EUV continuum radiation. A similar situation also holds for other Fe ions (e.g., Fe IV, Fe V, Fe VII). Fortunately the strongest lines do not lie at similar wavelengths so that it is possible to derive a quantitative Fe abundance even in stars where the strong winds broaden that spectral feature so much that individual lines cannot be discerned.

4. Progress Towards Non-LTE Line Blanketed Atmospheres

The principal problem computing non-LTE line blanketed model atmospheres is as follows: To determine the atmospheric structure it is necessary to solve the transfer equation at thousands of frequencies simultaneously with the statistical equilibrium equations. For complex atoms such as iron this is a non trivial task. For example, in Fe IV the first 3 configurations ($3d^5$, $3d^4 4s$, and $3d^4 4p$) alone contain 280 levels (108 terms) and over 8000 transitions! In most objects, several ionization stages of each species (Fe IV through Fe VIII in W-R stars) must be included simultaneously.

Progress in non-LTE blanketed models has been facilitated by several recent developments:

1. Improved methods for solving the statistical equilibrium equations in conjunction with the radiative transfer equations. The most popular of these is the accelerated lambda iteration technique (see reviews by Rybicki 1991, Hubeny 1992)
2. The availability of atomic data. While the Opacity Project (Seaton 1987) has supplied some of the necessary data, there is still an urgent need for further data.
3. Increases in computational power and memory.
4. Better methods for handling level dissolution, and line overlap near series limits (e.g., Hubeny et al. 1994)
5. The concept of super-levels to minimize the number of atomic states whose populations must be solved. This idea was pioneered by Anderson (1989) and is discussed below.

The construction of non-LTE blanketed model atmospheres for hot stars was pioneered by Anderson (1985, 1989). To expedite the construction Anderson used the approach of super-levels. In this approach levels with similar atomic properties (e.g., energies, transition rates) are grouped together and treated as a single *super* level. In the production of the model atmospheres only the populations of the super-levels are solved for. To compute synthetic spectra, populations of individual levels are found using the assumption that the levels making up a given super-level all depart from LTE in an identical fashion. Alternatively, more detailed calculations can be performed for the ion of interest. The code of Anderson has been used to construct non-LTE blanketed model atmospheres for both B and late-type O stars (Grigsby et al. 1992). Anderson also utilized several other approximations, but it is the concept of super-levels which has facilitated the rapid growth in non-LTE blanketed atmosphere codes.

While the basic concept of super-levels is straight forward, different approaches have been adopted for its implementation. In part these implementations simply reflect different philosophies; in part they reflect the different needs of models in different parts of the H-R diagram.

Dreizler and Werner (1993) used the super-level approach to construct the first metal line blanketed model atmospheres for hot stars. Due to the similarity of the atomic properties of the iron group elements they created a generic ion representing each ionization stage of the iron group. Within each ionization stage they used the super-level concept, with each generic ion containing only 6 to 7 super-levels. To treat the multitude of transitions between a pair of super-levels they used an opacity sampling technique. With this procedure they were able to account for line blanketing due to over 1 million lines from Sc through Ni in 5 ionization stages. Their models have been used to interpret the spectra of white dwarfs, and sub-luminous O stars, and allowed them to deduce, for the first time, a reliable Fe abundance and Fe/Ni abundance ratio in the sdO star BD $+28^{o}4211$ (Haas et al. 1996)

Hubeny and Lanz (1995) have also adopted the super-level approach in conjunction with a hybrid linearization/accelerated lambda iteration method to construct non-LTE model atmospheres for hot stars. To minimize the number of frequencies Hubeny and Lanz used the concept of opacity distribution functions (ODF's). By careful choice of super-levels, transitions between 2 super-levels will occur in a relatively narrow spectral window. Within this spectral window ($\delta\lambda/\lambda < 0.3$) the exact location of individual transitions is generally unimportant. Thus they redistribute the line opacity within this window so that it varies smoothly, and as a consequence they can use relatively few frequencies to sample the opacity variation.

A possible problem with their approach is that the overlap of lines arising from different pairs of super-levels (within the same species, or from a different species) will not be treated correctly. The effect, however, can be minimized by using smaller bands, and/or special treatment of important overlaps (e.g., regions around Ly α). There is also no natural extension to moving atmospheres, where the velocity causes an individual transition to have an influence over a much larger frequency band than its natural bandwidth.

In order to model luminous blue variables (LBVs), W-R stars and other similar objects, Hillier and Miller (1997; Hillier 1996) included another variant of the super-level approach into the non-LTE atmosphere code of Hillier (1990). In their approach, super-levels are utilized only as a means of facilitating the solution of the statistical equilibrium equations, and not the transfer equations.

As in earlier studies, Hillier and Miller group individual levels together to form a super-level, and within this super-level, the departure coefficients are assumed to be identical. No other assumptions are made. In particular,

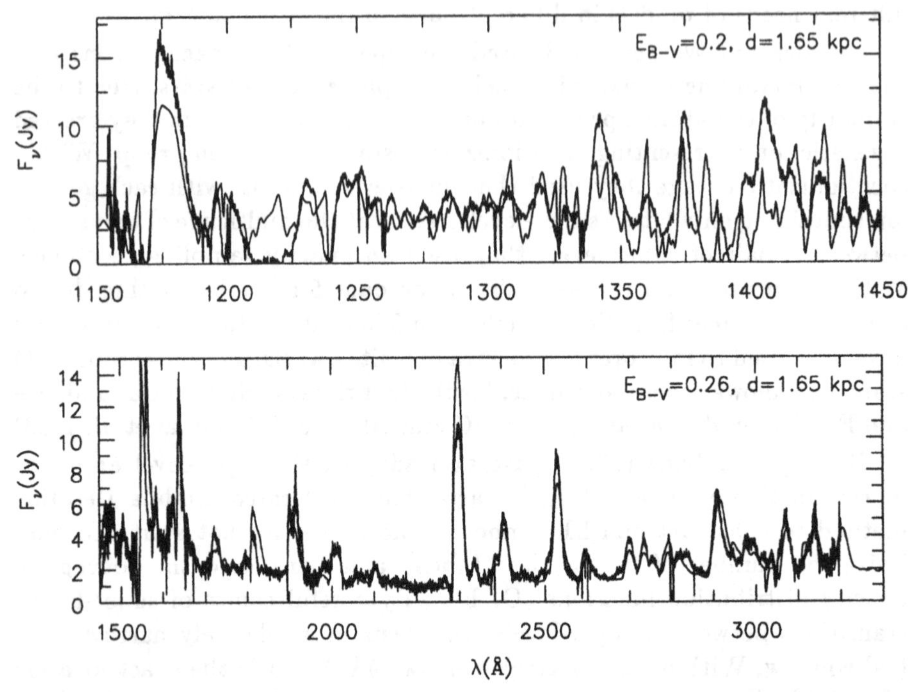

Figure 1. **Comparison of the synthetic spectrum computed using a non-LTE line blanketed model with the IUE spectrum of the galactic WC5 star HD 165763. In this spectral region the strongest spectral features are C III 1175, C IV 1550, He II 1640, and C III 2296. Much of the complex emission/absorption spectrum shortward of 1500Å is due to Fe V and Fe VI.**

individual transitions are treated at their correct wavelengths allowing line interactions, and the effect of velocity fields, to be correctly treated. An advantage of this technique is that LTE is recovered exactly at depth.

To illustrate progress in constructing W-R model atmospheres we compare a model for the galactic WC5 star HD 165763 with IUE observations in Fig. 1. While there are still discrepancies, the overall agreement is excellent. In particular, the models can produce the Fe emission/absorption complex between 1250Å and 1500Å. More importantly, the inclusion of Fe blanketing has helped to alleviate some of the discrepancies found with earlier unblanketed models.

Finally we note that a very different approach towards line blanketing calculations has been taken by Schmutz. Schmutz has utilized the Monte-Carlo code of Abbott and Lucy (1985) to estimate wavelength dependent blocking factors which characterize the influence of line blanketing. These blocking factors are then used to mimic the influence of line blanketing in "normal" non-LTE calculations (e.g., Schmutz 1991, Schmutz et al. 1990). The principal drawback with this approach is that the atomic populations used in the Monte-Carlo code are not necessarily consistent with the radiation field.

5. Novae and Supernovae

While these objects are not "hot" stars their analysis encounters many of the same problems as for W-R stars and LBV's. In particular, they possess atmospheres in which the effects of extension, non-LTE, and velocity fields are crucial for any spectroscopic analysis.

Extensive progress towards non-LTE blanketed modeling of SN has been made by Hauschildt and Ensman (1994), Eastman and Pinto (1993) and Pauldrach et al. (1996). Type II supernovae are simpler to model than Type I, and are more akin to hot stars. Type I are much more difficult to model as they are H/He deficient, and consequently metals dominate both continuum and line opacity. An additional complication of modeling supernovae is that it requires relativistic effects to be incorporated into the radiative transfer.

For novae, Hauschildt and collaborators have made enormous progress over the past 5 years (e.g., Hauschildt et al. 1992, 1994, 1995). In particular they are able to match the UV continuum in novae which is completely determined by the effects of Fe II blanketing. Their calculations indicate that at least 10,000 lines (and perhaps as many as 100,000) must be included if reliable synthetic spectra are to be computed (e.g., Hauschildt et al. 1994). For his models, Hauschildt has used a combination of both LTE and non-LTE. Species such as H are treated in non-LTE while many of the more important blanketing species (e.g., Fe, molecules) are treated in LTE. Later models treat Fe in non-LTE.

6. Conclusion

New computational methods, and the availability of atomic data now allow non-LTE blanketed models to be computed for hot stars, supernovae, and novae. In these models it is possible to handle over tens of thousands of transitions and thousands of levels. These new models should greatly improve spectroscopic analyses, allowing more concise constraints to be placed on stellar and galactic evolution.

References

Abbott, D.C., Lucy, L.B. 1985, ApJ, 288, 679
Anderson, L.S., 1985, ApJ, 298, 848
Anderson, L.S., 1989, ApJ, 339, 558
Auer, L.H., Mihalas, D., 1969, ApJ, 158, 641
Dreizler, S., Werner, K., 1993, A&A, 278, 199
Eastman, R.G., Pinto, P.A., 1993, ApJ, 412, 731
Grigsby, J.A., Morrison, N.D., Anderson, L.S, 1992, ApJS, 78, 205
Haas, S., Dreizler, S., Heber, U., Jeffery, S., Werner, K., 1996, A&A, 311, 669
Hauschildt, P.H., Ensman, L.M. 1994, ApJ, 424, 905
Hauschildt, P.H., Wehrse, R., Starrfield, S., Shaviv, G., 1992, ApJ, 393, 307
Hauschildt, P.H., Starrfield, S, Austin, S., et al. 1994, ApJ, 422, 831
Hauschildt, P.H., Starrfield, S, Shore, S.N., Allard, F., Baron, E., 1995, ApJ, 447, 829
Herrero, A., Kudritzki, R.P., Vilchez, J.M., et al. 1992, A&A, 261, 209
Hillier, D.J. 1990, A&A, 231, 116
Hillier, D.J., 1996, in Wolf-Rayet Stars in the Framework of Stellar Evolution, proc. 33rd Liege International Astrophys. Col. (Institut d'Astrophysique, Liege), p. 509
Hillier, D.J., Miller, D.L., 1997, submitted to ApJ
Hubeny, I., 1992, in lecture Notes in Physics, 401, The Atmospheres of Early Type Stars, eds. U. Heber, C.S. Jeffery (Belin: Springer-Verlag), p377
Hubeny, I., Hummer, D.G., Lanz, T., 1994, A&A, 282, 151
Hubeny, I., Lanz, T., 1995, ApJ, 439, 875
Hummer, D.G., Abbott, D.C., Voels, S.A., Bohannan, B., 1988, ApJ, 328, 704
Kudritzki, R.P., Hummer, D.G., 1990, ARA&A, 28, 303
Kudritzki, R.P., Hummer, D.G., Pauldrach, A.W.A., et al. 1992, A&A, 257, 655
Lanz, T., De Koter, A., Hubeny, I., Heap, S.R. 1996, ApJ, 465, 359
Lucy, L.B., Abbott, D.C., 1993, ApJ, 405, 738
Pauldrach, A.W.A, Duschinger, M., Mazzali, P.A., et al. 1996, A&A, 312, 525
Rybicki, G.B, 1991, in *Stellar Atmospheres: Beyond Classical Models*, NATO ASI series C, V341, eds. L. Crivellari, I. Hubeny, D.G. Hummer, (Kluwer, Dordrecht), p. 1
Schmutz, W., 1991, in *Stellar Atmospheres: Beyond Classical Models*, NATO ASI series C, V341, eds. L. Crivellari, I. Hubeny, D.G. Hummer, (Kluwer, Dordrecht) p.191
Schmutz, W., Abbott, D. C., Russell, R. S., et al. 1990, ApJ, 355, 255
Seaton, M. J. 1987, J. Phys. B, 20, 6363

DISCUSSION

Would you comment about models that include wind blanketing?

JOHN HILLIER: Detailed model calculations can be and have been used to provide estimates of wind blocking for plane-parallel static models. However, because of feedback effects and the necessity to explicitly allow for the wind, such methods are always approximate. Ultimately, unified spherical models in which the effects of wind blanketing are explicitly allowed for must be used.

NORBERT LANGER: You mentioned that the winds of WR stars are driven through iron opacity. What is the evidence for this?

JOHN HILLIER: There is no direct evidence. Iron is expected to be the dominant species responsible for the driving because of its large abundance and because its different ionization stages have a large number of transitions around the flux maximum. This is confirmed by the calculations of Lucy and Abbott.

PROGRESS ON MODEL ATMOSPHERES AND LINE DATA

ROBERT L. KURUCZ
Harvard-Smithsonian Center for Astrophysics, 60 Garden St., Cambridge, MA 02138, USA

Abstract. I discuss errors in theory and in interpreting observations that are produced by the failure to consider resolution in space, time, and energy. I discuss convection in stellar model atmospheres and in stars. One dimensional convective models can never work well, but the errors in predicted diagnostics for temperature, gravity, and abundances can be calibrated. I discuss the variation of microturbulent velocity with depth, effective temperature, and gravity. These variations must be dealt with in computing models and grids and in any type of photometric calibration. I describe the state of the art in computing a model atmosphere and complete spectrum for stars with arbitrary abundances. I have made no significant progress on atomic or molecular line data for the last two years, but I will soon begin large scale production. I will continue to distribute the results on CD-ROMs, and I will make them available on the World Wide Web.

1. Introduction

For the past two years I have been preoccupied with convection because the model atmospheres are now good enough to show shortcomings in the convective treatment. Here I will outline what I have learned. I will mainly list the conclusions I have come to from examining individual convective models and from examining grids of convective models as a whole. Eighteen figures illustrating the points made here can be found in Kurucz (1996a).

Every observation, measurement, model, and theory has seven characteristic numbers: resolution in space, in time, and in energy, and minimum and maximum energy. Many people never think about these resolutions. A low resolution physics cannot be used to study something in which the physical process of interest occurs at high resolution unless the high resolution effects average out when integrated over the resolution bandpasses.

T.R. Bedding et al. (eds.),
***Fundamental Stellar Properties: The Interaction between Observation and Theory*, 217–226.**

What does the sun, or any convective atmosphere, actually look like? We do not really know yet. There is a very simplified three-dimensional radiation-hydrodynamics calculation discussed in the review by Chan, Nordlund, Steffen, and Stein (1991). It is consistent with the high spatial and temporal resolution observations shown in the review by Topka and Title (1991). Qualitatively, there is cellular convection with relatively slowly ascending, hot, broad, diverging flows that turn over and merge with their neighbors to form cold, rapidly descending, filamentary flows that diffuse at the bottom. The filling factor for the cold downward flowing elements is small. The structure changes with time. Nordlund and Dravins (1990) discuss four similar stellar models with many figures. Every one-dimensional mixing-length convective model is based on the assumption that the convective structure averages away so that the emergent radiation depends only a one-dimensional temperature distribution.

There is a solar flux atlas (Kurucz, Furenlid, Brault, and Testerman 1984) that Ingemar Furenlid caused to be produced because he wanted to work with the sun as a star for comparison to other stars. The atlas is pieced together from eight Fourier transform spectrograph scans, each of which was integrated for two hours, so the time resolution is two hours for a given scan. The x and y resolutions are the diameter of the sun. The z resolution (from the formation depths of features in the spectrum) is difficult to estimate. It depends on the signal-to-noise and the number of resolution elements. The first is greater than 3000 and the second is more than one million. It may be possible to find enough weak lines in the wings and shoulders of strong lines to map out relative positions to a few kilometers. Today I think it is to a few tens of kilometers. The resolving power is on the order of 522,000. This is not really good enough for observations made through the atmosphere because it does not resolve the terrestrial lines that must be removed from the spectrum. (In the infrared there are many wavelength regions where the terrestrial absorption is too strong to remove.) The sun itself degrades its own flux spectrum by differential rotation and macroturbulent motions. The energy range of the atlas is from 300 to 1300 nm, essentially the range where the sun radiates most of its energy.

This solar atlas is of higher quality than any stellar spectrum taken thus far but still needs considerable improvement. If we have difficulty interpreting these data, it can only be worse for other stars where the spectra are of lower quality by orders of magnitude.

To analyze this spectrum, or any other spectrum, we need a theory that works at a similar resolution or better. We use a plane parallel, one-dimensional theoretical or empirical model atmosphere that extends in z through the region where the lines and continuum are formed. The one-dimensional model atmosphere represents the space average of the convective structure over the whole stellar disk (taking account of the center-to-limb variation) and the time average over hours. It is usually possible to compute a model that matches the observed energy distribution around the flux maximum. However, to obtain the match it is necessary to adjust a number of free parameters: effective temperature, surface gravity, microturbulent velocity, and the mixing-length-to-scale-height-ratio in the one-dimensional convective treatment. The microturbulent velocity parameter also produces an adjustment to the line opacity to make up for missing lines. Since much of the spectrum is produced near the flux maximum, at depths in the atmosphere where the overall flux is produced, averaging should give good results. The parameters of the fitted model may not be those of the star, but the radiation field should be like that of the star. The sun is the only star where the effective temperature and gravity are accurately known. In computing the detailed spectrum, it is possible to adjust the line parameters to match all but the centers of the strongest lines. Since very few lines have atomic data known accurately enough to constrain the model, a match does not necessarily mean that the model is correct.

2. Convective bullets

From plots of the convective flux and velocity for grids of models I have identified three types of convection in stellar atmospheres:

- normal strong convection where the convection is continuous from the atmosphere down into the underlying envelope. Convection carries more than 90% of the flux. Stars with effective temperatures 6000K and cooler are convective in this way as are stars on the main sequence up to 8000K. At higher temperature the convection carries less of the total flux and eventually disappears starting with the lowest gravity models. Intermediate gravities have intermediate behavior. Abundances have to be uniform through the atmosphere into the envelope. The highly convective models seem to be reasonable representations of real stars, except for the shortcomings cited below.

- atmospheric layer convection where. as convection weakens, the convection zone withdraws completely up from the envelope into the atmosphere. There is zero convection at the bottom of the atmosphere. Abundances in the atmosphere are decoupled from abundances in the envelope. For mixing-length models the convection zone is limited at the top by the Schwarzschild criterion to the vicinity of optical depth 1 or 2. The convection zone is squashed into a thin layer. In a grid, this layer continues to carry significant convective flux for about 500K in effective temperature beyond the strongly convective models. There is no common-sense way in which to have convective motions in a thin layer in an atmosphere. The solution is that the Schwarzschild criterion does not apply to convective atmospheres. The derivatives are defined only in one dimensional models. A real convective element has to decide what to do on the basis of local three-dimensional derivatives, not on means. These thin-layer-convective model atmospheres may not be very realistic.

• plume convection. Once the convective flux drops to the percent range, cellular convection is no longer viable. Either the star becomes completely radiative, or it becomes radiative with convective plumes that cover only a small fraction of the surface in space and time. Warm convective material rises and radiates. The star has rubeola. The plumes dissipate and the whole atmosphere relaxes downward. There are no downward flows. The convective model atmospheres are not very realistic except when the convection is so small as to have negligible effect, i.e. the model is radiative. The best approach may be simply to define a star with less than, say, 1% convection as radiative. The error will probably be less than using mixing-length model atmospheres.

Using a one-dimensional model atmosphere to represent a real convective atmosphere for any property that does not average in space and time to the one-dimensional model predictions produces systematic errors. The Planck function, the Boltzman factor, and the Saha equation are functions that do not average between hot and cold convective elements. We can automatically conclude that one-dimensional convective models must predict the wrong value for any parameter that has strong exponential temperature dependence from these functions.

Starting with the Planck function, the ultraviolet photospheric flux in any convective star must be higher than predicted by a one-dimensional model (Bikmaev 1994). Then, by flux conservation, the flux redward of the flux maximum must be lower. It is fit by a model with lower effective temperature than that of the star.

The following qualitative predictions result from the exponential falloff of the flux blueward of the flux maximum:

• the Balmer continuum in all convective stars is higher than predicted by a one-dimensional model;

• in G stars, including the sun, the discrepancy reaches up to about 400nm;

• in K stars it reaches to about 500nm. This flux "distortion" may be responsible for Short and Lester's (1994) problems with the ultraviolet flux of Arcturus;

• in M stars the whole Paschen continuum is higher;

• all ultraviolet photoionization rates at photospheric depths are higher in real stars than computed from one-dimensional models;

• flux from a temperature minimum and a chromospheric temperature rise masks the increased photospheric flux in the ultraviolet;

• the spectrum predicted from a one-dimensional model for the exponential falloff region, and abundances derived therefrom, are systematically in error;
• limb-darkening predicted from a one-dimensional model for the exponential falloff region is systematically in error;
• convective stars produce slightly less infrared flux than do one-dimensional models.

The Boltzman factor is extremely temperature sensitive for highly excited levels:
• the strong Boltzmann temperature dependence of the second level of hydrogen implies that the Balmer line wings are preferentially formed in the hotter convective elements. A one-dimensional model that matches Balmer line wings has a higher effective temperature than the real star;
• the same is true for all infrared hydrogen lines.

The Saha equation is safe only for the dominant species:
• neutral atoms for an element that is mostly ionized are the most dangerous because (in LTE) they are much more abundant in the cool convective elements. When Fe is mostly ionized the metallicity determination from Fe I can be systematically offset and can result in a systematic error in the assumed evolutionary track and age.
• in the sun convection may account for the remaining uncertainties with Fe I found by Blackwell, Lynas-Gray, and Smith (1995);
• the most striking case is the large systematic error in Li abundance determination in extreme Population II G subdwarfs. The abundance is determined from the Li I D lines which are formed at depths in the highly convective atmosphere where Li is 99.94% ionized (Kurucz 1995b);
• K giant abundances also have systematic errors that arise from similar mechanisms.
• molecules with high dissociation energies such as CO are also much more abundant in the cool convective elements. The CO fundamental line cores in the solar infrared are deeper than any one-dimensional model predicts (Ayres and Testerman 1981) because the cooler convective elements that exist only a short time have more CO than the mean model.

Given all of these difficulties, how should we proceed? One-dimensional model atmospheres can never reproduce real convective atmospheres. The only practical procedure is to compute grids of model atmospheres, then to compute diagnostics for temperature, gravity, abundances, etc., and then to make tables of corrections. Say, for example, in using the Hα wings as a diagnostic of effective temperature in G stars, the models may predict effective temperatures that are 100K too high. So if one uses an Hα temperature scale it has to be corrected by 100K to give the true answer. Every temperature scale by any method has to be corrected in some way.

Unfortunately, not only is this tedious, but it is very difficult or impossible because no standards exist. We do not know the energy distribution or the photospheric spectrum of a single star, even the sun. We do not know what spectrum corresponds to a given effective temperature, gravity, or abundances. The uncertainties in solar abundances are greater than 10%, except for hydrogen, and solar abundances are the best known. It is crucial to obtain high resolution, high signal-to-noise observations of the bright stars.

Another problem arises from using a one-dimensional model as a smooth background for some other physical process when the real star actually has large scale convective motions:

• strong convection in an atmosphere and pulsation, as in a Cepheid, are physically inconsistent. Pulsation and convection must be treated together using three-dimensional radiation-hydrodynamics to investigate whether both are possible at the same time and how the pulsating star would behave if they are.

• weak, plume convection is consistent with pulsation. When the star is expanding, the convective plumes, which carry only a small percentage of the flux, and which cover only a small percentage of the area at any given time, can add to the expansion velocity and become supersonic. Shocks can form lower in the atmosphere than expected from pulsation models that ignore convection. The shock region would be three-dimensional and complex, probably with overlapping bumps. This would produce a heating and an increase in broadening velocities that would be apparent in the spectrum.

3. Microturbulent velocity

In the sun the microturbulent velocity is empirically found to vary from 0.5 km-s^{-1} at the temperature minimum up 1.8 km-s^{-1} in the deepest layers we can see. Solar spectra and predicted fluxes are computed using this depth-dependant microturbulent velocity.

This is qualitatively the behavior we expect from convective motions. The maximum convective velocity in the solar atmosphere is about 2.0 km/s at optical depth about 4, and the (radial) velocity slows to zero at the "surface" because there is no mass loss. If, ad hoc, we set the "surface" equal to the temperature minimum, and if we assume that the microturbulent velocity arises from, and is in some way proportional to this convective velocity, then models should incorporate depth-dependent microturbulent velocity. Because of the change in line width, all convective one-dimensional model atmospheres would then have less opacity near the surface and more at depth than is now assumed.

Presuming that all strongly convective models have similar behavior, I propose simply scaling the solar Vturb-τ_{Ross} relation by the maximum convective velocity found in the atmosphere. I have tabulated this quantity for my grids of mixing-length convective models (Kurucz 1993b). It varies smoothly from 0.2 km-s^{-1} for an M dwarf to 3 km-s^{-1} for an F dwarf, to 4 km-s^{-1} for an M supergiant, and to >10 km-s^{-1} for an F supergiant. I will compute a grid where Vturb varies with Teff, log g, and τ_{Ross} according to this prescription to see what happens to the models and colors. This will result in a grid in which giants automatically have high microturbulent velocities and main sequence stars have low microturbulent velocities. And in which purely radiative model atmospheres have no microturbulent velocity. Stars with plume convection would be radiative over almost all the surface so they would have bimodal line profiles with a thermal doppler core and a small blueshifted tail from the plumes. I have not yet had a chance to construct such a grid to see how it quantitatively and qualitatively differs from my current grid with constant Vturb. I should get to it later this year.

In the sun the microturbulent velocity increases in the chromosphere above the temperature minimum because of waves. The model atmosphere grids available now are model photospheres so they ignore both the heating and the increased Vturb.

Since pulsation tracks vary in Teff and gravity, models that represent pulsating stars must have Vturb and line opacities that vary with phase. In interpreting the pulsation, grids of models must be used that take into account this variation.

4. Models and spectra for individual stars

ATLAS12 is an opacity sampling program that I originally wrote in 1992 to allow computation of models with individual abundances. It is described in Kurucz (1993b). The reader should consult that paper for details about ATLAS12 and about my existing grids of models computed with scaled solar-abundance distribution-function line opacities. That paper has figures showing sample calculations. Castelli and Kurucz (1994) present an ATLAS12 model for Vega.

When actually working with ATLAS12 sampled fluxes, Castelli and I found that, while quite accurate for predicting the total flux, the fluxes are not accurate in intermediate or narrow bandpass intervals because the sample size is too small. I wrote a special stripped version of the spectrum synthesis program SYNTHE to generate the surface flux for the converged model using the line data from CD-ROMs 1 and 15 (Kurucz 1993a;d). Plots showing sample calculations are given in Kurucz (1995a), the complete spectrum of Arcturus with contributions by individual molecules in Kurucz

(1994), and the complete spectrum for Sirius in Kurucz (1996b). As most of the line positions are predicted, the computed flux spectrum is not realistic when compared to a high resolution observed spectrum. It is possible to select only the subset of lines with accurate wavelengths, but many lines will be missing. This difficulty can be removed only by improving laboratory spectrum analyses. When more energy levels are known, more line positions can be computed accurately.

The spectrum is computed in pieces, typically at resolving power 500000. It is actually computed as 17 intensity spectra spread across the disk of the star from center to limb. These pieces are merged into 17 continuous intensity spectra. Then rotationally broadened flux spectra are computed for a number of values of v sin i, still at a resolving power of 500000, by interpolating and integrating over the disk. Alternatively, the pieces can be rotationally broadened and then merged. In these calculations all information about individual lines is thrown away. To get the complete information, the calculation must be made with the full SYNTHE program. The intensity spectra from different models can be combined to mimic surface features. Of course it is not necessary to compute the whole spectrum, but only the part of interest. Any model can be used; the spectrum calculation is not coupled to ATLAS12. These high resolution spectra can be compared directly to high resolution observed spectra or they can be instrumentally broadened to compare to low resolution spectra or photometry. It is now feasible to compute grids of color indices by computing sections of spectra for every model in a grid.

ATLAS12 can be used to produce improved models for Am and Ap stars. It should be very useful for investigating diffusion effects in atmospheres. It can be used to model exciting stars for H II regions with abundances consistent with those of the H II region. These programs and line files will be distributed on CD-ROMs.

5. Line data

The last CD-ROM I published is Kurucz CD-ROM 23 Atomic Line Data (Kurucz and Bell 1995), which is in collaboration with Barbara Bell. It has all the laboratory and computed line data with good wavelengths sorted into one file, and then also divided into 10 nm or 100 nm blocks for convenience. It also has hyperfine splittings for the iron group.

I have made no significant progress on the line data in the last two years because of problems with office space and computers. I now have an Alpha and several Vaxes with 60 GB of disk, a CD writer, and a web server. There are directories for every atom and diatomic molecule. I will soon begin large scale production of atomic and molecular line data. I will continue to distribute the results on CD-ROMs, and I will make them available on the World Wide Web.

References

Ayres, T.R. and Testerman, L. 1981, ApJ 245, 1124-1140.
Bikmaev, I. 1994, personal communication.
Blackwell, D.E., Lynas-Gray, A.E., and Smith, G. 1995, A&A, 296, 217.
Castelli, F. & Kurucz, R. L. (1994) A&A 281, 817-832.
Chan, K.L., Nordlund, Å, Steffen, M., Stein, R.F. (1991) *Solar Interior and Atmosphere*, A.N. Cox, W.C. Livingston, and M. Matthews, eds. (Tucson: U. of Arizona Press) 223-274.
Kurucz, R. L. (1993a) Atomic data for opacity calculations. Kurucz CD-ROM No. 1.
Kurucz, R. L. (1993b) in *Peculiar versus Normal Phenomena in A-type and Related Stars*, M. M. Dworetsky, F. Castelli, and R. Faraggiana, eds., A.S.P. Conference Series, 87-97.
Kurucz, R. L. (1993c) ATLAS9 Stellar Atmosphere Programs and 2 km/s grid. Kurucz CD-ROM No. 13.
Kurucz, R. L. (1993d) Diatomic Molecular Data for Opacity Calculations. Kurucz CD-ROM No. 15.
Kurucz, R. L. (1994) in *Molecules in the Stellar Environment*, U. G. Jørgensen, ed., Springer-Verlag, Berlin, 282-295.
Kurucz, R. L. (1995a) in *Highlights of Astronomy*, Vol. 10, I. Appenzeller, ed., 407-409.
Kurucz, R. L. (1995b) Ap.J.,452, 102-108.
Kurucz, R. L. (1996a) in ASP Conf. Series Vol. 108, *Model Atmospheres and Stellar Spectra* (ed. S. Adelman, F. Kupka, and W.W. Weiss) pp. 2-18.
Kurucz, R. L. (1996b) in ASP Conf. Series Vol. 108, *Model Atmospheres and Stellar Spectra* (ed. S. Adelman, F. Kupka, and W.W. Weiss) pp. 160-164.
Kurucz, R.L. and Bell, B. (1984) Atomic Line Data Kurucz CD-ROM No. 23.
Kurucz, R.L, Furenlid, I., Brault, J., and Testerman, L. (1984) Solar Flux Atlas from 296 to 1300 nm, (Sunspot, N.M.: National Solar Observatory)
Nordlund, Å. and Dravins, D. (1990) A&A 228, 155.
Short, C.I. and Lester, J.B. (1994) ApJ 436, L365.
Topka, K.P. and Title, A.M. (1991) in *Solar Interior and Atmosphere*, A.N. Cox, W.C. Livingston, and M. Matthews, eds. (Tucson: U. of Arizona Press) 727-747.

DISCUSSION

CLAUDE MEGESSIER: You mentioned the effects that affect the Balmer line wings and lead to errors in stellar effective temperature determinations through the Balmer line fitting. The advantage of the Infrared Flux Method is to take into account the integrated flux over the whole wavelength range and it does not consider some features in the spectra. Then the near-infrared region used by the method plays an important part and it is essential to know the physical phenomena that could affect it.

ROBERT KURUCZ: The Infrared Flux Method may be the only 'safe' method, although there is a systematic error. Because of flux conservation, an increased UV flux would be balanced by a decreased IR flux. This would probably be a few degrees error in effective temperature.

JOHANNES ANDERSEN: From your talk and the previous one, I gather that the useful days of 1-D atmosphere models are numbered. Do you think their shortcomings are the reasons for the problems with synthetic colours ('Hyades anomaly,' etc.) and abundances from temperature-sensitive lines (Li I, O I triplet)? And, as a comment, some years ago Dravins and Nordlund derived the kind of different convection patterns in F and later-type stars that you mentioned from an analysis of line bisectors in such stars.

ROBERT KURUCZ: There should be an inconsistency between ultraviolet and red or IR colors. And yes, there should be effects in abundance analysis.

UNKNOWN: Although we cannot do 3-D with good resolution yet (but soon), 2-D hydrodynamic simulations of stellar convection give higher velocities than mixing length theory. This is because mixing length is local, and plumes are accelerated over their path. Thus, microturbulence velocities inferred from mixing length should be used with some skepticism.

DIFFERENT UNKNOWN: A depth-dependent microturbulence would go part of the way in reducing the high Fe I abundance obtained by the Oxford group.

ROBERT KURUCZ: Yes, I think they did a test with depth-dependent microturbulent velocity.

ATMOSPHERE MODELS FOR VERY LOW MASS STARS, BROWN DWARFS AND EXOPLANETS

I. BARAFFE
Ecole Normale Supérieure, CRAL, 69364 Lyon Cedex 07, France

AND

F. ALLARD
Dept. of Physics, Wichita State University, Wichita, KS 67260-0032

1. Introduction

Over the past decade considerable effort, both observational and theoretical, has been directed towards a more accurate determination of the stellar lower main sequence and of the sub-stellar domain covered by Brown Dwarfs and Planets. Astronomers have been looking for brown dwarfs for more than a decade, either with standard astronomical technics or with microlensing experiments. A breakthrough in the search for brown dwarfs was very recently achieved with the discovery of the first cool brown dwarf GL 229B (Nakajima et al. 1995). At the same epoch, the search for planets blossomed with the discovery of a Jupiter - mass companion of the star 51 Pegasi (Mayor and Queloz 1995). Now, the number of faint, cool stars and substellar objects is rising rapidly.

The most reliable way to identify the nature of observed objects is a direct comparison of observed and synthetic spectra, to determine the effective temperature and metallicity. The mass requires interior models and is determined from consistent evolutionary models (cf. §2). Until recently, little about the atmospheres and spectral characteristics of such cool objects was understood.

T.R. Bedding et al. (eds.),
***Fundamental Stellar Properties: The Interaction between Observation and Theory*, 227–233.**

2. Characteristics of cool dwarf model atmospheres

Very low mass stars (VLMS) or M-dwarfs are characterized by effective temperatures from $\sim$ 5000 K down to 2000 K and surface gravities $log\, g \approx 3.5 - 5.5$, whereas brown dwarfs (BD) and extra-solar giant planets (EGP) can cover a much cooler temperature regime, down to some 100 K. Such low effective temperatures allow the presence of stable molecules (H_2, H_2O, TiO, VO, CH_4, NH_3,...), whose bands constitute the main source of absorption along the characteristic frequency domain. Such particular conditions are responsible for strong non-grey effects and significant departure of the spectral energy distribution from a black body emission. Tremendous progress has been made within the past years to derive accurate non-grey atmosphere models by several groups over a wide range of temperatures and metallicities (Plez et al. 1992; Saumon et al. 1994; Brett 1995 Allard and Hauschildt 1995,1997). A detailed description of the progress in the field is given in Allard et al. (1997).

The current grids of atmosphere models now available do not however include the condensation of molecules into grain, which should affect the atmosphere structure and the spectra of cool objects below $T_{eff} \sim 2500$ K (Tsuji et al. 1996a). A recent breakthrough was achieved by Tsuji et al. (1996a) who first included the grain formation and the grain opacities in non-grey atmosphere models for M-dwarfs and brown dwarfs. New generations of atmosphere models taking into account molecular condensation are now essential for a correct description of substellar objects and are now underway (e.g. Burrows et al. 1996, Allard and Alexander 1997).

Another difficulty inherent to cool dwarf atmopsheres is due to the presence of convection in the optically thin layers. This particularity is due to the molecular hydrogen recombination ($H+H \rightarrow H_2$) which lowers the adiabatic gradient and favors the onset of convective instability. Since radiative equilibrium is no longer satisfied in such atmospheres, the usual procedure based on $T(\tau)$ relationships to construct grey atmosphere models and to impose an outer boundary condition for the evolutionary models is basically incorrect (cf. Chabrier and Baraffe 1997; Chabrier, this conference). An accurate surface boundary condition based on *non-grey atmosphere models* is therefore required for evolutionary models.

3. Theory versus observation

3.1. M-DWARFS

The main improvement performed in the modelling of cool model atmospheres is essentially due to more accurate molecular opacity treatment (e.g opacity sampling method, line by line treatment) and improved molec-

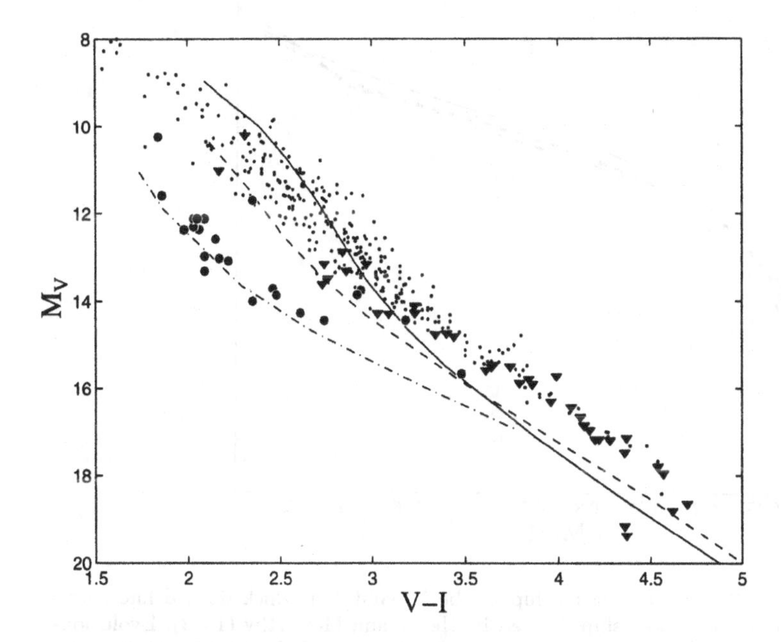

Figure 1. Color - Magnitude diagrams for different metallicity: [M/H]=-1.5 (dash-dot), [M/H]=-0.5 (dash) and [M/H]=0 (solid) from Baraffe et al. (1995). Subdwarf halo field stars from Monet et al. (1992) are indicated by full circles, as well as disk stars of Monet et al. (1992) (full triangles) and Dahn et al. (1995) (dots)

ular data for the main absorbers as H_2O and TiO (Jorgensen 1994; Miller et al. 1994; etc...).

The consistent coupling between non-grey atmosphere models and evolutionary models and the use of synthetic spectra to derive the colors provided recently the clue to reproduce the disk and halo field stars in a color-magnitude diagram (cf. Fig. 1). Same agreement is now reached with HST observations of Globular Cluster Main Sequences of low metallicity (Baraffe et al. 1997; see Chabrier 1997, this conference), based on the most recent generation of atmosphere models (Allard et al. 1997). These successful results suggest that the present stellar models are now sufficiently accurate to derive reliable mass-luminosity relationships and thus mass functions, although for solar metallicity a problem remains for the very bottom of the Main Sequence ($\leq 0.1 M_{\odot}$).

For solar metallicity, an empirical mass-luminosity relationship has been derived by Henry and McCarthy (1993, HMC93) based on observations of binary systems. Improvement in the molecular linelists of TiO and H_2O now leads to a good agreement of the models with the observed relationship in the optical (V-band) and in the near IR (K-band). Figure 2 shows such agreement for stellar models based on recent atmosphere models of Allard

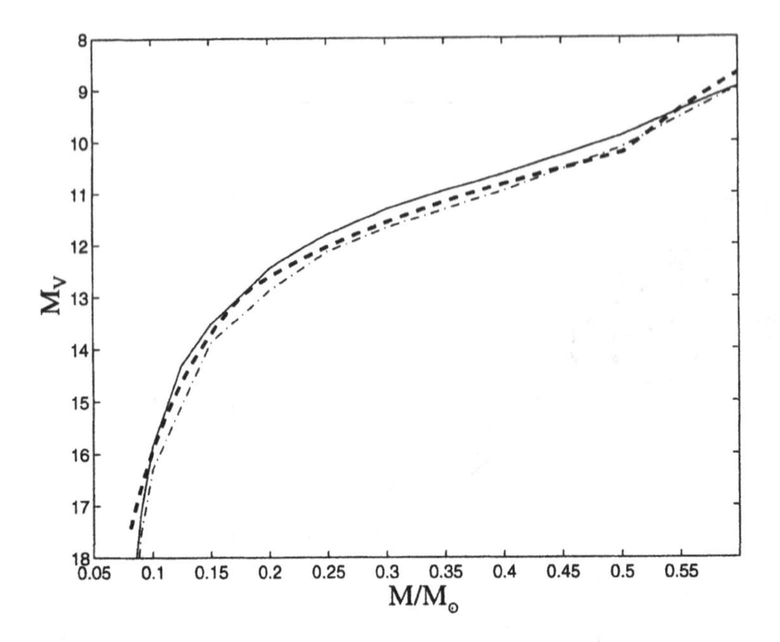

Figure 2. Mass - Magnitude relationship in the V-band. The thick dashed line corresponds to the empirical relationship derived by Henry and McCarthy (1993). Evolutionary models (Chabrier et al. 1996) are based on the Allard and Hauschildt (1997 , solid line) and the Brett-Plez (dash-dot) non-grey atmosphere models.

and Hauschildt (1997) and Brett-Plez (Brett 1995; Chabrier et al. 1996). Note that the change of slope in the HMC93 fit at $M_V \sim 11$ is exagerated and stems from a linear interpolation between remote objects (see Fig. 2 of HMC93).

The comparison of synthetic spectra with observed spectral distribution of M-dwarfs now reaches a good agreement in the IR (cf. Leggett et al. 1996; Allard et al. 1997) for atmosphere models including the new water linelist of Miller et al. (1994). Disagreement still remains in the optical colors (V-I, R-I) and for the latest-type M-dwarfs ($T_{eff} \leq 3500$ K) where the models systematically overestimate the water band strengh. Such discripancies may be due either to remaining shortcomings in the TiO (optical) and H_2O (IR) line lists or to the effect of grain formation, as discussed below.

3.2. SUBSTELLAR OBJECTS

As the effective temperature decreases below $\sim$ 2000 K, the IR spectrum is dominated mainly by H_2O, CH_4 and NH_3. The presence of methane is a clear signature of the substellar nature of an object. The most stringent test for extremely cool models was provided recently by the brown dwarf

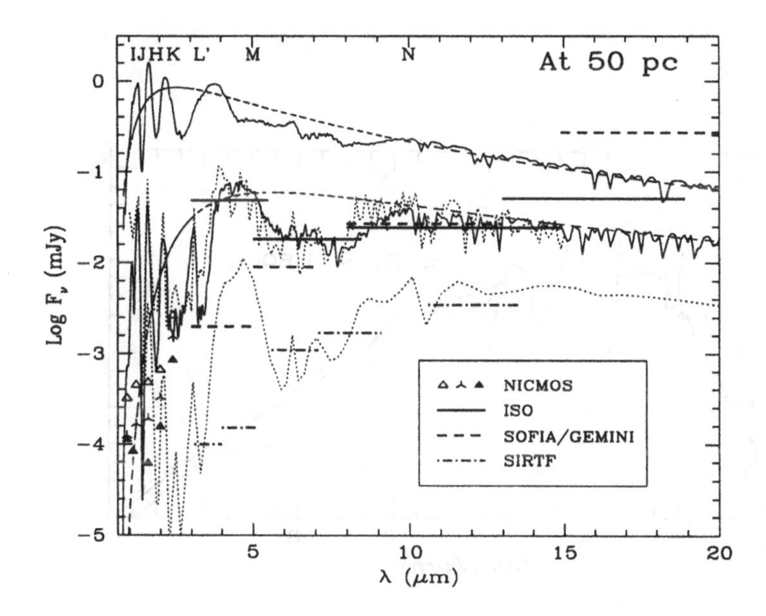

Figure 3. Predicted absolute fluxes of BD or EGP at 50 pc compared to the sensitivity of ground and space-based platforms. Models of both Allard et al. (1996) (solid) and Marley et al. (1996) (dotted) are shown which simulate (i) a brown dwarf near the hydrogen burning limit (topmost spectrum $T_{eff} = 2000K$), (ii) an evolved brown dwarf similar to GL 229B (central spectra: $T_{eff} = 900K$ and 960K), and (iii) an extremely cool object (lowermost spectrum: $T_{eff} = 500K$). Dashed curves give the corresponding black-body.

Gl 229B (Nakajima et al. 1996). The synthetic spectra calculated by different groups for an effective temperature of $\sim 1000K$ yield to a remarkable agreement with the observed spectrum, reproducing the main molecular absorption features (Allard et al. 1996, Tsuji et al. 1996b, Marley et al. 1996).

The predicted absolute fluxes of BD or EGP (Allard et al. 1997) are illustrated in Fig. 3 and compared to the sensitivity of ground and space-based observing platforms, as given by Saumon et al. (1996). The predictions are shown for different effective temperatures and compared to a black body emission at the same T_{eff}. Strong departure from the black body emission illustrates the necessity to derive synthetic spectra for this type of objects. As shown in Fig. 3, the flux peaks in the 4.5 - 5 μm window, which seems to be the best region for the search of cool BD or EGP.

Despite the success of modelling GL 229B, the remaining challenge for atmosphere modellers in the low temperature regime is the grain formation. Although the effects of grain absorption appear more subtil in currently available IR spectra of Gl229B (cf. Tsuji et al. 1996b; Allard et al.

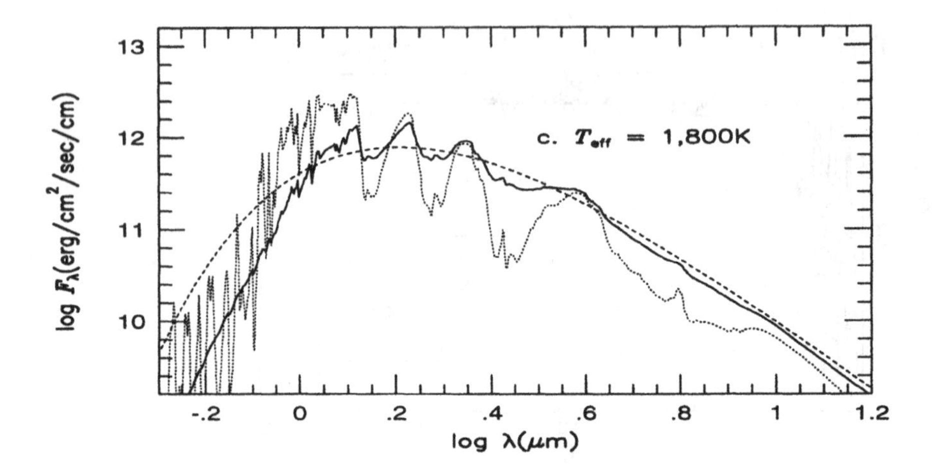

Figure 4. Predicted spectral energy distribution for dusty (solid line) and dust-free (dotted line) models with T_{eff} = 1800K (from Tsuji et al. 1996a). The black body curve corresponds to the dashed line.

1996, Marley et al. 1996), the inclusion of this effect seems necessary to explain the IR spectra of very late-type M-dwarfs. As shown by Tsuji et al. (1996a) and illustrated in Figure 4, kindly provided by T. Tsuji, dust opacities tend to reduce the molecular absorption feature and could explain the overestimation of water absorption found in dust-free model atmospheres.

4. Conclusion

The last few years have shown impressive improvement in the modelization of cool model atmospheres. The main efforts in both theoretical and observational directions lead to a much better understanding of the bottom of the Main Sequence and the substellar regime. Remarkable agreements between theory and observations are now reached on different fronts: color - magnitude diagram, mass - luminosity relationships, spectrum. The success of the theory is most encouraging and shows that it has finally come to maturity.

References

Allard, F., and Hauschildt, P. H. (1995): ApJ, 445, 433
Allard, F., Hauschildt, P.H., Baraffe, I., Chabrier, G. (1996): ApJl, 465, L123
Allard, F., and Hauschildt, P. H. (1997): in preparation
Allard, F., and Alexander D. R. (1997): in preparation
Allard, F., Hauschildt, P.H., Alexander D. R., Starrfield, S. (1997): Ann. Rev. A&A, 35, 137
Baraffe, I., Chabrier, G., Allard, F., Hauschildt P. (1995): ApJl, 446, L35
Baraffe, I., Chabrier, G., Allard, F., Hauschildt P. (1997): A&A, in press
Brett, J.M. (1995): A&A, 295, 736
Burrows. A. et al. (1996): Nucl. Phys. B, 51B, 76
Chabrier, G., Baraffe, I., Plez, B. (1996): ApJ, 459, L91
Chabrier, G., Baraffe, I. (1997): A&A, in press
Dahn, C.C., Liebert, J., Harris, H.C., Guetter, H.H. (1995): *The bottom of the main-sequence and below*, Ed. C. Tinney
Jorgensen, U. G.. (1994): ApJ, 284, 179
Henry , T.D., and McCarthy, D.W.Jr. (1993): AJ, 106, 773
Leggett. S.K., et al. (1996): ApJ, 104, 117
Marley et al. (1996): Science, 272, 1996
Mayor, M., Queloz, D. (1996): Nature, 378, 355
Miller, S., Tennyson, J., Jones, H.R.A, Longmore, A.J. (1994): in *Molecules in the Stellar Environment*, ed. U.G Jorgensen, Lecture Notes in Physics.
Monet D.G. et al. (1992): Astron. J., 103, 638
Nakajima et al. (1995): Nature, 378, 463
Plez, B., Brett, J.M., Nordlund, A. (1992): A&A, 256, 551
Saumon, D., Bergeron, P., Lunine, L.I., Hubbard, W.B., Burrows, A. (1994): ApJ, 424, 333
Saumon, D. et al. (1996): ApJ. 460, 993
Tsuji, T., Ohnaka, K., Aoki, W. (1996a): A&A, 305, L1
Tsuji, T., Ohnaka, K., Aoki, W., Nakajima. T. (1996b): A&A, 308, L29

DISCUSSION

BENGT GUSTAFSSON: Isn't the dilemma with ascribing the apparent saturation of water vapour bands in late M dwarfs that if you ascribe it to dust you have to explain why there are no similar problems for the brown dwarf Gl 229B. Or, is there no visible dust there? If so, why not?

ISABELLE BARAFFE: The spectrum of Gl 229B shows no evidence for the presence of dust. Tsuji has shown that dusty models cannot match the observed spectra. Therefore it seems that grains have settled down below the photosphere and their presence cannot be detected. Another possibility is that clouds formed. The question is open.

ANDRE MAEDER: About the inclusion of dust in the atmosphere models, do you mean including dust in the equation of state, in thermodynamics and opacities?

ISABELLE BARAFFE: Yes.

PARAMETER IDENTIFICATION – A NEW WAY OF ASSESSING ERRORS IN STELLAR ATMOSPHERE ANALYSIS

R. WEHRSE AND PH. ROSENAU

Institut f. Theoretische Astrophysik,
Tiergartenstraße 15, D 69121 Heidelberg

Interdisziplinäres Zentrum f. Wissenschaftl. Rechnen,
Im Neuenheimer Feld 368, D 69120 Heidelberg

Abstract. For the consistent and objective determination of stellar parameters and corresponding errors a constrained minimization approach is described which allows to take additional non-spectroscopic information into account and assures that the atmospheric structure equations are always fulfilled.

1. Introduction

Parameter identification and estimation is a new and fast growing field in numerical mathematics. Its results are applied in the modelling and optimization of complex systems. It has been applied very successfully e.g. in biology, robotics and chemical kinetics.

We want to adapt this approach to the analysis of stellar spectra since error estimates for stellar parameters are urgently needed e.g. for the interpretation of HR and color-color diagrams or for the assessment of metallicity-age relations and of metallicity gradients in the Galaxy. Unfortunately, up to now error estimates for the effective temperatures, the gravities and in particular the element abundances appear to be quite inaccurate and inconsistent (and in many cases even rather subjective) since the various quite intricate dependencies of the emergent flux on the basic parameters have not been fully taken into account (cf. Wehrse, 1990).

T.R. Bedding et al. (eds.),
***Fundamental Stellar Properties: The Interaction between Observation and Theory*, 235–238.**

2. Available information and assumptions

From observations we obtain the flux $F^{obs}(\lambda_i)$ at n wavelengths λ_i and the variances $\sigma(\lambda_i)$. We assume that the errors in the fluxes are normally distributed and statistically independent. In many cases, there is the additional information (e.g. from spectral classification, interferometry, or astrometry) that the gravity has to be within the range

$$g^l \le g \le g^u \tag{1}$$

and that the effective temperature is bracketed

$$T^l_{eff} \le T_{eff} \le T^u_{eff}. \tag{2}$$

In essentially all cases it is required that the abundances obey the inequalities

$$0 \le \epsilon_m \le \epsilon_{He} \le \epsilon_H, \; \epsilon^l_m \le \epsilon_m \le \epsilon^u_m. \tag{3}$$

On the other hand, one has to assume a model which can be considered to be true. For illustration, we discuss here the most simple case and assume that the star for which the spectral analysis is carried out can be described as a spherical gas mass in hydrostatic equilibrium and that the stellar atmosphere is the geometrically thin outer layer in radiative and local thermodynamical equilibrium so that it can fully be described by the parameter vector $\mathcal{P} = (T_{eff}, g, \epsilon_i, \ldots)$.

The resulting set of equations is given by the radiative transfer equation

$$\mu\frac{d}{dz}I(\lambda,z,\mu,\mathcal{P}) = -\chi(\lambda,z,\mu,\mathcal{P})(I(\lambda,z,\mu,\mathcal{P}) - S(\lambda,z,\mathcal{P})), \tag{4}$$

the definition of the model flux

$$F^{mod}(\lambda,z,\mathcal{P}) = 2\int_{-1}^{1}\mu I(\lambda,z,\mu,\mathcal{P})d\mu, \tag{5}$$

the energy equation

$$\frac{d}{dz}\int_0^\infty F^{mod}(\lambda,z,\mathcal{P})d\lambda = 0, \tag{6}$$

the hydrostatic equation

$$\frac{d}{dz}P(z,\mathcal{P}) = -g\rho(z,\mathcal{P}), \tag{7}$$

and the equation of state as well as expressions for the absorption and scattering coefficients.

3. Parameter and error estimation

We consider a parameter vector $\mathcal{P}^0$ to contain *the parameters of the star* if it minimizes the *objective function*

$$Z(\mathcal{P}) = \sum_{i=1}^{n} \left(\frac{F^{obs}(\lambda_i) - F^{mod}(\lambda_i, \mathcal{P})}{\sigma(\lambda_i)} \right)^2 \tag{8}$$

and if simultaneously the expressions (1) – (7) are fulfilled as constraints; i.e. we require the best possible agreement between the observed and calculated spectra, the *exact fulfilment* of the equations for the structure of the atmosphere and the spectrum formation, and the obedience of the parameters to the limits provided by the non-spectroscopic information. In mathematical terms, the parameters are obtained by a minimization constrained by a boundary value problem.

The accuracy of the resulting parameters can now be estimated from the *error ellipsoid* that contains the true solution with a probability $1 - \alpha$ (e.g. 95 %). It can be shown that for our set of assumptions the ellipsoid is approximated by the point set

$$G_L(\alpha) \;=\; \{\, \mathcal{P}^0 + \delta\mathcal{P} \;\big|\; J_c \delta\mathcal{P} = 0,\; \|J_1 \delta\mathcal{P}\|_2^2 \leq \gamma(\alpha) \,\} \tag{9}$$

with J_c and J_1 being the Jacobi matrices of the constraints and the objective function, resp., and $\gamma(\alpha)$ being essentially the quantile of the Fisher distribution. $\|...\|_2^2$ denotes the square of the Euclidean norm.

The basic idea for the approach described here dates back to C.F. Gauss. Although minimization problems with algebraic equations as constraints have already been solved by J.L. Lagrange, it was only learnt more than a century later how to treat inequalities (Karush, 1939; Kuhn and Tucker, 1951). Bock (1981) proposed an algorithm for the boundary value problem constrained case where the solution of the boundary value problem and the minimization are performed simultaneously. Our approach is based on that of Bock and involves a multiple shooting discretisation of the boundary value problems (e.g. for eqs. (4) and (7)). The resulting (in-)equality constrained minimization problem is then solved by a generalized Gauss-Newton iteration with line search, where the Jacobi matrices of the constraints are computed by internal numerical differentiation (Bock, 1985).

4. Discussion

It is evident that our approach involves the problems that it is hardly possible to estimate the errors resulting from uncertainties in the model and that it is not clear to what extent the assumptions on the flux errors are justified (empirical checks seem to be difficult). These problems are in fact inherent in all approaches for parameter determinations. In addition, compared to unconstrained minimization schemes (cf. Press et al., 1992) our method is quite complex and therefore the implementation is rather costly (but it has to be done only once). We think, however, that the problems are by far outweighed by the advantages: (i) the numerics is highly efficient. It is expected that – independent of the number of free parameters n – typically 10 to 20 iterations are needed to achieve a relative accuracy of $\approx 10^{-3}$; a corresponding determination on the basis of a model grid and unconstrained minimization would in most cases require the calculation of 10^n atmosphere models; (ii) more importantly, the parameter determination is objective, i.e. non-spectroscopic information and empirical knowledge can be introduced consciously but are not creeping in implicitly (and perhaps unintentionally); (iii) a fully consistent error ellipsoid is obtained without much additional cost; (iv) parameters are indicated that cannot or can hardly be determined simultaneously; (v) since all Jacobi matrices are available this approach can provide help for future observations as e.g. it can show why some colors are good indicators of a parameter and others are not; (vi) although the algorithm is not guaranteed to converge to the global minimum, experience has shown that in most cases it does so. The importance of this aspect can be seen e.g. from the discussion of the nature of the carbon star G77-61, cf. Gass et al. (1988); (vii) the algorithm is flexible, e.g. one of the first applications will be the empirical determination of the temperature structure of M dwarfs from ISO spectra; and (viii) last but not least the approach has a sound mathematical basis.

At the time of writing the project is still in progress but it is expected that numbers for actual stars will be available in near future.

Acknowledgements

This work has been supported by the Deutsche Agentur f. Raumfahrtangelegenheiten DARA (project 50 OR 96117).

References

Bock, H.-G. 1981, in: Modelling of Chemical Reaction Systems (Ebert, Deuflhard, Jäger, eds.), Springer Series Chemical Physics **18**, 102

Bock, H.-G. 1985, PhD. thesis, Bonn University

Gass, H., Liebert, J., Wehrse, R. 1988, AA **189**, 194

Karush, W. 1939, Master's Thesis, Department of Mathematics, University of Chicago

Kuhn, H. W., Tucker, A. W. 1951, in: Proc. of the Second Berkeley Symposion on Mathematical Statistics and Probability, J. Neyman (ed.), University of California Press

Press, W.H., et al. 1992, Numerical Recipes, Cambridge Univ. Press

Wehrse, R. (ed.) 1990, Accuracy of Element Abundances from Stellar Atmospheres, Lecture Notes in Physics **359**, Springer Verlag

ATMOSPHERIC MOTIONS AND GRANULATION IN COOL STARS - OBSERVATIONS AND THEORY

P.L. COTTRELL
Mount John University Observatory
Dept of Physics & Astronomy
University of Canterbury, Christchurch, New Zealand

Abstract.
This review presents a number of aspects in our quest for an understanding of the detailed structure of cool star stellar atmospheres through the study of atmospheric motion. It includes the observational tools (e.g. direct imaging, detailed analyses of line profiles) used to determine the type and amount of motion, as well as the modelling techniques (e.g. simulated images, velocity and pressure fields) used to interpret and describe these motions.

Linkages will be made through the solar-stellar connection to provide insight into the structure of stars where detailed imaging is not possible, and to the determination of the fundamental parameters of stars, such as effective temperature, surface gravity and elemental abundances.

1. Introduction

As telescopes have been made larger and larger and computers have become more powerful, astronomers and astrophysicists have endeavoured to understand the finer details in the structure of the outer layers of stars through a variety of techniques. So whether one is undertaking a high resolution imaging or spectroscopic observation, or a large scale hydro-dynamical calculation of the motions of a star, one is doing science which is just as fundamental as that which seeks answers to the questions about the large scale structure of the universe. Indeed, as we all know only too well, the basis for our determination of a more precise size of the universe depends critically upon the study of the structure of stars, in particular their mass, size, temperature and energy output.

T.R. Bedding et al. (eds.),
***Fundamental Stellar Properties: The Interaction between Observation and Theory*, 239–244.**

In the context of this review I will consider only non-oscillatory atmospheric motion. This is thus referring to the motions due to granulation in the stellar surface layers at the top of the convection zone. The term *granulation(s)*, or *granules*, is as originally applied by Dawes (1864), where he was endeavouring to introduce a more applicable term for the observations that he made of the solar surface. Older terms included *willow-leaves* and *rice-grains*, but Dawes was seeking a description which 'assumes nothing as to exact form or precise character; and (he) venture(d) to hope that the term will be generally adopted.' A reasonable legacy for the Reverend Dawes.

Granules are considered to be the rising gas in the surface layers and have a characteristic dimension of Mm (millions of metres). They are ~20% brighter than the material which is moving back down into the star's atmosphere and the timescale for them to rise and disperse is a few minutes. The term 'macro-turbulence' is a term which incorporates granulation and other processes whereby a spectral line is Doppler broadened. A simple visualization of granulation could be considered as a two stream model, one moving upwards and hot and the other downwards and cool, where there is some distribution of rising and falling velocities, which may not be the same, and the granulation velocity goes to zero at the limb of the star.

There have been some links that have been made (Stein & Nordlund 1991) between the convective motions associated with granulation and the periodic oscillations, but that is beyond the scope of the present review. This review will concentrate on the specific observational and modelling methods that have been and can be developed to obtain a more fundamental understanding of the structure of these stars. This provides input for both stellar interior and stellar atmosphere codes which are needed to deduce the fundamental parameters (radius, temperature, internal structure) of stars.

2. Observational Tools and Results

The most obvious technique is to make images of the 'surface' layers. To date, this has only been accomplished for the sun and we await new techniques which will enable other stars to be imaged in a similar way. Nonetheless, it is important that these solar images are combined with a review of this type as only in this way can we provide insights into the type of effects that we are endeavouring to interpret.

There have been many centuries of observations of the solar surface, predating the work of Dawes mentioned above. Many of these earlier works provided an incomplete picture as they were affected by the disturbing motion of the earth's atmosphere. However, with the advent of solar telescopes at sites with stable and excellent seeing (e.g. the Swedish Vacuum

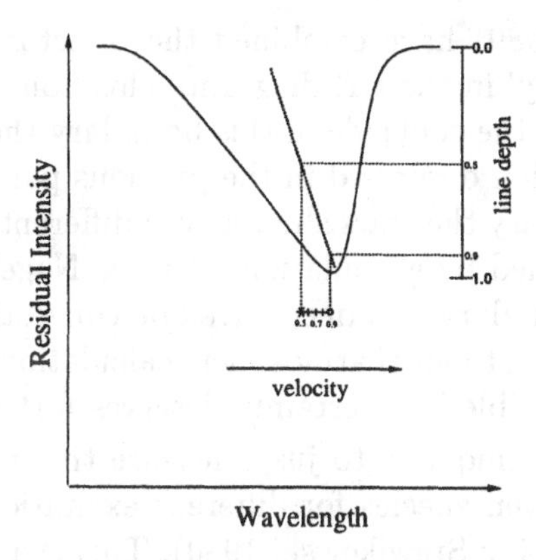

Figure 1. Schematic line bisector for a stellar spectral line. The labeling corresponds to depths in the line profile, where 0.0 is at the continuum and 1.0 is the core of the line (from Wallerstein et al. 1992).

Solar Telescope on La Palma, see Title & Berger 1996) and the use of (initially) photographic techniques, excellent imaging of the solar disk has shown accomplished and the two major granulation components have been clearly identified. These are the hot rising granules with the cooler falling regions at the margins of these granules.

Given that there is motion in these surface layers spectroscopic techniques can be employed to analyse this motion. Examples of spectroscopic observations of the sun can be found in Collados et al. (1996). To understand what these observations mean requires good instrument calibration, followed by innovative analysis techniques. David Gray provides examples of these techniques (see Gray 1988, particularly Lecture 4).

The tool that is the most useful in this work is to examine the line bisector (see Fig. 1) of a range of spectral lines. For an individual spectral line there is a contribution to the profile from a wide region in the star's atmosphere and when combined with lines of a range of elements, with different excitation and ionisation potential, one has a probe of the outer structure of a star's atmosphere. This includes the temperature and pressure structure and through stellar model atmosphere techniques the effective temperature and surface gravity of the star. The characteristic shape derived from the solar (and some other main sequence stars) observations is a 'C-shape' (see Gray 1988, Fig. 4-9), which includes a blue-shifted and larger contribution from the brighter rising granule and a red-shifted smaller component from the less bright falling intergranular region (Gray 1988, Fig. 4-1). In most other main sequence, and all giant stars, this characteristic shape is a truncated 'C', where the lower portion of the 'C' is no longer observed (Gray 1988, Fig. 4-25).

Gray & Nagel (1989) have combined the bisector data to delineate a 'granulation boundary' in the HR diagram. This boundary runs from F0V to G0Ib. For stars on the cool side of this boundary the line bisectors show the effect of granulation described in the previous paragraph, while on the hot side of the boundary the stars show a very different set of line bisectors, which are not explained by granulation. Gray & Nagel hypothesised some link between this boundary and deep envelope convective. From discussion with respect to the latest model atmosphere calculations at this Symposium (see Kurucz) this possible link certainly deserves further investigation.

An alternative technique is to just measure the 'mean line shift' for a range of lines of a given species for different excitation potential and line strength (see Glebocki & Stawikowski 1980). This is a single parameter fit, which is more applicable to situations where the acquisition of detailed line profiles is not possible.

3. Modelling Techniques

Gray (1988) used a simple two-stream model to compare with his bisector measurements and although he admits to it being unphysical the model provided insights into the nature of the rising and falling elements in the atmospheres of the sun and stars.

The important hydrodynamic modelling of the surface of the sun has been undertaken by numerous researchers (see Spruit et al. 1990). It relates to both the imaging of the surface layers, with the clear delineation of hot and cool regions, and the motion of these regions. Considerable progress has been made in this area through these models which, although only over relatively small *boxes*, clearly show the motions which are the cause of the *granules*.

In the following paragraphs, recent 2D and 3D hydrodynamical calculations of the granulation will be described.

Gadun & Vorob'yov (1995) have developed 2D hydrodynamic, time dependent models. These use a system of equations for a compressible, radiatively coupled, gravitationally stratified non-magnetic medium the equation of radiative energy transfer using gray Rosseland opacities. Their results cover a 2,000 km (~900 km is the atmosphere) by 4,000 km region with parameters calculated at 35 km spacings and at 0.3 s time steps. Their results with respect to the velocity and temperature profiles (see in particular their Fig. 1) show that they can reproduce the appropriate scales for the variation in distance, time and temperature. For example, granules with a linear dimension of 1,000 to 2,000 km and several thousand degree contrast in temperature changing over a few minutes seems typical of their calculations.

Nordlund & Stein (1991) and Stein & Nordlund (1991) have produced a series of larger scale, more realistic 3D hydrodynamical calculations. Their results have provided some examples of degraded resolution images which show good similarity to the types of structures observed in solar images (see Fig. 1 of Nordlund & Stein 1991). Their detailed calculations also provide excellent velocity field and temperature information (see Fig. 3 of Stein & Nordlund 1991). The velocity field simulations can be compared directly with the solar spectral observations. Because of the 3D work they are also able to give good qualitative agreement between the surface structure and deeper layers (see Fig. 1 of Stein & Nordlund 1991) and hence a better understanding of the overall structure of the outer layers of stellar envelopes.

These models are obviously only the first steps in a better understanding of the solar atmosphere, which can then be used to map structures in other stars where the imaging data are currently non-existent and the velocity data has much poorer spatial, and generally spectral and time resolution.

4. Prospects

The above discussion has highlighted some of the 'tools of the trade' that have been used to understand the detailed structure of stars. There are almost as many questions raised as answered in the observational field as the acquisition of high quality observations is a difficult process and requires very careful calibration of the instrumental system, as well as careful analysis techniques. In the modelling regime, there is a need for more extensive stellar envelope and atmosphere 3D hydrodynamical calculations to compare with the observational data.

However, in spite of the 'well-worn' phrase, "Further observations and models are required", there have been some great successes.

There has been excellent progress in the acquisition of ultra-high spatial and spectral resolution images, as well as polarimetry, of the sun's surface layers. One question raised here is: "What is the smallest scale structure that can be observed and can it be reproduced in the numerical models?"

With the use of very-high spectral resolution of cool stars, what implications are there to make connections with our current understanding of the solar situation? And with temporal coverage and links with other stellar parameters a more complete picture of the interrelation between the different observations (e.g., magnetic effects, luminosity variations - see Gray et al. 1996).

Considerable progress is being made in our understanding of the effects of motions on the determination of the temperature structure of cool stars, the effective surface gravity of these stars and the effect that atmospheric motion has on the determination of the abundances derived for these cool stars. Under- and over-excitation and ionisation effects will also be better understood with a more complete picture of the structure of the outer layers of these atmospheres.

The next generation of observations and models will surely provide some further progress in this area and enable astronomers to make significant advances in this area of stellar astrophysics.

References

Collados, M., Rodriguez Hidalgo, I., Ballesteros, E., Ruiz Cobo, B., Sanchez Almeida, J. & del Toro Iniesta, J.C. 1996 *A&A Supp.*, **115**, 367-77.

Dawes, W.R. (1964) *MNRAS*, **24**, 161.

Gadun, A.S. & Vorob'yov, Yu.Yu. (1995) *Solar Physics*, **159**, 45-51.

Glebocki, R. & Stawikowski, A. (1980) in *Stellar Turbulence*, (eds D.F. Gray & J.L. Linsky) IAU Coll. **51**, p55.

Gray, D.F. (1988) in *Lectures on spectral-line analysis: F, G and K stars*, Lecture 4 - Stellar Granulation.

Gray, D.F. & Nagel, T. (1989) *ApJ*, **341**, 421-6.

Gray, D.F., Baliunas, S.L., Lockwood, G.W. & Skiff, B.A. (1996) *ApJ*, **465**, 945-50.

Nordlund, Å. & Stein, R.F. (1991) in 'Challenges to theories of the structure of moderate-mass stars' (eds D. Gough & J Toomre) *Lecture Notes in Physics*, **388**, 141.

Paxman, R.G., Seldin, J.H., Lofdahl, M.G., Scharmer, G.B. & Keller, C.U. (1996) *ApJ*, **466**, 1087-99.

Spruit, H.C., Nordlund, Å. & Title, A.M. (1990) *Ann. Rev. Astron. Astrophys.*, **28**, 263.

Stein, R.F. & Nordlund, Å(1991) in 'Challenges to theories of the structure of moderate-mass stars' (eds D. Gough & J Toomre) *Lecture Notes in Physics*, **388**, 195.

Title, A.M. & Berger, T.E. (1996) *ApJ*, **463**, 797-807.

Wallerstein, G., Jacobsen, T.S., Cottrell, P.L., Clark, M. & Albrow, M.D. (1992) *MNRAS*, **259**, 474-88.

DISCUSSION

O'Mara: In your work have you looked at the hill vortex model developed by Roger Ulrich?

Cottrell: No. Rather than looking at individual models for convection, I was seeking full hydrodynamical descriptions of the convection/granulation problem.

Bell: When interpreting very accurate radial velocity observations of Cepheids, which show different velocities for different elements, should we consider that the Gray and Nagel effect is simply superimposed on the Cepheid oscillation?

Cottrell: I would suggest that one is going to require full hydrodynamical non-LTE models (incorporating the stellar envelope and atmosphere) to interpret the line level effects that are observed in Cepheid stars. Using any form of 'add-on' effect would not be a very satisfactory approach.

CHROMOSPHERES, ACTIVITY AND MAGNETIC FIELDS

YU. N. GNEDIN
The Central Astronomical Observatory at Pulkovo, 196140, St.Petersburg, Russia

Abstract. This brief review presents the current state of observations of stellar activity effects including the fluxes of chromospheric emission lines: CaII H+K, MgII h+k, SiII 1812 Å multiplet, CIV, as well as radio and X-ray fluxes versus B-V colours and luminosity classes, rotation periods, Rossby number and especially versus the mean magnetic flux density $\langle fB \rangle$. Results of stellar magnetic field measurements are presented.

1. Stellar Activity: What Is It?

Stellar activity means a set of phenomena that occur when the basic assumptions of a classical stellar atmosphere are not valid. The last ones are namely as follows:

1. Significant part of energy is transported from a star to space only by radiation and convection. The inevitable conversion of mechanical energy in a stellar core to heat is presumed not to alter significantly the thermal structure of the stellar atmosphere.
2. A good approximation is the atmosphere to be in hydrostatic equilibrium. That means there are no noticeable systematic flows such as winds, shocks, high-speed circulations, etc.
3. Magnetic fields are quite weak. They do not control the flow of matter and energy balance in the stellar atmosphere.

These classical assumptions lead to the situation when the mass, age, and initial chemical composition of a star determine its effective temperature, radius and gravity and hence its position on the HR diagram. The Sun is a very useful prototype of an active star. That is true, its activity is sufficiently weak as compared to other stars of late spectral types. On such stars rapid flares in X-ray, UV and microwave spectral ranges can be $10^3 - 10^5$ times more energetic than solar flares. What are the stellar pa-

T.R. Bedding et al. (eds.),
***Fundamental Stellar Properties: The Interaction between Observation and Theory*, 245–252.**

rameters that control the activity? It is well known that the surface fluxes of chromospheric emission lines (such as CaII H+K, MgII h+k, the SiII 1812 Å multiplet and CIV emission line) as well as X-ray and radio fluxes as a function of rotation period, B-V colour and luminosity class are indicators of stellar activity. These surface fluxes of active stars can be as much as 200 times larger than the basal stellar flux.

2. Stellar Parameters that Control Activity

Let us start with rotation rate (or rotation period). Stellar rotation decreases with age on the main sequence. It means that age and rotation rate are statistically interchangeable parameters and one requires additional information. Basri (1987) has proved from his investigation of RS CVn systems that rotation rather than age must be the controlling parameter. Fig. 1 shows X-ray luminosity versus stellar equatorial rotation velocity for stars of various types (see Bouvier 1990). The same correlation occurs also for the relationship between radio luminosity and rotation rate (Seaquist 1995). The next important empirical relation for stellar activity is connected with Ca II H+K and MgII h+k chromospheric line fluxes. Bouvier (1990) showed the strong correlation between both chromospheric line fluxes (Fig.2) and weaker but rather good correlation between H and CaII K fluxes. Recent ROSAT and HST observations show a good correlation between X-ray and CIV line fluxes (Ayres et al. 1993). There is correlation between the HST/FOS CIV fluxes and ROSAT/PSPC detections of cluster stars and field F-G dwarfs carried out during the ROSAT/IUE All-Sky Survey. The power law connecting the F9-G2 stars has a slope of 2.0.

3. Physical Parameters that Control Activity

The close connection between stellar rotation and chromospheric emission can be presented in terms of general stellar dynamo models. In the dynamo theory a dimensionless parameter, namely the dynamo number, characterizes the model behavior. The dynamo number NR is essentially proportional to the inverse square of the Rossby number R_0, which is the ratio of the stellar rotation period P to the convective turnover time τ_c: $NR \propto R_0^{-2}; R_0 = P/\tau_c$.

In practice, however, our knowledge of stellar convection is too limited to calculate correct convective turnover times. The reason is that the characteristic length scales, as well as the velocities, are not very well known. Recently Kim & Demarque (1996) have estimated the convective turnover timescales for Sun-like stars in the pre-main sequence and early post-main sequence phases of evolution. These estimates have been based on up-to-date physical input for the stellar models as well as on use of the micro-

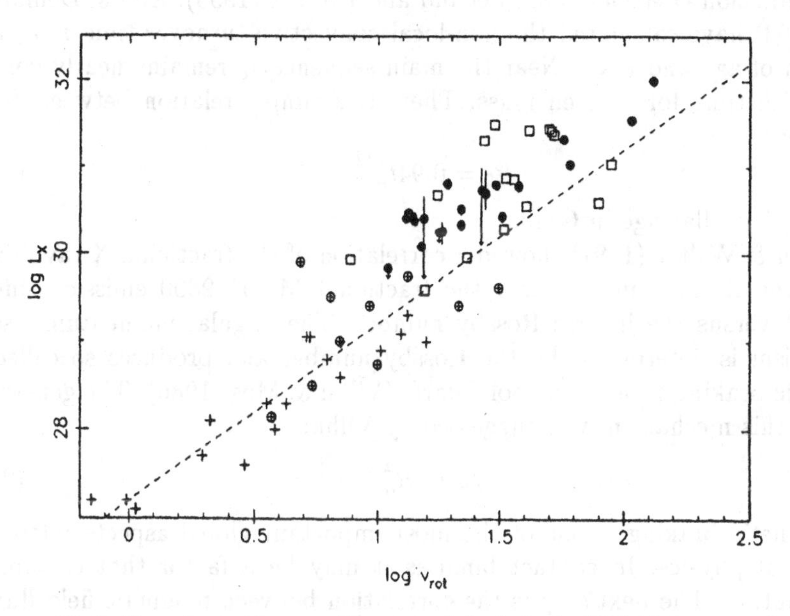

Fig.1. X-ray luminosity vs. stellar equatorial velocity for TTS (•), late-type dwarfs (+), dK_e - dM_e stars (⊕) and RS Cvn systems (□) from Bouvier, 1990.

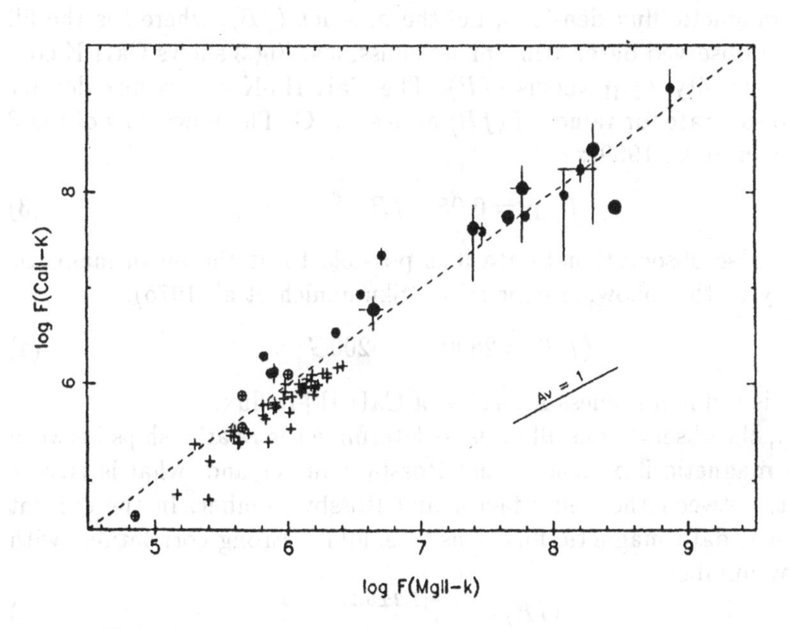

Fig.2. CaII K line flux vs. MgII k line flux from Bouvier, 1990.

scopic diffusion coefficients of Michaud and Proffitt (1993). Kim & Demarque (1993) have calculated the non-local convective turnover time τ_c as a function of age and mass. Near the main sequence τ_c remains nearly constant with time for a given mass. There is a simple relation between R_0 and age:

$$R_0 = 0.94 t_9^{1/2} \quad (1)$$

where t_9 is stellar age in Gyrs.

Vilhu & Walter (1987) showed a correlation of the fractional X-ray flux versus the Rossby number and the fractional Mg II 2800 emission line strength versus the inverse Rossby number. The angular-momentum-loss mechanism is determined by the Rossby number and produces so called magnetic braking process in cool dwarfs (Vilhu & Moss 1986). The general form of this mechanism was suggested by Vilhu:

$$\dot{J}/J \sim R_0^{-2} \quad (2)$$

Magnetic braking is one of the most important global aspects related to coronal physics. In contact binaries it may be a factor that controls the evolution. The next step is the correlation between magnetic field flux and CaII, MgII and X-ray fluxes (Marcy 1984). To study the relationship between magnetic field and stellar activity Saar (1988) has compared photospheric magnetic flux densities, i.e. the product $\langle fB \rangle$, where f is the fill factor, with observed outer atmosphere emissions. Fig.3 shows CaII K core excess flux density I_{CaII} versus $\langle fB \rangle$. The CaII H+K excess flux density appears to saturate for values of $\langle fB \rangle$ above 10^3 G. The power law of Fig.3 fits (Schrijver et al. 1989):

$$I_{CaII} = 0.055 \, \langle fB \rangle^{0.62} \quad (3)$$

Using these observational data it is possible to fit the mean magnetic flux density to the following expression (Skumanich et al. 1975):

$$\langle fB \rangle = 2800 R_C - 200G \quad (4)$$

where R_C is a dimensionless intensity of CaII H+K flux.

Finally, the observations allow us to determine the relationships between the mean magnetic flux density and Rossby number and, what is is very important, between the filling factor and Rossby number. In the current observational data magnetic flux density exhibits strong correlations with the Rossby number:

$$\langle fB \rangle \sim R_0^{-(1.2 \pm 0.1)} \quad (5)$$

and with the stellar rotation (see Haisch et al. 1994, Linsky & Saar 1987, Saar 1989)

$$\langle fB \rangle \sim \Omega^{1.3 \pm 0.1} \quad (6)$$

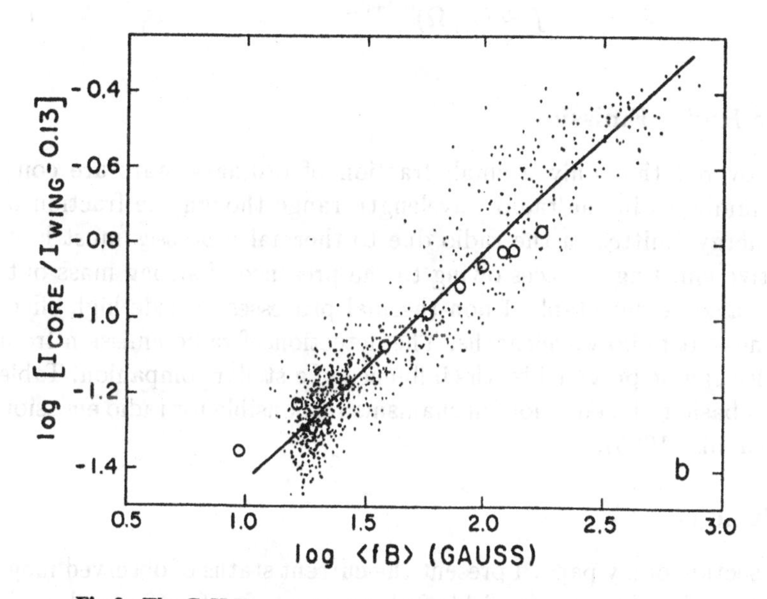

Fig.3. The CaII K core wing intensity ratio vs. the value of the mean magnetic flux density <fB>.

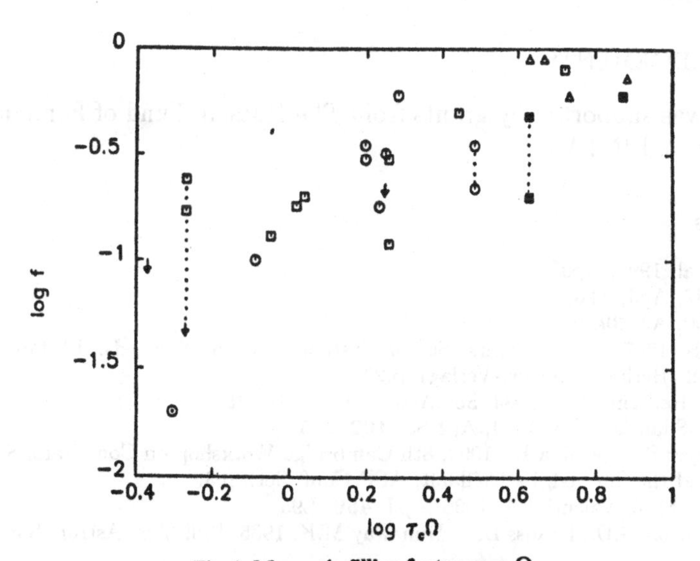

Fig.4. Magnetic filling factor vs. $\tau_c\Omega$.

The magnetic filling factor appears to be the dominant magnetic parameter controlling rotation $\sim$ activity and rotation $\sim$ age relations (Fig.4, see also Haish et al. 1994, Giampapa 1987):

$$f \sim (\tau_c/\Omega)^{0.9\pm0.2} \tag{7}$$

4. Stellar Radio Emission

It was discovered that only a small fraction of ordinary stars are comparatively luminous in the radio wavelength range though the fraction of the total energy emitted in the radio due to thermal processes must have large effective emitting surfaces owing to the presence of strong mass outflows. Stars also be detectable if non-thermal processes provide high effective brightness at radio wavelengths. The detection of radio emission from close binaries can be provided by the influence of a stellar companion. Table 1 represents basic radio emission mechanisms responsible for radio emission of stars (Seaquist 1995).

5. Conclusions

In the last section of my paper I present the current status of observed magnetism in stars of various types. Table 2 shows type of a star (first column), directly measured or indirectly estimated stellar magnetic field magnitude (second column) and radiation mechanism or method of a measurement (third column).

ACKNOWLEDGEMENTS

This paper was supported by grants from The Russian Fund of Fundamental Research and INTAS.

References

Ayres T.R. et al. 1993, ApJS
Basri G.S. 1987, ApJ, **316**, 377
Bouvier J. 1990, AJ, **99**, 946
Giampapa M.S. 1987, in Cool Stars, Stellar Systems and the Sun, eds. J.L.Linsky & R.E.Stencel (Berlin: Springer -Verlag), p.236
Gnedin Yu.N., Red'kina N.P. 1984, Sov.Astron. Lett., **10**, 255
Gnedin Yu.N., Silant'ev N.A. 1984, Ap&SS, **102**, 375
Haish B., Bowyer S. & Malina R., 1994, 8th Cambridge Workshop on Cool Stars, Stellar Systems and the Sun, ed. by Caillault, ASP Conf. Ser., p.3.
Johns-Krull Ch.M. & Valenti J.A. 1996, ApJ, **450**, L95
Kemp J.C., Henson G.D., Krauss D.J., Dunaway M.K. 1985, Bull. Am. Astron. Soc., **19**, 752
Kim Y-C & Demarque P. 1996, ApJ, **457**, 340

TABLE 1. Radio Emission Mechanisms

Mechanisms	Source size	T_B	Circul. polar.	Time Var.	Stars where observed
Thermal bremsstrahlung	large $R \gg R_s$	low $\sim 10^4$	low (~ 0)	low (years)	Sun, OB stars, K, M giants
Gyroresonance emission	large $R > R_s$	10^7	low	low	Sun, AM Her quiescent comp αM_e
Gyrosynchrotron or synchrotron	moderate $R \leq R_s$	$10^8 - 10^{10}$	mod. $< 30\%$	mod. (min,hr)	Sun, αM_e, RS Cvn, OB, B_P, A_P(CP)
Cyclotron maser	small $R \ll R_s$	high 10^{20}	$\sim 100\%$	high	Sun, αM_E flare AM Her-outburst
Plasma radiation+plasma-maser instability	small $R \ll R_s$	10^{17}	$(10 \div 90)\%$	high	Sun, αM_E fl, AM Her-outburst

Linsky J.L. & Saar S.H. 1987, in Cool Stars, Stellar Systems and the Sun, eds. J.L.Linsky and R.E.Stencel (Berlin: Springer -Verlag), p.44

Maheswarand M. & Cassinelli J.P. 1988, ApJ, **335**, 931

Marcy G.M. 1984, ApJ, **276**, 286

Michaud G. & Proffitt C.R. 1993, in Proc. IAU Colloq. 137, Inside the Stars, eds. A.Baglin and W.W.Weiss (San Francisco: ASP), 246

Nadeau R., Bastien P. 1986, ApJ, **305**, L5

Saar S.H. 1987, in Cool Stars, Stellar Systems and the Sun, eds. J.L.Linsky and R.E.Stencel (Berlin: Springer -Verlag), p.10

Saar S.H. 1988, ApJ, **276**, 286

Schrijver C.J., Cote J., Zwaan C., Saar S.H. 1988, ApJ., **337**, 964

Seaquist E.R. 1995, Rep.Progr.Phys., **3**

Skumanich A., Smythe G., Frazier E.W. 1975, ApJ., **200**, 747

Underhill A.B. 1984, ApJ, **276**, 583

Vilhu O. & Moss D. 1986, AJ, **92**, 1178

Vilhu O. & Walter F.M. 1987, ApJ, **321**, 958

White R.L. 1985, ApJ, **289**, 698

TABLE 2.

Type of Star	Magnetic Field Magnitude	Radiation Mechanism or Measurement Method
Neutron Stars (Radio pulsars)	$(10^{12} \div 10^{13})$G for Her X-1 $5 \times 10^{12} G (\sim 10^{12} G)$	Cyclotron Lines in X-ray (Radio-emission Pattern) $L \sim B_s^2 R_s^6 \Omega^4$
White Dwarfs	$(10^6 \div 10^8)G$	circular Polarimetry Zeeman Splitting in a Strong Magnetic Field
Magnetic Ap Stars	$(10^3 \div 10^4)G$	Zeeman Splitting
RS Canum Venaticorum Expanded Halo	$\sim 10^3 G$ $\sim 200G$	High Resolution Spectrum Circular Polarimetry (Kemp et al, 1985) Thermal Gyrosynchrotron
Flare Stars	$\geq 10^3 G$ for AD Leonis (dM3.5e)$3.5 \times 10^3 G$	High Resolution Infrared Spectroscopy (Johns-Krull, Valenti, 96)
WR Stars Nonthermal Radio-emission W-R ~1/6 of thermal ones or 1/3 OB stars	$\sim 1500G$ $\sim 100G$ $\sim 10G$	Magnetic Stellar Wind (Maheswaran and Casinelly, 1988): $B_r(R) \leq M^{1/2} U^{3/2} / \Omega R^2)$ Gyroresonance Radiation from surface of hot stars (Underhill, 1994) Shock Acceleration (White, 1985)
T Tau	$(10^3 \div 10^4)G$	Circular Polarimetry (Nadeu and Bastien, 1986) Linear Polarization with Faraday Rotation (Gnedin and Red'kina, 1984)
Be Stars	$(10 \div 100)G$	Magnetic Stellar Wind (Maheswaran & Cassinelly, 1988) Linear Polarization with Faraday Rotation Gnedin and Silan'tev, 1984)
Early-Type Stars ($\xi Per, \lambda Cep$)	$(70 \div 200)G$	Magnetic Stellar Wind
Cool Dwarfs, G Dwarfs	$(20 \div 300)G$	Equipation CaIIH+K Fluxes

PULSATING STELLAR ATMOSPHERES

DIMITAR D. SASSELOV
Harvard-Smithsonian Center for Astrophysics, Cambridge, MA 02138, USA

Abstract. We review the basic concepts, present state of theoretical models, and the future prospects for theory and observations of pulsating stellar atmospheres. Our emphasis is on radially pulsating cool stars, which dynamic atmospheres provide a general example for the differences with standard static model atmospheres.

1. Introduction

Most stars in the upper part of the Hertzsprung-Russell diagram pulsate. This pulsation involves large-scale cyclic motion of their envelopes, very often in a simple radial mode. Most often the stellar atmosphere has little effect on the driving and acoustics of the pulsation, which is in essence an envelope phenomenon. However, apart from providing the upper boundary condition, the atmosphere is what we observe. Except for the period of pulsation, all other stellar parameters we derive are affected by the ever changing state of the atmosphere.

There is a very large range in pulsation amplitudes – from the pervasive, but very calm, solar-type oscillations to radial pulsation of up to a fifth of the stellar radius in RR Lyrae stars. A large number of giants and supergiants exhibit cyclic variations of low amplitude which are most likely due to pulsation (Rao et al. 1993). Even stars which are often considered to be non-variable standards, like γ Cyg and α Per, show such variability when subjected to scrutiny (Butler 1997). There are specific classes of pulsating stars of great interest to distance measurement and stellar evolution, e.g. Cepheids, RR Lyrae, Miras, and post-AGB stars, where understanding their pulsating atmospheres is often crucial to the applications.

IAU Symposium 189, Sydney, 1997.

T.R. Bedding et al. (eds.),
Fundamental Stellar Properties: The Interaction between Observation and Theory, 253–260.

2. Concepts

In the talks and discussions of this session, the physics of stellar atmospheres was discussed explicitly or implicitly in the framework of some basic approximations. Hydrostatic equilibrium [HSE] is one of them, and a very well justified one at that. Here the basic physics of pulsating atmospheres can be illustrated with the help of juxtapositions and differences.

2.1. HSE VS. NON-HSE

The atmosphere of a pulsating star is not in HSE by definition. However, there is no general and sharply defined state at which HSE breaks down as a viable approximation. One obvious reason for this is that the amplitude of a global radial or nonradial oscillation is not the only parameter involved in reshaping the atmosphere – the period (i.e. velocity gradients) and the state of the atmosphere are equally important. The latter will determine the acoustic and hydrodynamic properties of the atmosphere and its response to the oscillation at its inner boundary.

Pulsation (and departure from HSE) leads to **extension** of the stellar atmosphere. This change in atmospheric structure is obvious, though not trivial (e.g. Bowen 1988; Cuntz 1989). Extension alone affects spectral line formation, invalidates the plane-parallel approximation, and facilitates dust formation and mass loss, as we describe below.

The development of extension in an atmosphere undergoing radial pulsation is shown in Figure 1. The model is for a bright giant or supergiant of $T_{\rm eff} \approx 5500K$. The pulsation period (75 days) and amplitudes (0.5 to 2 $km\ s^{-1}$) match those observed in stars like γ Cyg and α Per. The model is a 1-D radiation hydrodynamics calculation with H, He, and CaII in non-LTE (using code HERMES by Sasselov & Raga 1992). Two effects are illustrated: (1) the general extension of the atmosphere, and (2) the strong shock waves forming in the upper atmosphere when velocity gradients are too weak to induce resonant line cooling in the higher-density photosphere. Shock dynamics is discussed in the next section.

Atmospheric extension affects the formation of **photospheric lines**. With the accompanying changes in temperature and pressure structure, extension broadens and complicates the line forming region (Figure 2). This does not always cause departures from local thermodynamic equilibrium [LTE] – the transition in Figure 2 (taken from the Cepheid model of Sabbey *etal.* 1995) is formed very close to LTE in both the HSE and non-HSE versons of the same model. However, the resulting spectral line becomes more sensitive to depth-dependent perturbations.

The **plane-parallel** approximation fails when atmospheric extension exceeds ≈5% of the stellar radius (Schmid-Burgk & Scholz 1975). The equa-

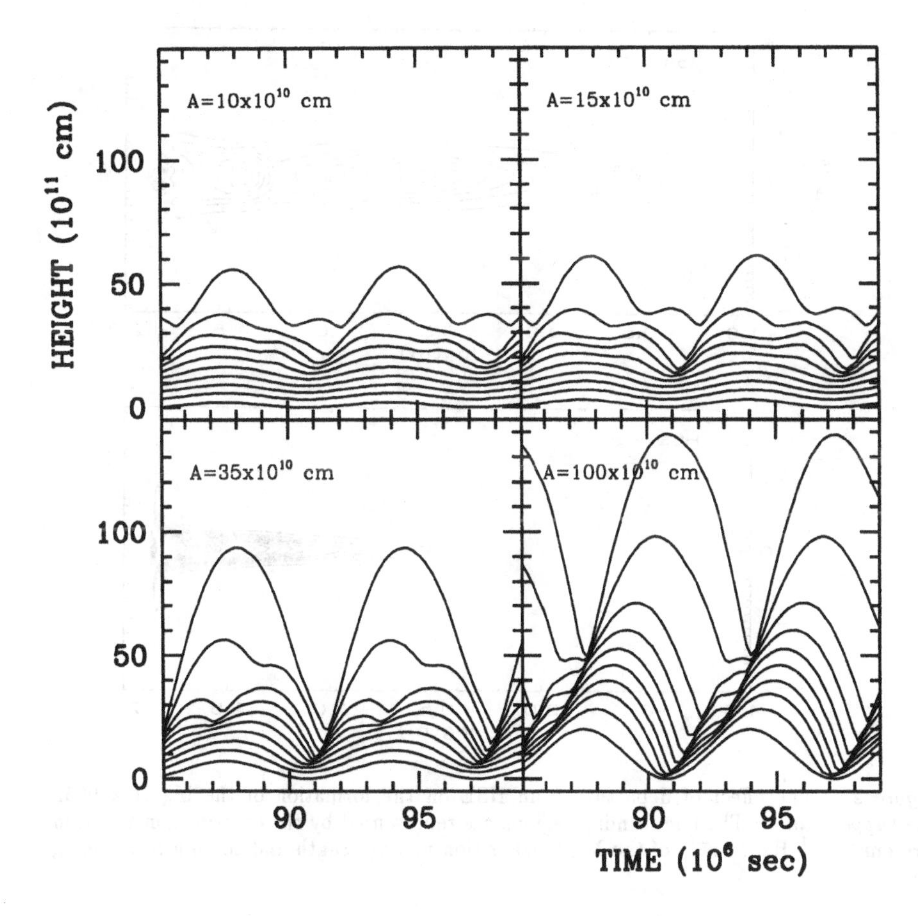

Figure 1. The effect of increasing the amplitude of pulsation in an atmosphere initially in HSE. The motion of selected mass zones in the pulsating atmosphere of a bright giant for four different amplitudes of radial pulsation. The atmosphere extends by a factor of 2 in the illustrated range of piston amplitudes; however, the relation is not linear.

tion of radiative transfer has to be solved in spherical geometry. The radiation field is diluted throughout the atmosphere, the temperature gradient becomes steeper, and the emergent flux distribution changes (it becomes flatter from UV to IR). Hence colour indices between blue, red, and near-IR bandpasses are underestimated (Fieldus & Lester 1990). The line spectrum changes too: a steeper source function tends to enhance most weak absorption lines.

With the drop in both temperature and effective gravity as a function of height, atmospheric extension creates better conditions for the formation of **dust** (Gauger, Sedlmayr, & Gail 1990) and for **mass loss** (Hoefner *etal.* 1996).

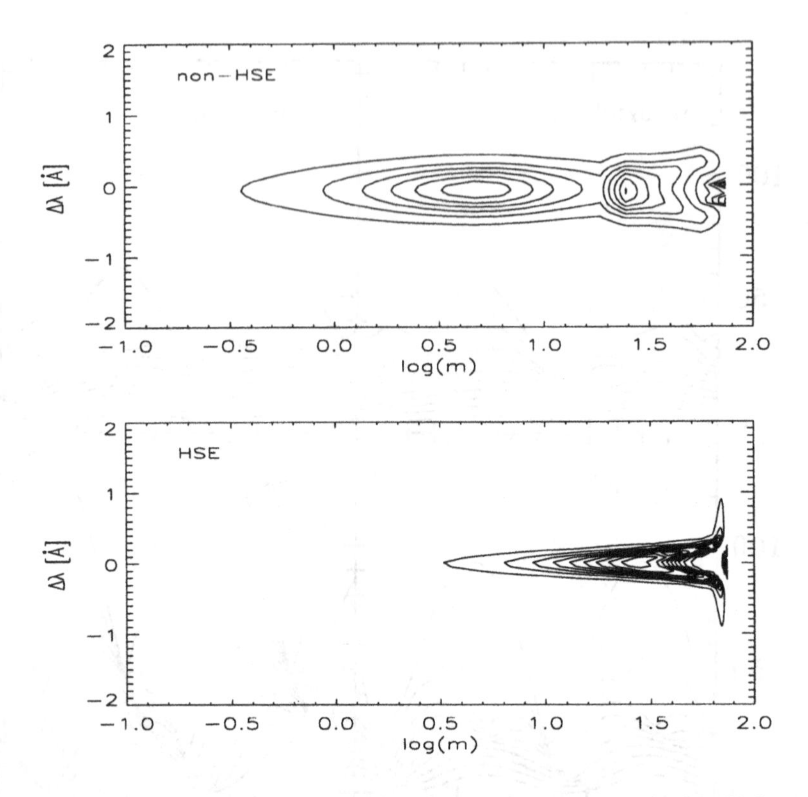

Figure 2. The effect of departure from HSE on the formation of the Mg II λ10952 line (*upper panel*). The line forming regions are represented by the contribution function (erg cm^{-2} s^{-1} Hz^{-1} sr^{-1}) of the Mg II transition in wavelength and column mass, m (g cm^{-2}).

2.2. RADIATIVE EQUILIBRIUM VS. SHOCK WAVES

Radiative equilibrium [RE] is unlikely in a dynamic and extended stellar atmosphere. Neither is convective equlibrium (see Kurucz, § 2, this volume), although both radiative and convective energy transport remain the most important. Dynamics and the development of shock waves introduce complexities in the atmosphere which affect both spectral lines and continuum emission.

The mechanical flux deposited into the atmosphere by the pulsation (and the dynamics/shocks induced by it) will affect its thermodynamical state. In cool star atmospheres this heating could go into ionization of H and He, and be lost radiatively in resonance lines, etc. It is thus important to treat the solution in non-LTE (Carlsson & Stein 1992). Emergent continuum radiation will be less affected than spectral lines due to the non-locality of the source function and the larger range of depths involved.

The large-scale dynamics resulting from the pulsation is strongly coupled and affected by the **small-scale nonthermal velocity fields** present in the atmosphere. The origin and nature of the small-scale velocity fields are probably unrelated to the large-scale radial pulsation motion. Instead, the solar-stellar connection points towards convective motions and high-frequency waves as being responsible (see Cottrell, this volume). Traditionally, photon pathlengths have been used to define the difference between large-scale and small-scale velocity fields in a stellar atmosphere. These are called macroturbulent and microturbulent velocity, respectively – *ad hoc* parameters not to be identified with hydrodynamic turbulence, but with extremes of spectral line broadening parametrization (Mihalas 1978). In a standard static atmosphere model the turbulent velocity is an empirical constant parameter.

The need for a *depth-dependent* microturbulent velocity (also, time-dependent) arises clearly in pulsating atmospheres, when hydrodynamics and radiative transfer are coupled (Sasselov & Lester 1994). This is important for synthesizing realistic line profiles, but also because more microturbulent velocity broadening in cool stars reduces atmospheric opacity by reducing the gas pressure contribution. As regards macroturbulent velocity, there has been no need for introducing it in HERMES models of Cepheids, because such broadening is consistently computed in terms of the large-scale motions induced by the pulsation. The same set of hydrodynamic equations should also be adequate to compute the small-scale velocity fields if not for: (1) limited grid resolution (compared to photon mean free paths), and (2) lack of a mechanism for energy transport between large and small scales (e.g. turbulent cascade) in a 1-D model. Undertanding the mechanism for energy transport could solve the problem even for 1-D codes. For example, assuming that small-scale velocity fields are well represented by a turbulent spectrum one could model the depth-dependent microturbulent velocity with a simple self-consistent model of turbulence in the spirit of Canuto, Goldman, & Mazzitelli (1996). Thus one would treat both the subgrid cascade, as well as the nonlinear interaction of the pulsation shock waves with the small-scale field.

3. Current State of Theory

3.1. MODELS

Computing pulsating stellar atmospheres across the HR diagram poses challenges of different nature in different types of stars. In some the requirements towards the hydrodynamics are higher, in others – the radiative transfer is more important. Therefore, currently there are three loosely defined types of models for pulsating atmospheres. (We limit references to most recent models only.)

Envelope models have an inner boundary at a few 10^6K and calculate the driving and acoustics of the pulsation with a first-order hydro scheme. The pulsating atmosphere is calculated (in LTE or non-LTE) after the run of the envelope model from selected snapshots. Such models are useful for both hotter pulsators (like Cepheids) and cooler ones (like Miras), and especially for the latter (e.g. Bessell, Scholz, & Wood 1996; Luttermoser & Bowen 1990). For Cepheids and similar stars – models calculations by Fokin (1991); Albrow & Cottrell (1996). All these models are in 1-D.

Gray atmosphere models make use of a piston at their inner boundary (at $\approx 10^4$K) to introduce the pulsation motion. At their outer boundary the radiative transfer is calculated assuming constant opacity (gray approximation). Despite the strong limitations of the gray approximation, these codes are useful for studies of the bulk dynamics of very cool pulsators in 1-D, like Miras (Bowen 1988; Feuchtinger, Dorfi, & Hoefner 1993 - who use the Eddington approximation).

Multi-level non-LTE hydro models also make use of a piston at their inner boundary (at $\approx 10^4$K). They use second order accurate (no artificial viscosity) hydro schemes with multi-level non-LTE radiative transfer for cooling and heating terms in the energy equation. Standard assumptions in such models are: (1) that non-LTE does not affect the hydro solution for the density and velocity; and (2) time dependence for bound-free transitions (assuming that the bound-bound rates are much larger). Models based on the method of characteristics were developed by Cuntz (1989). Sasselov & Raga (1992) use the Godunov method and multi-level atoms of H, He, Ca, and Mg. These codes are also in 1-D.

3.2. CHALLENGES & PROBLEMS

All problems of standard stellar model atmospheres (see Gustafsson, Kurucz, this volume) apply also to pulsating stellar atmospheres. In addition, there are some important bits of physics which are still left out of pulsating model atmospheres.

Line blanketing is not yet incorporated fully (or at all) and all current experience from static model atmospheres points to its importance. This means that we are currently unable to judge the systematic errors in the use of colors and color calibrations to pulsating stars. There is no unsurmountable obstacle in building a line-blanketed 1-D raditive hydrodynamics model for hotter variables (like Cepheids). This should be done, given the recent interest in metallicity dependence of Cepheid distances, which relies strongly in understanding colors and color changes.

Important coolants are still missing from the coupled non-LTE hydrodynamics models, in particular - FeI and FeII. While Fe species are included very roughly in the background opacities, short of a full line blanketing calculation, non-LTE coupling of Fe may be as important as that of He, Ca, and Mg, which are currently used. Including Fe in non-LTE models is also within reach, given the successes of novae and supernovae models.

Shock precursors and the treatment of strong shocks in general require significant improvement. This is a difficult problem, as it touches upon the isssues discussed above, as well as the physically consistent treatment of shock-turbulence interaction. On one hand, the recurrence of shocks in a pulsating atmosphere could lead to enhancement of the small-scale nonthermal velocity fields – a typical shock strength in a Cepheid model would cause up to a factor of 2 increase in the turbulent kinetic energy. In a fluid dynamics sense the problem has been studied by e.g. Rotman (1991); some observational evidence comes from e.g. Breitfellner & Gillet (1994). Another issue is the effect these small-scale velocity fields have on the front of the shock wave – under most stellar atmosphere conditions the shock front will be corrugated, thus affecting significantly the energetics and the precursor solution. In 1-D models this can be handled approximately by using simple relations for the development of Richtmeyer-Meshkov [RM] instabilities in the linear regime. One can treat RM instabilities in the spirit of the Rayleigh-Taylor intability with the shock as an instantaneous acceleration (Mikaelian 1991). However, it is highly questionable whether this approach is justified in a 1-D model, where also the small-scale velocity fields have to be calculated with the same underlying approximation. Clearly a 2-D or 3-D model is the only consistent solution.

4. The Future

Many of the advances in understanding pulsating stellar atmospheres will follow or parallel advances in standard atmospheres and the Sun. Among them are the physics of small-scale nonthermal velocity fields and the use of 2-D and 3-D models. Currently there are no outstanding theoretical problems in coupling radiative transfer and hydrodynamics in 3-D. There has been considerable progress in radiative transfer techniques – to mention a few: the ALI and the MALI methods (Rybicki & Hummer 1991; Auer *etal.* 1994). Fast or refined hydro schemes – e.g. PPM and Godunov methods (Zachary *etal.* 1994) are also available.

The ultimate model for a pulsating stellar atmosphere – the *unified pulsation model*, has been a goal for decades. Its completion (even in a 1-D) will most likely take us beyond the year 2000. The task is complex – a model which couples consistently a realistic pulsating envelope and a realistic non-LTE line-blanketed atmosphere requires a solution to at least three outstanding problems. The first problem is time-dependent convection. The second problem is the smooth transition between table opacities (in the envelope) and multi-level radiative transfer (in the atmosphere). This transition occurs in a very sensitive location in the envelope-atmosphere interface (with effects on the acoustic cavity and the convection zone). In order to tackle this problem we need a "smarter" adaptive grid method which can manage efficiently the often opposing demands of the hydrodynamics and the non-LTE radiative transfer. Building such an adaptive grid is the third outstanding problem – it is not simply an issue of enough resolution (which is sorely needed!), but rather of being able to handle the approximate physics in the envelope-atmosphere transition region, and keep the two together. This will be, in the words of Sir Walter Scott, "*The silver link, the silken tie, Which heart to heart, and mind to mind, In body and in soul can bind*".

References

Auer, L., Bendicho, F., & Bueno, T. 1994, A&A, 292, 599
Albrow, M., & Cottrell, P. 1996, MNRAS, 280, 917
Bessell, M., Scholz, M., & Wood, P. 1996, A&A, 307, 481
Bowen, G. W. 1988, ApJ, 329, 299
Breitfellner, M., & Gillet, D. 1993, A&A, 277, 553
Butler, P. 1997, preprint
Canuto, V. M., Goldman, I., & Mazzitelli, I. 1996, ApJ, 473, 550
Carlsson, M., & Stein, R. 1992, ApJ, 397, L59
Cuntz, M. 1989, PASP, 101, 560
Feuchtinger, M., Dorfi, E., & Hoefner, S. 1993, A&A, 273, 513
Fieldus, M., & Lester, J. 1990, in *6th Cambridge Workshop*, ASP 9, 79
Fokin, A. B. 1991, MNRAS, 250, 258
Gauger, A., Sedlmayr, E., & Gail, H.P. 1990, A&A, 235, 345
Hoefner, S., *et al.* 1996, A&A, 314, 204
Luttermoser, D., & Bowen, G. 1990, in *6th Cambridge Worksh.*,ASP 9,491
Mikaelian, K. 1991, Phys. Fluids A, 3, 2625
Rybicki, G., & Hummer, D. 1991, A&A, 245, 171
Rao, L., *et al.* 1993, in *Luminous High – latitude Stars*, ASP 45, 300
Rotman, D. 1991, Phys. Fluids A, 3, 1792
Sabbey, C., Sasselov, D., Fieldus, M., Lester, J., Venn, K., & Butler, P. 1995, ApJ, 446, 250
Sasselov, D., & Raga, A. 1992, in *7th Cambridge Workshop*, ASP 26, 549
Sasselov, D., & Lester, J. B. 1994, ApJ, 423, 795
Schmid-Burgk, J., & Scholz, M. 1975, A&A, 41, 41
Zachary, A., Malagoli, A., & Colella, P. 1994, SIAM J. Sci. Comp., 15, 263.

Discussion of this paper appears at the end of these Proceedings.

WHAT DO WE DO WHEN MODELS DON'T FIT? ON MODEL ATMOSPHERES AND REAL STELLAR SPECTRA

BENGT GUSTAFSSON
Uppsala Astronomical Observatory
Box 515, S-751 20 Uppsala, Sweden
(Bengt.Gustafsson@astro.uu.se)

Abstract. Developments in the modelling of stellar atmospheres and results from the confrontation between calculated and observed fluxes and spectra are discussed. It is argued that, although impressive advances in the study of O-type, WR and cool star spectra have recently occurred, significant improvements should be possible with already existing methods in the analysis of, e.g., solar-type stars.

1. Introduction

The art of interpreting stellar spectra and fluxes relies on models of the flux-emitting layers, the stellar atmospheres. However, these models are never perfect, and often unsatisfactory. The question in the title is therefore relevant in almost all analyses of stellar radiation. An answer to this question, as given in practical work, is often the following: First, we change the fundamental parameters of the models to get a good fit. If such a fit is obtained with a certain parameter set, these specific parameter values are dubbed "determinations" of, e.g., effective temperature or abundances. If the variation of parameters does not resolve the mis-match, we question the input physics, such as gf values or opacities. Also, the observations may be scrutinised. If these efforts cannot explain away our problems we blame the basic assumptions of the models, like LTE or hydrostatic equilibrium. This procedure may well be adequate in many cases, but – in spite the well-tested scientific prescription to always first try the simplest explanations, which goes back to medieval theology ("Occam's razor") – may also lead us astray.

T.R. Bedding et al. (eds.),
***Fundamental Stellar Properties: The Interaction between Observation and Theory*, 261–276.**

There is also a reverse, theoretical and more critical tradition in this field of research. Here, attempts to improve the internal physical consistency of the models are in focus. If this can be achieved (if not, the discussion may turn rather frustrating for practitioners interested in applications and results more than in philosophical principles), the predicted spectra of the new models are compared with the old ones and observations are used to support the claim that the new models are indeed more realistic.

However, these two ways towards a better understanding of stellar spectra and stellar atmospheres are rarely so simple and the difference between them not often so clear-cut. Instead, practice in this field today is developing towards mixed approaches, with close interaction between the two different ways of procedure. This partly reflects the methodological advances in the last decades both in observations and in theory. On the observational side, the enormous improvement in the possibility of getting high S/N spectral data, even at high spectral resolution and extending from the extreme ultraviolet to the far infrared spectral region, has enabled studies of fundamentally new aspects of stellar atmospheres and to check models to an extent far beyond what could be contemplated before. On the theoretical side, new numerical methods now make it possible to abstain from some, or even most of the assumptions behind classical model atmospheres, such as LTE or mixing-length convection. Also as regards basic atomic and molecular data the advance is great, which has put both direct interpretation of stellar spectral line measurements and the modelling of the indirect effects of line blanketing on a much safer ground.

So, today stellar spectroscopic analysis is a field with close and intricate interaction between the two approaches discussed above. This brings new progress and new complications. Sometimes theoretical and more empirical approaches are intertwined so intimately that it is difficult to see how new evidence would affect the conclusions of a study. Also, the great number of different observational criteria that are accessible today for an object may tempt to compromises in model fits or model construction that can be hard to describe or reproduce. In the following some examples will be given of such complex situations, as well as more classical confrontation between the two approaches. I shall start with a presentation of a highly successful fit of standard models, then turn to examples of the great progress in recent years, mainly from theoretical modelling of hot stars, then go to the other extreme in the HR diagram, the cool stars, for which significant advances have recently been made due to improvements in molecular data, and finally end up with stars of intermediate temperatures for which the situation at present seems more frustrating.

My comments are not meant to be comprehensive – the reader is advised to consult the conference proceedings edited by Crivellari et al. (1991)

and Strassmeier and Linsky (1996) for reviews on progress in atmosphere modelling; see also Gustafsson and Jørgensen (1994) as regards models for late-type stars. Reference to reviews on more specific topics will be made below.

2. Vega: Success of classical models (?)

The classical model atmospheres are based the assumptions of plane-parallel stratification, hydrostatic and radiative equilibrium (if need be modified by mixing-length convection) and local thermodynamic equilibrium (LTE). An extensive grid of such models, with detailed consideration of line blanketing, has been produced by Bob Kurucz, and such models were compared in detail with observations of the standard star Vega (AOV) by Castelli and Kurucz (1994). They found that models with $T_{\rm eff}$ in the interval 9550–9650 K and logarithmic gravity of at least 3.95 (the exact values depending on $E(B-V)$ and the helium abundance) give the optimal fit to observations. With reddening and He abundance known the observed visual fluxes would enable a determination of $T_{\rm eff}$ to 50 K (0.5%) and the Hγ profile a determination of $\log g$ to 0.05 dex. In the ultraviolet below 200 nm the computed flux is higher than the observed by typically 4%, and other localised differences appear, probably caused by errors in, or totally missing strong lines. In the visual, the agreement is excellent within the small observational errors ($< 1\%$), while in the infrared the observed flux is higher than the calculated (e.g., by 3% at 1040 nm). Also, the observed Paschen lines are much broader than the computed ones; the reason for this needs further exploration, but may well turn out to be trivial.

The excellent agreement found by Castelli and Kurucz certainly lends credibility to the validity of the basic assumptions behind classical models. These assumptions may be valid in this case because the atmosphere is not convective and the gravity is high enough to keep the atmosphere well stratified. Certainly, departures from LTE are significant for atoms like Fe, Ba and N (Gigas (1986), Gigas (1988), Takeda (1992), Lemke and Venn (1996), Rentzsch-Holm (1996) and references therein) and these effects, mainly reflecting the hot non-local radiative field in the ultraviolet, must be considered in abundance analysis. Yet, the basic photospheric structure is probably not very much affected in the continuum flux forming layers (cf. Frandsen (1974)). However, models with a relatively complete non-LTE blanketing should be calculated, and compared to corresponding LTE models before one can be sure that the neglect of departures from LTE in the model atmospheres is insignificant. One should also note that Vega has been found to be rapidly rotating but seen pole-on (Gulliver et al. (1994)) which means that $T_{\rm eff}$ and $\log g$ vary considerably across its surface.

3. The virtue of handling radiation properly

Very impressive progress has recently been made in the analyses for hot stars, as reviewed by John Hillier in these Proceedings. On the theoretical side this is due essentially to three different developments:

Unified models of atmospheres with radiatively driven winds have been developed, in particular by the Munich group (cf. Pauldrach et al. (1993) for a review). This progress will be further illustrated below.

Major advances in techniques for solving radiative transfer problems for complex atoms now make it possible to calculate non-LTE model atmospheres with consideration of all important atomic and molecular transitions. The Accelerated Lambda Iteration (ALI) method, based on the method of Scharmer (1981), has been developed and speeded up, and combined with the statistical approach of Anderson (1989) to handle complex atoms with opacity sampling (Anderson (1991), Dreizler and Werner (1993), and references cited therein). An alternative method is the hybrid method by Hubeny and Lanz (1995), combining the advantage (global convergence) of the classical Complete Linearization Method of Auer and Mihalas with the great speed and flexibility of the ALI. Hubeny and Lanz also develop a statistical treatment of the complex atoms, inspired by Anderson (1989). These major advances are applicable for stars in all parts of the HR diagram, but as yet, they have mainly been applied for the hotter stars. A major remaining problem is the lack of basic physics data, in particular cross sections for inelastic collisions with electrons, atoms and (for the coolest stars) molecules. Another basic restriction is that complete redistribution is assumed within the transitions. If partial redistribution should also be allowed for, further methodological development will be important. In this case, smaller radiative losses are expected from strong lines for both static and dynamic atmospheres, in particular in low-density atmospheric regions (cf. Huenerth and Ulmschneider (1995), Avrett (1996)).

A third area where major improvements have taken place during the latest decade is in atomic absorption data. Thus, the Opacity Project (Seaton et al. (1994)) and the impressive line lists calculated by Kurucz (1995) and other atomic data bases (e.g., Piskunov et al. (1995), Bell et al. (1994)) have enabled a much more realistic treatment of line blanketing.

These advances have especially been applied in the modelling of O-type stars and Wolf-Rayet stars. For the O-type stars the Munich group has constructed spherically extended NLTE-models where the photosphere and the radiation-driven wind are unified in a self-consistent way (Gabler et al. (1989)). The models are successful in reproducing a number of important observational phenomena – e.g. the long standing "Zanstra-discrepancy" in the different temperature estimates from H and He II lines. Gabler et al. (1992) also demonstrated that the increased EUV flux in these models, as compared with plane-parallel ones, make massive O stars possible candidates for providing enough nebular excitation for the He II emission from high-excitation H II regions. For central stars in planetary nebulae the EUV flux is, however, still higher than predicted in a number of cases, which Gabler et al. (1991) have ascribed to shock generation of extra EUV photons.

Pauldrach et al. (1994) studied the UV spectra of winds of hot luminous stars with a very detailed non-LTE treatment of most important ionization stages. Photospheric line blocking was considered but not completely self-consistently, and with an approximate calculation of the shock emission. Higher mass loss rates resulted than with previous standard models, and a generally very good agreement between observed and calculated spectra was obtained for ζ Pup and Melnick 42 – the brightest star in the LMC.

Schaerer and Schmutz (1994) showed, in their first line-blanketed models of spherically expanding atmospheres of hot stars (with simultaneous solution of the equations of motion, the non-LTE problem of H and He and radiative transfer in the line-blanketed atmosphere), that even "purely" photospheric lines, on which spectroscopic determinations of stellar parameters are based, are strongly affected by the velocity field in the transition zone between the photosphere and the wind, and that this does not only depend on the mass loss rate but on the velocity structure of the wind. Problems still remain in comparisons with observations, e.g. for ζ Pup, for spectral lines dependent on the wind structure.

For O-type subdwarfs (sdO stars), which are immediate progenitors to the white dwarfs, fully line-blanketed models have recently been constructed by Haas et al. (1996), using the method of Dreizler and Werner (1993). Haas et al. included a detailed study of the line blanketing effects of iron-group elements but found that the effects on the main optical diagnostics, lines of H and He and other light elements, are relatively small.

For Wolf-Rayet stars the so-called standard model – based on the assumption of a spherically symmetric homogeneous expanding atmosphere with a stationary flow – has been elaborated with detailed non-LTE model atmospheres for several elements, and, which is found important, with proper treatment of the frequency redistribution of line photons by electron scattering (cf. reviews by Hamann (1995), Hillier (1995)).

In analyses of observed WR spectra the parameters of the standard model are estimated with accuracies of typically 10% for $T_{\rm eff}$ and the velocity parameter, of 20% for the radius parameter and about 30% for the mass-loss rate.

The different shortcomings of the standard model, such as predicted too strong P-Cygni absorption of He I lines, or too strong electron scattering wings, or the observational indications of polarisation, suggesting departures from spherical symmetry, and line-profile variations, suggesting wind inhomogeneities, and strong X-ray fluxes from many stars, are further commented on by Hillier ((1995)).

An important and quite difficult step to take for the hot stars is modelling of clumps or other departures from spherical symmetry. It is possible that this step will be needed in order to decrease the errors in fundamental parameters much below those of today.

A basic link is still missing for the WR stars: the velocity structure of the wind is not calculated from first principles. Also, not the least for these stars, departures from spherical symmetry are known to occur, and these require further detailed exploration.

4. Cool giant opacities: just a beginning

The progress in atomic line data is matched for cool stars by corresponding improvements for molecules. Very extensive calculations and compilations of molecular line data have recently been made (cf. Gustafsson and Jørgensen (1994) and Jørgensen (1997) for recent reviews) and these have led to a much more realistic treatment of line blanketing as well as improved calculations of these heavy line-blocked spectra and colours. In spite of these achievements, still more improvements are needed (cf., e.g., Gustafsson (1995) for a summary of current opacity needs for cool star modelling).

For the M and C stars only LTE models with mixing-length convection exist until now. The most realistic model grid for non-Mira high-luminosity M stars is the opacity sampling grid of Plez et al. (1992). When comparing these models to observed fluxes these authors found a fair agreement with the IR fluxes of three M giants with $T_{\rm eff}$ in the interval 3200 K-4000 K. The visual fluxes are predicted to be too strong, probably because of missing opacity. In the visual and near IR, the model TiO bands come out too strong, seemingly with a discrepancy increasing with decreasing temperature. Thus for g Her ($T_{\rm eff}$ = 3235 K) they found the reduction needed of the band oscillator strengths to be more than a factor of two. A possibility that the mismatch reflects non-LTE effects, over-ionization in Ti, keeping the abundance of TiO molecules below its thermal values, should be investigated further, but may not seem very probable in view of the results reported by Allard et al. (1994). Considerably revised measured band oscillator strengths of TiO have recently been introduced in synthetic spectra by Plez (cf. Bessell et al. (1997)) and these tend to lead to a much better agreement.

Plez et al. (1992) also found a satisfactory agreement between calculated and observed V, R, I, J, K, and L colours. A new much more extensive comparison between calculated and observed colour-colour relations and temperature-colour relations for A-M stars is being made by Bessell, Castelli and Plez (1997)).

Grids of blanketed LTE-model atmospheres of cool carbon stars have been published most recently by Jørgensen et al. (1992). These models are spherically symmetric and include polyatomic absorption by HCN, C_2H_2, shown to have very great structural effects by Eriksson et al. (1984). The models used by Lambert et al. (1986) in their abundance analyses for 30 bright N stars are plane parallel but otherwise fairly similar to those of Jørgensen et al. Opacity Sampling is used for all recent N star models.

Comparison of observations of spectra of these models with current temperature scales, partly based on angular diameters and partly on the Infrared flux method or colours, show problems with the polyatomic bands from HCN and C_2H_2 which come out too strong (Lambert et al. (1986), Jørgensen (1989)). Systematic revisions of the effective temperatures by about 200 K upwards would be needed to remedy this discrepancy. Another problem in the abundance analysis of Lambert et al. – based on high resolution IR FTS spectra – were the CH vibration-rotation lines being significantly weaker than calculated. This mismatch has not been understood and needs further exploration. E.g., an increase in effective temperature by about 500 K would be needed to remedy this. An astonishing fact was that the nitrogen abundances derived mainly from CN lines were only solar or less. Many consistency checks come, however, out positively in the analysis of Lambert et al.; among those were $^{12}C/^{13}C$ ratios in fair agreement from different criteria. These were, however, recently questioned by Ohnaka and Tsuji (1996) who got considerably lower values from lines in the crowded near-IR region. These authors suggest that severe differences between the model atmospheres used should be the explanation – this, however, does not seem to be the case. The major difference between the grids is that the Ohnaka and Tsuji models include a turbulent pressure different from zero, which the models of Lambert et al. do not. This, however, may be treated just as a gravity shift, which does not change the isotopic ratios significantly. We note that the effective temperatures of Ohnaka and Tsuji, based on the Infrared flux method, are several hundred K in excess of those of Lambert et al. and systematically higher than current angular diameter measurements. Further analyses of carbon star spectra, with systematic consistency checks using different criteria, must be pursued.

It is not known to which extent discrepancies like those mentioned above could be ascribed to departures from LTE. In fact, for the cool stars this is an almost totally unexplored area.

Dust probably plays an important role in all cool star atmospheres, as an opacity source and for the transfer of momentum and mass, as well.

Future models must include this, which will require improvements in the theory of dust formation, and dynamics.

Velocity fields in the atmospheres of red giant stars show quite interesting structures, indicating the inadequacy of the standard micro/macro-turbulence concepts. Important results indicating this were obtained by Tsuji (1991) in his study of CO VR lines in high-resolution spectra of M giants. Line shifts and bisectors reveal complicated velocity fields, possible to interpret as granular motions. (Especially, standard abundance determinations based on saturated lines, e.g., the 2.3 micron CO bands, may lead to systematic errors.) 3D simulations of convection should also be attempted for these stars (see further below). Also Mira models with a simultaneous detailed treatment of pulsation, radiation and dust formation should be developed. No doubt, a much closer interaction between detailed modelling and systematic observing will be needed for significant further progress in this field.

5. Cool dwarfs: a field in rapid development

A major step forward in the analysis of M dwarfs is the work by France Allard, Peter Hauschildt, Hugh Jones, and collaborators. Allard and Hauschildt (1995) published an extensive grid for M dwarfs with $T_{\rm eff}$ down to 1500 K, and different gravities and metallicities. A great number of molecular bands and atomic lines were handled in opacity sampling, however, the most important line opacities, water vapour and TiO, were still only treated with rough statistical methods. These models represent a significant improvement in consistency, and in their fit to observed near-infrared spectra. However, severe discrepancies prevail in the infrared beyond 1.3 μm which is seen both in low resolution spectra and in IR two-colour diagrams. This is probably mainly due to over-estimated water vapour absorption. The calculated spectra demonstrate a number of interesting and important properties which are partly counter-intuitive, e.g. the strengthened metal hydride absorption with decreasing metallicity. Moreover, the overall flux is not shifted very much towards longer wavelengths when the effective temperature is decreased, due to the increased water vapour absorption and, at cooler temperatures, the growing collision-induced absorption (CIA) by H_2, centred around 2 μm.

Later, Jones et al. (1995) studied the behaviour of the H_2O absorption bands in cool dwarfs around 3 μm with similar models but with a more detailed line-by-line treatment of TiO and with data for H_2O in the synthetic spectra based on recent ab-initio calculations. From matching the spectra the authors find effective temperatures in fair agreement with other determinations, however, with errors on the order of 7%. The authors warn that more definite determinations must wait until the water-vapour opacity calculations are extended to wavelengths below 2 μm. A better understanding of molecular line broadening by damping is also needed – the results by Jones et al. (1995) may indicate that the damping constants have been overestimated when chosen according to prescriptions for atomic lines.

Jones et al. (1996) explored the capacity of the models of Allard and Hauschildt (1995) to reproduce M dwarf spectra in the J band. The region studied contains strong KI, Fe I, Mg I lines and FeH bands. A reasonable, though not yet satisfactory, agreement is found – suggesting metallicities from -2.5 to $+0.5$ on a logarithmic scale relative to the Sun. The problems with the fit of the Fe I lines are at least partially due to too great $\log gf$ values.

The same models were also used by Leggett et al. (1996) in discussing infrared low-resolution spectra of 16 M dwarfs (M0-M6.5). They found that the spectra give consistent effective temperature estimates, although discrepancies remain between theory and observation such as an excess model flux in V, problems with hydrides such as FeH and with reproducing IR fluxes to the 10% level. The effective temperatures derived are hotter than the commonly used temperature scale (by Berriman and Reid (1987)) would suggest by on average 130 K. With these temperatures new radii were estimated. The $T_{\rm eff}$ and radii thus obtained, as well as their dependence on metallicity, seem to agree well with most recent stellar models, which resolves a long-standing discrepancy. A relatively high $T_{\rm eff}$ scale for earlier M-type dwarfs is also supported by the study, based on independent models, of Brett and Plez (1993).

Tsuji et al. (1996) (see also references therein) have explored the results of dust-formation in M dwarf model atmospheres. The basic effect is a heating of the photosphere by several hundred Kelvin for stars with $T_{\rm eff} < 2500$ K, counteracting the dust condensation. This may keep the surface layers just at the condensation temperatures of the most abundant grain species. The effect on the spectrum is a general weakening of molecular bands, due to the stronger dust absorption and less molecules in the surface layers. The application of these dusty models in determining effective temperatures from spectra tend to favour a fairly low scale as compared with similar determinations with non-dusty models. For the interesting continuation of these studies to the very cool end, the brown dwarfs, see Baraffe's contribution in the present Proceedings.

The models for M dwarfs are now advanced enough to be tried also in detailed abundance analyses. Recent instruments (such as the Kitt Peak Phoenix IR spectrometer) admit such work for many stars. Analyses of M dwarfs in binaries or other systems with hotter components with known compositions are important as tests. Further studies of the role of stellar activity for the structure of M star atmospheres must also be carried out.

6. R CrB stars: progress and a worrying failure

This interesting but relatively small group of stars, with their H poorness, carbon richness and sudden declines of brightness, have earlier only been analysed with un-blanketed model atmospheres. Those models did not produce both high-excitation features, like He I or C II lines, and low excitation features such as C_2 lines which often are present in the same stellar spectra. We have calculated a grid of blanketed LTE models for these stars (Asplund et al. (1997)) and found that blanketing is so strong, since the continuous absorption is severely reduced by the reduction of hydrogen, that the temperature gradient is raised such that both the high- and low-excitation features show up at roughly the right strength. The strong temperature gradient and the He I absorption at depth even bring the stars across the effective Eddington instability border where radiative acceleration is taking over gravitation. Asplund and Gustafsson (1996) suggest this to be an important factor behind the sudden declines, ascribed to occultation by dust formed in gas clouds ejected from the stars.

However, when using the models in detailed abundance analyses we, in collaboration with David Lambert and Kameswar Rao, came across an interesting problem. The abundant C I lines in the stellar spectra are all considerably weaker than those predicted, corresponding to an abundance reduction by about a factor of 4 (Lambert et al. (1997), Gustafsson and Asplund (1996)). The discrepancy cannot be removed by a simple abundance change – the continuous opacity is also mainly due to C I bound-free absorption, originating from levels only slightly higher than those of the lower levels of the lines. Therefore, a temperature change is also not a realistic explanation, nor a gravity change or a change in microturbulence parameter. Non-LTE effects are also found to be small, and errors in gf values or bound-free absorption coefficients are less probable. The most probable explanation is a drastic temperature structure modification, e.g. related to the unstable region in the deep atmosphere. Radiation-hydrodynamic simulations should be used to explore the stability and the structure of objects of this character, with a super-Eddington zone in the deep photosphere. However, it is worrying that this – in principle very straightforward and simple – test case of the classical model atmospheres come out with a great inconsistency. In other more normal stars this inconsistency would have been masked as an abundance effect.

7. Solar-type stars: home-work to do!

A great step forward in stellar physics is the construction of reasonably realistic models of convection and granulation. Pioneering 3D convection models with a rather coarse spatial resolution and assuming an an-elastic fluid were constructed by Nordlund (1982), Nordlund and Dravins (1990), and compared with observations of spectral-line profiles by Dravins and Nordlund (1990a), (1990b). Compressible 2D models were made by Steffen and collaborators (Freytag et al. (1996), and references therein). (Magneto)hydrodynamic simulations of fully compressible convection can now be made in 3D, with a relatively high spatial resolution and a rather detailed consideration of radiative transfer (Rast et al. (1993), Nordlund and Stein (1995), and references therein), a possibility not yet exploited for other stars than the Sun.

Attempts to fit the solar spectrum and limb-darkening with more classical flux-constant plane-parallel model atmospheres have given conflicting results. Gustafsson et al. (1975) found their blanketed solar model to be too bright in the ultraviolet by e.g. 25% around 350 nm and ascribed this to a missing opacity source, probably a veil of weak metal lines. Kurucz later, using a more extensive atomic line list found a good agreement (Kurucz (1992)) which was verified by Edvardsson et al. (1993) with essentially the same list but a different model atmosphere program (OSMARCS). (Note, however, the critical remarks made by Bell et al. (1994)). The Kurucz (1992) model (ATLAS9) and the OSMARCS model are different as regards prescriptions for convection. In particular, the ATLAS9 model has an *ad hoc* consideration of convective overshoot, which gives the temperature structure a characteristic hump around $\tau(\mathrm{cont}) = 0.6$ of about 200 K, and an even more extended effect for metal-poor models (cf. Fig 11 in Gratton et al. (1996)). This difference in structure makes the ATLAS9 model reproduce solar limb-darkening observations better than the OSMARCS model (Blackwell et al. (1995)), but none of them are as successful as the classical Holweger-Müller empirical model in this respect.

The Hβ wings of the ATLAS9 model are, however, far too bright, and convective overshoot must be switched off and the mixing length parameter α lowered to 0.5 in order for a good agreement with observations to result (van't Veer-Menneret and Megessier (1996), Fuhrmann et al. (1993)). Similar results were found for Procyon (F5 IV-V). In OSMARCS models the standard choice of $\alpha = 1.5$ leads to a good fit, which is since the value of the structure parameter y in the mixing-length formalism (cf Henyey et al. (1965)) is chosen to 0.076 instead of 0.5. Similar results were obtained by Fuhrmann et al. (1993), see also Fuhrmann et al. (1994), who stress the significance of the Balmer line profiles as tests of the model atmospheres.

Obviously, none of the two theoretical models give satisfactory results in both tests. This is presumably because the real Sun can not be represented by a flux-constant 1D model with mixing-length convection. It would be of great interest to check the predicted limb-darkening and spectrum, not the least that of the Balmer lines, of the Nordlund 3D models.

What models should then be used for solar-type stars? Gratton et al. (1996) showed that the differences between the OSMARCS models of Edvardsson et al. (1993) and the ATLAS9, as well as the more recent Kurucz models from 1995 with certain errors corrected, are very considerable, the latter being several hundred K hotter in the flux-forming layers due their "convective overshoot". This introduces a considerable uncertainty in the effective temperature scale. In a detailed study of the temperature scale for solar-type dwarfs, as derived from Balmer line profiles and various colour indices, Fuhrmann et al. (1994) found significant inconsistencies and scatter between different criteria which casts doubt on scales based on broad-band colours when used for individual stars as well as on calibrations by means of plane-parallel models in general. Gratton et al. (1996) also derived a semi-empirical temperature calibration of colours for metal poor late-type stars based on Kurucz ATLAS9 models. This calibration is significantly different from that of Edvardsson et al. (1993) and Nissen et al. (1994) which is based on calculated $(b-y)$ colours of OSMARCS models. The difference is ascribed by Gratton et al. to the differences in model structures - this requires further exploration. Gratton et al. check their scale by comparing calculated to observed Hα profiles – in view of the problems encountered with Kurucz's models, Hβ profiles should also be checked.

Obviously, the situation for metal-poor dwarf atmospheres is far from satisfactory - uncertainties by 400 K and 0.4 dex could be present in $T_{\rm eff}$ and [Fe/H], respectively! Some of the effect may be compensated for by the fact that convective overshoot will also affect the line forming atmospheric layers. For the line strengths of many lines, however, the temperature gradient is what matters and this is drastically different in the different models.

Studies based on spectra of 3D models of convection for solar-type dwarfs of different metallicities are thus urgently needed. Simulations of this type, combined with detailed spectroscopic observations, up along the sub-giant and red-giant branch are also important. Moreover, the advance in non-LTE blanketing already exploited for hotter stars should be applied for solar-type stars. Until this has been done one should systematically use weak lines, formed in layers close to the continuum forming layers, in abundance analyses and also explore the virtues of semi-empirical model atmospheres in detailed abundance analyses. There have been only scattered attempts to derive such models for late-type stars, from individual line profiles or other criteria (see Gustafsson and Jørgensen (1994) for a review; see also Morossi's presentation at this meeting).

The solar-type stars are of key significance in the study of galactic evolution. Their ages may be determined and range all across the life-span of the Galaxy, and their rich visual spectra enable the determination of abundances which may be assumed to be characteristic of the gas from which the stars once formed. In fact, much of our understanding of the evolution of the Galaxy emanates from studies of these stars (cf., e.g., Edvardsson et al. (1993). They may also, with available techniques, be used as tracers of the chemical evolution in nearby dwarf galaxies (cf. Ardeberg et al. (1997)). Given this significance, it is astonishing that the great progress made in hydrodynamics and in non-LTE modelling, has hardly been exploited in analyses of solar-type stars. A particularly significant group of stars for which 3D simulations should now be explored (and compared with profiles of hydrogen, and other strong lines and bisectors of metal lines) and detailed non-LTE analyses ought to be made systematically, is the metal-poor dwarfs. In view of the significance of their abundances for cosmology and for the study of the evolution of the early Galaxy, this must be regarded a high-priority project.

References

Allard, F. and Hauschildt, P.H.. (1995) ApJ **445**, 433
Allard, F., Laylor, L., Alexander, D.R., Hauschildt, P.H. (1994) BAAS **187**, 10311
Anderson, L.S. (1989) ApJ **339**, 558
Anderson, L. (1991), in Crivellari et al. (1991), p. 29
Ardeberg, A., Gustafsson, B., Linde, P., Nissen, P.E. (1997), A&A, in press
Asplund, M. and Gustafsson, B. (1996) *Hydrogen-deficient stars*, C.S. Jeffery & U. Heber (eds.), ASP Conf. Ser. **Vol. 96**, 39
Asplund, M., Gustafsson, B., Kiselman, D., Eriksson, K. (1997) A&A **318**, 521
Avrett, E.H. (1996) in Strassmeier & Linsky (1966), p. 503
Bell, R.A., Paltoglou, G. and Tripicco, M.J. (1994) MNRAS **268**, 771
Berriman, G.B. and Reid, N. (1987) MNRAS **227**, 315
Bessell, M., Castelli, F. and Plez, B. (1997), in preparation
Blackwell, D.E., Lynas-Gray, A.E., Smith, G. (1995) A&A **296**, 217
Brett, J.M. and Plez, B. (1993) Proc. ASA **10**, 250
Castelli, F. and Kurucz, R.F. (1994) A&A **281**, 817
Crivellari, L., Hubeny, I., Hummer, D.G. (eds.) (1991) *Stellar Atmospheres: Beyond Classical Models*, NATO ASI Series, **Vol. 341**
Dravins, D. and Nordlund, Å. (1990a) A&A **228**, 184
Dravins, D. and Nordlund, Å. (1990b) A&A **228**, 203
Dreizler, S. and Werner, K. (1993) A&A **278**, 199
Edvardsson, B., Andersen, J., Gustafsson, B. et al. (1993) A&A **275**, 101
Eriksson, K., Gustafsson, B., Jørgensen, U.G., Nordlund, Å. (1984) A&A **132**, 37
Frandsen, S. (1974) A&A **37**, 139
Freytag, B., Ludwig, H.-G. and Steffen, M. (1996) A&A **313**, 497
Fuhrmann, K., Axer, M., Gehren, T. (1993) A&A **271**, 451
Fuhrmann, K., Axer, M., Gehren, T. (1994) A&A **285**, 585
Gabler, R., Gabler, A., Kudritzki, R.P., Puls, J. and Pauldrach, A. (1989) A&A **226**, 162
Gabler, R., Kudritzki, R.P. and Mendez, R.H. (1991) A&A **245**, 587
Gabler, R., Gabler, A., Kudritzki, R.P. and Mendez, R.H. (1992) A&A **265**, 656

Gigas, D. (1986) A&A **165**, 170
Gigas, D. (1988) A&A **192**, 264
Gratton, R.G., Caretta, E., Castelli, F. (1996) A&A **314**, 191
Gulliver, A.F., Hill, G., Adelman, S.J. (1994) ApJ **429**, L81
Gustafsson, B. (1995) *Astrophysical Applications of Powerful New Databases*, S.J. Adelman & W.L. Wiese (eds.), ASP Conf. Ser. **Vol. 78**, 347
Gustafsson, B. and Asplund, M. (1996) *Hydrogen-deficient stars*, C.S. Jeffery & U. Heber (eds.), ASP Conf. Ser. **Vol. 96**, 27
Gustafsson, B., Bell, R.A., Eriksson, K., Nordlund, Å. (1975) A&A **42**, 407
Gustafsson, B. and Jørgensen, U.G. (1985) *Calibration of Fundamental Stellar Quantities*, Proc. IAU Symp. **Vol. 111**, D.S. Hayes et al. (eds), D. Reidel Publ. Co.,p. 303
Gustafsson, B. and Jørgensen, U.G. (1994) A&AR **6**, 19
Haas, S., Dreizler, S., Heber, U., Jeffery, S. and Werner, K. (1996) A&A **311**, 669
Hamann, W.-R. (1995) *Wolf-Rayet Stars: Binaries, Colliding Winds, Evolution*, Proc. IAU Symp. **Vol. 163**, K.A. van der Hucht, P.M. Williams (eds.), Kluwer, p. 105
Henyey, L., Vardya, M.S., Bodenheimer, P. (1965) ApJ **142**, 841
Hillier, D.J. (1995) *Wolf-Rayet Stars: Binaries, Colliding Winds, Evolution*, Proc. IAU Symp. **Vol. 163**, K.A. van der Hucht, P.M. Williams (eds.), Kluwer, p. 116
Hubeny, I. and Lanz, T. (1995) A&A **439**, 875
Huenerth, G and Ulmschneider, P. (1995) A&A **293**, 166
Jørgensen, U.G. (1989) ApJ **344**, 901
Jørgensen, U.G. (1997) *Molecules in Astrophysics*, E. van Dieshoeck (ed.), Kluwer Academic Publ., in press
Jørgensen, U.G., Johnson, H.R., Nordlund, Å. (1992) A&A **261**, 263
Jones, H.R., Longmore, A.J., Allard, F. et al. (1995) MNRAS **277**, 767
Jones, H.R., Longmore, A.J., Allard, F., Hauschildt, P.H. (1996) MNRAS **280**, 77
Kurucz, R.L. (1992) Rev. Mex. Astron. Astrof. **23**, 181
Kurucz, R.L. (1995) *Astrophysical Applications of Powerful New Databases*, S.J. Adelman & W.L. Wiese (eds.), ASP Conf. Ser. **Vol. 78**, 205
Lambert, D.L., Gustafsson, B., Eriksson, K., Hinkle, K. (1986) ApJS **62**, 673
Lambert, D.L., Rao, K.W., Asplund, M., Gustafsson, B. (1997), in preparation
Leggett, S.K., Allard, F., Berriman, G. et al. (1996) ApJS **104**, 117
Lemke, M., Venn, K.A. (1994) A&A **309**, 558
Nissen, P.E., Gustafsson, B., Edvardsson, B., Gilmore, G. (1994) A&A **285**, 440
Nordlund, Å. (1982) A&A **107**, 1
Nordlund, Å. and Dravins, D. (1990) A&A **228**, 115
Nordlund, Å. and Stein, R.F. (1995), in Stellar Evolution: What Should Be Done, Proc. 32nd Liege Int. Astroph. Coll, p. 75.
Ohnaka, K.. and Tsuji, T. (1996) A&A **310**, 933
Pauldrach, A., Feldmeier, A., Puls, J. and Kudritzki, R. (1993) *Space Sci. Rev.* **66**, 105
Pauldrach, A., Kudritzki, R.P., Puls, J. et al. (1994) A&A **283**, 525
Piskunov, N.E., Kupka, F., Ryabchikova, T.A. et al. (1995) A&AS **112**, 525
Plez, B., Brett, J.M., Nordlund, Å. (1992) A&A **256**, 551
Rast, M.P., Nordlund, Å., Stein, R. and Toomre, J. (1993) ApJ **408**, L53
Rentzsch-Holm, I. (1996) A&A, **305**, 275
Schaerer, D., Schmutz, W. (1994) A&A **288**, 231
Scharmer, G.B. (1981) ApJ **249**, 720
Seaton, M.J., Yu Yan, Mihalas, D. and Pradhan, A.K. (1994) MNRAS **266**, 805
Strassmeier, K.G. and Linsky, J.L. (1996) *Stellar Surface Structure*, Proc. IAU Symp. **Vol. 176**, Kluwer Academic Publ., p. 501
Takeda, Y. (1992) PASJ **44**, 649
Tsuji, T. (1991) A&A **245**, 203
Tsuji, T., Ohnaka, K., Auki, W. (1996) A&A **30**, 1
van't Veer-Menneret, C., and Megessier, C. (1996) A&A **309**, 879

DISCUSSION

MONIQUE SPITE: Is it possible now with the new MARCS models to have a good representation of the centre-to-limb variation of the solar lines?

BENGT GUSTAFSSON: Concerning spectral lines, I don't know yet. The limb darkening in the continuum is, however, still not very well reproduced (c.f. Blackwell et al., 1995, A&A 314, 191).

MIKE BESSELL: What was the difference in the lithium abundance from the hydrodynamic simulation and the p-p atmosphere?

BENGT GUSTAFSSON: For LTE models in the solar case it seems to be less than 0.1 dex. It could be worse in more metal-poor stars.

ROGER BELL: Since you showed the effect of including more and more H_2O lines, is it necessary to include lines of H_2O^{17} and H_2O^{18}?

BENGT GUSTAFSSON: Yes, it might be necessary for detailed models.

MIKE BESSELL: The problem with the R Cor Bor stars may be related to that seen in other carbon-rich objects. The carbon-rich dwarf G77−61 has been determined to have a Mg abundance 100 times lower than expected from its space motion. Similarly, Lamers has analysed Pop I A-F supergiants in advanced evolutionary stages which are C-rich. He finds Fe abundances as low as 1% solar, although again, one would expect near solar abundances. It seems that in C-rich atmospheres metal lines are computed much too strong, as though an important opacity source is missing.

BENGT GUSTAFSSON: Yes, that is an interesting possibility. The question is what this opacity could be due to.

ROBERT KURUCZ: The convective models have systematic errors. The physical basis for the errors is understood. Here is how to proceed: make ad hoc corrections to the predictions of each grid; tabulate the systematic error in $T_{\rm eff}$ as a function of $T_{\rm eff}$ from methods using wings of Hα, Hβ, etc.; tabulate corrections to a V magnitude as a function of $T_{\rm eff}$; tabulate corrections to UV and visible flux; then determine which lines species are sensitive to convective structure and do not use them in abundance analyses.

BENGT GUSTAFSSON: I think one should put a strong emphasis on developing and trying as physically self-consistent approaches as possible before adopting ad-hoc procedures - essentially introducing extra free parameters.

CLAUDE MEGESSIER: Could you comment more on the problem of carbon in the R Cor Bor stars. Why do you think that NLTE would not improve the situation? The work by Hubeny and Lanz shows that the NLTE treatment of C has a large effect on the $T(\tau)$ temperature law.

BENGT GUSTAFSSON: In the much cooler R Cor Bor stars (as compared with the white dwarfs in the work of Hubeny and Lanz), carbon has been found to be close to LTE by detailed statistical-equilibrium calculation by Asplund and Ryde at Uppsala. This was obtained for a given model atmosphere - I do not think, however, the coupling between the C I excitation/ionization and the model structure to be such that this situation would change considerably in a self-consistent model.

RAINER WEHRSE: There is another side concerning the progress caused by the introduction of the accelerated Λ iteration: the accelerated Λ iteration is simply a Jacobi iteration, which is well known in mathematics for more than 100 years and is considered obsolete for more than 30 years. There are now much more efficient methods available, as e.g., multi-grid or Krylow space methods.

BENGT GUSTAFSSON: Maybe, but in spite of this, the (re)invention of the ALI has really been a major step forward in this field of research. It is also, not the least in combination with the complete linearization as proposed by Hubeny and Lanz, a very flexible and efficient method. A comparison of disadvantages and advantages of that method with multi-grid and Krylow space methods would be interesting.

ROGER BELL: Where did you get your value of 3% for the C/He ratio?

BENGT GUSTAFSSON: We used values of 0.3, 1 and 3% for the C/He ratio, mainly inspired by determinations for EHe stars, which we think are evolutionarily related to the R CrB stars. At such high ratios, He mainly acts as an inert element in the R CrB photospheres.

9. STELLAR OSCILLATIONS AND PULSATIONS

OBSERVING SOLAR-LIKE OSCILLATIONS

HANS KJELDSEN
Teoretisk Astrofysik Center, Danmarks Grundforskningsfond, Aarhus University, DK-8000 Aarhus C, Denmark

AND

TIMOTHY R. BEDDING
School of Physics, University of Sydney 2006, Australia

Abstract. We review techniques for measuring stellar oscillations in solar-type stars. Despite great efforts, no unambiguous detections have been made. A new method, based on monitoring the equivalent widths of strong lines, shows promise but is yet to be confirmed. We also discuss several subtleties, such as the need to correct for CCD non-linearities and the importance of data weighting.

1. Why search for solar-like oscillations?

Given the tiny amplitudes of oscillations in the Sun and the obvious problems in detecting similar oscillations in other stars, we should first ask whether the effort is justified. Oscillation frequencies give information about the sound speed in different parts of the stellar interior. They can be measured much more precisely than can any of the other fundamental parameters which have been discussed at this meeting. Accuracies of 10^{-3}–10^{-4} have been achieved for "classical" multi-periodic pulsators stars such as δ Scuti stars, rapidly oscillating Ap (roAp) stars and β Cephei stars. These stars pulsate with amplitudes typically 1000 times greater than seen in the Sun, so why are we not satisfied with observing them?

One reason is that the classical pulsating stars are only found in restricted areas of the HR diagram (the instability strips). Since oscillations in the Sun are thought to be excited by convective turbulence near the surface, all stars with an outer convective zone should undergo similar oscillations. This makes it possible, at least in principle, to perform seismic studies on *all* stars with spectral type later than about F5.

T.R. Bedding et al. (eds.),
***Fundamental Stellar Properties: The Interaction between Observation and Theory,* 279–284.**

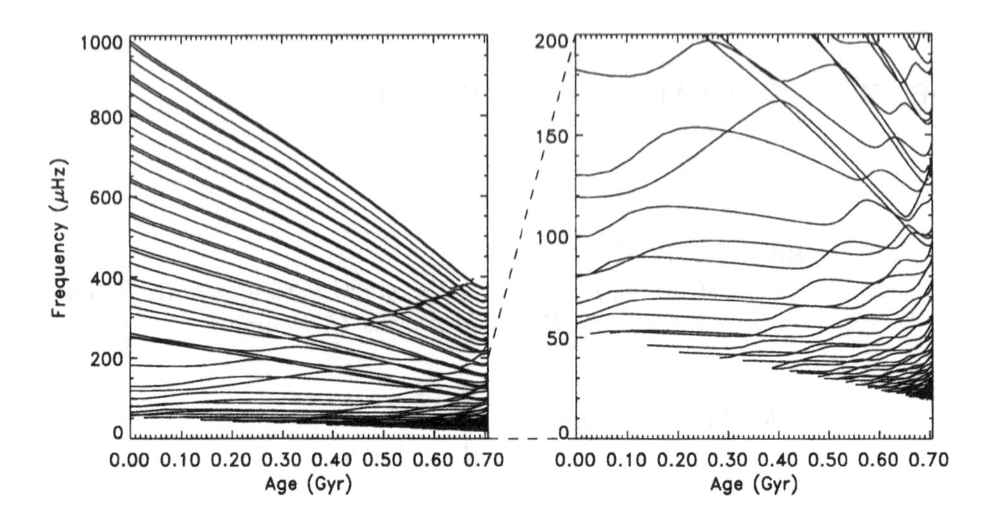

Figure 1. Evolution of oscillation frequencies in a $2.2\,M_\odot$ star, from model calculations by J. Christensen-Dalsgaard. Only modes with $\ell = 0, 1, 2$ and $n \leq 10$ are shown.

A second reason for studying solar-like oscillations is that the modes are easy to identify. There is little point in knowing the frequency of an oscillation mode unless you also know in which part of the star that mode is trapped. An oscillation mode is characterized by three integers: n (the radial order), ℓ (the angular degree) and m (the azimuthal order)[1]. These specify the shape of the eigenfunction, which in turn determines the sensitivity of the oscillation frequency to the internal structure of the star.

Figure 1 shows the oscillation frequencies of a non-rotating star (mass $2.2\,M_\odot$) as it evolves. At any instant during the star's evolution, a vertical cross-section through this figure shows the frequencies of oscillation modes with $\ell = 0$, 1 and 2 (which are most easily observed in an unresolved star). However, in multi-periodic δ Scuti and β Cephei stars, only the lowest frequency modes are found to be excited to an observable level, presumably due to the details of the excitation process (the so-called κ mechanism). We are therefore forced to identify modes in the crowded lower region of the diagram. To further complicate matters, these stars tend to be rapid rotators, which causes a splitting of frequencies (analogous to Zeeman splitting). Finally, a given star is only observed to oscillate in a seemingly random subset of possible modes. Until reliable mode identification is achieved, it will be impossible to apply asteroseismology to these "classical" pulsating stars.

In contrast, it is easy to identify the modes of solar-like oscillations. At least in the Sun, all modes in a broad frequency range are excited.

[1]In a star with no rotation or magnetic field, frequencies do not depend on m.

Furthermore, these modes approximately satisfy an asymptotic relation, with modes of fixed ℓ and differing n having regularly spaced frequencies separated by the so-called large separation, $\Delta\nu$. The resulting comb-like structure is clearly seen in the upper part of Figure 1 and allows modes to be identified directly from the oscillation spectrum.

Measuring $\Delta\nu$ provides an estimate of the stellar density. Moreover, the small differences between observed frequencies and those predicted by the asymptotic relation give crucial information about the sound speed deep inside the star.

2. Sensitivities of detection methods

Velocity In the Sun, the strongest modes have velocity amplitudes of about 25 cm/s, which corresponds to a wavelength variation ($\delta\lambda/\lambda$) of less than 10^{-9}, or 4.2 μÅ at 5000 Å. Detecting such miniscule Doppler shifts in other stars is extremely difficult. Spectrographs cannot be made with absolute stabilities of 10^{-9}, so one must simultaneously monitor the wavelength of a stable reference (e.g., a Na or K resonance cell, an I_2 absorption cell or telluric absorption features). The noise levels at present are down to about 0.5 m/s, which is a factor of two higher than the solar signal.

Radius Given that solar periods are around 5 min, the change in radius is only about 12 m or 17 microarcseconds. Astrometry of the solar limb using SoHO/MDI has recently revealed the oscillations (J. Kuhn et al., Proc. IAU Symp. 181, in press), but such observations will surely be impossible for other solar-like stars.

Intensity The solar oscillations have been observed as variations in total intensity, with amplitudes of about 4 ppm (parts per million). Open clusters are a natural target for differential CCD photometry and the lowest noise level so far achieved is 5–7 ppm, from observations by Gilliland et al. (1993) of twelve stars in M 67 using six telescopes (2.5 m to 5 m) during one week. This is an interesting noise level, less than a factor of two away from solar photometric amplitude.

Ground-based photometric observations are severely hampered by atmospheric scintillation. Several space missions have been proposed, but only one has so far been launched: the EVRIS experiment, on board the Russian Mars96 probe, which ended in the Pacific Ocean.

Temperature Since the change in radius during solar oscillations is insignificant, the intensity fluctuations observed in the Sun must result from local temperature changes in the atmosphere of about 6 mK ($\delta T_{\rm eff}/T_{\rm eff} \approx 10^{-6}$). It has been suggested that these temperature changes can be measured by their effect on spectral absorption lines (Kjeldsen et al. 1995;

Bedding et al. 1996). For example, the Balmer lines in the Sun should show variations in equivalent width of about 6 ppm. As discussed below, the equivalent-width method has so far attained noise levels in other stars of 2–3 times the solar peak amplitude.

3. Some subtleties

Achieving low noise levels demands care during both observing and data analysis. One major requirement is high efficiency, in order to get as many photons as possible (photon counting statistics are a fundamental limitation). This requires optical systems with high transmissions, detectors with high Q.E. (i.e., CCDs) and observations with a high duty cycle. This may force one to observe under quite unusual conditions. For example, in the case of photometry these requirements mean observing defocussed stars in order to avoid saturating the CCD.

Linearity of the system is another important issue. Measuring oscillations at the ppm level requires that the detector be linear to the level of 10^{-3} or better. This is certainly not trivial and our tests of different CCDs and controllers often reveal deviations from linearity of up to a few per cent. Unless correction is made for these effects, the extra noise will destroy any possibility of detecting oscillations.

Each step in the data reduction procedure must be tested to establish how much noise it adds to the time series. It also helps if, as well as measuring the parameter which is expected to contain the oscillation signal (magnitude, velocity or the line strength), one also monitor extra parameters. For example, by correlating measured magnitudes with seeing variations, one has a chance to remove the influence of seeing simply by subtracting that part of the signal which correlates with seeing. Of course, this assumes that the real oscillations do not correlate with the seeing. This process of decorrelation, which can be repeated for other parameters (total light level, position on detector, etc.), is very powerful but can also be quite dangerous if not done with care.

Once a time series has been extracted, the search for oscillation frequencies is done by calculating the power spectrum. The simplest method is to Fourier transform the time series and take the squared modulus. The resulting spectrum shows power as a function of frequency, and a significant peak in this spectrum implies a periodic signal in the time series data. However, the standard Fourier transform treats all data points as having equal weight. In reality, data quality can vary significantly within a data set, due to variable weather conditions or even because data are being combined from different telescopes. The power spectrum is very sensitive to bad data points – the final noise level will be dominated by the noisiest parts of the time series. One should therefore calculate a weighted power spectrum, with each data point being allocated a statistical weight according to its quality (e.g., Frandsen et al. 1995). Unfortunately, this procedure is not widely used and many published power spectra have higher noise than necessary.

4. Recent reults

Attempts to detect solar-like oscillations have been reviewed by Brown & Gilliland (1994) and Kjeldsen & Bedding (1995), and here we only discuss more recent results. Most efforts have concentrated on subgiants, since these are expected to have higher oscillations amplitudes than the Sun.

η Boo is the brightest G-type subgiant. We observed this star over six nights with the 2.5-m Nordic Optical Telescope (Kjeldsen et al. 1995). Using the equivalent-width method, we claimed a detection of solar-like oscillations with amplitudes at the expected level and frequencies that were subsequently shown to be consistent with models (Christensen-Dalsgaard et al. 1995; Guenther & Demarque 1996). However, a search for velocity oscillations in η Boo by Brown et al. (1997) has failed to detect a signal, setting limits level below the value expected on the basis of the Kjeldsen et al. result.

Some support for the equivalent-width method was given by Keller et al. (1997), who detected the 5-minute oscillations in the Sun from measurements of H-beta equivalent widths. However, they have subsequently had difficulties in reproducing these results (Keller, priv. comm.).

α Cen A is the brightest G-type main-sequence star. We obtained Hα spectra over six nights in April 1995 using the 3.9-m AAT (UCLES) and the 3.6-m ESO (CASPEC). Data reduction using the equivalent-width method was hampered by a variability of the continuum, which seems to be due to some kind of colour term in scintillation at a level of about 10^{-4} per minute (well below the normal photometric scintillation).

Procyon is the brightest F-type subgiant. Recent results from Doppler-shift measurements are: (i) Bedford et al. (1995), using a narrow-band filter, have retracted an earlier possible detection; and (ii) Brown et al. (1996), using an echelle spectrograph, have not detected a signal. We observed Procyon for several hours per night during the 1995 run mentioned above. Preliminary analysis reveals excess power at the expected amplitude and frequency, but sparse sampling prevents a definite measurement of the frequency splitting. A recent campaign on Procyon in Jan–Feb 1997 by several members of SONG (see below) should produce results soon.

Arcturus and similar red giants are variable in both velocity (e.g., Hatzes & Cochran 1996 and references within; Merline 1996) and intensity (e.g., Edmonds & Gilliland 1996), but the presence of solar-like oscillations has not yet been established.

5. Conclusion

In the last few years, the precision in velocity and photometric measurements has not been significantly improved. The new equivalent-width method is far from being fully developed and no confirmation of the claimed signal in η Boo has been made. Hopefully, the formation of SONG (Stellar Oscillations Network Group; see `http://www.noao.edu/noao/song/`), which aims to do joint research in this field, will soon produce robust detections of oscillation signals.

Space would be a wonderful place to do photometry. Although COROT has been selected, for now we will have to continue using ground-based facilities. It is important to remember that we are only about a factor of two from producing noise levels equal to the solar oscillation signal, and that some stars are expected to oscillate with higher amplitudes than our own Sun. A network of 10-m class telescopes should provide scintillation levels low enough for detection of oscillations in M 67 (Gilliland et al. 1993), but getting a week on each of these big telescopes will not be easy.

We still await real asteroseismic results for solar-type stars. However, twenty-five years ago we were in a similar situation concerning oscillations in the Sun. First, people had to believe that these oscillations actually existed. Next, they had to measure their frequencies accurately. Finally, we have reached a stage where we truly see the Sun as a physics laboratory. The same will one day be true for other stars. It might take longer than twenty-five years, but it could also happen much faster.

Acknowledgements This work was supported in part by the Danish National Research Foundation through its establishment of the Theoretical Astrophysics Center. TRB is grateful for funding from the University of Sydney Research Grants Scheme and the Australian Research Council.

References

Bedding, T. R., Kjeldsen, H., Reetz, J., Barbuy, B., 1996, MNRAS 280, 1155
Bedford, D. K., Chaplin, W. J., Coates, D. W., et al., 1995, MNRAS 273, 367
Brown, T. M., Gilliland, R. L., 1994, ARA&A 33, 37
Brown, T. M., Kennelly, E. J., Noyes, R. W., et al., 1996, BAAS 188, 5902
Brown, T. M., Kennelly, E. J., Korzennik, S. G., et al., 1997, ApJ 475, 322
Christensen-Dalsgaard, J., Bedding, T. R., Kjeldsen, H., 1995, ApJ 443, L29
Edmonds, P. D., Gilliland, R. L., 1996, ApJ 464, L157
Frandsen, S., Jones, A., Kjeldsen, H., et al., 1995, A&A 301, 123
Gilliland, R. L., Brown, T. M., Kjeldsen, H., et al., 1993, AJ 106, 2441
Guenther, D. B., Demarque, P., 1996, ApJ 456, 798
Hatzes, A. P., Cochran, W. D., 1996, ApJ 468, 391
Keller, C. U., Harvey, J. W., Barden, S. C., et al., 1997, PASP (submitted)
Kjeldsen, H., Bedding, T. R., 1995, A&A 293, 87
Kjeldsen, H., Bedding, T. R., Viskum, M., Frandsen, S., 1995, AJ 109, 1313
Merline, W. J., 1996, BAAS 28, 860

Discussion of this paper appears at the end of these Proceedings.

CONSTRAINTS ON STELLAR INTERIOR PHYSICS FROM HELIOSEISMOLOGY

J. CHRISTENSEN-DALSGAARD
Teoretisk Astrofysik Center, Danmarks Grundforskningsfond, and Institut for Fysik og Astronomi, Aarhus Universitet, DK-8000 Aarhus C, Denmark

1. Introduction

Traditional observations of the properties of stars generally provide tests of only the gross aspects of stellar structure and evolution. The limitation lies in the amount and precision of the available data of relevance to the structure of the stellar interior, *i.e.*, the determination of stellar effective temperatures, surface composition, luminosities and, in a few cases, masses. Additional constraints on the observed stars, such as the common age and composition normally assumed for stars in clusters or multiple systems, clearly increase the information. However, detailed information on the physics and processes of stellar interiors requires more extensive data, with a dependence on stellar structure sufficiently simple to allow unambiguous interpretation. Such data are offered by observations of stellar pulsation frequencies: they can be observed with great accuracy and their dependence of stellar structure is generally well understood. In particular, the richness and precision of the observed frequencies of solar oscillation are now offering a detailed view on the interior properties of a star.

Reviews on solar oscillations were provided by, *e.g.*, Gough & Toomre (1991). The modes are characterized by a degree l measuring, approximately, the number of wavelengths in the stellar circumference. For each l, there is a set of possible modes of oscillation, characterized by the radial order n, and with angular frequencies ω_{nl}. In a spherically symmetric model the frequencies are independent of the azimuthal order m. This degeneracy is lifted by rotation or other departures from spherical symmetry.

In the Sun, modes are observed at each l between $l = 0$ and at least 2000, with cyclic frequencies $\nu = \omega/2\pi$ between about 1000 and 5000 μHz. The relative standard deviations are less than 5×10^{-6} in many cases, making

T.R. Bedding et al. (eds.),
***Fundamental Stellar Properties: The Interaction between Observation and Theory,* 285–292.**

these frequencies by far the most accurately known properties of the Sun. The observed modes are essentially standing acoustic waves, propagating between a point just below the photosphere and an inner turning point at a distance r_t from the centre such that $c(r_t)/r_t = \omega/\sqrt{l(l+1)}$, c being the adiabatic sound speed. Over the observed range of modes r_t moves from very near the centre of the Sun to just below the solar surface. This variation of the region to which the mode frequencies are sensitive permits inverse analyses to determine localized properties of the solar interior.

The analysis of the observed frequencies is generally aimed at determining differences between the Sun and reference solar models and hence inferring the errors in the assumed physics or other properties of the models. The quality of the helioseismic data has inspired considerable efforts to improve the solar model computations, by including as far as possible known processes and by using the most precise description of the physics available. In particular, diffusion and gravitational settling, which have often been neglected in the past, have a substantial effect on the models at this level of precision and hence must be included.

Here I shall use as reference the so-called Model S of Christensen-Dalsgaard *et al.* (1996), which includes diffusion and settling of helium and heavy elements; OPAL equation of state (Rogers, Swenson & Iglesias 1996) and opacity (Iglesias, Rogers & Wilson 1992) were used, as well as nuclear reaction rates largely from Bahcall & Pinsonneault (1995). This, as well as other models discussed here, were calibrated to present solar luminosity and radius, as well as to the observed surface ratio $Z_s/X_s = 0.0245$ between the heavy-element and hydrogen abundances, by adjusting the initial composition and the mixing-length parameter.

2. Results of helioseismic inversion

The differences between the observed and model frequencies are small, of order 0.3 % or less. This motivates analysis in terms of linearized relations between the frequency differences and differences in suitable sets of model variables. Here I use (c^2, ρ), ρ being density. It is possible to form linear combinations of the frequency differences, in such a way as to obtain a measure of the sound-speed difference $\delta c/c$, localized to a small region of the Sun, while suppressing the influence of the difference in ρ, of uncertainties in the modelling of the near-surface region of the Sun and of observational errors (for details, see for example Basu *et al.* 1996). Differences between the Sun and Model S obtained in this manner are shown in Fig. 1, based on frequencies resulting from a combination of observations from the BiSON network (Chaplin *et al.* 1996) and the LOWL instrument (Tomczyk *et al.* 1995). As indicated by the horizontal bars, the points provide averages of

$\delta c^2/c^2$ over limited regions in the Sun, from the centre to near the surface. Furthermore, the estimated random errors in the result, based on the quoted errors in the observed frequencies, are minute, less than 2×10^{-4} in most of the solar interior. Thus it is in fact possible to measure a property of the solar interior as a function of position, with great precision.

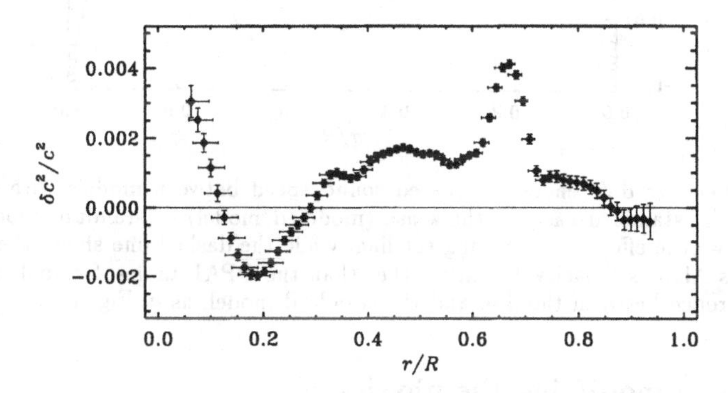

Figure 1. Inferred difference in squared sound speed, in the sense (Sun) - (model). The horizontal error bars mark the first and third quartile points of the averaging kernels, whereas the vertical error bars show 1-σ errors, as progated from the errors in the observed frequencies. From Basu *et al.* (1997).

It is evident, also, that at a superficial level the agreement between the Sun and the model is excellent: we have been able to predict the solar sound speed with a precision of better than 0.2 %. This has required improvements in the modelling inspired by the high accuracy of the observations; however, the calculation involves no adjustment of parameters to fit the model to the data. On the other hand, the difference between the Sun and the model is highly significant, given the very small error in the inferred difference. Thus, in that sense, the model is hardly satisfactory.

The oscillations depend essentially only on the dynamical properties of the Sun, *e.g.*, pressure, density and sound speed. Since, approximately, $c^2 \propto T/\mu$, where T is temperature and μ the mean molecular weight, helioseismology constrains T/μ but not T and μ separately. This has ramifications for the possibility of helioseismic constraints on the solar neutrino production (*e.g.* Antia & Chitre 1995; Christensen-Dalsgaard 1997).

The dependence of the oscillation frequencies on azimuthal order m carries information about the solar internal angular velocity. Inversion of this dependence shows that in the convection zone rotation varies with latitude approximately as on the surface, with modest dependence on r. Near the base of the convection zone there is a rapid transition, over less than about $0.1R$, to rotation depending little on r and latitude in much of the radiative interior (*e.g.* Thompson *et al.* 1996; Kosovichev 1996).

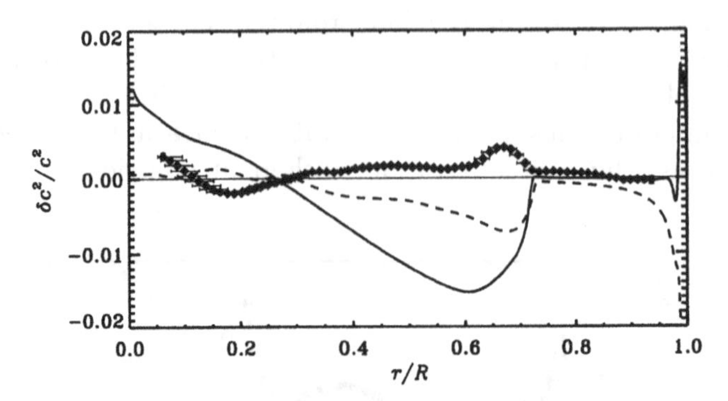

Figure 2. Relative differences in squared sound speed between models with modified physics and the standard case, in the sense (modified model) – (standard model). The solid line shows the effect of neglecting settling, while the dashed line shows the effect of using the Los Alamos Opacity Library rather than the OPAL tables. Symbols show the inferred difference between the Sun and the standard model, as in Fig. 1.

3. Effects of modifying the physics

To evaluate the significance of the comparatively close agreement between the "standard" Model S and the Sun it is necessary to consider other models with differing assumptions or physics. An important example is the inclusion of settling and diffusion. Figure 2 compares the difference between a model without these effects, but using otherwise the same physics, and Model S, with the difference between the Sun and Model S. It is evident that neglect of settling would very considerably worsen the agreement between the Sun and the models (see also Cox, Guzik & Kidman 1989; Christensen-Dalsgaard, Proffitt & Thompson 1993; Bahcall *et al.* 1997). The figure also shows the effect of replacing the OPAL opacities with the older Los Alamos Opacity Library (Huebner *et al.* 1977). Clearly the revision of the opacity has improved the agreement between the model and the Sun considerably, although less so than the inclusion of settling. This also suggests that the apparent improvement brought about by including settling is not compromised by opacity errors, as suggested by Elliott (1995).

Very considerable effort has gone into work on the equation of state (*e.g.* Däppen 1992), with corresponding improvements in the agreement between the resulting models and the observed frequencies (*e.g.* Christensen-Dalsgaard, Däppen & Lebreton 1988). Detailed analyses have demonstrated the ability of the helioseismic data to probe subtle properties of the thermodynamic state of the solar plasma (*e.g.* Christensen-Dalsgaard & Däppen 1992; Vorontsov, Baturin & Pamyatnykh 1992; Elliott 1996); this allows the use of the solar convection zone, where the structure depends largely on the equation of state, as a laboratory for plasma physics.

4. What is wrong with the solar model?

Despite the improvements in solar modelling and in the agreement between the model and the Sun, the remaining highly significant discrepancy between the model and the Sun indicates a lack in our understanding of stellar evolution; although modest in the solar case, the error could quite possibly have more substantial effects in other stars where conditions are more extreme than in the Sun.

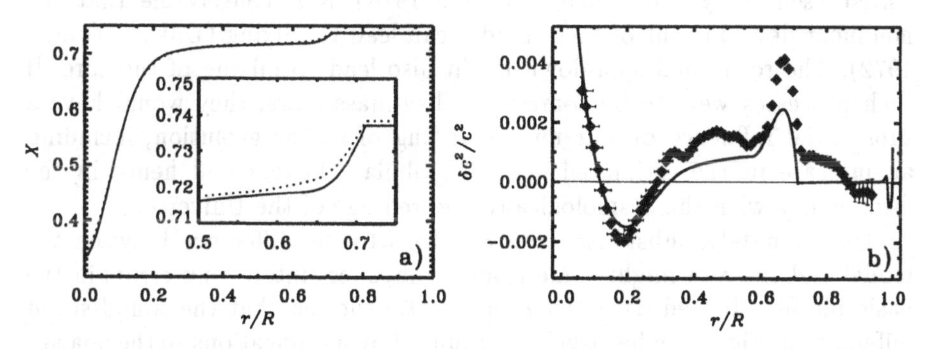

Figure 3. (a) Profiles of the abundance X by mass of hydrogen. The solid line shows the profile in Model S of Christensen-Dalsgaard *et al.* (1996), whereas the dotted line shows a modified profile aimed at trying to match the sound-speed difference shown in Fig. 1 between the Sun and the model. (b) Difference in squared sound speed between the model with modified X-profile and Model S. The symbols show the inferred difference between the Sun and Model S, as in Fig. 1. Adapted from Bruntt (1996).

The dominant features in the sound-speed difference occur in regions of the model with strong composition gradients (*cf.* Fig. 3), resulting from nuclear burning in the core or helium settling from the convection zone. These gradients would be affected by "non-standard" processes causing mixing in convectively stable regions. Mixing in the core could increase the hydrogen abundance at the centre of the model while reducing it at the edge of the core; the central sound speed would therefore be increased, and the sound speed at the edge of the core reduced, as required by Fig. 1. Similarly, weak mixing beneath the convection zone would increase the hydrogen abundance and sound speed in this region, again potentially according for the observed bump. As a toy model of such processes, Fig. 3 also shows an artificially modified hydrogen profile and the corresponding change in the sound speed, confirming that redistribution of hydrogen can in fact largely account for the observed behaviour.

Such suggestions evidently require physical mechanisms for the mixing. Just beneath the convection zone the steep gradient in the helioseismically inferred rotation rate is likely to be associated with circulation which could cause mixing (Spiegel & Zahn 1992; Gough *et al.* 1996; Elliott 1997). Mixing might also be caused by instabilities associated with the spin-down of the Sun from the usually assumed initial state of rapid rotation (*e.g.* Chaboyer *et al.* 1995), or by penetration of convection beyond the unstable region. Independent evidence for mixing beneath the convection zone is provided by the destruction of lithium and beryllium (*e.g.* Chaboyer *et al.* 1995).

There appears to be no similarly simple explanation of potential core mixing. However, the Sun has been shown to be unstable to low-order, low-degree g modes, at least during earlier phases of its evolution (*e.g.* Christensen-Dalsgaard, Dilke & Gough 1974); it is conceivable that the nonlinear development of these modes can lead to mixing (Dilke & Gough 1972). The rotational spin-down might also lead to mixing of the core. If such processes were to be common to low-mass stars, they would have a substantial influence on our understanding of stellar evolution, including an increase in the estimated ages of globular clusters and hence in the discrepancy with the cosmologically inferred age of the Universe.

Unfortunately, substantial contributions to the difference between the model and the Sun might come from perhaps less interesting errors in the basic physics. Indeed Tripathy *et al.* (1997) showed that the sound-speed difference in Fig. 1 can be largely reproduced by modifications to the opacity of less than about 5 %. While this is certainly smaller than the generally assumed uncertainty in current opacity calculations, it remains to be seen whether the specific change required is physically plausible.

Finally, I note that there is evidence for errors in the equation of state in and below the helium ionization zone (*e.g.* Dziembowski, Pamyatnykh & Sienkiewicz 1992), even when using the OPAL equation of state (*e.g.* Basu & Christensen-Dalsgaard 1997). The effect is small in the Sun, but it might be substantial in lower-mass stars where non-ideal plasma effects could be stronger.

5. Relation to stellar astrophysics

The helioseismic results clearly give some confidence in modelling of stellar evolution. However, in part this relative success of the solar models undoubtedly stems from the fact that the Sun is a comparatively simple type of star: for example, at only slightly higher mass than solar the problems of a convective core would play a major role. The ability to cover a broad range of parameters makes investigations of other stars, "classical" as well as seismological, an essential complement to the solar studies, even though they can never be as detailed and precise as those obtained for the Sun. For example, discrepant period ratios in models of double-mode Cepheids and δ Scuti stars led to the prediction of a substantial increase in opacities (*e.g.* Simon 1982; Andreasen & Petersen 1988), at temperatures in the range $10^5 - 10^6$ K; this falls within the solar convection zone and hence would have no effect on solar structure. The opacity increase in the OPAL tables has in fact largely resolved this discrepancy (*e.g.* Moskalik, Buchler & Marom 1992). Similarly, properties of convective cores, including the important but highly uncertain question of mixing beyond the region of instability, might well be studied from observations of solar-like oscillations in other stars (Kjeldsen, these proceedings) or observation of sufficiently detailed spectra of oscillations in, for example, δ Scuti or β Cephei stars.

Independent stellar information may also help to compensate for the non-uniqueness in the physical interpretation of the helioseismic data, illustrated in the preceding section. For example, investigations of element abundances may provide further insight into the physics of mixing beneath stellar convection zones (*e.g.* Baglin & Lebreton 1990).

6. Concluding remarks

Major advances in helioseismology will result from the extensive new data from the GONG network (*e.g.* Harvey *et al.* 1996) and from the instruments on the SOHO satellite (*e.g.* Scherrer *et al.* 1996), as well as from the continued observations from other ground-based instruments. As a result, we shall be able to investigate in more detail solar structure, not least in the core and the convection zone, as well as rotation and other aspects of solar internal dynamics. In parallel with this, major advances in the study of stellar oscillations may lead to the definite detection of solar-like oscillations in other stars and a detailed analysis of oscillations in δ Scuti stars and other "classical" pulsators. Finally, the new very large telescopes and advances in stellar-atmosphere modelling (Gustafsson, these proceedings) are likely to lead to major improvements in our knowledge about stellar composition and the processes that control it, for a variety of stars.

In this way we shall obtain much firmer tests of stellar modelling, reaching beyond the fundamental properties of stars to the even more fundamental questions of their physical basis.

Acknowledgements

I thank the authors of Basu *et al.* (1997) for permission to quote results in advance of publication. This work was supported in part by the Danish National Research Foundation through its establishment of the Theoretical Astrophysics Center.

References

Andreasen G. K. & Petersen J. O., 1988. *Astron. Astrophys.*, **192**, L4
Antia H. M. & Chitre S. M., 1995. *Astrophys. J.*, **442**, 434

Baglin A. & Lebreton Y., 1990. *Proc. IAU Colloquium No 121, Inside the Sun*, p. 437 eds Berthomieu G. & Cribier M., Kluwer, Dordrecht.
Bahcall J. N. & Pinsonneault M. H., 1995. *Rev. Mod. Phys.*, **67**, 781
Bahcall J. N., Pinsonneault M. H., Basu S. & Christensen-Dalsgaard J., 1997. *Phys. Rev. Lett.*, **78**, 171
Basu S. & Christensen-Dalsgaard J., 1997. *Astron. Astrophys.*, submitted.
Basu S. *et al.*, 1997. *Mon. Not. R. astr. Soc.*, submitted.
Basu S., Christensen-Dalsgaard J., Pérez Hernández F. & Thompson M. J., 1996. *Mon. Not. R. astr. Soc.*, **280**, 651
Bruntt H., 1996. *Batchelor thesis*, Aarhus University.
Chaboyer B., Demarque P., Guenther D. B. & Pinsonneault M. H., 1995. *Astrophys. J.*, **446**, 435
Chaplin W. J. *et al.*, 1996. *Mon. Not. R. astr. Soc.*, **282**, L15
Christensen-Dalsgaard J., 1997. In *Proc. 18th Texas Symposium on Relativistic Astrophysics*, eds Olinto A., Frieman J. & Schramm D., World Scientific Press, in press.
Christensen-Dalsgaard J. & Däppen W., 1992. *Astron. Astrophys. Rev.*, **4**, 267
Christensen-Dalsgaard J., Däppen W., Ajukov S. V. *et al.*, 1996. *Science*, **272**, 1286
Christensen-Dalsgaard J., Däppen W. & Lebreton Y., 1988. *Nature*, **336**, 634
Christensen-Dalsgaard J., Dilke F. W. W. & Gough D. O., 1974. *Mon. Not. R. astr. Soc.*, **169**, 429
Christensen-Dalsgaard J., Proffitt C. R. & Thompson M. J., 1993. *Astrophys. J.*, **403**, L75
Cox A. N., Guzik J. A. & Kidman R. B., 1989. *Astrophys. J.*, **342**, 1187
Däppen W., 1992. *Proc. Workshop on Astrophysical Opacities, Revista Mexicana de Astronomia y Astrofisica*, **23**, eds Lynas-Gray C., Mendoza C. & Zeippen C., 141
Dilke F. W. W. & Gough D. O., 1972. *Nature*, **240**, 262
Dziembowski W. A., Pamyatnykh A. A. & Sienkiewicz R., 1992. *Acta Astron.*, **42**, 5
Elliott J. R., 1995. *Mon. Not. R. astr. Soc.*, **277**, 1567
Elliott J. R., 1996. *Mon. Not. R. astr. Soc.*, **280**, 1244
Elliott J. R., 1997. *Astron. Astrophys.*, submitted.
Gough D. O. & Toomre J., 1991. *Ann. Rev. Astron. Astrophys.*, **29**, 627
Gough D. O., Kosovichev A. G., Toomre J. *et al.*, 1996. *Science*, **272**, 1296
Harvey J. W., Hill F., Hubbard R. P. *et al.*, 1996. *Science*, **272**, 1284
Huebner W. F., Merts A. L., Magee N. H. & Argo M. F., 1977. *Astrophysical Opacity Library*, Los Alamos Scientific Laboratory Report LA-6760-M.
Iglesias C. A., Rogers F. J. & Wilson B. G., 1992. *Astrophys. J.*, **397**, 717
Kosovichev A. G., 1996. *Astrophys. J.*, **469**, L61
Moskalik P., Buchler J. R. & Marom A., 1992. *Astrophys. J.*, **385**, 685
Rogers F. J., Swenson F. J. & Iglesias C. A., 1996. *Astrophys. J.*, **456**, 902
Scherrer P. H., Bogart R. S., Bush R. I. *et al.*, 1996. *Solar Phys.*, **162**, 129
Simon N. R., 1982. *Astrophys. J.*, **260**, L87
Spiegel E. A. & Zahn J.-P., 1992. *Astron. Astrophys.*, **265**, 106
Thompson M. J., Toomre J., Anderson E. R. *et al.*, 1996. *Science*, **272**, 1300
Tomczyk S. *et al.*, 1995. *Solar Phys.*, **159**, 1
Tripathy S. C., Basu S. & Christensen-Dalsgaard J., 1997. In *Poster Volume; Proc. IAU Symposium No 181*, eds Schmider F.X. & Provost J., Nice Observatory, in press.
Vorontsov S. V., Baturin V. A. & Pamyatnykh A. A., 1992. *Mon. Not. R. astr. Soc.*, **257**, 32

Discussion of this paper appears at the end of these Proceedings.

BULGE δ SCUTI STARS IN THE MACHO DATABASE

D. MINNITI[1], C. ALCOCK[1], D.R. ALVES[1], T.S AXELROD[2], A.C BECKER[3], D.P. BENNETT[4], K.H. COOK[1], K.C. FREEMAN[2], K. GRIEST[5], M.J. LEHNER[5], S.L. MARSHALL[1], B.A. PETERSON[2], P.J. QUINN[6], M.R. PRATT[3], A.W. RODGERS[2], C.W. STUBBS[3], W. SUTHERLAND[7], A. TOMANEY[3], T. VANDEHEI[5], D. WELCH[8],

[1]*LLNL;* [2]*MSSSO;* [3]*U. Washington;* [4]*Notre Dame;* [5]*UC San Diego;* [6]*ESO;* [7]*Oxford U.;* [8]*McMaster U.;*

1. Introduction

We describe the search for δ Scuti stars in the MACHO database of bulge fields. Concentrating on a sample of high amplitude δ Scutis, we examine the light curves and pulsation modes. We also discuss their spatial distribution and evolutionary status using mean colors and absolute magnitudes.

The 61" telescope at Mt. Stromlo Observatory observes once a night several bulge or LMC fields in two bandpasses. While the MACHO system is optimized for microlensing, it has also had great success detecting variable stars (Cook et al. 1995). We have searched for δ Scuti stars in the bulge fields (these variable stars are beyond the detection limit in the LMC fields). The δ Scutis have very short periods, typically between 1 hr and 6 hr (e.g. Breger 1995, Rodriguez et al. 1994), which may be challenging using a once-a-day observing routine. Amazingly enough, the light curves of δ Scu stars phase very well for periods as short as 0.08 days. Because subsequent observations are separated by 5-15 pulsation periods, the phases are randomized, giving a rather uniform light curve coverage. Also, because we have so many datapoints per star ($\sim$ 500 over four seasons), aliasing is not a significant problem. The quality of the light curves (Figure 1) say more than a thousand words...

T.R. Bedding et al. (eds.),
***Fundamental Stellar Properties: The Interaction between Observation and Theory*, 293–298.**

The δ Scu stars are selected using the period-amplitude diagram for all stars with $P < 0.20^d$, and amplitudes $A_V > 0.1$ mag in the MACHO blue band, taking for comparison the catalogue of Rodriguez et al. (1994). We demand that $P_V = P_R$ to within 2%. The δP criterion implies that we keep mainly the variables that have good quality light curves, and that are not multi-mode pulsators. The lower cut in the amplitudes is imposed to secure good data, even though small amplitude δ Scu stars will also be neglected. We also demand that $A_R < 0.8 A_V$, in order to eliminate eclipsing binaries which have $A_V \approx A_R$. Finally, the identification of these large amplitude δ Scu candidates is confirmed by visual inspection of the light curves.

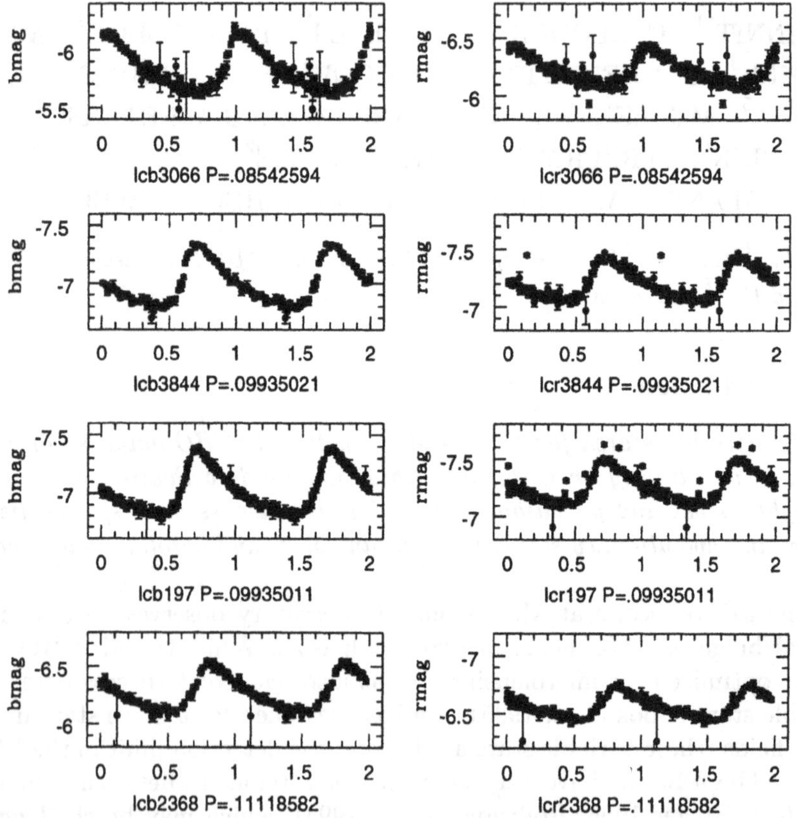

Fig. 1 - Typical blue (left panels) and red (right panels) δ Scu light curves.

Thus, we do not consider the low amplitude, the short period, and most double or multi-mode δ Scu stars. Even though the sample is incomplete, it is adequate to examine the extension of the existing P-L-Z relations into the metal-rich domain, and to determine the evolutionary status of these stars in the bulge.

2. Light Curves: Fundamental, 1st Overtone, or 2nd Overtone?

The light curve shapes of the δ Scu stars in our sample are similar to short-period versions of RR Lyrae type ab and type c variables. Theoretical light curves for δ Scu stars pulsating in the fundamental, first overtone, and second overtone modes have been recently computed by Bono et al. (1997). In particular, they predict for the first time the occurrence of stable second overtone pulsators. While a more detailed exploration of the full parameter space is needed, these models (Bono et al. 1997, their Figures 3–7) reproduce the observed light curves of our sample very well.

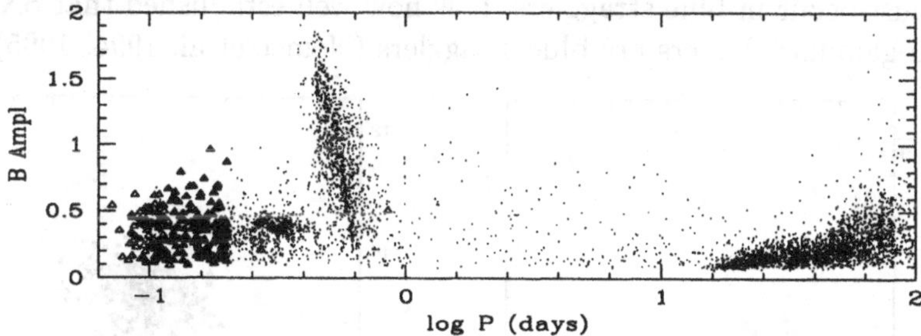

Fig. 2 - Period-amplitude diagram for pulsating bulge variables with P< 100^d. The δ Scu are plotted with triangles. RR Lyrae stars type ab, c, and e are also seen at *log P* ≈ −0.25, −0.5, and −0.6 days, respectively.

Few fundamental δ Scu pulsators would be included in the sample, since these would have $P > 0.12^d$ and $A \leq 0.1$ mag. Based on these periods and amplitudes, it appears that the majority of the sample stars would be second overtone pulsators, with a few first overtone pulsators.

Bono et al. (1997) find that, contrary to different RR Lyrae pulsators, the second overtone δ Scu pulsators show larger amplitudes, in spite of having smaller periods than the first overtone or the fundamental pulsators. This is in direct contradiction with the assumption that, like RR Lyrae and Cepheid stars, the higher modes have lower amplitudes (e.g. McNamara 1995). Furthermore, the theoretical light curves of Bono et al. (1997) show that fundamental δ Scu pulsators should have sinusoidal light curves, and that first or second overtone pulsators should have asymmetric light curves. This also is in direct contradiction with the assumption that, like RR Lyrae stars, the asymmetric light curves are indicative of fundamental mode pulsation, with sinusoidal light curves present in first –and maybe also in second– overtone pulsators (e.g. Nemec et al. 1995, McNamara 1995).

3. The Evolutionary Status of δ Scu Stars in the Bulge Fields

The reddening in the bulge fields is very inhomogeneous. From the VR photometry it is convenient to use the reddening independent magnitude

W_V, defined as $W_V = V - 3.97 \times (V - R)$, which assumes a standard extinction law for the bulge fields. Most of the δ *Scu* stars in our sample belong to the Galactic bulge; their magnitudes peak at $W_V = 15.9$, which places them at about 8 kpc.

The δ Scu are core hydrogen burning stars located inside the instability strip, on the main sequence or just above the main sequence (Breger 1995, Rodriguez et al. 1994). For a population older than about 10^{10} yr, the instability strip lies bluewards of, and it is brighter than the main sequence turn-off. Thus, in old populations, the pulsating variables in the instability strip must occur in blue stragglers. It is now well established that SX Phe stars in globular clusters are blue stragglers (Nemec et al. 1994, 1995).

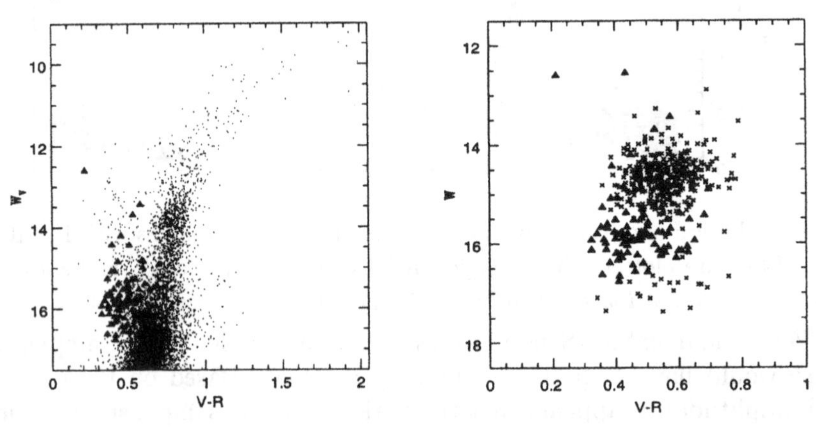

Fig. 3 - Color-magnitude diagrams of bulge δ Scuti stars, compared with Baade's window stars (left panel), and bulge RR Lyrae stars (right panel).

Are the bulge δ Scu stars also blue stragglers? In order to answer this question, Figure 3a shows the MACHO W_V *vs* $V - R$ color-magnitude diagram of Baade's window (only 5000 stars are plotted). This color-magnitude diagram shows a prominent red giant branch (RGB), a red horizontal branch (or RGB clump) at $W_V = 13.5 - 14$, and the blue disk main sequence with $V - R < 0.5$. The bulge turn-off would be located at the faint limit of our photometry in this field. Note, however, that we reach the bulge turn-off in other MACHO fields which are less crowded than Baade's window. The δ Scu stars with $0.08^d < P < 0.20^d$ from all the bulge fields are also plotted (filled triangles) in Figure 3a. From this Figure we conclude that their colors and reddening-independent magnitudes are consistent with bulge blue stragglers. This is confirmed also by the location of δ Scuti stars (triangles) with respect to bulge RR Lyrae type ab (crosses, from Alcock et al. 1997) in the color magnitude diagram shown in Figure 3b.

4. The Distribution along the Line-of-sight

Figure 4a shows the W_V magnitude distribution of δ Scu and RR Lyr stars. The δ Scu distribution is very peaked, with $FWHM = 0.6 \pm 0.05$ mag. This argues for δ Scu stars pulsating in a single dominant mode. The observed magnitude scatter is consistent with the line-of-sight depth of the Galactic bulge, and it would be larger if pulsators in two or more different modes are represented. Also note that the observed FWHM is slightly smaller than that of RR Lyrae type ab stars. This means that the metallicity dependence of the magnitude $\Delta M_V/\Delta[Fe/H]$ is smaller, either because the RR Lyrae have a wider metallicity distribution, or because their spatial distribution is slightly different.

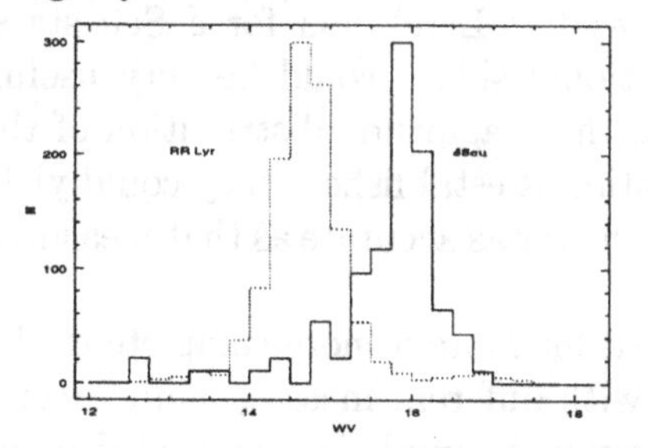

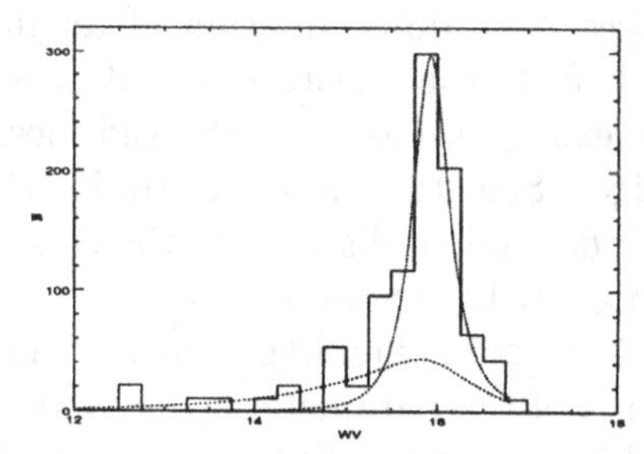

Fig. 4 - Luminosity distribution of δ Scu stars (arbitrary normalization).

We assume that the line of sight distributions of δ Scu and RR Lyrae stars are similar. In the sky, these stars are concentrated towards the inner fields, as expected if they follow the bulge radial density profile rather than the disk, which would show a more uniform, shallower distribution.

In order to decide if the stars in the present sample belong to the disk or the bulge, one can assume that all δ Scu stars have roughly the same magnitude, and compute the expected magnitude distribution for these two Galactic components integrated along the line of sight. Figure 4b shows the observed magnitude distribution, along with the line of sight density distribution expected for the bulge and the disk (upper and lower dotted curves, respectively). The disk is assumed to be a double exponential of the form $\rho_d \propto exp(-r/h_r)exp(-z/h_z)$, with scale-height $h_z = 0.3$ kpc, and scale-length $h_r = 3$ kpc, and the bulge is assumed to be approximated by a power law of the form $\rho_b \propto r^{-n}$, with $n = 3.5$. The counts have been normalized arbitrarily. If all the δ Scu stars from Table 1 were disk stars, the FWHM of their observed magnitude distribution would be ~ 2 mag, as seen in Figure 4b. We conclude that the majority of the stars in our sample are bulge δ Scu stars. However, about 10% of the δ Scu stars shown in Figure 4 are significantly brighter than the mean. These could be δ Scu stars in the foreground disk, or merely blends.

5. Conclusions

We have analyzed a sample of δ Scu variable stars within a narrow period range, $0.08^d < P < 0.20^d$, identified in the MACHO bulge database. The colors and magnitudes of these stars are consistent with them being bulge blue stragglers. A comparison of the observed light curves with the recent theoretical models of Bono et al. (1997) suggests that the present sample consists mainly of second overtone pulsators.

We note that adopting the P–L–Z relations of Fernie (1992), McNamara (1995), or Nemec et al. (1995), different mean absolute magnitudes can be obtained. Distances measured using these δ *Scu* stars are presently very uncertain, because of the unknown metallicity dependence and pulsational stages. The determination of an improved P–L relation for δ Scu stars, both from the theoretical and observational sides, would be very useful. For example, based on the tightness of the magnitude distribution of the bulge δ Scu stars, when a firm P–L relation is established, they could yield an independent distance to the Galactic center as accurate as that measured using RR Lyrae stars.

We would then like to stress the need for more a more complete exploration of the parameter space (models with different masses, temperatures and luminosities). Spectroscopy of the present sample is also needed in order to determine the abundances and masses of these stars, and to decide if δ Scu stars are the mere metal-rich extension of SX Phe stars.

Acknowledgements

We would like to thank T. Bedding for useful suggestions, and to all the organizers for a great meeting. We are always very grateful for the skilled support by the technical staff at MSO. Work at LLNL is supported by DOE contract W7405-ENG-48. Work at the CfPA is supported NSF AST-8809616 and AST-9120005. Work at MSSSO is supported by the Australian Department of Industry, Technology and Regional Development.

References

Alcock, C., et al. (The MACHO Collaboration) 1997, ApJ, 474, 217
Bono, G., Caputo, F., et al. 1997, A&A in press (astro-ph/9609087)
Breger, M. 1995, Baltic Ast., 4, 423
Cook, K. H., et al (The MACHO Collaboration) 1995, ASP No. 83, p. 221
Fernie, J. D. 1992, AJ, 103, 1647
McNamara, D. H. 1995, AJ, 109, 1571
Nemec, J. M., Mateo, M., Burke, M., & Olszewski, E. 1995, AJ, 110, 1186
Nemec, J. M., Nemec, A. F. L., & Lutz, T. E. 1994, AJ, 108, 222
Rodriguez, E., Lopez de Coca, P., Rolland, A., et al. 1994, A&AS, 106, 21

PROPERTIES OF CEPHEIDS AND LONG-PERIOD VARIABLES

P.R. WOOD
Mount Stromlo and Siding Spring Observatories
Private Bag, Weston Creek PO, ACT 2611, Australia
wood@mso.anu.edu.au

Abstract. Some recent theoretical results concerning Cepheids and long-period variables are presented and compared with observations. Nonlinear pulsation calculations for select bump Cepheids are shown capable of yielding very accurate fundamental properties of these stars. The implications of the derived parameters are discussed. For the long-period variables, recent observational and theoretical results pertaining to the long-standing problem of the mode of pulsation are presented. The conflicting results point to some fundamental problem with current interpretations.

1. Cepheid variables

There are a number of ways in which the fundamental parameters of Cepheid variables (mass M, luminosity L, effective temperature T_{eff}, abundance) can be determined. Firstly, abundances must always be determined by spectroscopic means, or adopted, and will not be considered further. T_{eff} can be determined by comparing colours and spectra with model atmospheres while luminosity can be obtained if a distance is known, for example, from cluster or external galaxy membership. Then the relation between M, L, T_{eff} and pulsation period P given by *linear* pulsation theory can be used to determine M. This method, which yields the so-called pulsation mass, has been known for a long time but suffers from observational uncertainty in distance modulus and, most importantly, uncertainty in the calibration of T_{eff} (the derived pulsation mass varies as ${T_{\text{eff}}}^{5.1}$ - Chiosi, Wood & Capitanio 1993).

Another way in which fundamental properties of Cepheids can be derived from *linear* pulsation theory is by examining Cepheids which exhibit multiple modes of pulsation, the beat Cepheids. The observed period ratios

T.R. Bedding et al. (eds.),
Fundamental Stellar Properties: The Interaction between Observation and Theory, 299–304.

of the modes in the beat Cepheids can only be reproduced theoretically by Cepheids which obey a specific M-L relation (Petersen 1973; Moskalik, Buchler & Marom 1992; Christensen-Dalsgaard & Petersen; Buchler et al. 1996). This M-L relation comes from stellar evolution theory and is quite uncertain due to the uncertainties in the amount of convective core overshoot that occurs during main-sequence evolution. One might hope that the Cepheids can actually calibrate the M-L relation and hence, indirectly, the amount of main-sequence convective core overshoot.

Another feature of Cepheids related to multiple modes is the existence of a bump on the light curve. The bump is associated with a resonance between the fundamental and second overtone modes and occurs when $P_0/P_2=2$ (Simon & Schmidt 1976). Fourier analysis of light curves of various periods shows that the resonance condition is satisfied exactly when $P_0 \approx 10$ days for Galactic and LMC Cepheids. Once again, assuming a M-L relation, the characteristic M-L relation for bump Cepheids can be derived from linear pulsation theory (Moskalik et al. 1992; Buchler et al. 1996).

2. Nonlinear models for bump Cepheids

As mentioned above, in the bump Cepheids, it is generally assumed that the ratio P_0/P_2 is identically equal to 2.0. However, this identity can not be satisfied exactly except in rare stars. A glance at a compendium of Cepheid light curves (Gaposhkin 1970) shows that, at a given P, there are light curves with a wide range of amplitudes and shapes. Similarly, the light curve Fourier decomposition parameters scatter about a mean trend line in plots such as ϕ_{21} vs $log P$ (Welch et al. 1995). Both these situations can be explained by the fact that the instability strip has a finite width, so that at a given P, Cepheids exist with a range in M, $T_{\rm eff}$ and L, and hence light curve shape and P_0/P_2.

A method of obtaining very accurate parameters for a Cepheid, is to carry out nonlinear modelling of the lightcurve. Figure 1 shows the result of such modelling for the bump Cepheid HV905 in the LMC (Wood, Arnold and Sebo 1997). The only assumption made during this modelling is the abundance of the Cepheid. In particular, no evolutionary M-L relation is assumed. The nonlinear models are forced to fit the V light curve shape and amplitude only. The offset in V-I of the models from the observations gives the reddening. The properties derived for this Cepheid are: $M = 5.20 \pm 0.2 M_\odot, L = 4897 \pm 150 L_\odot, E_{B-V} = 0.105$ and distance modulus 18.53 ± 0.04. The distance modulus derived is in excellent agreement with the distance modulus to the LMC derived from SN1987A(Panagia et al. 1991). The mass is much lower than expected for the luminosity according to standard stellar evolution calculations which predict a mass of $\sim 7.6 M_\odot$.

Stellar evolution calculations that include moderate amounts of convective core overshoot during main-sequence evolution predict a mass of $\sim 6.8 M_{\odot}$. The result derived for HV905 suggests that a large amount of main-sequence convective core overshoot is required in stars with $M \sim 5 M_{\odot}$. The linear pulsation analysis of LMC bump Cepheids yields similar results regarding the M-L relation (Buchler et al. 1996).

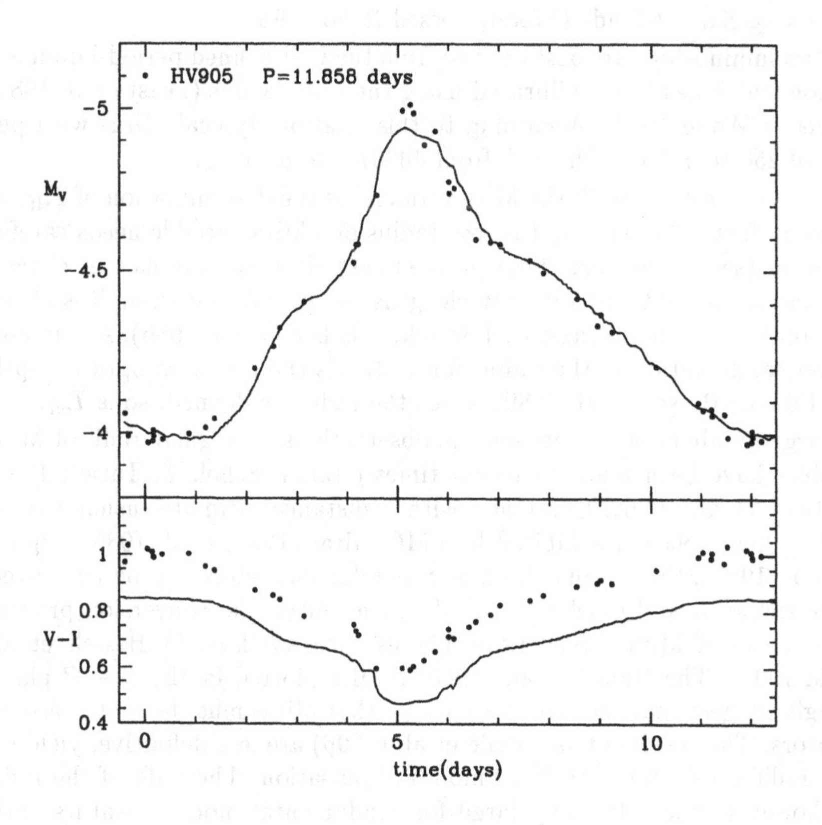

Figure 1. The V and $V - I$ light curves of the LMC bump Cepheid HV905 (dots) compared with theoretical simulations (continuous lines).

3. The long-period variables

The fundamental parameters for long-period variables (LPVs) are much less well known than those of the Cepheids. Here, discussion will be confined to the large amplitude LPVs known as Mira variables. Masses are known only in a statistical sense from studies of Galactic kinematics and scale heights (Feast 1963; Jura & Kleinmann 1992). The mean (initial) Mira mass in the solar vicinity is $\sim$1-1.2$M_{\odot}$, but since Miras occur in metal-rich globular clusters such as 47 Tuc, initial masses as small as 0.8 $M_{\odot}$ must occur. Because of mass loss, the Mira mass is less than the initial mass, but it has a lower bound corresponding to the mass of the degenerate core, which is $\sim$0.6$M_{\odot}$ for Miras with $L \sim 5000L_{\odot}$ (Wood & Zarro 1981). On the other hand, large-amplitude LPVs with masses up to 5-7$M_{\odot}$ are known to occur in the Magellanic Clouds (Wood, Bessell & Fox 1983).

Mira luminosities are best derived from the well-defined period-luminosity relation which has been calibrated using the LMC Miras (Feast et al. 1989; Hughes & Wood 1990). According to this relation, typical Miras with periods of 250 to 400 days have L from $4000L_{\odot}$ to $6200L_{\odot}$.

The real problem with the Mira variables is the determination of $T_{\rm eff}$, or radius. In fact, what is meant by the radius of a Mira variable needs careful definition (see M. Scholz, these proceedings) since the radius can change by a factor of 2 at different wavelengths or by 25% between Rosseland mean optical depth 0.01 and 1.0 (Bessell, Scholz & Wood 1996). A common theoretical definition of the radius for a Mira is the radius at optical depth $\frac{2}{3}$ or 1.0 (see Bessell et al. 1989). Once the radius is defined, so is $T_{\rm eff}$.

Large numbers of interferometric observations of angular radii of Mira variables have been made in recent times (Haniff, Scholz & Tuthill 1995; van Belle et al. 1996). Combined with a distance estimate, usually from the $K - logP$ relation exhibited by LMC Miras (Feast et al. 1989; Hughes & Wood 1990), these monochromatic angular diameters can be converted to the radius R at Rosseland optical depth unity. The conversion process involves use of Mira model atmospheres such as those in Bessell et al. (1989; 1996). The Haniff et al. (1995) results plotted in the $R - P$ plane strongly suggest that the radii are so large that Miras must be first overtone pulsators. The results of van Belle et al. (1996) are less definitive, yielding some radii consistent with both modes of pulsation. The bulk of the radii are, however, uncomfortably large for fundamental mode pulsators with M$\sim M_{\odot}$.

In spite of these results, there are good reasons to believe that the Mira variables are really fundamental mode pulsators with radii only $\frac{2}{3}$ as large as suggested by the direct angular diameter measurements.

The first problem with the large radii is a theoretical one. Mira variables are observed to have radial velocity pulsation amplitudes of $\sim$25 km s^{-1} (Hinkle, Scharlach and Hall 1984), or >30 km s^{-1} when corrected for limb darkening. With the large radii suggested by the results of Haniff et al. and van Belle et al., the gravities are too small to produce the observed pulsation velocity amplitudes, no matter how hard the pulsation in the outer layers is driven (Bowen 1988; Bessell et al. 1996).

The second reason to suspect that Miras are first overtone pulsators is provided by comparison of the $K - logP$ relations for Mira and small amplitude semi-regular LPVs in the LMC. Wood & Sebo (1996) have shown that the Miras and semi-regulars follow parallel $K - logP$ sequences that are separated by ~0.35 in $logP$. This is just as expected if the Miras are fundamental mode pulsators and the semi-regulars are first (or second) overtone pulsators. Furthermore, calculations of the pulsation periods of stars on the LMC giant branch predict that the fundamental mode periods are very similar to those observed while the first overtone periods are much shorter.

Obviously, there is still a serious conflict between the various pieces of observational/theoretical evidence regarding the pulsation mode, or radii, of the Miras. The Wood & Sebo results are based on a small number of stars and need confirming with a larger sample. But, if the Miras are indeed fundamental mode pulsators, then the large angular diameters measured by the interferometric techniques will need an explanation. Perhaps the solution will be improved model atmospheres for converting the observed monochromatic radii into "theoretical" radii.

References

Bessell, M.S., Brett, J.M., Scholz, M., Wood, P.R. 1989, A&A, 213, 209
Bessell, M.S., Scholz, M., Wood, P.R. 1996, A&A, 307, 481
Bowen, G.H. 1988, ApJ, 329, 299
Buchler, J.R., Kollath, Z., Beaulieu, J.P. and Goupil, M.J. 1996, ApJ, 462, L83
Christensen-Dalsgaard, J. & Petersen, J.O. 1995, A&A, 299, L17
Feast, M.W. 1963, MNRAS, 125, 367
Feast, M.W., Glass, I.S., Whitelock, P.A. & Catchpole, R.M. 1989, MNRAS, 241, 375
Gaposhkin, S. 1970, Smithsonian Astrophysical Observatory, Special report 310
Haniff, C.A., Scholz, M. & Tuthill, P.G. 1995, MNRAS, 276, 640
Hinkle, K.H., Scharlach, W.W.G. & Hall, D.N.B. 1984, ApJS, 56, 1
Hughes, S.M.G & Wood, P.R. 1990, AJ, 99, 784
Jura, M. & Kleinmann, S.G. 1992, ApJS, 79, 105
Panagia, N., Gilmozzi, R., Macchetto, F., Adorf, H.-M. & Kirshner, R.P. 1991, ApJ, 380, L23
Petersen, J.O. 1973, A&A, 27, 89
Moskalik, P., Buchler, J.R. & Marom, A. 1992, ApJ, 385, 685
Simon, R.R. & Schmidt, E.G. 1976, ApJ, 205, 162
van Belle, G.T., Dyck, H.M., Benson, J.A. & Lacasse, M.G. 1996, AJ, 112,

2147
Welch, D.L. et al. 1995, ASP Conf. Series, 83, 232
Wood, P.R., Arnold, A. & Sebo, K.M. 1997, in preparation
Wood, P.R., Bessell, M.S. & Fox, M.W. 1983, ApJ, 272, 99
Wood, P.R. & Sebo, K.M. 1996, MNRAS, 282, 958
Wood, P.R. & Zarro, D.M. 1981, ApJ, 247, 247

DISCUSSION

ANDRE MAEDER: You have used $Y = 0.25$ for Cepheids of about $5M_{\odot}$ in the LMC. Is there no He enrichment by a few percent due to dredge up? This may significantly influence the pulsation properties. I also wonder whether an overshooting with $\Lambda = 1$ would not just suppress the blue loops and the Cepheids as well.

PETER WOOD: The Cepheid calculations shown were for $Y = 0.25$, a few percent too low perhaps following first dredge-up. However, the calculations were also done with $Y = 0.30$ and yield almost identical results.

You are correct that more overshoot on the main-sequence suppresses blue loops. However, adding more undershoot in the envelope can counteract this, as shown by the Padova group. The Cepheids can calibrate the M-L relation. It is then up to stellar evolution theory to reproduce this.

DAVID ARNETT: Please expand on your interesting comment that the matter doesn't have time to fall back.

PETER WOOD: Each pulsation cycle, a shock in the LPV atmosphere imparts an outward velocity of $\sim 10\,\mathrm{km/s}$ to mass particles there. If followed ballistically in the low gravity, large radius case, the particles don't have time to return before the next shock, one pulsation cycle later. This could lead to extremely large mass loss rates, but these are not observed in Miras with $P \sim 300$-400 days.

RR LYRAE VARIABLES

G. BONO
Trieste Astronomical Observatory, Via G.B. Tiepolo 11, 34131 Trieste, Italy, bono@oat.ts.astro.it
AND
M. MARCONI
Dept. of Physics, Univ. of Pisa, Piazza Torricelli 2, 56100, Pisa, Italy, marcella@astr1pi.difi.unipi.it

1. Historical Overview

On the basis of a historical overview we outline the relevant results which allowed to approach on firm basis the physical mechanisms which govern the radial pulsation of RR Lyrae variables, revealing the astrophysical parameters which account for their limiting amplitude pulsational behavior.

1.1. THEORETICAL ROUTE

After the pioneering papers by Eddington and Schwarzschild, the theoretical approach to the pulsational properties and modal stability of RR Lyrae variables finds its cornerstone in the seminal investigation by Christy (1966). In a series of thorough papers Christy (1968 and references therein) first showed that nonlinear radiative models can reproduce the main observational properties of RR Lyrae. He found, in close agreement with observational data, that first overtone (FO) pulsators are located at mean effective temperatures larger than the fundamental (F) ones. However, one of the main drawbacks of Christy's approach was that the bulk of the computed envelope models were followed in time only for a small number of periods and therefore they could supply only plausible guesses concerning the approach to limit cycle stability of radial motions.
This problem was solved by Stellingwerf (1974) and von Sengbusch (1975) who developed a relaxation method for evaluating the limiting amplitude modal behavior based on a linear perturbation analysis (*Floquet coeffi-*

T.R. Bedding et al. (eds.),
***Fundamental Stellar Properties: The Interaction between Observation and Theory*, 305–310.**

cients) at each limit cycle and at the origin. A further problem properly settled by Stellingwerf (1975) was the method for handling, through the artificial viscosity pressure, the development and the propagation of shocks both in the ionization zones and in the deeper layers of stellar envelopes. At the same time he was also able to put on a more secure physical basis the destabilization due to both hydrogen and helium ionization regions.

All along the seventies, although several relevant agreements, both linear and nonlinear calculations could not account for the quenching of the pulsation instability close to the red edge of the instability strip. Baker & Kippenhahn (1965), on the basis of linear nonadiabatic models including a mixing-length theory of convection, had earlier suggested that the damping of the pulsation was induced by convection. However, the linear and nonlinear calculations for evaluating F and FO red edges were at odds with observational data (Baker & Gough 1979). This thorny and long-standing problem was solved by Deupree (1979) and Stellingwerf (1982) who developed two different formulations to account for the coupling between radial pulsation and convective motions. In particular, Stellingwerf (1984 and references therein) by adopting a 1-D, nonlinear, nonlocal and time-dependent treatment of convection succeeded in evaluating, at fixed astrophysical parameters, the pulsation quenching caused by convection and therefore the location in the HR diagram of cool RR Lyrae instability boundaries. Subsequent improvements and refinements of this approach confirmed the overall theoretical scenario (Gehmeyr 1993), casting light on the role played by limiting amplitude calculations on modal behavior and pulsational amplitudes (Bono & Stellingwerf 1994) and their dependence on astrophysical parameters (Bono et al. 1997a).

1.2. OBSERVATIONAL ROUTE

The discovery and the first systematic searches for RR Lyrae variables in Galactic globular clusters date back to the observational investigations provided by Shapley and Bailey. However, Baade first recognized the paramount role played by *cluster-type variables* (this was the name originally adopted for RR Lyrae) for tracing the spatial distribution in the Galaxy of what we now call population II stars. One of the first important results obtained by Baade (Osterbrock 1996 and references therein) was that RR Lyrae were not an intrinsic property of globular clusters since in his search for RR Lyrae in three different clusters he found several variables of the same type in the field well outside the clusters. On the basis of the constancy of the mean absolute magnitude of RR Lyrae previously discovered by Shapley, he also suggested that these objects do not belong to the Galactic plane and that their "distances are comparable with those of the globular clusters".

RR Lyrae and globular clusters were the crown witnesses of a spherical and high space velocity stellar population, whereas the eclipsing variables were the tracers of a stellar population characterized by a flat distribution and low space velocities. The kinematics properties of these objects provided by Lindblad and Oort strengthen the observational scenario which led Baade to the pivotal discovery of two different stellar populations (Baade 1944).

Another seminal result obtained by Baade was the discovery of a large number of RR Lyrae in a region of the Galactic bulge that we now call *Baade's window* (BW). At the Vatican Conference on "Stellar Populations", he pointed out a "very curious" observational evidence concerning this group of variables. The period distribution of RR Lyrae in the BW showed a peak at P=0.33 days and, in contrast with variables in globular clusters characterized by the same periods, they presented asymmetrical light curves and canonical pulsational amplitudes (Baade 1958).

The Galactic RR Lyrae observational scenario was partially stirred up by the photometric and spectroscopic data collected by Preston (1959). In this investigation Preston introduced the ΔS parameter as a metallicity indicator and applied it to field RR Lyrae. The data presented in that paper show that there are at least three important "departures" from the common "mean" properties of RR Lyrae known at that time. In fact, he found that field short-period RRab (fundamental) variables are characterized by low ΔS values -i.e. high metal contents-, that they are distributed close to the Galactic plane and present small radial motions relative to the sun, in contrast with metal-poor variables which resemble the "pure population II" halo stars. The subtle consequence of this and of the subsequent thorough investigations is that the pulsational behavior and the rotational kinematic of RR Lyrae in different Galactic components (disk, bulge, halo) present substantial variations and therefore can be adopted as fundamental diagnostic tools of the Galactic structure (Suntzeff et al. 1991; Layden 1995 hereinafter L95).

In the last few years the modal behavior and both the evolutionary and pulsational properties of metal-poor RR Lyrae in globular clusters have been widely discussed (see e.g. Sandage 1993; Bono et al. 1997a). Nevertheless, only recently a detailed theoretical analysis of the evolutionary and pulsational characteristics of metal-rich RR Lyrae has been provided (Bono et al. 1997b,c and references therein).

The main aim of this paper is to discuss the comparison between theoretical predictions and observational data in the period metallicity plane and in the Bailey diagram (pulsational amplitude vs period). The reader interested to a more detailed analysis of both F and FO pulsators in the BW and in the Galactic field is referred to Bono et al. (1997c). In the final section we briefly outline some future developments.

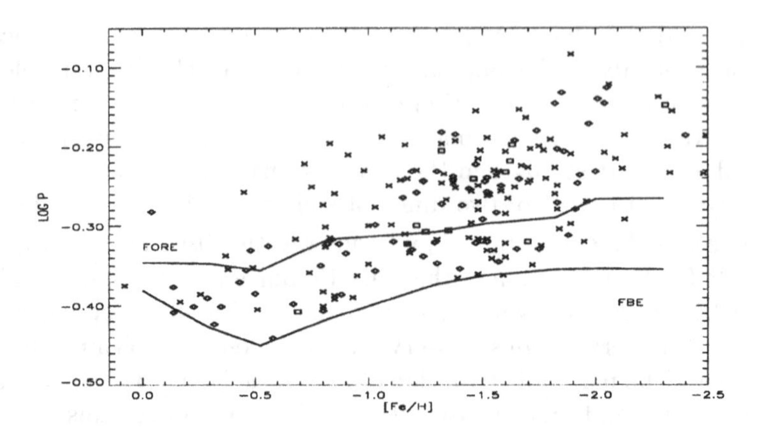

Figure 1. Pulsational periods as a function of metal abundance for Galactic field RRab variables collected by L95. The solid lines show the fundamental blue edge and the first overtone red edge. Crosses and squares are referred to variables with uncertain or unknown blue amplitudes and uncertain ΔS values respectively.

2. Field RR Lyrae Variables

As a first step in disclosing the evolutionary and pulsational properties of RRab variables we take into account the variation of the pulsational period as a function of the metal content. Figure 1 shows in this plane the data collected by L95. The metallicity indicators provided by the quoted author have been transformed into the Zinn & West (1985) metallicity scale by adopting the relation suggested by Suntzeff, Kinman & Kraft (1991). In this figure the Fundamental Blue Edge (FBE) and the First Overtone Red Edge (FORE) are also plotted. The theoretical periods have been evaluated by convolving, at selected metal contents, the stellar mass and the luminosity level predicted by ZAHB evolutionary calculations with the FBE and the FORE constructed by adopting the same input parameters (M, L, χ).
The agreement between theoretical predictions and observational data is quite satisfactory and some features of this figure are worth noting: 1) field variables are affected by the Oosterhoff dichotomy and indeed metal-poor variables ($[Fe/H] < -1.4$) are characterized by longer mean periods since the transition takes place close to the FORE, thus resembling Oosterhoff type II clusters. On the other hand, at higher metal contents the mean periods decrease since the transition takes place at the FBE, thus resembling Oosterhoff type I clusters. 2) Moving toward higher metal abundances the period at the FBE decreases. In fact, in agreement with observational data and with the Baade's "very curious" evidence, we find that at $[Fe/H] \approx -0.6$ the period at the FBE is approximately equal to 0.35 days. The subsequent increase at higher metal contents is mainly due to

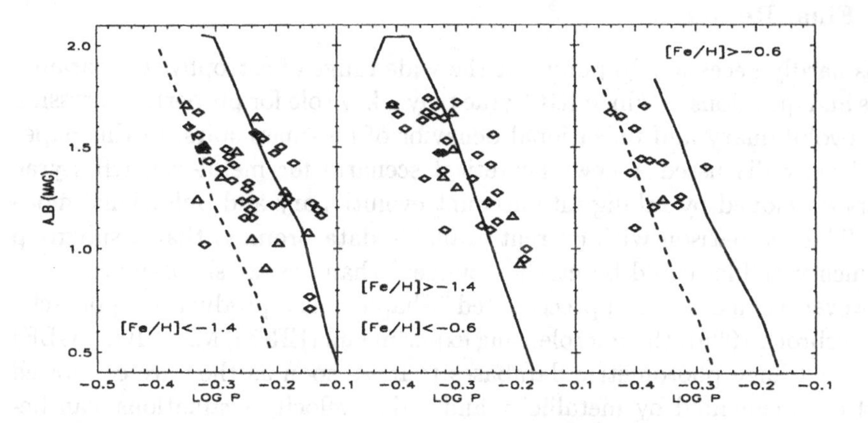

Figure 2. Bailey diagram, B amplitude versus period, for Galactic field RRab variables collected by L95. Each panel is referred to a different range of metallicity, see labeled values. The lines plotted in each panel show the theoretical amplitudes. Triangles are referred to variables with uncertain blue amplitudes.

the increase in the helium abundance since at these metallicities the range of stellar masses which populate the instability strip narrows and therefore the ZAHB luminosity levels attain quite similar values. This finding is quite interesting since the comparison between predicted and observed periods can provide useful constraints on the value of the enrichment ratio between helium and heavy elements in low-mass stars (Carigi et al. 1995).
Figure 2 shows the comparison into the Bailey diagram between predicted and observed field RRab amplitudes. The bolometric amplitudes have been transformed by adopting Kurucz's atmosphere models (1992). This figure shows quite clearly that the overall agreement is much less satisfactory. In fact, the theoretical amplitudes for metal-poor variables are in reasonable agreement with observational data, whereas for metal-intermediate and metal-rich variables the predicted amplitudes are systematically larger. In order to account for this effect we cannot rule out that the transformation from bolometric into B amplitudes through static atmosphere models could be affected by uncertainties and/or by systematic errors. Nevertheless, the large discrepancy (≈ 0.5 mag) we find at fixed period for metal-rich variables cannot be only due to transformation errors, since the same effect should be also present among metal-poor pulsators. However, let us note that this odd group of variables can be properly explained by assuming "young" metal-rich pulsators (i.e. $t \approx 1 - 2$ Gyr) since in this context the theoretical period-amplitude relation moves toward shorter periods. Finally it is worth underlining that quite recently a similar evidence of "young" low-mass stars in the solar neighborhood has been suggested by Gonzalez (1997) in his spectroscopic investigation of a sample of "51 Pegasi" stars.

3. Final Remarks

It is hardly necessary to point out the wide range of astrophysical parameters and questions in which RR Lyrae play a key role for properly addressing the evolutionary and pulsational behavior of low-mass stars. In this paper we briefly discussed a new theoretical scenario for metal-rich RR Lyrae stars developed by taking into account evolutionary and pulsational models. The comparison with current available data prompts that a subgroup of such variables could be much "younger" than previously assumed.
However, thanks to an unprecedented "Shapley mass production approach" (Osterbrock 1996), the microlensing experiments (EROS, MACHO, OGLE) collected huge photometric databases that, as soon as they are calibrated and complemented by metallicity and radial velocity evaluations, can become a fundamental benchmark for testing theoretical models. At the same time, a substantial improvement in the evaluations of astrophysical parameters based on RR Lyrae pulsational properties could be soon achieved.

References

Baade, W. 1944, ApJ, 100, 147
Baade, W. 1958, in Stellar Populations, ed. D.J.K. O'Connell, (Amsterdam: North Holland), 303
Baker, N. H., & Gough, D. O. 1979, ApJ, 234, 232
Baker, N. H., & Kippenhahn, R. 1965, ApJ, 142, 868
Bono, G., Caputo, F., Castellani, V., & Marconi, M. 1997a, A&AS, 121, 327
Bono, G., Caputo, F., Cassisi, S., Castellani, V., Incerpi, R., & Marconi, M. 1997b, ApJ, 478, April 10
Bono, G., Caputo, F., Cassisi, S., Incerpi, R., & Marconi, M. 1997c, ApJ, 283, July 10
Bono, G., & Stellingwerf, R. F. 1994, ApJS, 93, 233
Carigi, L., Colin, P., Peimbert, M., & Sarmiento, A. 1995, ApJ, 445, 98
Christy, R. F. 1966, ApJ, 144, 108
Christy, R. F. 1968, QJRAS, 9, 13
Deupree, R. G. 1979, ApJ, 234, 228
Gehmeyr, M. 1993, ApJ, 412, 341
Gonzalez, G. 1997, MNRAS, 285, 403
Kurucz, R. L. 1992, in IAU Symp. 149, The Stellar Populations of Galaxies, eds. B. Barbuy & A. Renzini (Dordrecht: Kluwer), 225
Layden, A. C. 1995, AJ, 110, 2288 (L95)
Osterbrock, D. E. 1996, Jour. for the History of Astron., 27, 301
Preston, G. W. 1959, ApJ 130, 507
Sandage, A. F. 1993, AJ, 106, 719
Stellingwerf, R. F. 1974, ApJ, 192, 139
Stellingwerf, R. F. 1975, ApJ, 195, 441
Stellingwerf, R. F. 1982, ApJ, 262, 330
Stellingwerf, R. F. 1984, ApJ, 284, 712
Suntzeff, N. B., Kinman, T. D., & Kraft, R. P. 1991, 367, 528
von Sengbusch, K. 1975, Mem. Soc. Roy. des Sci. de Liege, 6, VIII, 189
Zinn, R., & West, M. J., 1984, ApJS, 55, 45

10. STELLAR MODELS VERSUS OBSERVATIONS

THE EVOLUTION OF MASSIVE STARS

A. MAEDER
Geneva Observatory, CH—1290 Sauverny, Switzerland

1. Introduction

Massive stars are the crossroads of many important astrophysical problems and thus a proper understanding of their evolution is very needed. They are the main sources of UV radiation, by heating the interstellar dust they produce the far-IR luminosities of galaxies. They are the precursors of Supernovae and also the main sources of nucleosynthesis. They are visible in distant galaxies and the recent observations of starbursts have shown their major role in the spectral and chemical evolution of galaxies. They begin to be observed in the galactic center and in regions of star formation around galactic nuclei.

The properties of massive stars have been extensively studied over the last decade (cf. Chiosi & Maeder, 1986; Maeder & Conti, 1994). In Geneva (for example !), the following grids have been made:

- Pre–MS evolution with accretion (Bernasconi & Maeder, 1996; Bernasconi, 1996)
- MS and post–MS evolution in the range of 0.8 to 120 $M_\odot$ for:

Z = 0.10	Y = 0.48	Mowlavi *et al.*, 1997	(paper VIII)
Z = 0.04	Y = 0.34	Schaerer *et al.*, 1993b	(paper IV)
Z = 0.02	Y = 0.30	Schaller *et al.*, 1992	(paper I)
Z = 0.008	Y = 0.264	Schaerer *et al.*, 1993a	(paper II)
Z = 0.004	Y = 0.252	Charbonnel *et al.*, 1993	(paper III)
Z = 0.001	Y = 0.243	Schaller *et al.*, 1992	(paper I)

- Models of stars with M > 15 $M_\odot$ for the above (Z,Y) values have been calculated with mass loss rates increased by a factor of 2 (Meynet *et*

T.R. Bedding et al. (eds.),
Fundamental Stellar Properties: The Interaction between Observation and Theory, 313–322.

al., 1994, paper V).

- Models for Horizontal Branch (HB), post–HB and AGB stars have been constructed for various (Z,Y), (cf. Charbonnel *et al.*, 1996, paper VI).
- Models for low mass stars in the range 0.4 to $1\,M_\odot$ for $Z = 0.02$ and $Z = 0.001$ have been made (cf. Charbonnel *et al.*, 1997, paper VII).
- Complete tables of isochrones for the various Z and for ages from 2×10^6 yr to 2×10^{10} yr.
- A code for calculating isochrones with various color indices, synthetic open clusters, population synthesis is kindly provided on request.

These data are available at the CDS in Strasbourg or by e-mail on request (georges.meynet@obs.unige.ch).

2. Pre–MS evolution

New pre–MS models have been made (cf. Bernasconi & Maeder, 1996; Bernasconi, 1996). They follow the accretion scenario firstly proposed by Palla and Stahler (1991, 1992, 1993; see also Beech & Mitalas, 1994). The basic idea is the following one. Molecular clouds can become gravitationally unstable and produce a large inflow of matter, which will be accreted in the associated star formation process. In the central volume, matter organizes itself in a spatially thin and very extended disk (several 10^2 AU) from which matter is progressively accreted onto the central protostellar core. The pre–MS evolution is very different according to stellar mass. For a $1\,M_\odot$ star the accretion period is terminated at sufficiently early stages, so that the Hayashi phase occurs like in models without accretion and the pre–MS stage lasts the classical 3×10^7 yr (Palla & Stahler, 1993). For stellar masses larger than $7 - 10\,M_\odot$, the accretion time becomes longer than the Kelvin–Helmholtz time-scale, and massive stars continue to accrete matter hidden within their molecular cloud, while they are already centrally contracted and enter their hydrogen burning phase.

A basic parameter is the mass accretion rate. In our models, the equilibrium equation of cloud models are solved with the assumption that the clouds have both a thermal pressure and a non–thermal contribution supporting them. The non–thermal effect is based on the velocity dispersion-size relation by Larson (1981). This way, accretion rates are obtained; they are growing function of the already accreted stellar mass.

Fig. 1a shows two birthlines in the accretion scenario. The birthline is the path followed by a continuously accreting star. Depending on the moment when the accretion becomes negligible, the star will stop its rising evolution on the birthline. Various mass ranges on the birthline of a massive star can be distinguished (Bernasconi & Maeder, 1996).

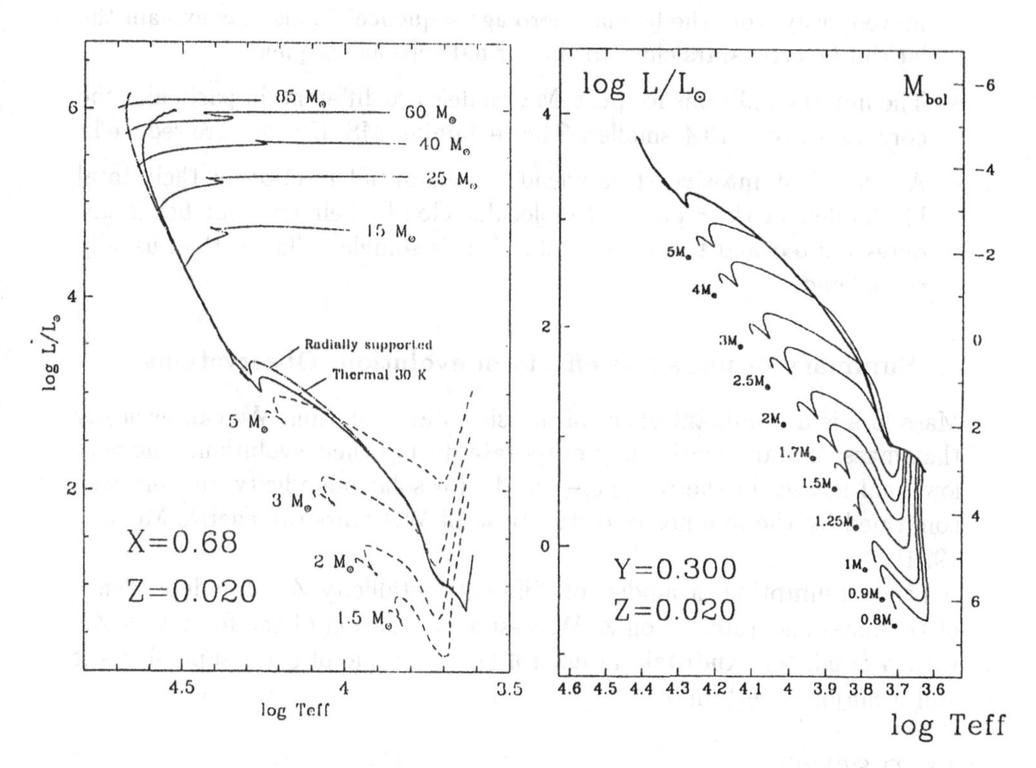

Fig.1a and 1b. Pre–MS evolutionary tracks with accretion. Fig. 1a: Standard pre–MS tracks are given without accretion (broken line); the continuous lines for 15 to 85 $M_\odot$ are post–MS tracks. The two upwards going lines are birthlines for two different hypotheses on the cloud support. Fig. 1b: Detailed pre–MS tracks with accretion, the thicker line is the birth line, which is followed by massive pre–MS stars (cf. Bernasconi & Maeder, 1996; Bernasconi, 1996).

In particular, at about 8 $M_\odot$, the star reaches its minimum radius, which we define as the point where the star reaches the ZAMS. This very moment is preceded by the slow ignition of CN burning and by some ^{3}He spikes due to the successive back and forth motions of the convective core. Fig. 1b shows the details of pre–MS tracks leaving the birth line, which is the path followed by massive pre–MS stars.

These new models show several interesting features:

- The pre–MS lifetimes for massive stars with accretion are much larger (i.e. about 1.3×10^6y for 25 $M_\odot$) than for standard evolution (i.e. 7.4×10^4yr for a 25 $M_\odot$ star).
- At the time it becomes visible, a newly formed star with $M \geq 40\,M_\odot$ has already burnt some of its central hydrogen and consequently has already

moved away from the formal "zero age sequence". This may explain the lack of O–type stars close to the formal zero age sequence.

- The initial conditions for post–MS models are different; in particular the core sizes are 5–10% smaller. The remaining MS lifetimes are reduced.
- As the most massive stars spend a substantial fraction of their total life hidden in their parental molecular cloud, their true number is underestimated and the slope of the IMF is somehow flatter than usually considered.

3. Summary of mass loss effects in evolution. Observations

Mass loss is a dominant effect of massive star evolution. We can even say that massive stars nearly fully evaporate during their evolution. The very low final masses in the range $5 - 10\ M_\odot$ for solar metallicity are very well confirmed by the luminosity distribution of WR stars (Maeder & Meynet, 1994).
A basic assumption in models of different metallicity Z is the dependence of the mass loss rates $\dot{M}$ on Z. We assume a relation of the form $\dot{M} \sim Z^\alpha$, with $\alpha \simeq 0.5$ (cf. Kudritzki *et al.*, 1991). The value of α for supergiants is still a major uncertainty.

3.1. O–STARS:

During MS evolution, the effects of mass loss at the observed rates are rather small:

- The sizes of the convective core are reduced and thus their luminosities as well.
- However, the core mass fractions are larger than in a star of the same actual mass, but evolving at constant mass. The stars evolving with mass loss are also over-luminous with respect to their actual masses.
- The MS lifetimes undergo little increases ($< 10\%$).
- A certain MS widening occurs around $60\,M_\odot$ for solar composition and at lower masses for higher Z (cf. Fig. 2).
- Mass loss at the observed rate is unable to produce an increase of the N/C and N/O ratios during the MS for stellar masses below about $70\,M_\odot$.

Concerning the observations, the most interesting result is that by Herrero *et al.* (1992), who shows that fast rotating O–stars exhibit He–enhancements. This observations, joined to the existence of ON–stars, suggest that at least a fraction of O–stars is experiencing additional mixing processes, either due

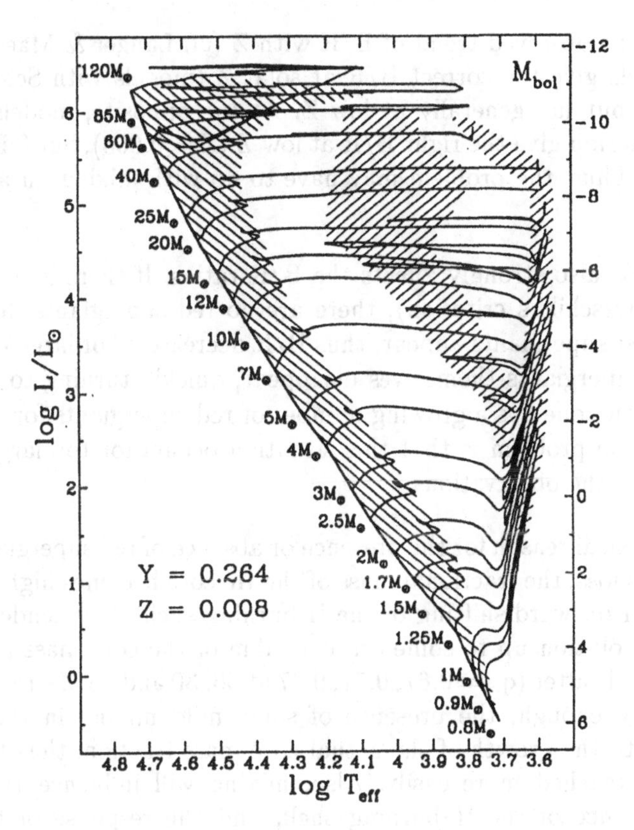

Figure 2. Evolutionary tracks appropriate to LMC, composition with indications of the slow phases of nuclear burning.

to rotation and/or to tidal mixing in binaries. Another point, possibly related to the previous one, is the fact that the upper MS of young clusters is more elongated than predicted (cf. Meynet *et al.*, 1993). This could be the cause of the absence of a visible gap at the end of the MS.

3.2. SUPERGIANTS, THE B/R RATIO:

At the present stage, the problems of blue, yellow and red supergiants are more intricate. The models, on the whole, correctly predict the observed ratios of red supergiants to MS stars for galactic clusters in the range of 3×10^6 yr, (cf. Meynet, 1993). However, the study of clusters in the SMC, LMC and Galaxy show that the number ratio B/R of blue-to-red supergiants is an increasing function of metallicity (see Langer & Maeder, 1995 for recent ref.). Clusters in inner galactic regions have much larger B/R ratios than in the SMC, where large numbers of red supergiants are present as in NGC 330. Up to now, no sets of models are able to correctly reproduce the observed trend of B/R with Z (cf. Langer & Maeder, 1995). Some models give the correct B/R at solar Z (models with Schwarzschild criterium), but fail generally at low Z. At the opposite, models with the Ledoux criterion give the right B/R at low Z ($Z = 0.003$), but fail generally at solar Z. Thus, the problem may have to do with mixing, in addition to mass loss.

Mass loss also strongly affects the B/R ratios. If there is no mass loss (and Schwarzschild's criterion), there are no red supergiants; for growing $\dot{M}$–rates, red supergiants appear, thus B/R decreases. For large enough $\dot{M}$–rates, red supergiants themselves disappear, quickly turning to WR stars. In this regime, one has a growing number of red supergiants for lower Z, as observed. The problem is that this situation occurs for too large $\dot{M}$–rates compared to the observations.

The physical reason for the presence or absence of red supergiants is that due to mass loss, the fractional mass of the He core becomes higher and also produces an outward shifting of the H-burning shell. This tendency favors redwards evolution up to some critical value of the core mass fraction q_c and bluewards after ($q_c = 0.67, 0.77, 0.97$ at 60, 30 and 15 $M_\odot$ respectively). Interestingly enough, the presence of some mild mixing in the star will contribute to the growth of the actual core mass fraction, thus the critical q_c will be reached more easily. Also, mixing will influence the location and importance of the H–burning shell, and the response of the star is then more complex with in particular the presence or absence of blue loops (cf. Langer, 1991). Models in progress with rotational mixing may help to solve this longstanding problem. This possibility is also supported by the evidences of N/C and He enhancements in blue and yellow supergiants (Venn, 1995), which are not predicted for non rotating models.

Very often, star models in the supergiant stage are close to a neutral equilibrium between a blue and red location. Processes affecting the envelope structure may give the final kick to the blue or the red. Such uncertainties are of course a difficulty to interprete the populations of blue and red supergiants, particularly at low metallicities. However, these uncertainties have very little consequences on the subsequent evolution, since further mass loss is quickly removing the outer envelopes, with or without additional mixing, to let finally bare cores identified as Wolf–Rayet stars. The evolution of WR stars are dominated by mass loss and the effects of previous differences are washed out.

3.3. WOLF–RAYET STARS

WR stars have an average mass below $10\,M_\odot$, they are over-luminous with respect to main-sequence stars. They fit well the mass–luminosity relation of bare–core models (Smith & Maeder, 1989). More generally, bare cores containing no hydrogen (roughly corresponding to WNE and WC stars) fit a $M - L - \dot{M} - R - T_{eff}$ – chemistry relation (cf. Schaerer & Maeder, 1992). To get R and T_{eff}, a proper account for the optically thick wind has to be made. A simple procedure has been used in Schaller *et al.*, (1992) and by Schaerer and Maeder; despite its simplicity it is consistent with the results of the complete stellar models, including the interior, atmosphere and winds (cf. Schaerer *et al.*, 1996).

The mass loss rates in the WR stages without hydrogen seem to have little or no dependence on metallicity. The main parameter is the actual mass of the WR stars, and a relation of the form $\dot{M}_{WR} \sim M_{WR}^{\alpha}$ with a value $\alpha \approx 2$ has been proposed by Langer (1989), while Moffat (1995) suggests a rather linear dependence.

At present, the following filiations can be distinguished for massive stars (cf. Chiosi & Maeder, 1986; Crowther *et al.*, 1995; Langer *et al.*, 1994).

$M \geq 60\,M_\odot$
O — Of — WNL + abs — WN7 — (WNE) —WCL — WCE — SN
At low Z:... WN7 —WCE —SN

$M \simeq 40 - 60\,M_\odot$
O — Of — LBV — WN8 — WNE — WCE — SN

$M \simeq 25 - 40\,M_\odot$
O — (BSG) —RSG —(BSG) —WNE — (WCE) — SN

$M \simeq 25\,M_\odot$
O — (BSG) —RSG — BLUE ⟶ RSG⟶ SN
YELLOW
SUPG ⟶ SN

Several uncertainties remain, e.g. whether stars with $M \geq 60\ M_\odot$ avoid the LBV (Luminous Blue Variables), or whether the LBV turns into WN8 as supposed by Crowther *et al.* Differences in the mass limits and in the exact sequences may also depend on Z. For example, for $Z \geq 0.02$, we likely have WNL → WCL, avoiding the WNE stage. For $Z < 0.02$, WNL → WCE → WO, mostly avoiding the WCL stage, and WNE do not lead to WC stars.

In WN stars, CNO ratios are in agreement with equilibrium values, thus CNO abundances are a test for nuclear cross-sections rather than for model assumptions. Surface helium contents range from $Y_s = 0.4$ (gen. in WNL)

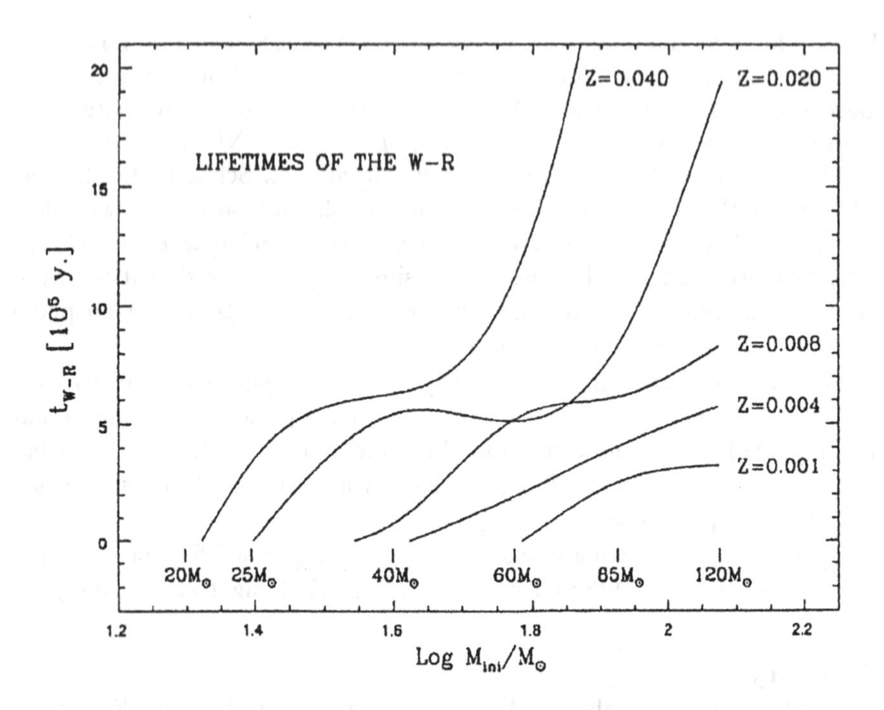

Figure 3. WR lifetimes as a function of M and Z.

to $Y_s = 0$ (gen. in WNE). The Y_s vs. logL diagram is a key diagram for WN stars (cf. Langer *et al.*, 1994; Maeder 1995). In this diagram it seems, however, difficult to account for the WNE stars of low luminosity with non-zero Y_s. This might be a signature of mixing.

For WC stars, the sequence WC9, WC8 ... WC4, W0 seems to be a sequence of increasing (C+O)/He (cf. Smith & Hummer, 1988; Smith & Maeder, 1991). This ratio is preferable to C/He, which does not vary monotonously during evolution. WC abundances are basically a test of model assumptions, since they are products of partial He-burning. The models made at various Z allow to understand the main observed properties of WC stars in galaxies (cf. Smith & Maeder, 1991): 1) WCL stars are preferentially created at high Z, 2) at a given Z, WCL stars are brighter than WCE stars, 3) stars of a given WC subtype are brighter at low Z.

The number ratios WR/O and WC/WN change very much according to galaxies and locations in galaxies (cf. Maeder & Conti, 1994). To carefully discuss this important property, we must distinguish:

a) galaxies and areas in galaxies, where the assumption of constant star formation rate (SFR) over the last 10^7 yr is valid and

b) starburst regions and starburst galaxies, where the SFR has greatly varied in recent times.

For case a), the main effect responsible for changes in massive star population is metallicity Z. Fig. 3 shows the lifetimes of WR stars as a function of M and Z for models with enhanced $\dot{M}$-rates (cf. Maeder & Meynet, 1994). When integrated over the current mass spectrum and with the hypothesis of constant SFR, these lifetimes well reproduce the observed number ratios WR/O in nearby galaxies (cf. also Maeder & Conti, 1994).

It does not seem necessary to advocate the contribution of mass transfer in many binaries. Case b) of starbursts is quite different. There it appears that the observations pick up these objects at a specific time after a very short starburst, so that they show very high WR/O ratios. Theses ratios may be up to an order of magnitude larger than in case a). The situation has been discussed by Arnault *et al.* (1989), Maeder & Conti (1994) and by Meynet (1995), who has made very complete calculations.

The WR stars are most useful objects, even observable in the integrated spectrum of galaxies; thus WR stars may give us an access to the properties (age, SFR, mass spectrum etc.) of distant starbursts.

References

Arnault, P., Schild, H. (1989) *A&A* **224**, 73

Beech, M., Mitalas, R. (1994) *ApJS* **95**, 517

Bernasconi, P. A. (1996) *A&A* **120**, 57

Bernasconi, P., Maeder, A. (1996) *A&A* **307**, 829

Charbonnel, C., Meynet, G., Maeder, A., Schaller, G., Schaerer, D. (1993) *A&AS* **101**, 415 (paper III)

Charbonnel, C., Meynet, G., Maeder, A., Schaerer, D. (1996) *A&AS* **115**, 339 (paper VI)

Charbonnel, C., D=E4ppen, W., Bernasconi, P., Maeder, A., Meynet, G., Schaerer, D., Mowlavi, N. (1997) *A&AS*, in press (paper VII)

Chiosi, C., Maeder, A. (1986) *A&A* **24**, 329

Crowther, P. A., Smith, L. J., Hillier, D. J., Schmutz, W. (1995) *A&A* **293**, 427

Herrero,, A., Kudritzki, R. P., Vichez, J. M., Kunzer, D., Butler, K. (1992) *A&A* **261**, 209

Kudritzki, R. P., Pauldrack, A., Puls, J., Voels, S. R.. (1991), *IAU Symp. 148*, Ed. R. Haynes and D. Milne, p. 279

Langer, N. (1989) *A&A* **220**, 135

Langer, N. (1991) *A&A* **252**, 669

Langer, N., Hamann, W. R., Lennon, M., Najarro, F., Pauldrach, A. W. A., Puls, J. (1994) *A&A* **290**, 819

Langer, N., Maeder, A. (1995) *A&A* **295**, 685

Larson, R. B. (1981) *MNRAS* **194**, 809

Maeder, A. (1995) in "Wolf-Rayet stars: Binaries Colliding Winds, Evolution" *IAU Sump. 163*, Eds. K. van der Hucht and P.M. Williams p. 280

Maeder A., Conti, P. (1994) *ARAA* **32**, 227

Maeder A., Meynet, G. (1994) *A&A* **287**, 803

Meynet, G. (1993) in "The Feedback of Chemical Evolution on the Stellar Content of Galaxies", Eds. D. Alloin, G. Stasinska, Obs. de Paris, p. 40
Meynet, G. (1995) *A&A* **298**, 767
Meynet, G., Maeder, A., Schaller, G., Schaerer, D., Charbonnel, C. (1994) *A&AS* **103**, 97 (paper V)
Meynet, G., Mermilliod, J.-C., Maeder, A. (1993) *A&AS* **98**, 477
Moffat, A. (1995) in "Wolf-Rayet stars: Binaries Colliding Winds, Evolution" *IAU Symp. 163*, Eds. K. van der Hucht and P.M. Williams p. 213
Movlawi, N., Schaerer, D., Meynet, G., Bernasconi, P.A., Charbonnel, C., Maeder, A. (1997) *A&AS*, in press (paper VIII)
Palla, F., Stahler, S. W. (1991) *ApJ* **375**, 288
Palla, F., Stahler, S. W. (1992) *ApJ* **392**, 667
Palla, F., Stahler, S. W. (1993) *ApJ* **418**, 414
Schaerer D., Maeder, A. (1992) *A&A* **263**, 129
Schaerer, D., Meynet, G., Maeder, A., Schaller, G. (1993a) *A&AS* **98**, 523 (paper II)
Schaerer, D., Charbonnel, C., Meynet, G., Maeder, A., Schaller, G. (1993b) *A&AS* **102**, 339 (paper IV)
Schaerer, D., de Koter, A., Schmutz, W., Maeder, A. (1996) *A&A* **312**, 475
Schaller, D., Schaerer, D., Meynet, G., Maeder, A. (1992) *A&A* **96**, 269 (paper I)
Smith, L., Hummer, D.G. (1988) *MNRAS* **230**, 511
Smith, L., Maeder, A. (1989) *A&A* **211**, 71
Smith, L., Maeder A. (1991) *A&A* **241**, 77
Venn, K. (1995) *ApJS* **99**, 659

DISCUSSION

BENGT GUSTAFSSON: I was impressed with your good agreement between the predicted number of Wolf Rayet Stars relative to the number of O stars for different metallicities and the observed ratio for different galaxies. Are there any free parameters in that fit?

ANDRE MAEDER: The good agreement between the observed and theoretical number ratios suggests that the fraction of WR owing their existence to binary mass transfer is small. This is certainly one free parameter in the comparisons.

STEVE KAWALER: How many free parameters are used in your formulations for rotational mixing?

ANDRE MAEDER: We have been working over the last two years on the Richardson criterion, meridional circulation, thermal instabilities, etc., in order to have a theory as consistent as possible. I believe there is no interest nowadays to build 'parametrized' theories.

JOHN NORRIS: Is the nitrogen produced by your rotating models primary or secondary?

ANDRE MAEDER: The nitrogen enrichment we are getting is secondary during the O-type star and super-giants phases. However, it is quite possible that rotational mixing may produce some primary nitrogen during the WR phases, when some freshly produced ^{12}C is diffused in H-burning regions. We must check that quantitatively.

EVOLUTION OF INTERMEDIATE MASS STARS

CESARE CHIOSI
Department of Astronomy
Vicolo dell'Osservatorio 5, 35122 Padova, Italy

Abstract. We summarize the evolution of intermediate-mass stars, calling attention to the uncertainties related to mixing in convectively unstable regions, and to recent developments in the theory of AGB stars with envelope burning. Finally, we briefly report on the distribution of C-stars of the Large Magellanic Cloud in the M_{bol} -log(age) plane.

1. Summary of IMS evolution

Definition. Intermediate mass stars (IMS) are those that ignite helium non degenerately, but following central He-exhaustion develop an electron degenerate C-O core, and as AGB stars experience He-shell flashes or thermal pulses (so-called TP-AGB). This places both the lower limit M_{HeF} (which is also the upper limit for low mass stars) and an upper limit M_{up} on the initial mass of such stars. In classical models $M_{HeF} = 1.8 - 2.2 M_{\odot}$ and $M_{up} = 8 - 9 M_{\odot}$ depending on the chemical composition.

Core and shell H-burning phases. In IMS the core H-burning phase is characterized by the formation of a convective core, a steady increase in luminosity and radius, and a decrease of the T_{eff}. The size of the convective core is customarily fixed by the Schwarzschild criterion $\nabla_R = \nabla_A$ (the so-called classical models). After central H-exhaustion, IMS evolve rapidly to the red giant region, burning hydrogen in a thin shell above a rapidly contracting and heating He core. As they approach the Hayashi line, a convective envelope develops whose base extends inward until it reaches layers in which H has been converted into He and C into N via the CNO cycle (changes in surface abundances, *1st dredge-up*).

Core He-Burning Phase. In IMS core He-burning ignites in non-degenerate conditions. This requires a minimum core mass of 0.33 $M_{\odot}$. Slow core He-burning of IMS takes place in two distinct regions of the HRD, a first near the Hayashi line and a second at higher T_{eff}s and luminosities, therefore extended loops in the HRD may occur, whose precise modelling depends on M, chemical composition, nuclear reaction rates, size of the convective

T.R. Bedding et al. (eds.),
***Fundamental Stellar Properties: The Interaction between Observation and Theory*, 323–330.**

core, opacity, mass loss along the RGB, inward penetration of the outer convection during the RGB stages, and other physical details. The blue band of the core He-burning models may intersect the instability strip of the Cepheid stars. The core He-burning phase of IMS (toward the lower mass end) is affected by two types of convective instability: in early stages by a semi-convective mixing (He-semi-convection) similar to that encountered by massive stars (H-semi-convection), and in late stages by the so-called breathing convection. While He-semi-convection has a negligible effect on IMS, breathing convection moderately increases t_{He} (about 20%) but gives origin to much larger C-O cores. After core He-exhaustion, the structure of the stars is composed of a C-O core, a He-burning shell, and an H-rich envelope at the base of which an H-burning shell is active. During the RGB phase, little mass is lost by stellar wind according to current mass loss rates (Reimers 1975).

AGB Phase. Following the core He-exhaustion IMS evolve through the AGB phase, which splits in two parts: early (E-AGB) and TP-AGB. In brief, the He-exhausted core contracts and heats up while the H-rich envelope expands and cools and the H-burning shell extinguishes. In the HRD the stars evolve running almost parallel to the RGB, and once again the base of the convective envelope penetrates inward. There is a limiting mass above which external convection eventually reaches layers processed by the CNO cycle. This means that fresh helium and fresh nitrogen are brought to the surface (*2nd dredge-up*). Eventually, the expansion of the envelope is halted by its own cooling and the envelope re-contracts, the luminosity decreases, and matter at the base of the convective envelope heats up. Ultimately, the H-burning shell is re-ignited, forcing the envelope convection to move outward in mass ahead of the H-burning shell. This terminates the E-AGB. In the meantime, the matter in the C-O core becomes electrons-degenerate. and nearly isothermal, while neutrino cooling carries away the liberated gravitational energy. Therefore, the temperature in the core tends to equal the temperature of the He-burning shell (about 10^8 K), well below the threshold value for C-ignition. Following the re-ignition of the H-burning shell, nuclear burning in the He-shell becomes thermally unstable. The H- and He-burning shells alternate as the major source of energy. During this phase, material processed into the intershell region can be brought into the outer convective envelope and exposed to the surface. The so-called *3rd dredge-up* can then take place. In AGB stars of large C-O core mass (hence with large initial mass) the dredge-up can occur easily. But in AGB stars of small C-O core mass (hence with small initial mass) this is possible only if extra mixing is forced into the inter-shell region. The goal is achieved either by means of semi-convection (Iben & Renzini 1982) induced by the more opaque C-rich material deposited in the intershell region by the tiny convective shell ahead of the flashing He-burning shell or by crude overshoot of convective elements from the convective shell itself. (Hollowell & Iben 1989). In both cases, C-rich material is deposited in more external layers where it can be easily engulfed by the external convection during the subsequent cycle. This is the basic mechanism to convert an M giant into a C-star. To first approximation, along the TP-AGB phase the luminosity of the star increases linearly with the mass of the H-exhausted core (Paczynski 1970) and the star brightens in M_{bol} at a constant rate (Iben & Renzini 1983). TP-AGB evolution occurs in presence of substantial mass loss terminating the AGB phase. In the classical scenario, the bulk of mass removal occurs via a fast wind (cf. Iben & Renzini 1983 and Chiosi et al. 1992 for all details). C-ignition in highly degenerate conditions requires a C-O core mass of 1.4 $M_{\odot}$ which implies, considering the effect of mass loss, a minimum initial mass of the star, M_W, in the range 4 to 6 $M\odot$. Stars lighter than the above limit will fail C-ignition and, by losing the H-rich envelope will become C-O white dwarfs.

2. Convection: the major uncertainty

Although the above evolutionary scenario substantially agrees with the observational data, a closer scrutiny reveals that there are many points of severe uncertainty, those related to convection in particular. In the classical approach, the Schwarzschild criterion provides the simplest evaluation of the size of the convectively unstable regions and the MLT simplifies the complicated pattern of motions therein by saying that full, and instantaneous mixing of material takes place. In this scheme, well known inconsistencies are known to develop at the border of the convective regions leading to the problems of the He-semi-convection and breathing convection we have already mentioned. Various attempts have been made to cure the above difficulties, among which we recall overshoot and diffusion. Generally speaking the problem is cast as follows:

(1) What determines the extension of the convectively unstable regions (either core or envelope or both) together with the extension of the surrounding regions formally stable but that in a way or another are affected by mixing? In other words how far convective elements can penetrate into formally stable regions (overshoot)?

(2) What is the thermodynamic structure of the unstable and potentially unstable regions?

(3) What is the time scale of mixing? instantaneous or over a finite (long) period of time? What is the mechanism securing either full or partial homogenization of the unstable regions?

Over the past decade different answers to above questions have been suggested and in turn different types of stellar model have been calculated. Among others we recall: (a) The extension of the convective regions is set at the layer where the velocity rather than acceleration of convective elements vanishes. The overshoot region (beyond the Schwarzschild border) is adiabatic, and mixing is instantaneous. The real extension of the overshoot region is a matter of vivid debate. Nowadays it has settled to a sizable fraction of the local pressure scale hight. For the particular case of overshoot from the H-burning convective cores in stars of low mass, the additional (reasonable) assumption is made that it must vanish at decreasing convective core. This is indeed suggested the morphology of the turnoff in the CMD of old clusters (Bertelli et al. 1992). We will refer to stellar models of this type as those with straight overshoot (Bressan et al. 1981, Bertelli et al. 1985). (b) Within the same scheme, the attempt is made to take into account that mixing actually requires a a suitable time scale to occur. To this aim straight mixing is abandoned, and the more appropriate diffusive approach is adopted (cf. Deng et al. 1996). The efficiency of diffusion (or equivalently the time scale of it) seeks to incorporate physical processes known to occur in laboratory hydrodynamics, such as intermittence and stirring, and varies as function of the local properties of the overshoot region. In this context, the thermodynamics structure of these layer plays a secondary role even if a radiative stratification ought to be preferred. Other more physically grounded but by far more complicated formulations of the problem (cf. Xiong 1986; Grossman et al. 1993) have not yet been included in stellar model calculations.

Stellar models with straight overshoot. The core H-burning phase of all stars possessing a convective core on the zero-age main sequence ($M \geq M_{con} \simeq 1.2 M_{\odot}$) is affected by convective overshoot. The models run at higher luminosities and live longer than the classical ones. They also extend the main sequence band over a wider range of T_{eff}s, this trend increasing with stellar mass (e.g. Maeder & Meynet 1991). The over-luminosity caused by overshoot during the core H-burning phase still remains during the shell H- and core He-burning phases because of the larger size of the He-exhausted core. As a consequence of it, t_{He} gets shorter in spite of the larger mass of the convective core. This, combined with the longer H-burning lifetime, t_H, makes the ratio t_{He}/t_H fairly low (from 0.12 to 0.06 when the stellar mass varies from 2 $M_{\odot}$ to 9 $M_{\odot}$). The lifetime ratio is about a factor of 2 to 3 lower than in classical models of the same mass. Models with core overshoot alone produce luminosity functions of main sequence stars that agree much better with the observational data for rich clusters (Chiosi et al. 1989), however they hardly match the extension of the blue loops observed in the same clusters because they possess less extended blue loops in the HRD (Alongi et al. 1993). However extended loops are reinstated when the effect of overshoot from the bottom of the convective envelope during the RGB phase is taken into account (Alongi et al. 1991). Convective envelopes extending deeper inside a star than predicted by the classical models are also suggested by the surface abundances of RGB stars, and the bump in the RGB luminosity function. Finally, Due to the larger masses of the He and C-O cores left over at the end of core H- and He-burning phases, respectively, the critical masses M_{up} and M_{HeF} are about 30% smaller than in classical models (Barbaro & Pigatto 1984; Bertelli et al. 1985; Bertelli et al 1986).

Stellar models with diffusion. Diffusive models of IMS have been recently calculated by Deng et al. (1996). These models share the properties of the classical semi-convective models and those with straight overshoot. In brief, the more extended convectively unstable cores yield lifetimes, lifetime ratios, and limiting masses M_{up} and M_{HeF} much alike to those of the straight overshoot models, whereas the partial diffusive mixing in the overshoot regions induces very extended loops in the the HRD. These models have not yet been applied to studies of the CMD and luminosity functions of real star clusters.

3. Why different kinds of mixing ? Observational hints

The many uncertainties in extant theories of convective overshooting and mixing reflect onto the variety of solutions and evolutionary models that have been proposed over the years, and has spurred many studies aimed at assessing the soundness of the proposed alternative by comparing parameterized models with observations. Among the various tests, three of them are particularly relevant.

Old open clusters. As first pointed out by Barbaro & Pigatto (1984), the interpretation of the CMD of old open clusters (e.g. M67, NGC 2420, NGC 3680, IC 4651) in terms of the classical models encounters some difficulties that can be solved by invoking a certain amount of convective overshoot during the main sequence core H-burning phase and hence older ages with respect to those from classical models. The main signatures are the detailed shape of the main sequence turnoff, the shape of the RGB, the clump of red stars (most likely core He-burners), and the number of stars brighter than the main sequence at the beginning of the subgiant branch with respect to the main sequence stars. Another type of evidence comes from small samples of stars for which good determinations of mass, radius, luminosity, and abundances are available (Andersen et al. 1990; Napiwotzki et al. 1991; Nordstrom et al. 1996), falling near the turnoff of some of these clusters. It seems that their position in the CMD is best accounted for by models with significant overshoot.

Young LMC clusters. The young rich clusters of the Large Magellanic Cloud (LMC) are classical templates to which the results of stellar evolution theory for intermediate-mass stars are compared. A powerful workbench is NGC 1866, which is well populated throughout the various evolutionary phases, exhibits an extended loop of giant stars, and is rich in Cepheids (Mateo et al. 1990). For the observed luminosity of the giants, there are too many stars above the predicted main sequence turnoff, whose number is a significant fraction of the number of giant stars. Furthermore, he predicted ratio of post main sequence stars to the main sequence stars is about four times the observed one. Making use of the integrated luminosity function of the main sequence stars normalized to the number of giants (NILF), which simply reflects the ratio of core He- to H-burning lifetimes, Chiosi et al. (1989) showed that models with substantial core overshoot reproduced the observed NILF, whereas classical models failed. Similar studies by Vallenari et al. (1992) on other clusters of the LMC reached the same conclusions.

Cepheids. The above arguments in favour of convective overshoot were also reinforced by the study of the Cepheid stars in the LMC cluster NGC 2157 by Chiosi et al. (1992), where it was shown that the use of overshoot models brings into agreement the evolutionary and pulsational mass of these stars. See also Wood (1997, this conference).

4. Mass loss from RGB and AGB stars

Mass loss during the RGB and AGB phases bears very much on the evolution of IMS. In RGB stars the question is whether or not the mass loss rate depends on the metallicity Studying the stars in the red clump of the old open cluster M67 with nearly solar metallicity, Tripicco et al. (1993) argue that the mass of these stars in much lower than commonly assumed and that the rate of mass loss along the RGB increases with the metallicity above the value holding for Globular Clusters and predicted by the classical Reimers (1975) relation. Their conclusion has been questioned by Carraro et al. (1996) who showed that fits of the CMD of M67 (and position of the clump stars in particular) are possible, in which the classical value of the mass for the stars in question is recovered.

In AGB stars, it is long known that the classical Reimers (1975) mass loss rate does not remove sufficient mass as suggested by the properties of PN stars. In the past, the discrepancy has been cured invoking the occurrence of superwind during the latest TP-AGB stages (cf. Iben & Renzini 1983). In recent models of TP-AGB stars (see below), the semi-empirical formalism of Vassiliadis & Wood (1993) is used to evaluate the mass-loss rate by stellar wind as a function of the pulsation period of variable AGB stars. It is derived from observational determinations of mass-loss rates for

Mira variables and pulsating OH/IR stars both in the Galaxy and LMC. The notable feature of this prescription is the onset of superwind which develops naturally on the AGB, instead of the artificial sudden transition that is needed if a Reimers-like law is used.

5. AGB & Carbon stars: recent results

The most recent studies of these topics are by Marigo et al. (1996a,b; 1997) who have investigated the TP-AGB phase of low and IMS with the aid of a semi-analytical model following the TP-AGB evolution from the first pulse till the complete ejection of the envelope by stellar wind (cf. also Renzini & Voli 1981, Groenewegen & de Jong 1993), paying particular attention to the changes in the chemical composition of the envelope due to (a) intershell nucleosynthesis and convective dredge-up; (b) envelope burning in the most massive AGB stars ($M \geq 3 - 4M_{\odot}$); (c) mass loss by stellar wind.

Third dredge-up. The analytical treatment of the 3rd dredge-up involves two parameters: M_c^{min}, the minimum core mass for convective dredge-up, and λ the fractionary core mass increment during the previous interpulse period dredged up to the surface. The calibration ($M_c^{min} = 0.58M_{\odot}$ and $\lambda = 0.65$) is constrained on the luminosity function of C-stars in the LMC.

Envelope burning. In agreement with previous studies (Boothroyd & Sackmann 1992 and references therein), in massive TP-AGB stars ($M \geq 3-4M_{\odot}$) with deep and hot-bottom convective envelopes ($T_{\rm B} \geq 60-100 \times 10^6 K$) the evolution of the core is not de-coupled from that of overlying layers. At high luminosities, the standard core mass - luminosity ($M_{\rm c} - L$) relation breaks down and the stars rapidly get much higher luminosities. This would anticipate the onset of the super-wind phase, thus favouring the ejection of the residual envelope.

Chemical abundances. As far as the chemical surface abundances are concerned, the rapid conversion of ^{12}C into ^{13}C and then into ^{14}N via the first reactions of the CNO cycle, can delay and even prevent the formation of C-stars. Moreover, the production of ^{7}Li possibly occurs by means of electron captures on ^{7}Be nuclei convected from the hot regions of the envelope into cooler layers ($T < 3 \times 10^6 K$) before the reaction $^{7}{\rm Be}(p,\gamma)^{8}{\rm B}$ proceeds.

AGB & C-stars in LMC. Examining the available data for AGB stars in the LMC clusters, Marigo et al. (1996b) address the question about the mass interval of low- and intermediate-mass stars which eventually evolve into C-stars during the TP-AGB phase. They combine the data compiled by Frogel et al. (1990) – near infrared photometry and spectral classification for luminous AGB stars in clusters – with the ages for individual clusters derived from independent methods (Girardi et al. 1995). The resulting distribution of C-stars in the $M_{\rm bol}$ – log(age) plane evidences that the upper and the lower limits of the mass range for the formation of C-stars cannot be derived from cluster data. The explanation of this resides in the presence of two different periods of quiescence in the cluster formation history of the LMC, shaping the age (and progenitor mass) distribution of C-stars. The most recent of these quiescence episodes could also explain the lack of very luminous AGB stars (with $-6 > M_{\rm bol} > -7$) in the clusters, contrary to what observed in the field. Finally, they compare the distribution of C-stars in the $M_{\rm bol}$ – log(age) diagram with models of AGB evolution which were previously constrained to reproduce the observed luminosity function of C-stars in the field. These models provide a good description of the relative frequency of M- versus C-stars.

References

Alongi M., Bertelli G., Bressan A., Chiosi C., 1991, A&A, 244, 95
Alongi M., Bertelli G., Bressan A., et al., 1993, A&AS, 97, 851
Andersen J., Nordstrom N., Clausen J.V., 1990, ApJ 363, L33
Barbaro G., Pigatto L., 1984, A&A, 136, 355
Bertelli G., Bressan A., Chiosi C., 1992, ApJ
Bertelli G., Bressan A., Chiosi C., 1985, A&A, 150, 33
Bertelli G., Bressan A., Chiosi C., Angerer K., A&AS, 66, 191
Bressan A., Bertelli G., Chiosi C., 1981 A&A, 102, 25
Boothroyd A.I., Sackmann I.J., 1992, ApJ 393, L21
Carraro G., Girardi L., Bressan A., Chiosi C., 1996, A&A 305, 849
Chiosi C., Bertelli G., Bressan A., ARA&A, 30, 235
Chiosi C., Bertelli G., Meylan G., Ortolani S., 1989, A&A 219, 167
Chiosi C., Wood P.R., Bertelli G., Bressan A., Mateo M., 1992, ApJ, 385, 205
Deng L., Bressan A., Chiosi C., 1986, A&A, 313, 145; 159
Frogel J.A., Mould J., Blanco V.M., 1990, ApJ, 352, 96
Girardi L., Chiosi C., Bertelli G., Bressan A., 1995, A&A, 298, 87
Groenevegen M.A.T., de Jong T., 1993, A&A, 267, 410
Grossman S.A., Narayan R., Arnett D., 1993, ApJ, 407, 284
Iben I.Jr., Renzini A., 1982, ApJ 259, L79
Iben I.Jr., Renzini A., 1983 ARA&A 21, 271
Hollowell D.E., Iben I.Jr., 1989, ApJ 340, 966
Maeder A., Meynet G., 1991. A&AS, 89, 451
Marigo P., Bressan A., Chiosi C., 1996a, A&A, 313, 545
Marigo P., Bressan A., Chiosi C., 1996b, A&A, 316, L1
Marigo P., Bressan A., Chiosi C., 1997, A&A, to be submitted
Mateo M., Olszewski E.W., Madore B., 1990, ApJ 107, 203
Napiwotzki R., Schoenberner D., Weidmann V., 1991, A&A 243, L5
Nordstrom B., Andersen J., Andersen M.I., 1996, A&AS, 118, 407
Paczynski B., 1970, Acta Astron. 20, 47; 287
Reimers D., 1975, Mem. Soc. R. Sci. Liege, 6 (8), 369
Renzini A., Voli M., 1981, A&A 94, 175
Tripicco M.J., Dorman B., Bell R.A., 1993, AJ, 106, 618
Vallenari A., Chiosi C., Bertelli G., Meylan G., Ortolani S., 1992, AJ, 104, 1100
Vassiliadis E., Wood P.R., 1993, ApJ, 413, 641
Xiong D.R., 1986, A&A 167, 239

Discussion of this paper appears at the end of these Proceedings.

THEORY OF LOW MASS STARS, BROWN DWARFS AND EXTRA-SOLAR GIANT PLANETS

GILLES CHABRIER AND ISABELLE BARAFFE
C.R.A.L., Ecole Normale Supérieure, 69364 Lyon Cedex 07, France
chabrier@ens-lyon.fr; ibaraffe@ens-lyon.fr

1. Introduction

Accurate modeling of the mechanical and thermal properties of very-low-mass stars (VLMS), Brown Dwarfs (BD) and Extra-solar Giant Planets (EGP) is of prior importance for a wide range of physical and astrophysical problems, from the fundamental physics point of view to the astrophysical and cosmological implications. They provide natural laboratories to test the different theories, equations of state, nuclear reaction rates, model atmospheres aimed at describing the physics of dense and cool objects. They represent the largest stellar population in the Galaxy, and thus provide a substantial contribution to the Galactic (disk) mass budget. Finally they represent one of the most intriguing questions in our understanding of the formation of star-like objects: are planet and star formation processes really different ? Is there, and if so what is, a minimum mass for the formation of star-like objects ? This field has blossomed recently with the discovery of several brown dwarfs (Nakajima et al. 1995; Rebolo et al., 1995) and numerous exoplanets since 51 Pegasi (Mayor and Queloz 1995; Mayor, this conference), which provide important information to challenge the theory.

2. Theoretical improvements

VLMS, or M-dwarfs, defined hereafter as objects with masses below $0.6M_\odot$, are compact objects, with characteristic radii in the range $\sim 0.1 - 0.6\,R_\odot$. Their central densities and temperatures are respectively of the order of $\rho_c \approx 10 - 10^3$ gcm^{-3} and $T_c \approx 10^6 - 10^7$ K, so that the stellar interior represents a strongly correlated plasma. BD's are defined as objects not

T.R. Bedding et al. (eds.),
Fundamental Stellar Properties: The Interaction between Observation and Theory, 331–340.

massive enough for their central temperature to sustain hydrogen burning. This characteristic stems from the onset of degeneracy in the contracting protostar, which prevents an increase of the temperature as contraction proceeds. A BD therefore never reaches thermal equilibrium and cools down for its whole life. The hydrogen-burning minimum mass (HBMM) is $\sim 0.07 - 0.09 M_{\odot}$, depending on the initial composition from Z=Z$_{\odot}$ to Z=0, respectively (Chabrier & Baraffe, 1997). The minimum mass depends on stellar formation theory and is still uncertain (Hubbard, 1994). Brown dwarfs bridge the gap in the observed mass-distribution of astrophysical objects between the lowest-mass star and the largest solar planet, Jupiter ($M_J = 0.001\, M_{\odot}$). The distinction between BD and Giant Planets is based on different formation scenarios, but their inner structure (apart from the central rocky core in planets) and spectral signatures are both governed by the same physics.

The effective temperatures of these low-mass objects are below $T_{eff} \approx$ 5000 K, down to $\sim$ 100 K for BDs and EGPs, and surface gravities $g = GM_{\star}/R_{\star}^2$ are in the range log $g \approx 3.5 - 5.5$. The low effective temperature allows the presence of stable molecules in the atmosphere (see Baraffe & Allard, these proceedings). The presence of these bands complicates tremendously the treatment of radiative transfer, not only because of the numerous transitions to include in the calculations, but also because the molecular absorption coefficients strongly depend on the frequency. Moreover molecular recombination in the interior ($2H \rightarrow H_2$) leads to a decrease of the adiabatic gradient so that convection penetrates deeply into the optically-thin atmospheric layers. Because of these physical processes, the grey-approximation is no longer valid below $T_{eff} \sim$ 5000 K, as shown by several authors (Allard, 1995; Saumon et al., 1994; Chabrier et al., 1996; Chabrier & Baraffe, 1997).

These conditions show that the modeling of low-mass objects requires a correct description of non-ideal effects in the interior, as well for the equation of state (EOS) as for the screening factors of the nuclear rates, a derivation of accurate models for dense and cool atmospheres, where molecular opacity becomes eventually the main source of absorption, and consistent (non-grey) boundary conditions between the atmosphere and the interior along evolution.

A new EOS, devoted to the description of low-mass stars and giant planets, has been derived recently, which presents a consistent treatment of pressure ionization (Saumon, Chabrier & VanHorn 1995; SCVH), altough based on the so-called ideal-volume mixing between hydrogen and helium (see SCVH). Improvement in the field of cool atmosphere models has blossomed within the past few years with the work of Allard and Hauschildt (1995; 1997), Brett (1995), and Saumon et al. (1994). Non-grey atmosphere models now extend down to $T_{eff} = 900$ K, since the discovery of the first cool BD GL 299B offered a stringent test for such models (Allard et al. 1996; Tsuji et al. 1996a; Marley et al., 1996).

A detailed description of the recent improvement in the theory of low-mass objects (LMS and BDs), and a comparison with standard grey-like treatments, is given in a recent paper by Chabrier & Baraffe (1997). The modelization of VLMS, BDs and EGPs and the confrontation with observations is examined below.

2.1. VLM STARS

In spite of considerable progress in stellar theory - internal structure, model atmospheres and evolution - all the VLMS models so far failed to reproduce accurately the observed color-magnitude diagrams (CMD) of disk or halo stars below ~ 4000 K, i.e. $\sim 0.4 - 0.6\, M_{\odot}$, depending on the metallicity. All the models predicted too hot an effective temperature for a given luminosity, i.e. were too blue compared to the observations by at least one magnitude. Such a disagreement stemmed essentially from shortcomings both in the physics of the interior, i.e. equation of state (EOS) and thus mass-radius relationship and adiabatic gradient, and in the atmosphere, since all models were based on grey atmospheres and approximate outer boundary conditions. Important progress has been made recently in this field with the derivation of evolutionary models based on a *consistent treatment between the interior and the atmosphere profile* (Baraffe, Chabrier, Allard & Hauschildt 1995, 1997; Chabrier, Baraffe & Plez 1996). The effect of the outer boundary condition on the mass-T_{eff} relationship has been examined in detail by Chabrier & Baraffe (1997). These authors have made comparison with models based on different grey-like treatments and have shown convincingly that such treatments (which imply a $T(\tau)$ relationship as an external boundary condition) are incorrect, or at best highly unreliable as soon as molecular formation sets in, i.e. for any object below $T_{eff} \lesssim 5000$ K, i.e. $m \sim 0.5\, M_{\odot}$.

The LMS models based on the afore-mentioned updated physics and consistent (non-grey) boundary condition now reach quantitative agreement with observations for both the disk and the halo stellar population down to the bottom of the main sequence (Figure 1). For the *disk population* the models reproduce the observed color-magnitude diagrams both in the infrared (M_K vs (I-K)) and in the optical (M_V vs (V-I)) (Baraffe et al., 1997b), although below 0.1 $M_{\odot}$, $M_V \gtrsim 15$, the models are still too blue by $\sim 0.2 - 0.4$ mag in (V-I), as illustrated in Figure 1. This is likely to stem from the still inaccurate TiO line list which shapes the flux in the optical, or possibly from the onset of grain formation for solar metallicity (see

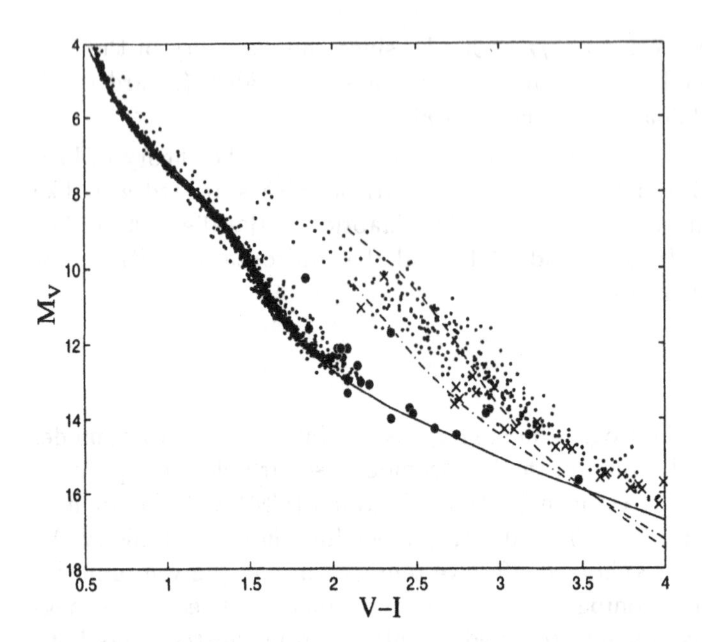

Figure 1. M_V-(V-I) diagram for different metallicities : [M/H]=-1.5 (solid line), [M/H]=-0.5 (dash-dot) and [M/H]=0 (dash). The Main Sequence of the Globular Cluster NGC6397 from Cool et al. (1996) is shown on the left part. Subdwarf halo field stars from Monet et al. (1992) are indicated by full circles, as well as disk M-dwarfs of Monet et al. (1992) (crosses) and Dahn et al. (1995) (dots on the right hand side)

Baraffe & Allard, these proceedings). More importantly, the models are in excellent agreement with the *observationally-determined* mass-magnitude relationship (Henry & McCarthy, 1993) both in the infrared and in the optical (Chabrier, Baraffe & Plez, 1996; Baraffe et al., 1997b) as shown in the present Figure 2 and in Figure 1 of Baraffe & Allard (these proceedings). For *metal-depleted populations*, the models are in remarkable agreement with the main sequences of globular clusters observed with the HST nearly down to the bottom of the main sequence and with the halo field subdwarf sequence, as shown in Figure 1 (Baraffe et al., 1997a). We stress that, since LMS are essentially fully convective below $\sim 0.4\, M_\odot$ (Chabrier & Baraffe, 1997), the models are not hampered by *any* adjustable parameter and the agreement between theory and observation reflects directly the reliability of the physics entering the models. These models yield the derivation of reliable mass-functions for the disk population (Méra, Chabrier & Baraffe, 1996) and for globular clusters and halo field stars down to the brown dwarf limit (Chabrier and Méra, 1997).

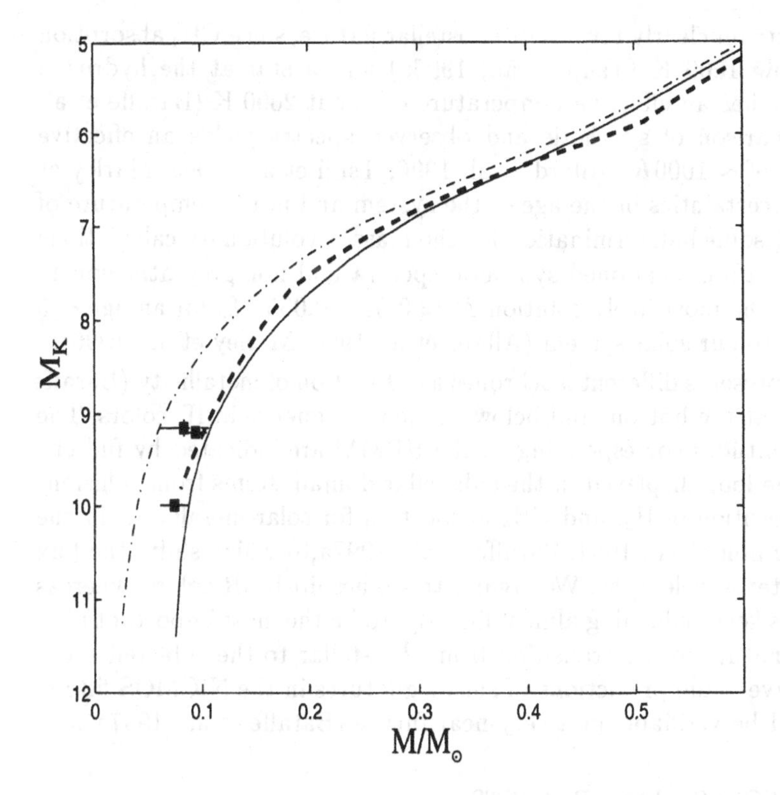

Figure 2. Mass-luminosity relationship in the K-band. The dash-line is the fit of the observationally-determined relation (Henry & McCarthy, 1993; HMc93), the solid line is based on the most recent models for solar metallicity (Baraffe et al., 1997b) for $t = 5$ Gyr, whereas the dash-dot line is the same models for $t = 0.1$ Gyr (the ZAMS age for $0.1\ M_\odot$ is about 0.5 Gyr). Note that the three lowest-mass objects observed by HMc93, shown on the figure with their error bars, all exhibit strong surface (H_α-emission) and coronal (X-emission) activity. Although this is not a proof of a young age, it is a strong hint and is supported by the present analysis.

2.2. BROWN DWARFS

The recent discovery of substellar objects enables us to test the reliability of the theory in the BD regime. The first young brown dwarfs were identified in the Pleiades star cluster (Rebolo et al. 1995), namely Teide 1 and Calar 3, which both have retained their initial Lithium abundance (Rebolo et al., 1996). Their extremely low luminosity ($\log L/L_\odot \sim -3.1$) combined with the presence of lithium yields masses $m \sim 0.05 M_\odot$, well below the hydrogen-burning minimum mass (Rebolo et al., 1996; Baraffe et al, 1997b). Given the large effective temperature of these two BDs ($T_{eff} \sim 2500K$), due to their young age ($\sim 10^8$ yrs), the most stringent test for extremely cool models is provided by Gl 229B (Nakajima et al. 1996). The presence of methane in its infrared spectrum clearly reveals its substellar nature, since CH_4 absorption appears below 1600 K (Tsuji et al., 1996a) and a star at the hydrogen burning limit has an effective temperature of about 2000 K (Baraffe et al. 1995). Comparison of synthetic and observed spectra yields an effective temperature of $\sim 1000K$ (Allard et al. 1996; Tsuji et al. 1996a; Marley et al. 1996). Uncertainties in the age of the system and in the temperature of Gl229B yield some indetermination for the mass. Evolutionary calculations based on the afore-mentioned synthetic spectra and non-grey atmosphere models yield the most likely solution $M \approx 0.04 - 0.055\ M_\odot$ for an age ~ 5 Gyr, similar to our solar system (Allard et al. 1996; Marley et al., 1996).

Figure 3 presents different isochrones as a function of metallicity (Baraffe et al. 1997b) at the bottom and below the main sequence in IR colors. The K-limit magnitudes corresponding to the HBMM are indicated by full circles. The blue loop displayed in the substellar domain stems from collision-induced absorption of H_2, and CH_4 absorption for solar-metallicity, in the K-band (Saumon et al., 1994; Baraffe et al., 1997a,b) which shifts the flux back to shorter wavelengths. We predict this blueshift in IR-colors, whereas optical colors keep reddening almost linearly, to be the most important photometric signature of the transition from the stellar to the sub-stellar domain. We have made predictions of such signatures in the NICMOS filters, which should be verifiable in a very-near future (Baraffe et al., 1997a).

2.3. EXTRASOLAR GIANT PLANETS

A general theory of EGPs, from 0.3 to 15 Jupiter masses, and the possibility of detection by existing and future observational projects, has been performed recently by Saumon et al. (1996). These calculations take root in the theory of solar giant planets derived previously (Chabrier et al., 1992; Guillot et al. 1995). In this first generation of calculations, the spectral emission of the EGP and of the parent star is approximated by a black body distribution. Although weakened by this approximation, the predictions made in Saumon et al. (1996) represent the first benchmark in search strategies for the detection of EGPs. They have been applied more specifically to the case of 51 PegA to demonstrate the stability of giant planets orbiting nearby stars (Guillot et al., 1996).

More recent calculations by Marley et al. and Allard et al. do now predict synthetic spectra in the substellar domain from $T_{eff} \sim 2000$ K down to ~ 300 K. The predicted absolute fluxes of BD or EGP (Allard et al. 1997) are displayed in Fig. 3 of Baraffe & Allard (these proceedings) and compared to the sensitivity of ground and space-based observing platforms.

An extra-degree of complication in the atmosphere of objects below $T_{eff} \sim 2600K$ is due to the onset of grain formation, as suggested by Tsuji

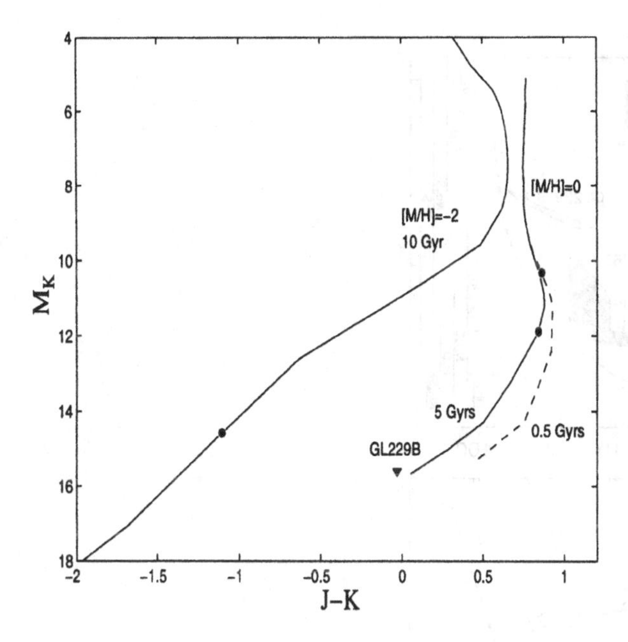

Figure 3. Color-magnitude diagram in the near IR for the metallicity and ages indicated on the figure. The full circles on the curve indicate the stellar/sub-stellar transition.

and collaborators (Tsuji et al. 1996b). This is illustrated in Figure 4 which displays the effective temperature and the radius of various exoplanets discovered recently with the domain of condensation (bars on the right hand side) of various compounds (Burrows et al., 1997).

3. Conclusion

The recent improvements in the description of the mechanical and thermal properties of cool, compact objects, and of their photometric signature now provide solid grounds to analyse the observations and make reliable predictions. The three essential inputs in the theory of these objects are an accurate EOS, with a reliable treatment of non-ideal effects and pressure ionization, synthetic spectra with accurate molecular absorption coefficients in the optical and the IR, and consistent evolutionary calculations with correct (non-grey) boundary conditions between the atmosphere and the interior profiles.

This improved theory of LMS and substellar objects allows now the derivation of reliable stellar mass-functions down to the brown dwarf limit and, depending on the rate of discovery of substellar objects in the near future, will yield eventually the brown dwarf mass function, in connection with microlensing observations. This will bring new insight on the still

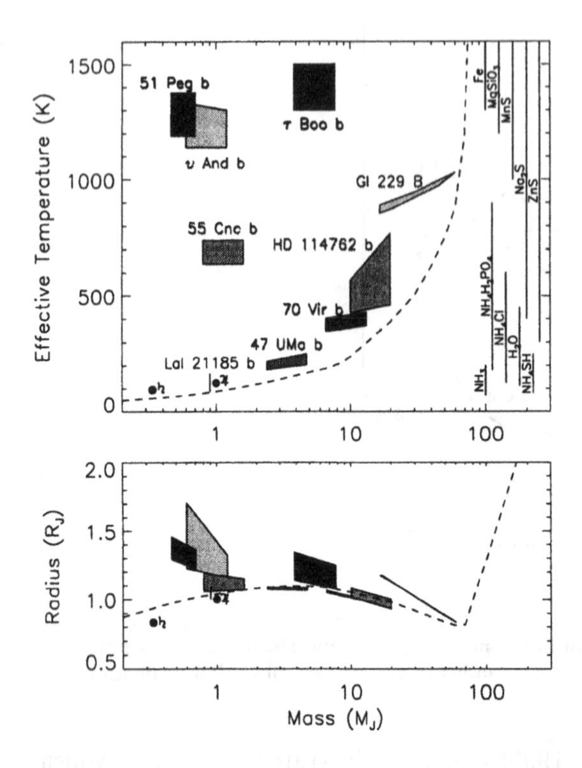

Figure 4. T_{eff} and radius as a function of mass for several EGPs (kindly provided by A. Burrows). Jupiter and Saturn are also indicated by their usual symbols. The bars on the right hand side show the domain of condensation of different species.

unsolved problem of stellar versus planetary formation.

The derivation of cool atmosphere models and synthetic spectra including grain formation represents the next substantial improvement. Work is already under progress in this field, both in the Tucson group and in the Lyon-Wichita group, and promises more exciting results in the near future.

References

Allard, F., 1995, *The bottom of the main-sequence and below*, Ed. C. Tinney, Springer Verlag
Allard, F., and Hauschildt, P. H., 1995, ApJ, 445, 433
Allard, F., Hauschildt, P.H., Baraffe, I., Chabrier, G., 1996, ApJ, 465, L123
Allard, F., Hauschildt, P.H., Alexander D. R., Starrfield, S., 1997, ARA&A, 35, 137
Baraffe, I., Chabrier, G., Allard, F., Hauschildt P., 1995, ApJ, 446, L35
Baraffe, I., Chabrier, G., Allard, F., Hauschildt P., 1997a, A&A, in press
Baraffe, I., Chabrier, G., Allard, F., Hauschildt P., 1997b, in preparation
Brett, J.M., 1995, A&A, 295, 736
Burrows A. et al., 1997, in preparation

Chabrier, G., Saumon, D., Hubbard, W.B. & and Lunine, J.I., 1992, ApJ, 391, 317
Chabrier, G., Baraffe, I., Plez, B., 1996, ApJ, 459, L91
Chabrier, G., Baraffe, I., 1997, A&A, in press
Chabrier, G. & Méra, D., 1997, A&A, in press
Cool, A. M., Piotto, G., King, I.R., 1996, ApJ, 468, 655
Dahn, C.C., Liebert, J., Harris, H.C., Guetter, H.H., 1995, *The bottom of the main-sequence and below*, Ed. C. Tinney
Guillot, T., Chabrier, G., Gautier, D., Morel P., 1995, ApJ, 450, 463
Guillot, T., Burrows, A., Lunine, J.I., Saumon, D. 1996, ApJ, 459, L35
Henry , T.D., and McCarthy, D.W.Jr, 1993, AJ, 106, 773
Hubbard. W. B., 1994: *the Equation of State in Astrophysics*, Ed. G. Chabrier and E. Schatzman, Cambridge University Press
Marley et al., 1996, Science, 272, 1919
Mayor, M., Queloz, D., 1995, Nature, 378, 355
Méra, D., Chabrier, G. and I. Baraffe, 1996, ApJ, 459, L87
Monet D.G. et al., 1992, Astron. J., 103, 638
Nakajima et al., 1995, Nature, 378, 463
Rebolo, R., Zapatero Osorio, M.R., Martin, E.L., 1995, Nature, 377, 129
Rebolo, R., Martin, E.L., Basri, G., Marcy, G.W. & Zapatero Osorio, M.R., 1996, ApJ, 469, L53
Saumon, D., Bergeron, P., Lunine, L.I., Hubbard, W.B., Burrows, A., 1994, ApJ, 424, 333
Saumon, D., Hubbard, W.B., Burrows, A., Lunine, J.I. and Chabrier, G., ApJ, 1996, 460, 993
Saumon, D., Chabrier, G., VanHorn, H.M., 1995, ApJS, 99, 713
Tsuji, T., Ohnaka, K., Aoki, W., Nakajima. T., 1996a, A&A, 308, L29
Tsuji, T., Ohnaka, K., Aoki, W., 1996b, A&A, 305, L1

DISCUSSION

PIERRE MAXTED: You suggested that two of the brown dwarfs seen in the Pleiades are binary stars since they sit so far above the isochrones in the HR diagram. Is there any independent observational evidence of binarity in these stars?

GILLES CHABRIER: The groups of Rafael Rebolo and Gibor Basri, who suggested the possible binarity, are examinating this possibility but, as far as I know, they don't have a definitive answer yet.

JØRGEN CHRISTENSEN-DALSGAARD: Would it be fair to ask what the MACHO objects are, if not brown dwarfs?

GILLES CHABRIER: As we've shown in a recent paper (ApJ 468 L21), they can be explained by halo white dwarfs, providing some stringent conditions are met. I will illustrate this point in my white dwarf talk.

FLAVIO FUSI PECCI: You have shown nice fits of observed faint main sequences with Galactic globulars and claim that the quality of the fits is much better than obtained so far, thanks to the improved quality of your models compared to previous ones. I have seen similar fits obtained, for instance, for NGC 6397 (HST Data) by Alexander et al. (1997 A&A, 317 90) using their own models - could you schematically explain which are the main differences between

the two sets and how you determine and evaluate the quality of the best fitting solution?

GILLES CHABRIER: A detailed answer is given in our papers (Chabrier & Baraffe 1997; Baraffe et al. 1997a). To summarize, there are five important differences:

1. Our models rely on consistent non-grey boundary conditions, whereas the Teramo models use a grey condition, based on $T(\tau)$ approximation. Even when modified (or adjusted), the grey approximation is basically *not* correct/reliable for LMS, because of the very physics characteristic of these objects (see the present review).
2. With the *same physics*, and *no* adjustable parameter, our models reproduce all the observed sequences from Fe/H = -2.2 (M 15) to -1.0 (ω Cen) and even -0.5 (47 Tuc), not just NGC 6397.
3. The Teramo models assume $[M/H] = [Fe/H]$ which is not correct for metal-depleted stars, because of the O/Fe enrichment, which yields $[M/H] = [O/H]$.
4. The Teramo models rely on previous (so-called "Base") colors by Allard & Hauschildt which overestimate the opacity (because of the Straight mean approximation, see Allard et al., 1997, ARA&A) and are now abandoned. Moreover the bolometric corrections have been shifted arbitrarily to recover the Kurucz values at high temperatures (see their §4).
5. Our models use consistent reddening corrections, calculated from Allard's synthetic spectra. These corrections (E(V-I)=0.22) are in excellent agreement with the ones used by the observers (E(V-I)=0.23; Cool et al., 1996). The ones used by the Teramo group (E(V-I)=0.19, see their Fig. 7) differ significantly. With the reddening and metallicity quoted by the observers, their models are too red w.r.t. the observations (see e.g. Fig 5 of Baraffe et al., 1997a).

11. STELLAR INTERIORS

ROTATION: A FUNDAMENTAL PARAMETER OF MASSIVE STARS

N. LANGER
Universität Potsdam, Institut für Theoretische Physik und Astrophysik, D-14415 Potsdam, Germany

AND

A. HEGER AND J. FLIEGNER
MPI für Astrophysik, D-85740 Garching, Germany

1. Introduction

Massive stars are rapid rotators. Equatorial rotation velocities span the range $v_{\rm rot} = 100-400\,{\rm km\,s^{-1}}$, with B stars rotating closest to their break-up speed $v_{\rm crit}$ (Howarth et al. 1997). During the last decade, many observations have revealed unusual surface abundances that may require additional internal mixing (beyond that of simple convection and overshooting) for their explanation, most important helium and nitrogen enrichment in main sequence O and B stars (Gies & Lambert 1992), in the SN 1987A progenitor (Fransson et al. 1989), and boron depletions in main sequence B stars (Venn et al. 1996). In particular the latter observations clearly point towards internal mixing and rule out a close binary origin of the abundance peculiarities (Fliegner et al. 1996). Altogether, the occurrence of some form of additional mixing responsible for altering the surface abundances in a large fraction, if not all massive stars appears to be beyond reasonable doubt, and mixing processes due to rotation are the most natural explanation.

Here we report on results obtained with a 1D implicit hydrodynamic stellar evolution code, which was modified to incorporate several effects of rotation. The angular momentum is treated as a local variable. The centrifugal force is included in latitude-averaged form, with non-spherical equipotential surfaces replacing the usual Lagrangian mass variable as independent spatial coordinate (Endal & Sofia 1978). The approximate constancy of all physical variables on equipotential surfaces is due to the action of the baroclinic instability (Zahn 1992). The transport of chemical elements

T.R. Bedding et al. (eds.),
***Fundamental Stellar Properties: The Interaction between Observation and Theory*, 343–348.**

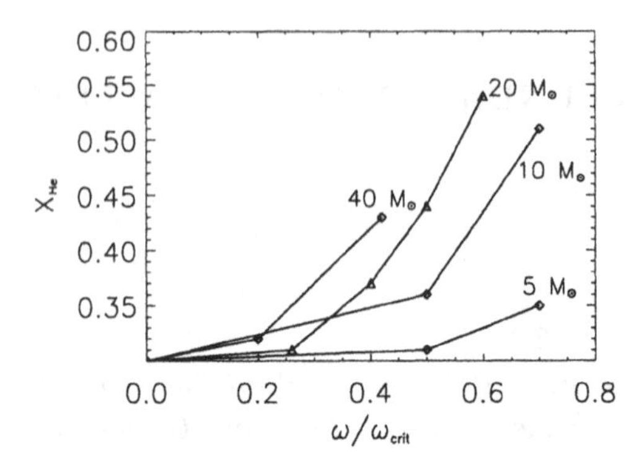

Figure 1. Helium surface mass fraction at core hydrogen exhaustion for models with an initial mass of 5, 10, 20 and 40 $M_{\odot}$, as function of the ratio of rotation to critical rotation at the zero age main sequence.

and angular momentum is performed in the diffusion approximation, with diffusion coefficients appropriate for the following instabilities: convection, semiconvection, dynamic and secular shear instability, Goldreich-Schubert-Fricke instability and Eddington-Sweet circulation (cf. also Maeder 1997). Mass loss due to stellar winds is taken into account, which is important to maintain a small but significant angular momentum gradient and thus efficient rotational mixing (Zahn 1992).

2. Rotationally induced mixing

2.1. ENVELOPE POLLUTION AND SURFACE ENRICHMENT

Fig. 1 shows the effect of rotationally induced mixing on the surface helium abundance of galactic (Z=2%) massive main sequence stars. While the non-rotating models ($\omega/\omega_{\mathrm{crit}} = 0$) maintain their initial helium surface abundance ($X_{\mathrm{He}} = 0.30$) until core helium exhaustion, rotating models develop a helium enrichment which is more pronounced for larger initial rotation rates. Fig. 1 shows also that, for a fixed value of $\omega/\omega_{\mathrm{crit}}$, stars with a higher initial mass obtain higher helium enrichment. This is due to the increasing contribution of radiation pressure at higher masses, which decreases the effect of mean molecular weight barriers on the mixing.

Like helium, all other abundances which are changed in the core due to hydrogen burning are altered at the stellar surface. This concerns in particular boron and carbon, which are depleted, and nitrogen, oxygen, sodium, and the radio-nuclide ^{26}Al, which are all enhanced. Furthermore, the enrichment of helium not only at the surface but throughout the radiative stellar

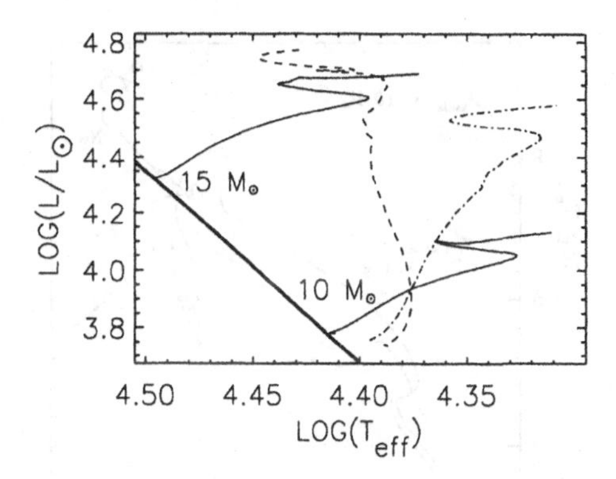

Figure 2. Evolutionary tracks of three $10\,M_\odot$ stars with different initial equatorial rotation velocities (v_{eq} = 0, 280, and 400 km/s; solid, dash-dotted, and dashed lines, respectively) during the core hydrogen burning phase. The track of a non-rotating $15\,M_\odot$ star is also shown. The thick solid line marks the ZAMS position of non-rotating stars.

envelope increase significantly the mean molecular weight of the star. This leads to a strong increase of the luminosity (cf. Fig. 2), consequently to a larger convective core and to a larger helium core mass at core hydrogen exhaustion, compared to the non-rotating case (Langer 1992). It is obvious that all these effects significantly alter the nucleosynthesis predictions of massive star models (cf. Langer et al. 1997, for details).

2.2. ISOCHRONES, INITIAL MASS FUNCTION AND SUPERNOVA RATE

Figure 2 shows the effects of rotation on the main sequence evolution in the HR diagram for stars with an initial mass of $10\,M_\odot$. Although the more rapid rotators are initially less luminous due to the reduced effective gravity, the increasing luminosity due to the mixing of helium (cf. Sect. 2.1) is soon the overwhelming effect. If the distribution of initial rotation velocities is not very strongly peaked, the tracks shown in Fig. 2 demonstrate that the concept of isochrones in the HR diagram changes its character. Due to the dependence of the tracks on the initial rotation rate, isochrones are no longer lines but become surfaces in the HR diagram.

A similar effect occurs for the mass-luminosity relation: the luminosity depends on mass *and* initial rotation rate. Stellar masses obtained from observed luminosities through standard M-L relations will in general result in too large masses. This has been found empirically by comparing evolutionary stellar masses to masses derived by other methods (e.g., Herrero et al. 1992). One consequence is that initial mass functions which are derived from observed luminosity functions using a mass-luminosity relation

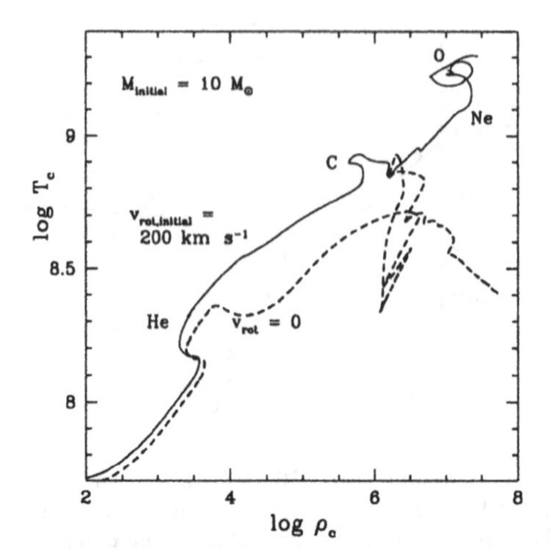

Figure 3. Evolutionary tracks of a rotating (solid line) and a non-rotating (dashed line) $10\,M_\odot$ star in the $\log \rho_c - \log T_c$–diagram. The various core burning stages are indicated along the track of the rotating model.

for non-rotating stars will systematically overestimate the number of the more massive stars.

Finally, also the fate of a massive star may depend on its rotation rate. Fig. 3 compares the evolution of a rotating and a non-rotating $10\,M_\odot$ star in the central temperature vs. central density diagram. Due to its larger helium core, the rotating star evolves through all the nuclear burning phases without being strongly affected by electron degeneracy, and explodes as a Type II supernova. The non-rotating star ignites carbon off-center, whereafter the core cools dramatically. Depending on the mass loss rates in the late stages, it may evolves into a ONeMg white dwarf rather than exploding. I.e., rotation decreases the critical initial mass for supernova explosion (which now is a function of the initial rotation rate) and therefore leads to a substantial increase of the supernova rate.

3. Mass loss of massive stars at the Ω-limit

Massive stars have radiation driven winds. Friend & Abbott (1986) showed that rotation at a rate $\Omega := v_{\rm rot}/v_{\rm crit}$ leads to an enhancement of the mass loss rate by the factor $(1/(1-\Omega))^{0.43}$ in O and B stars. This effect alone results in a dependence of the evolution of massive stars on their initial rotation rate.

Furthermore, it has been shown by Langer (1997) that massive main sequence stars may actually reach the Ω-limit $\Omega \simeq 1$. The coupling of mass

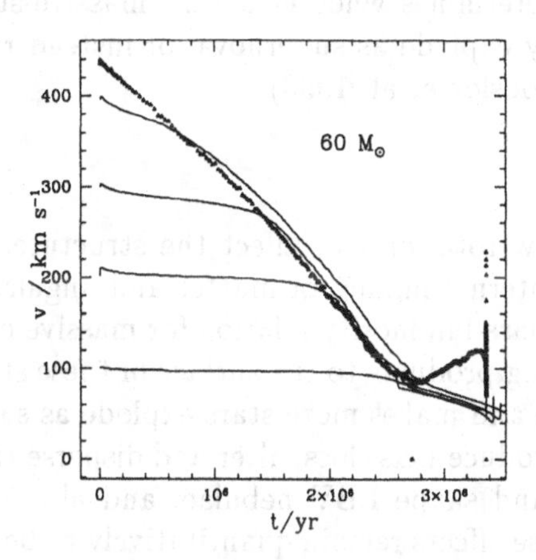

Figure 4. Evolution of the equatorial rotation velocity with time during the core hydrogen burning phase of four 60 $M_\odot$ stars with different initial rotation rates (see at $t = 0$). The thick line displays the evolution of the critical rotational velocity for the sequence with an initial rotation velocity of 100 km s^{-1}; it is very similar for the other sequences. The stars are assumed to maintain rigid rotation, and the evaluation of $v_{\rm crit}$ is uncertain (cf. Langer 1997, for details). For $v_{\rm rot} \simeq v_{\rm crit}$, the stars evolve at the Ω-limit.

and angular momentum loss results in a well defined limiting mass loss rate $\dot{M}_\Omega$: larger $\dot{M}$ would result in so much angular momentum loss that the star would evolves away from the Ω-limit. For a 60 $M_\odot$ star, Langer (1997) found $\dot{M}_\Omega \simeq 10^{-5}\ M_\odot\ {\rm yr}^{-1}$. Mass loss at the rate $\dot{M}_\Omega$ keeps the star at the Ω-limit until the critical rotational velocity is decreasing again with time, i.e. for decreasing effective temperature during the main sequence evolution. Fig. 4 illustrates that the time spent at the Ω-limit and therefore the total amount of mass lost depends strongly on the initial rotation rate. The stellar radiation field may not be strong enough to push all the mass lost at the Ω-limit to infinity; rather a ring or disk may form around the star, which may thus resemble a Be or B[e] star.

During the post main sequence evolution, massive stars may again hit the Ω-limit, but due to the much shorter evolutionary time scale $\dot{M}_\Omega$ may be much larger, mass and angular momentum loss decouple, and the mass loss becomes unstable. Garcia-Segura et al. (1997) showed that the assumption of Luminous Blue Variable eruptions being due to massive post main sequence stars hitting the Ω-limit naturally explains the bipolarity of most LBV nebulae and in particular the shape of the Homunculus nebula around η Carinae.

Clearly, the type of mass loss described here affects the initial-final mass relation of massive stars, which now depends on Ω. The initial angular momentum even determines whether a very massive star can lose enough mass to successfully explode as supernova, or instead rather collapses into a black hole (cf. Woosley et al. 1993).

4. Conclusions

We have shown how rotation can affect the structure, evolution and fate of massive stars. Internal mixing of matter and angular momentum shifts and disperses the mass-luminosity relation for massive main sequence stars, transports H-burning products to the surface of these stars, alters the character of isochrones, and makes more stars explode as supernovae. Rotation can enhance and produce mass loss, alter and disperse the initial-final mass relation, produce and shape LBV nebulae, and also lead to more supernovae. Many of these effects remain quantitatively to be explored. However, for most of them there is overwhelming observational evidence showing their qualitative significance. There is no doubt that the initial rotation rate is a truly fundamental parameter for massive stars, which is equally important as the initial mass and metallicity.

Acknowledgements

This work has been supported by the Deutsche Forschungsgemeinschaft through grant La 587/8-2.

References

Endal A.S., Sofia S. 1978, ApJ, 220, 279
Fliegner J., Langer N., Venn K.A. 1996, A& A, 308, L13
Fransson C., Cassatella A., Gilmozzi R., et al. 1989, ApJ, 336, 429
Friend D.B., Abbott D.C. 1986, ApJ, 311, 701
Herrero A., Kudritzki R.P., Vilchez J.M., et al., 1992, A& A 261, 209
Howarth I.D., Siebert K.W., Hussain G.A.J., Prinja R.K., 1997, M.N.R.A.S., in press
García-Segura G., Langer N., Mac Low M.-M. 1997, in *Luminous Blue Variables: Massive Stars in Transition*, eds. A. Nota, H. Lamers, ASP Conf. Ser., in press
Gies D.R., Lambert D.L. 1992, ApJ, 387, 673
Langer N. 1992, A& A, 265, L17
Langer N. 1997, in *Luminous Blue Variables: Massive Stars in Transition*, eds. A. Nota, H. Lamers, ASP Conf. Ser., in press
Langer N., Fliegner J., Heger A., Woosley S.E. 1997, in *Nuclei in the Cosmos IV*, ed. M. Wiescher, Nucl. Phys. A, in press
Maeder A. 1997, A& A, in press
Venn K.A., Lambert D.L., Lemke M., 1996, A& A, 307, 849
Woosley S.E., Langer N., Weaver T.A. 1993, ApJ, 411, 823
Zahn, J.P. 1992, A& A, 265, 115

Discussion of this paper appears at the end of these Proceedings.

THE EFFECT OF ROTATION ON RGB SURFACE ABUNDANCES

C. CHARBONNEL
Laboratoire d'Astrophysique de Toulouse - CNRS UMR 5772 - 14, av.E.Belin, 31400 Toulouse, France

1. Introduction

Pop II field and globular cluster giant stars (and, to a less extent, Pop I giants) exhibit chemical anomalies which are not predicted by standard stellar evolution theory. Two hypotheses have been proposed to explain these abundance variations, namely the primordial and the evolutionary explanations. A primordial origin for intracluster abundance anomalies (see e.g. Cottrel & Da Costa 1981) would be related to inhomogeneities in the cluster material due to pollution by a prior generation of massive stars. In the evolutionary hypothesis, abundance variations would be due to nuclear and mixing processes internal to the giant stars themselves. Many good reviews exist on the subject (see e.g. Briley et al. 1994a, Kraft 1994), in which observational evidence supporting both hypotheses are presented. In this conference, Da Costa recalls the most recent observational data, and some excellent poster contributions bring essential clues to the subject.

I will concentrate here on the evolutionary hypothesis. I will first recall the main observations which reveal that an extra-mixing process occurs in low mass stars while they evolve on the red giant branch (RGB), and more precisely between the completion of the standard first dredge-up and the onset of the helium flash. I will then show how rotation-induced mixing can simultaneously account for the observed behavior of carbon isotopic ratios and lithium abundances in low mass giants. This process also avoids large ^{3}He production by low mass stars in the Galaxy. New developments will be proposed.

2. Observational evidence supporting the evolutionary scenario

When the stars experience the first dredge-up, the deepening convective envelope brings up to the surface internal matter which was nuclearly-

T.R. Bedding et al. (eds.),
***Fundamental Stellar Properties: The Interaction between Observation and Theory*, 349–354.**

processed during the main-sequence evolution. This leads to modifications of the surface abundances by amounts that depend both on the stellar mass and initial metallicity. In low mass stars, the convective envelope reaches only regions where ^{12}C was processed in favor of ^{13}C and ^{14}N. So basically, the carbon isotopic ratio declines (from 90 to about 20-30), the carbon abundance drops (by about 30 %) and nitrogen increases (by about 80 %), but oxygen and all other element abundances remain unchanged. Still according to the standard scenario, the surface abundances then stay unaltered as the convective envelope slowly withdraws during the end of the RGB evolution.

However, observational data on the abundance variations of C, N, O, Na, Al in evolved stars reveal a different reality.

- Pop II field and globular cluster giants present $^{12}C/^{13}C$ ratios lower than 10, even down to the near-equilibrium value of 4 in many cases (Sneden et al. 1986, Smith & Suntzeff 1989, Brown & Wallerstein 1989, 1992, Bell et al. 1990, Suntzeff & Smith 1991, Shetrone et al. 1993, Briley et al., 1994b, 1997).
- In evolved halo stars, the lithium abundance continues to decrease after the completion of the first dredge-up (Pilachowski et al. 1993).
- A continuous decline in carbon abundance with increasing stellar luminosity along the RGB is observed in globular clusters such as M92 (Carbon et al. 1982; Langer et al. 1986), M3 and M13 (Suntzeff 1981), M15 (Trefzger et al. 1983), NGC 6397 (Bell et al. 1979, Briley et al. 1990), NGC 6752 and M4 (Suntzeff & Smith 1991)
- In some globular clusters (M92, Pilachowski 1988; M15, Sneden et al. 1991; M13, Brown et al. 1991, Kraft et al. 1992; Omega Cen, Paltoglou & Norris 1989), giants exhibit evidence for O$\rightarrow$N processed material.
- In addition to the O versus N anticorrelation, the existence of Na and Al vs N correlations and Na and Al vs O anticorrelations in a large number of globular cluster red giants has been clearly confirmed (Drake et al. 1992, Kraft et al. 1992, 1993, Norris & Da Costa 1995, Shetrone 1996).

These observations suggest that, while they evolve on the RGB, low mass stars undergo an extra-mixing in the region situated between the hydrogen burning shell (where the material is processed through the CN-cycle and possibly the ON-cycle) and the deep convective envelope. This extra-mixing adds to the standard first dredge-up to modify the surface abundances. Observations of $^{12}C/^{13}C$ ratios in M67 evolved stars (Gilroy & Brown 1991) strongly suggest that the extra-mixing process is only efficient when the hydrogen burning shell has crossed the discontinuity in molecular weight built by the convective envelope during the first dredge-up (Charbonnel 1994). Before this evolutionary point, the mean molecular weight gradient probably acts as a barrier to the mixing in the radiative zone. Above this point, no gradient of molecular weight exists anymore above the hydrogen burning shell, and extra-mixing is free to act.

3. Rotation-induced mixing on the red giant branch

Different mixing processes were proposed to explain the abundance anomalies in evolved stars. Sweigart & Mengel (1979) suggested that meridional circulation induced by stellar rotation on the RGB could lead to the low $^{12}C/^{13}C$ ratios observed in field giants. More recently, Charbonnel (1995), Wasserburg et al. (1995), and Denissenkov & Weiss (1996) reconsidered this idea in order to explain the carbon and oxygen isotope problems on the RGB and AGB. Moreover, if Na and Al were produced in the CN- and ON- processed region, the deep mixing scenario may also explain Na, Mg and Al anomalies (Denissenkov & Denissenkova 1990, Langer et al. 1993; Langer & Hoffman 1995, Denissenkov & Weiss 1996, Cavallo et al. 1996).

Rotation-induced mixing seems to be a very promising candidate. We investigated the influence of such a process on the RGB (Charbonnel 1995), by taking into account the most recent progress in the description of the transport of chemicals and angular momentum in stellar interiors : We used Zahn's (1992) consistent theory which describes the interaction between meridional circulation and turbulence induced by rotation. In this framework, the global effect of advection moderated by horizontal turbulence can be treated as a diffusion process. In our context, four important points must be emphasized : *1.* The resulting mixing of chemicals in stellar radiative regions is mainly determined by the loss of angular momentum via a stellar wind. *2.* Even in the absence of such mass loss, some mixing can take place wherever the rotation profile presents steep vertical gradients. *3.* Additional mixing is expected near nuclear burning shells. *4.* Due to the stabilizing effect of the composition gradients, the mixing will be efficient on the RGB only when the hydrogen-burning-shell will have crossed the chemical discontinuity created by the convective envelope during the first dredge-up. Since all these conditions are expected to be fulfilled during the non-homologous evolution of low mass stars on the RGB, we suggested that this process could be responsible for the extra-mixing we are looking for.

We estimated the effect of rotation-induced mixing on some surface abundances. Stellar evolutionary models were computed with the Toulouse code, a version of the Geneva stellar evolution code in which we have introduced the numerical method described in Charbonnel et al. (1992) to solve the diffusion equation. We restricted our study to the case where the stars undergo a moderate wind (see Zahn 1992).

Figures 1 and 2 show the influence of rotation-induced mixing on the

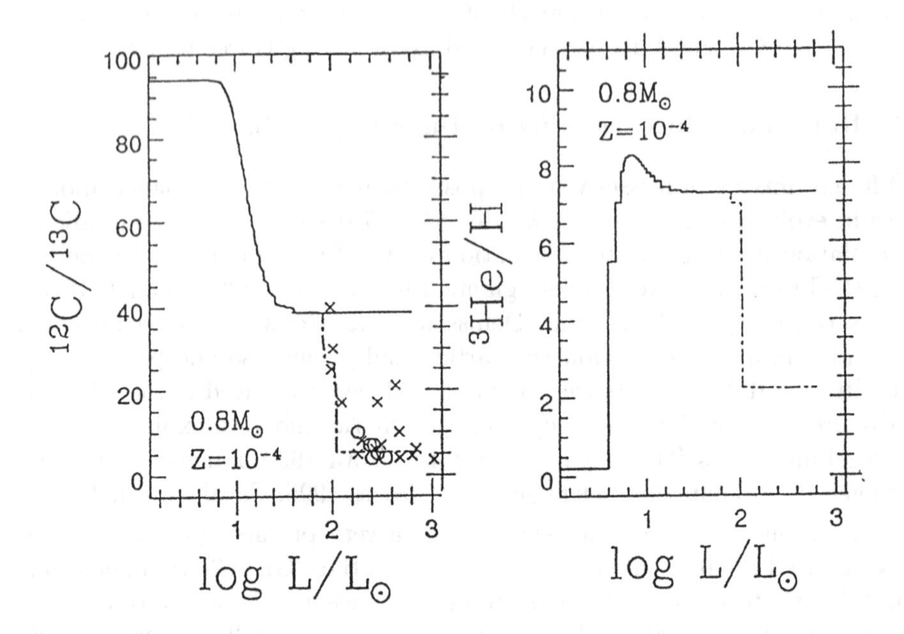

Figure 1. Theoretical behavior of $^{12}C/^{13}C$ and of $^3He/H$ (in units of 10^{-4} as a function of luminosity, for standard evolution (solid lines) and for the evolution including extra-mixing (dashed-dotted lines). Observations of the carbon isotopic ratio in field Population II (crosses; Sneden et al. 1986) and globular cluster M4 (circles; Smith & Suntzeff 1989) giant stars

surface values of $^{12}C/^{13}C$ ratio and ^7Li abundance in Pop II giants. In the standard case, the post-dilution values of the carbon isotopic ratio and of the ^7Li abundance remain constant and are substantially higher than observed in the most evolved RGB stars. However, when extra-mixing begins to act, the $^{12}C/^{13}C$ rapidly drops. The observed behavior of $^{12}C/^{13}C$ is well reproduced, and one reaches the low values currently observed in globular cluster giants, namely 3-8. Simultaneously, when ^7Li diffuses, it rapidly reaches the region where it is burned by proton capture. Due to rotation-induced mixing on the RGB, the surface abundance of lithium rapidly decreases down to the very low values observed in the halo giants. In addition, ^3He also reaches the region where it is nuclearly burned by the $^3He(\alpha,\gamma)^7Be$ reaction. This leads to a rapid decrease of the surface value of $^3He/H$, confirming the predictions by Hogan (1995). This result strongly modifies the actual contribution of low mass stars to the galactic evolution of ^3He.

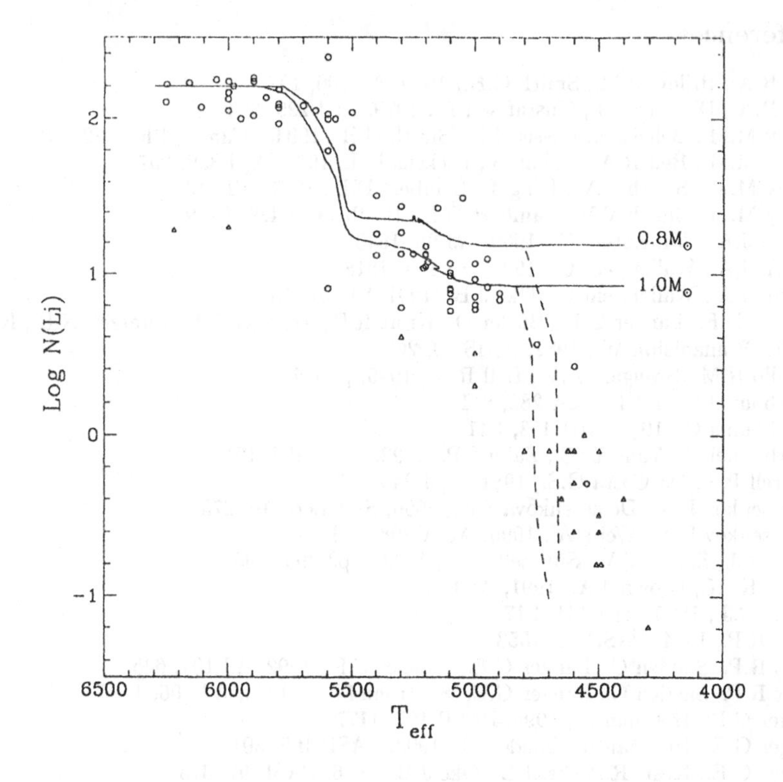

Figure 2. Theoretical behavior of the lithium abundance as a function of T_{eff}, for standard evolution (solid lines) and for the evolution including extra-mixing (dashed-dotted lines), for 0.8 and $1M_\odot$ models computed with $Z=10^{-4}$. The very low lithium abundances observed in the most evolved stars of the sample (halo giant stars from Pilachowski et al. (1993); open circles for real lithium detection, open triangles for upper limits) can only be reproduced when rotation-induced mixing is taken into account

4. Future developments

Preliminary results indicate that a realistic physical process, rotation-induced mixing, can simultaneously account for the observed behavior of carbon isotopic ratios and for the lithium abundances in Population II low mass giants. It also avoids large ^{3}He production by low mass stars in the Galaxy.

Detailed simulations, with a consistent treatment of the transport of matter and angular momentum, have now to be carried out for different stellar masses, initial metallicities, mass loss and rotational histories. The impact of this process on other chemical anomalies on the RGB (C $\downarrow$, Na $\uparrow$, O $\downarrow$, Al $\uparrow$, Mg O $\downarrow$) and on the precise yields of ^{3}He has to be investigated in details.

References

Bell R.A., Briley M.M., Smith G.H., 1990, AJ 100, 187
Bell R.A., Dickens R.J., Gustafsson B., 1979, ApJ 229, 604
Briley M.M., Bell R.A., Hesser J.E., Smith G.H., 1994a, Can. J. Phys. 72, 772
Briley M.M., Bell R.A., Hoban S., Dickens R.J., 1990, ApJ 359, 307
Briley M.M., Smith V.V., King J., Lambert D.L., 1997, AJ 113, 1
Briley M.M., Smith V.V., Lambert D.L., 1994b, ApJ 429, L119
Brown J.A., Wallerstein G., 1989, AJ 98, 1643
Brown J.A., Wallerstein G., 1992, AJ 104, 1818
Brown J.A., Wallerstein G., Oke J.B., 1991 AJ 101, 1693
Carbon D.F., Langer G.E., Butler D., Kraft R.P., Trefzger C.F., Suntzeff N.B., Kemper E., Romanishin W., 1982, ApJS 49, 207
Cavallo R.M., Sweigart A.V., Bell R.A., 1996, preprint
Charbonnel C., 1994, A&A 282, 811
Charbonnel C., 1995, ApJ 453, L41
Charbonnel C., Vauclair S., Zahn J.P., 1992, A&A 255, 191
Cottrell P.S., Da Costa G.S., 1981, ApJ 245, L79
Denissenkov P.A., Denissenkova S.N., 1990, SvA Lett. 16, 275
Denissenkov P.A., Weiss A., 1996, A&A 308, 773
Drake J.J., Smith V.V., Suntzeff N.B., 1992, ApJ 395, L95
Gilroy K. K., Brown J.A., 1991, ApJ 371, 578
Hogan C.J., 1995, ApJ 441, L17
Kraft R.P., 1994, PASP 106, 553
Kraft R.P., Sneden C., Langer G.E., Prosser C.F., 1992, AJ 104, 645
Kraft R.P., Sneden C., Langer G.E., Shetrone M.D., 1993, AJ 106, 1490
Langer G.E., Hoffman R., 1995, PASP 107, 1177
Langer G.E., Hoffman R., Sneden C., 1993, PASP 105, 301
Langer G.E., Kraft R.P., Friel E., Oke J.B., 1986, PASP 97, 373
Norris J.E., Da Costa G.S., 1995, ApJ 441, L81
Paltoglou G., Norris J., 1989, ApJ 336, 185
Pilachowski C.A., 1988, ApJ 326, L57
Pilachowski C.A., Sneden C., Booth J., 1993, ApJ 407, 713
Smith V.V., Suntzeff N.B., 1989, AJ 97, 1699
Sneden C., Kraft R.P., Prosser C.F., Langer G.E., 1991, AJ 102, 2001
Sneden C., Pilachowski C.A., VandenBerg D.A., 1986, ApJ 311, 826
Shetrone M.D., 1996, AJ
Shetrone M.D., Sneden C., Pilachowski C.A., 1993, PASP 195, 337
Suntzeff N.B., 1981, ApJS 47,1
Suntzeff N.B., Smith V.V., 1991 ApJ 381, 160
Sweigart A. V., Mengel J. G., 1979, ApJ 229, 624
Trefzger C.F., Carbon D., Langer G.E., Suntzeff N.B., Kraft R.P., 1983, ApJ 266, 144
Wasserburg G. J., Boothroyd A. I., Sackmann I. J. 1995, ApJ 447, L37
Zahn J.P., 1992, A&A 265, 115

THE IMPORTANCE OF HELIUM AND METALS DIFFUSION IN STARS

CHARLES R. PROFFITT
Computer Sciences Corporation and Catholic University of America, Code 680, Goddard Space Flight Center, Greenbelt MD 20771, USA

Abstract. Comparisons between models of the solar interior and sound speed profiles derived from inversions of helioseismic data have demonstrated that it is essential to include the effects of gravitational settling when calculating the structure and evolution of the Sun. Including settling should also be necessary for models of metal poor main-sequence stars and results in a substantial reduction in the ages derived for globular clusters.

In many cases it is clear that competing hydrodynamic processes, such as mass loss or rotationally driven mixing, will limit the effectiveness of gravitational separation of chemical elements. However, the quantitative details and even the relative importance of the different processes in various types of stars remains poorly understood.

1. Introduction

The chemical separation of elements in stellar atmospheres and interiors is driven by gravitational and radiative forces and is a basic physical process which should not be ignored in computations of stellar evolution. Preventing chemical separation requires the existence of competing hydrodynamic mechanisms, such as mixing or mass loss, which may have substantial consequences of their own for a star's structure and evolution.

The effects of diffusion are most obvious in white dwarfs and the chemically peculiar A and B stars, where extraordinarily large abundance anomalies occur. Detailed calculations are difficult for these stars because of the short timescales involved and the large radiative acceleration expected for the individual elements. In this review we will concentrate on the more

T.R. Bedding et al. (eds.),
***Fundamental Stellar Properties: The Interaction between Observation and Theory*, 355–360.**

subtle effects of diffusion on the interiors of low-mass main-sequence stars, especially the Sun and main-sequence metal-poor dwarfs.

2. Physics of Diffusion

Most work on chemical diffusion uses either the Chapman-Enskog procedure (Chapman & Cowling 1970) or the method of Burgers (1969) for deriving the transport properties from the Boltzmann equation. The method of Burgers is more easily generalized to multi-component plasmas, and has been adapted for the case of partially ionized plasmas by Geiss & Burgi (1986, 1987). For collisions between charged particles, the collision integrals computed numerically by Paquette *et al.* (1986) which assume screened Coulomb potentials are currently the most reliable. Caution should be used when substituting approximate analytic expressions.

How reliable are current estimates of diffusion rates? It should be noted that many of the assumptions implicit in the diffusion calculations are only marginally satisfied in the interiors of main-sequence stars (see the discussion in Michaud & Proffitt 1993). The Coulomb interaction energies of colliding particles are only a few times smaller than their kinetic energies, and there are typically only a few screening particles per Debye sphere. The effects of inelastic collisions on diffusion coefficients in ionization zones have also not been considered. While there is no reason to believe that any of these effects cause serious errors in calculations of the transport properties, a derivation done in the context of an equation of state which rigorously treats multi-particle interactions, such as the OPAL EOS (Rogers 1994), would be extremely valuable. In the absence of such work, it is difficult to assign quantitative errors to current treatments. It does seem likely that the stronger Coulomb coupling and more complicated electronic structures of heavy elements would result in uncertainties in diffusion rates substantially larger than for hydrogen and helium.

3. The Sun

In a star like the Sun, gravitational settling of the helium and metals will be the primary effect of atomic diffusion. Noerdlinger (1977) first studied the effects of the gravitational settling of helium in solar models, and found that surface helium decreased by a few percent over the lifetime of the Sun, and that there was a slight acceleration of the core evolution caused by central settling. These results were mostly ignored for a number of years. However, more recent work done with more accurate diffusion coefficients and more carefully matched to the observed solar parameters finds similar results (Cox, Guzik, & Kidman 1989, Proffitt & Michaud 1991, and Bahcall & Pinsonneault 1992 among others). These latter works find that the current mass fraction of helium in the solar convection zone has decreased by about 10%, ($\Delta Y \approx -0.03$), since the Sun formed. Models with gravitational settling also have slightly deeper convection zones and slightly more concentrated cores. There is also a steep gradient in the helium abundance just below the surface convection zone. Guzik *et al.* (1989), Proffitt (1994), and Bahcall & Pinsonneault (1995) also considered the effects of heavy element settling on solar structure, and found them smaller than the effects of helium settling alone, but still significant.

As discussed elsewhere in these proceedings (see the contribution by Christensen-Dalsgaard) helioseismology can be used to test models of the structure of the solar interior. Because of uncertainties in the superficial layers of the solar convection zone, comparison between theory and observation is best done by comparing the sound speed profile of computed models with a seismic model derived from an inversion of the observed frequencies (Christensen-Dalsgaard, Gough, & Thompson 1989). Inversion techniques were first used to test models that included helium settling by Christensen-Dalsgaard, Proffitt, and Thompson (1993). Additional comparisons using improved input physics and observational data have recently been done by Basu *et al.* (1996), Antia (1996), and Richard *et al.* (1996). They all find that models with settling reproduce the interior sound speed of the Sun to within 0.5% throughout the region where the observations are reliable, and are clearly superior to models which neglect gravitational settling. There is a suggestion that the calculated helium abundance gradient just below the convection zone is steeper than in the real Sun. This suggests that some modest amount of mixing has occurred below the solar convection zone, and is consistent with the mixing required to explain the observed solar lithium depletion. There are also some hints that the Sun's central helium profile is slightly flatter than that of the models, which might be evidence for a small amount of mixing in the core. However, it is not possible to unambiguously distinguish errors in the abundance profile from errors in the opacities.

Helioseismology observations can also be used to measure the helium abundance in the surface convection zone. By measuring the sound speed as a function of depth in the second helium ionization zone, the helium mass fraction can be inferred, provided the equation of state is sufficiently well understood. This has been done recently by Dziembowski *et al.* (1994), Hernandez & Christensen-Dalsgaard (1994), Basu & Antia (1995), and Richard *et al.* (1996). They find $Y \approx 0.24 - 0.25$ in the surface convection zone, which is substantially less than the initial helium abundance, $Y \approx 0.27 - 0.28$, needed to produce stellar interior models that match the current solar luminosity. This provides additional strong evidence that the predicted amount of helium settling has occurred.

4. Metal Poor Dwarfs

The material in the interiors of the metal poor G and K main-sequence and turn-off stars of the halo and the globular clusters has densities and temperatures similar to that of the solar interior. Since we appear to understand the equation of state, nuclear reaction rates, opacities, and diffusion rates well enough to build extremely accurate solar models, it might seem unlikely that there is any new physics needed to model these stars. The most significant difference is that many of these stars have substantially thinner surface convection zones than the Sun. This is especially true for the most metal-poor turn-off stars with $T_{eff} \approx 6000$ to 6400 K.

The effects of helium settling on the evolution of low mass globular cluster stars was first studied by Stringfellow *et al.* (1983), who found that settling in the center of the star accelerated the main-sequence evolution, thereby reducing the turn-off luminosity at a given age. From comparisons of calculated evolutionary tracks they suggested that this would reduce the estimated ages for globular clusters by 25%. However, newer studies using improved physics and, more importantly, directly comparing isochrones rather than evolutionary tracks (e.g., Proffitt & VandenBerg 1991), found that including helium diffusion reduces globular cluster ages as deduced from the luminosity of the main-sequence turn-off by about 10%. Ages estimated from the difference in luminosity between the turn-off and zero age horizontal-branch (ZAHB) (e.g. Chaboyer, Sarajedini, & Demarque 1992) are less affected, because helium settling in the interior results in a surface helium mass fraction after the first dredge up that is slightly lower than in models without diffusion. This leads to lower predicted ZAHB luminosities, partially offsetting the decrease in the turn-off luminosity. However, Sweigart (1997) has recently suggested that extra helium from the shell source might be mixed into the envelope during the first ascent RGB, *increasing* the luminosity of the ZAHB and decreasing the derived ages for globular clusters. This suggests that any theoretical calculation of the ZAHB luminosity should be avoided in measuring absolute ages for globular clusters. As noted above, calculations of main-sequence evolution can be tested using the Sun, and therefore observations of the turn-off and sub-giant luminosities combined with measures of the distance moduli not based on theoretical HB luminosities appear to be the most reliable way to measure the absolute ages of globular clusters.

It is, however, also clear that simple, uninhibited helium settling does not occur in the surface layers of metal poor dwarfs. Models with a reduced surface helium abundance have larger radii and cooler T_{eff}. Proffitt & VandenBerg (1991) found that the morphology of such isochrones is inconsistent with the observed color-magnitude diagrams of metal poor globular clusters such as M92. Helium settling is also accompanied by a similar reduction in the lithium abundance (which sinks at almost the same rate as helium), and this would lead to noticeable lithium depletion in the warmest halo turn-off stars which have the thinnest convective envelopes. However, observations have consistently shown no trace of this "Li dip", and instead suggest a slight increase of the lithium abundance with increasing T_{eff} (see Spite *et al.* 1996, and Spite 1997 in this volume).

The surface settling of lithium (and helium) must be moderated in such a way that a nearly uniform abundance is maintained in most main-sequence halo stars with T_{eff} between 6500 and 5700 K. Swenson (1995) and Vauclair & Charbonnel (1995) have demonstrated that models combining gravitational settling with mass loss rates of a few$\times 10^{-13}\, M_{\odot}\, \mathrm{yr}^{-1}$ can match the observed abundance pattern, but it is unclear if such a large and uniform mass loss rate is plausible, (the solar mass loss rate is $10^{-14}\, M_{\odot}\, \mathrm{yr}^{-1}$). Mixing processes related to rotation have also been proposed as a way to limit settling (e.g., Vauclair 1988, Chaboyer 1994). Preventing too much settling in the warm stars implies that at least the outer 1 to 2% of the stellar envelopes are well mixed, even in stars with surface convection zones much thinner than this. However, mixing that extends much beyond the outer 2% would be inconsistent with the apparent detection of ^{6}Li in at least one near turn-off halo star (Smith, Lambert, & Nissen 1994), as ^{6}Li is more easily destroyed by nuclear reactions than the more abundant isotope ^{7}Li. Mixing beyond the outer 3% of the mass would begin to destroy significant amounts of ^{7}Li.

5. Conclusions and Recommendations

We must reconcile the helioseismic observations of the Sun, which show that the the expected helium settling has been at most slightly moderated by mixing below the surface convection zone, with the lithium abundances observed in the halo stars, which suggest that the depth of surface mixing is independent of the depth of the surface convection zone. The observed lithium abundances strongly suggest that the mixed region in the globular cluster/halo stars is shallower than in the Sun. Since settling has had three times longer to act in these stars, it seems inevitable that a significant amount of helium settling has occurred. The surface helium mass fraction in these stars is likely to be in the range 0.1–0.2 rather than at the expected initial abundances of 0.23–0.25. This should cause a small redward shift (≈ 0.02 to 0.03 in $B - V$) in the predicted location of the F and early G-stars relative to the RGB, although uncertainties in convection theory and spectral synthesis models may prevent direct detection of this effect.

We strongly recommend that all future isochrone calculations for 0.6–

$1.2\,M_\odot$ stars include the gravitational settling of helium and heavy elements, along with a model of some other hydrodynamic process which prevents excessive settling that is inconsistent with the observed chemical abundances of lithium and other elements. The physics used in such calculations should always be checked against precise solar seismic models. It is probably premature to identify any particular mixing or mass-loss formulation as *the* mechanism that limits settling, and this regrettably limits the usefulness of parameter-free diffusion models. However, at least for metal poor dwarfs, available data places strong limits on amount of surface settling that has occurred, and this should enable models to be produced that are significantly superior to any that ignore gravitational settling.

References

Antia, H. M. 1996, *A&A*, **307**, 609

Bahcall, J. N. & Pinsonneault, M. H. 1992, *ApJ*, **395**, 119

Bahcall, J. N. & Pinsonneault, M. H. 1995, *Rev. Mod. Phys.*, **67**, 781

Basu, S. & Antia, H. M. 1995, *MNRAS*, **276**, 1402

Basu, S, Christensen-Dalsgaard, J., Schou, J., Thompson, M. J., & Tomczyk, S. 1996, *ApJ*, **460**, 1064

Burgers, J. M. 1969, *Flow Equations for Composite Gases*, New York: Academic Press

Chaboyer, B. 1994, *PASP*, **106**, 200

Chaboyer, B., Sarajedini, A., & Demarque, P. 1992, *ApJ*, **394**, 515

Chapman, S. & Cowling, T. G., 1970 *The Mathematical Theory of Non-Uniform Gases*, Cambridge University Press, 3rd ed.

Christensen-Dalsgaard, J., Gough, D. O., & Thompson, M. J. 1989, *MNRAS*, **238**, 481

Christensen-Dalsgaard, J. Proffitt, C. R. & Thompson, M. J. 1993 *ApJ*, **403**, 75

Cox, A. N., Guzik, J. A., & Kidman, R. B 1989, *ApJ*, **342**, 1187

Dziembowski, W. Q., Goode, P. R., Pamyatnikh, A. A., & Sienkiewicz, R. 1994, *ApJ*, **432**, 417

Geiss, J. & Bürgi, A. 1986, *A&A*, **159**, 1

Geiss, J. & Bürgi, A. 1987, *A&A*, **178**, 286

Hernandez, F. P., & Christensen-Dalsgaard, J. 1994, *MNRAS*, **269**, 475

Michaud, G. & Proffitt, C. R. 1993, in *Inside the Stars*, eds. W. W. Weiss, A. Baglin, IAU Coll. 137, PASP Conf. Ser. **40**, 246

Noerdlinger, P. D. 1977, *A&A*, **57**, 407

Paquette, C., Pelletier, C., Fontaine, G., & Michaud, G. 1986, *ApJ*, **61**, 177

Proffitt, C. R. 1994, *ApJ*, **425**, 849

Proffitt, C. R. & Michaud, G. 1991, *ApJ*, **380**, 238

Proffitt, C. R. & VandenBerg, D. A. 1991, *ApJS*, **77**, 473

Richard, O., Vauclair, S., Charbonnel, C., Dziembowski, W. A. 1996, *A&A*, **312**, 1000

Rogers, F. J. 1994, in IAU Coll. 147, *The Equation of State in Astrophysics*, ed. G. Chabrier, & E. L. Schatzman, Cambridge, Cambridge University Press, 16

Smith, V. V., Lambert, D. L., & Nissen, P. E. 1993, *ApJ*, **408**, 262

Spite, M., Francois, P., Nissen, P. E., & Spite, F. 1996, *A&A*, **307**, 172

Stringfellow, G. S., Bodenheimer, P., Noerdlinger, P. D., & Arigo, R. J. 1983, *ApJ*, **264**, 228

Sweigart, A. V. 1997, *ApJL*, **474**, 23

Swenson, F. J. 1995, *ApJL*, **438**, 87

Vauclair, S. 1988, *ApJ*, **335**, 971

Vauclair, S. & Charbonnel, C. 1995, *A&A*, **295**, 715

Discussion of this paper appears at the end of these Proceedings.

12. EVOLVED STARS

NON-VARIABLE HORIZONTAL-BRANCH STARS

ROBERT T. ROOD
University of Virginia

Abstract. For 25 years our ignorance of the physical basis of this mass loss process has been the barrier to progress in understanding horizontal branch morphology. I review some recent observational and theoretical results which may be giving us clues about the nature of the mass loss process.

1. Introduction

Much of the effort in horizontal branch (HB) research has been directed toward understanding its most obvious feature—the "horizontal" distribution of stars. How and why does HB morphology change from cluster to cluster. Early observations showed that in general HBs become redder as metallicity (usually measured by the iron abundance—[Fe/H]) increased. Faulkner (1966) showed how this arose naturally in core helium burning stars and led to there identification with the observed HB. However, even early data showed that more than one parameter appeared to be driving HB morphology. The quest to identify this second parameter still continues.

Early attempts to explain observed HBs (e.g. Iben & Rood 1968) were made without invoking mass loss in an effort to circumvent the introduction a free parameter. Unfortunately these efforts failed to reproduce either the centroid or breadth of observed HB log T_{eff} distributions. An alternate hypothesis was that the HB was a mass sequence rather than an evolutionary sequence. Because the mass of the helium core at the time of the helium flash was insensitive to total stellar mass, it was plausible to model the HB as sequence of stars with constant core mass (M_{core}) and varying total mass (M). The zero-age HB (ZAHB) of stars just beginning core He burning is populated with stars which have undergone some mean mass loss (ΔM) with dispersion around that mean [$\sigma(M)$](Iben & Rood 1970;

T.R. Bedding et al. (eds.),
Fundamental Stellar Properties: The Interaction between Observation and Theory, 363–368.

Faulkner 1972). Subsequent evolution carries stars away from their ZAHB location, but under most circumstances the primary factor affecting HB morphology is what part of the ZAHB is populated. Thus four mass parameters were required to model HBs: $M_{\rm RG}$, $M_{\rm core}$, ΔM, & $\sigma(M)$. Only, two of these, $M_{\rm RG}$ & $M_{\rm core}$, can be determined with evolutionary calculations. The other two are taken as free parameters—with $\Delta M \sim 0.25 M_{\rm RG}$ and a $\sigma(M) \sim 0.1 \Delta M$ Rood (1973) could mimic the observed CMDs of many clusters. HBs are still modeled in this fashion (e.g. Lee, Demarque, & Zinn 1990, 1994; Catelan 1993; Catelan & de Freitas Pacheco 1993)

Adopting this model, in addition to [Fe/H], age, helium abundance (Y), and the summed abundance of the CNO elements with respect to iron ([(C + N + O)/Fe]) were found to affect where stars fall on the HB. One could fix [Fe/H] and vary one the others to explore the second parameter problem, but this requires an assumption about mass loss, e.g. that the mean mass loss is a constant or that the mass loss was given by the Reimers' formula (Reimers 1977). Even the early work of Rood (1973) (see his Fig. 4) showed the potential pitfalls of such assumptions. However, age is the most attractive and tractable of the second parameter candidates, and in recent years many (most notably Lee, Demarque, & Zinn 1990, 94) have argued that dominant second parameter is age.

There are two factors which suggest that age differences are not the dominant driver of HB diversity at fixed [Fe/H]. The first of these is that as second parameter pairs are better studied, age differences seem to be disappearing rather than being confirmed. Perhaps the best studied second parameter pair for many years was NGC 288/NGC 362. The case for an age difference seemed particularly strong (Bolte 1989; Green & Norris 1990; Dickens et al. 1991). Yet in very detailed reanalysis Stetson, VandenBerg, & Bolte (1996) argue that the difference in HB morphology is not due to age and point out the difficulties in obtaining differential age measures. One crucial factor that is that the photometry for the brighter cluster stars and fainter turn-off stars be obtained and calibrated in *exactly the same way*. Ferraro et al. (1997) have obtained HST observations of the second parameter pair M3/M13 which satisfy this criterion. Again they find that no significant age difference which could account for the dramatic difference in HB morphology. Earlier, Catelan & de Freitas Pacheco (1995) had reached a similar conclusion. Likewise Richer et al. (1996) find that the M3/NGC 6752 pair have the same age. Indeed, the only clusters with certifiably different ages seem to be two small groups (including Rup 106, Arp 2, & IC 4499) associated with the accretion of other small galaxies by the Milky Way (Fusi Pecci et al. 1995; Richer et al. 1996).

The second factor causing me to doubt the ubiquity of age differences is the growing complexity of observed HB morphology. The first hint of what was to come was the CMD of NGC 6752 presented by Russell Cannon at the 1973 Frascati Globular Cluster workshop (Cannon 1985). It showed a very hot vertical HB with a very distinct *gap* in the stellar distribution. The ensuing years have given even more surprises. Strongly bimodal HBs are found in NGC 2808 (Ferraro et al. 1990) and NGC 1851 (Walker 1992). Clusters which we thought we knew, like M13, turn out to have gaps (Ferraro et al. 1997). There are additional BHB gaps in NGC 2808 (Piotto et al. 1997; Sosin et al. 1997). There are prominent blue extensions to the HBs of the metal rich clusters NGC 6388 and NGC 6441 (Rich et al. 1997; Sosin et al. 1997).

Neither blue HB gaps as found in NGC 6752 (Crocker, Rood, & O'Connell 1988) nor bimodal HBs as in NGC 2808 (Rood et al. 1993) can easily be accommodated into a simple HB scheme. One needs more than an age difference to turn a "classic" HB like that of M3 into a "NGC 6752" or a "NGC 2808." Since some factor other than age must be invoked for these extreme cases, perhaps it is a factor all of the time.

2. HB Evolution Does Not Produce Gaps & Bimodality

One might wonder whether HB evolution could naturally lead to the exotic HB morphologies described above. Indeed, it has been long know that there are families of HB track types, and Newell (1973) suggested that BHB gaps might arise in the region where the tracks were changing from the mode where evolution eventually carries the star back toward the RGB to the mode where all HB evolution takes place at high log T_{eff}. Our current results suggest that this is not the case (Rood, Whitney, & D'Cruz 1997; Whitney, Rood, & O'Connell 1997a). This is most clearly shown using Hess Diagrams (Hess 1924). A Hess Diagram represents the CMD as an "image" rather than a collection of points. The "intensity" of each pixel is the observed or theoretical density of stars in the CMD. Such a representation combines the advantages of a CMD with the temporal information contained in number counts and luminosity functions (LFs). Theoretical HB Hess Diagrams are constructed by weighting each pixel of the CMD image according to the amount of time evolving stars spend in that region of the CMD. The resulting image is a population density plot, indicating where evolving HB and post-HB stars spend their lifetimes, and thus, where they are most likely to be observed. Our point here is best illustrated in Figs 1–4 of Rood et al. (1997). In these a flat ZAHB mass distribution ($\mathcal{P}(M) = \text{constant}$) has been assumed. While this is unphysical it illustrates one very important point. The HB shown is analogous to a block of clay from which an observed HB must be sculpted. Any "observed" features like gaps which are not present in this block of clay must be introduced either by the transformation to the observed quantities or by sculpting "structure" using $\mathcal{P}(M)$. The conversion to colors and magnitudes is analogous to bending and stretching the clay block. It introduces only mild features. The resulting Hess diagram has neither BHB gaps nor strong HB bimodality. Thus, these structures are a product of the mass loss process.

3. Natural Sources of Bimodality

Our assumption is that HB structure like gaps and bimodality must arise naturally. It is quite possible to produce HB bimodality by assuming a bimodal mass distribution (and 4 free parameters), but that is a most unsatisfying approach (Rood et al. 1993). This structure must be telling us something about the mass loss process.

Extreme horizontal branch stars (EHB) are stars with such small envelope masses that the ensuing helium shell burning phase occurs at high log $T_{\rm eff}$—the so-called AGB-manqué and P-EAGB stars (Greggio & Renzini 1990; Dorman, Rood & O'Connell 1993). Some clusters have HB blue tails which extend into the EHB region. There is sometimes a gap at or near the beginning of EHB regions (e.g. M13—Ferraro et al. 1997 and NGC 2808—Piotto et al. 1997). One especially puzzling aspect of clusters with extensive EHB populations is that the EHB is populated by a very narrow range of HB masses. Surely some fine tuning of the mass loss process is required to funnel so many stars into such a small mass range

However, D'Cruz et al. (DDRO) showed that this was not the case. Simply by modeling the HB using $\mathcal{P}(\eta_{\rm ML})$, where $\eta_{\rm ML}$ is a Reimers-like mass loss parameter, rather than $\mathcal{P}(M_{\rm HB})$ one can produce EHB stars without fine tuning and produce a gap just above EHB transition. I conjecture that the lower BHB gap observed in many clusters is this EHB gap. Observations of the location of the EHB gap and the distribution of stars on the hot side of the gap will yield information on the mass loss mechanism.

Bimodality could also arise naturally from stellar interactions. Buonanno, Corsi & Fusi Pecci (1985) first suggested that cluster density, and thus stellar interactions, might play a role in HB morphology. More recently, Fusi Pecci et al. (1993) and Buonanno et al. (1997) have shown a correlation between the presence of long HB blue tails (BT) and cluster central density. Others (e.g. Bailyn et al. 1992) have argued that BT stars arise from stellar collisions and binary mergers.

There are several reasons to think this is not the case: (1) The observed stars have parameters more or less like those of single star models. One would expect that mergers and collisions produce a more diverse population. (2) There is no variation (at least in a gross sense) of BHB/EHB population with position in a cluster (Rich et al. 1997;

Whitney et al. 1997b). This is in contrast to the blue stragglers, which are generally considered to be certifiable interaction products and normally strongly concentrated toward the cluster center. (3) The correlations between BT stars and cluster density are not perfect. In most respects M3 & M13 are twins (with moderate central density), yet M13 has a prominent BT and M3 does not (Ferraro et al. 1997). The richest population of EHB/BT stars of any Galactic globular cluster is that in ω Cen a very "open" cluster. Thus, there has to be a mechanism for making BT stars not involving collisions.

So we are left with a tight, but not one-to-one, correlation between BTs and cluster density. Maybe this is telling us that density affects some parameter which affects mass loss. Basically, high density clusters seem to be able to "turn-on" some additional mass loss process. The mediating parameter must be affected by factors other than cluster density. The most likely candidates are stellar rotation (Peterson, Rood, & Crocker 1995) or helium mixing on the giant branch (Sweigart 1997a; Sweigart 1997b), which in turn may be tied to each other. The degree to which this extra driver of mass loss operates may determine whether there is another BHB gap near where the HB turns sharply downward in the V, $B-V$ plane.

4. Summary

- Gaps along the HB do not arise naturally from the models. $\Rightarrow$ HB gaps probably tell us something about mass loss.
- There are many strange HBs. There are no simple one-to-one correlations between HB morphology and other parameters. $\Rightarrow$ It is dangerous to assume HB morphology can be a measure of age.
- If one thinks in terms of $\mathcal{P}(\eta_{\rm ML})$ rather than $\mathcal{P}(M_{\rm HB})$, then (1) Fine tuning is not required to produce stars with very low envelope masses. (2) A gap (the lower, or EHB, gap in M13 and NGC 2808) can arise naturally. (3) As metallicity increases the mid-HB becomes difficult to populate. Bimodality is easy to explain in high metallicity systems. (4) It is just as easy to make EHB stars at high metallicity as at low.
- There may be two or more mass loss mechanisms.

This research was supported by NASA Long Term Astrophysics Grant NAGW-2596. My thoughts on this subject have been shaped by conservations with many of my colleagues, in this particular case, most strongly by Ben Dorman and Flavio Fusi Pecci.

References

Bailyn, C. D., Sarajedini, A., Cohn, H., Lugger, P. M., & Grindlay, J. E. 1992, AJ, 103,

1564
Bolte, M. 1989, AJ, 97, 1688
Buonanno, R., Corsi, C., and Fusi Pecci, F. 1985, A&A, 145, 97
Buonanno, R., Corsi, C., Bellazzini, M., Ferraro, F. & Fusi Pecci, F. 1997, AJ, 113, 706
Cannon, R. D. 1985, in Observational Tests of Stellar Evolution Theory, ed. A. Maeder & A. Renzini (Dordrect: Reidel), 123
Catelan, M. 1993, A&AS, 98, 547
Catelan, M. & de Freitas Pacheco, J. A. 1993, AJ, 106, 1858
Catelan, M., & de Freitas Pacheco, J. A. 1995, A&A, 297, 345
Crocker, D. A., Rood, R. T., & O'Connell, R. W. 1988 ApJ, 332, 236
D'Cruz, N. L., Dorman, B., Rood, R. T., & O'Connell, R. W. 1996, ApJ, 466, 359
Dickens, R. J., Croke, B. F. W., Cannon, R. J., & Bell, R. A. 1991, Nature, 351, 212
Dorman, B., Rood, R. T., & O'Connell, R. W. 1993, ApJ, 419, 596
Faulkner, J. 1966, ApJ, 144, 978
Faulkner, J. 1972, ApJ, 173, 401
Ferraro, F. R., Clementini, G., Fusi Pecci, F., Buonanno, R., & Alcaino, G. 1990, A&AS, 84, 59
Ferraro, F. R., Paltrinieri, B., Fusi Pecci, F., Cacciari, C., Dorman, B., Rood, R. T. 1997, submitted to ApJ Letters
Fusi Pecci, F., Ferraro, F.R., Bellazzini, M., Djorgovski, D.S., Piotto, G., Buonanno, R., 1993, AJ, 105, 1145
Fusi Pecci, F., Bellazzini, M., Cacciari, C., & Ferraro, F. R. 1995, AJ, 110, 1664
Greggio, L. & Renzini, A., 1990, ApJ, 364, 35
Green, E. M., & Norris, J. E. 1990, ApJ, 353, L117
Hess, R. 1924, in Probleme der Astronomie: Seeliger Festscrift, ed. H. Kienle (Springer: Berlin) p. 265
Iben, I., Jr., & Rood, R. T. 1968, ApJ, 154, 215
Iben, I., Jr., & Rood, R. T. 1970, ApJ, 161, 587
Newell, E. B. 1973, ApJS, 26, 37
Peterson, R. C., Rood, R. T., Crocker, D. A. 1995, ApJ, 453, 214
Piotto, G. et al. 1997, in Advances in Stellar Evolution, eds R. Rood & A. Renzini, (Cambridge: CUP), 84
Reimers, D. 1977, A&A, 57, 395
Rich, R. M. et al. 1997, ApJ, in press
Richer, H. B. et al. 1996, ApJ, 463, 602
Rood, R.T. 1973, ApJ, 184, 815
Rood, R. T., & Crocker, D. A. 1989, in The Use of Pulsating Stars in Fundamental Problems of Astronomy, ed. E. G. Schmidt (Cambridge: Cambridge University Press), 103
Rood, R. T., Crocker, D. A., Fusi Pecci, F., Ferraro, F. R., Clementini, G., & Buonanno, R. 1993, in The Globular Cluster-Galaxy Connection, ed. G. H. Smith, J. P. Brodie (San Francisco: ASP), 218
Rood, R. T., Whitney, J. & D'Cruz, N. L. 1997, in Advances in Stellar Evolution, eds R. Rood & A. Renzini, (Cambridge: CUP), 74
Sosin, C., Piotto, G., Djorgovski, S. G., King, I. R., Rich, R. M., Dorman, B., Liebert, J., & Renzini, A. 1997, in Advances in Stellar Evolution, eds R. Rood & A. Renzini, (Cambridge: CUP), 92
Stetson, P.B., Vandenberg, D.A., Bolte, M., 1996, PASP, 108,560
Sweigart, A. V. 1997, ApJ, 474, L23
Sweigart, A. V. 1997, in Faint Blue Stars III, ed. A. G. D. Philip (San Francisco: ASP)
Walker, A. R. 1992, PASP, 94, 1063
Whitney, J. H., Rood, R. T., & O'Connell, R. W. 1997, in preparation
Whitney, J. H., et al. 1997b, in preparation

NLTE ANALYSES OF PG 1159 STARS: CONSTRAINTS FOR THE STRUCTURE AND EVOLUTION OF POST-AGB STARS

S. DREIZLER, K. WERNER AND T. RAUCH
Institut für Astronomie und Astrophysik,
Universität Tübingen, Germany

U. HEBER
Dr. Remeis Sternwarte Bamberg,
Universität Erlangen–Nürnberg, Germany

I.N. REID
California Institute of Technology, USA

AND

L. KOESTERKE
Lehrstuhl Astrophysik, Universität Potsdam, Germany

1. Introduction

The majority of all stars ($M_i < 8M_\odot$) end their lives as white dwarfs (WD). On the tip of the Asymptotic Giant Branch (AGB) the star ejects its outer layers which become a Planetary Nebula and the stellar core evolves rapidly towards very high effective temperatures (T_{eff} >100 kK). When the nuclear burning in the H or He shell ceases the star enters the WD cooling sequence. The evolution starting from the AGB is separated in a H–rich and a H–deficient sequence where the latter contributes with a number fraction of ≈ 20%. In this paper we describe our recent effort in the analysis of one group of these stars, the PG 1159 stars.

PG 1159-035, the prototype of hot C– and O–rich pre–WDs was discovered in the Palomar Green Catalog (Green et al. 1986). It also became the prototype of a new class of variables, the GW Vir (=PG 1159-035) stars, when McGraw et al. (1979) discovered non–radial g–mode pulsations in PG 1159-035. First LTE analyses of Wesemael et al. (1985) revealed the very high T_{eff} (>100 kK) and the high surface gravity (log g~7). These stars represent corner stones in our understanding of the late phases of stellar

T.R. Bedding et al. (eds.),
***Fundamental Stellar Properties: The Interaction between Observation and Theory*, 369–372.**

evolution since many important effects of the post–AGB evolution can be studied here. Selected examples shall demonstrate how our analyses provide constraints for the structure and evolution of post–AGB stars.

2. Properties of PG 1159 stars

Presently 27 PG 1159 stars are known and most of them have been analyzed by our group (see Dreizler et al. 1995 and Werner et al. 1997 for latest reviews) using highly sophisticated NLTE model atmospheres (Dreizler & Werner 1993, Rauch 1997). Due to the enormous improvements in the model atmosphere techniques and also in the observation facilities we can now derive their effective temperatures, gravities, and surface abundances with high precision. These stars cover the hottest part in the post–AGB evolution in the transition between the Central Stars of Planetary Nebulae (CSPN) and the WDs with effective temperatures ranging from 65 kK to 180 kK. On the luminous end they have a broad overlap with O(He) and [WCE] CSPNe (Rauch et al. 1997, Koesterke & Hamann 1997), on the compact end with the hot He–rich DO WDs (Dreizler & Werner 1996).

The abundance pattern of PG 1159-035 (He/C/O=.5/.33/.17 by mass, H below the detection limit [H/He<1]; Werner, Heber & Hunger 1991, [WHH]) is found to be typical for PG 1159 stars. As already discussed by WHH these abundances can not be explained by standard evolutionary models which retain a H–rich envelope during the entire post-AGB phase. However, the surface abundances of PG 1159 stars are in qualitative agreement with the chemical composition in the deep intershell layers (Iben 1984) which led WHH to propose that PG 1159 stars are born–again post–AGB stars (Iben et al. 1983). In this scenario the star suffers a late He flash when it already left the AGB. The strong mass loss during the second (post–) AGB phase might be sufficient to erode the stellar surface down to the former intershell layers producing C and O rich stars. Despite the qualitative agreement between theory and observation the situation is still not satisfactory. Models consistently evolved through the numerically very difficult phase of the late He flash (Iben & MacDonald 1995) can not reproduce the observed abundance pattern since mixing processes produce significant admixtures of H and N which are not observed in this pattern in PG 1159 stars, e.g. N and H being more abundant than O. A very interesting point of view comes from recent work of Blöcker et al. (1997) who can produce the typical PG 1159 abundances already on the top of the He convection zone with a realistic treatment of the overshoot. Less drastic mass loss would be sufficient to exhibit these layers.

The GW Vir stars have also been subject to extensive photometric studies to derive asteroseismological parameters (Whole Earth Telescope,

Winget et al. 1991). The spectroscopic parameters like effective temperature and surface abundance serve as constraints for these analyses in order to reduce the huge parameter space or resolve ambiguities in the solution (Kawaler & Bradley 1994, Kawaler et al. 1995). Asteroseismology can then provide otherwise inaccessible information on the structure of these stars. It is also possible to derive much lower limits on the rotation periods and magnetic fields and to obtain independently derived masses and luminosities for PG 1159 stars. A comparison between the asteroseismological and spectroscopic masses is very satisfying for those stars with reliable pulsational data (Werner et al. 1996). It is found that PG 1159 stars have a mean mass of 0.6 $M_{\odot}$ which is close to the CSPN and WD mean masses.

Spectroscopic results are also important to investigate the driving mechanism of the GW Vir pulsations. Starrfield et al. (1985) proposed the $\kappa - \gamma$ mechanism of C and O to cause the pulsations. The chemical composition in the driving region, which is very close to the surface, is a very important parameter for the efficiency of this mechanism. According to Stanghellini et al. (1991) and Bradley & Dziembowski (1996) this layer must be H free. Also C and O must be even more enriched then in the atmosphere. In contrast, recent calculations of Saio (1996) and Gautschy (1997) show that pulsations are possible with the presence of H in the driving zone. We therefore started an intensive search for H in PG 1159 stars. This is a difficult task since all H lines are blended by much stronger He II lines. High resolution is necessary to resolve small rest wavelength differences e.g. at Hα. Due to the faintness of these stars large telescope are required. Observations at the 3.6 m ESO and 10 m Keck telescope of 4 PG 1159 stars could reduce the upper limit to 5 % by number (Werner 1996, Dreizler et al. in prep). On the other hand we detected 4 stars with typical spectra of PG 1159 stars but with Balmer lines clearly present (Napiwotzki & Schönberner 1991, Dreizler et al. 1996). They were therefore termed hybrid PG 1159 stars. According to the results of Dreizler et al., one of them (HS 2324+3944) lies in the GW Vir instability strip. Indeed, variability was detected by Silvotti (1996) and an analysis of the light curve of Handler et al. (1997) makes this star a good candidate for a member of the GW Vir group. A multi site campaign in order to corroborate this hypothesis by resolving more frequencies is planned. Another important constraint for the pulsation theory are the boundaries of the instability strip. We therefore obtained HST spectra of 9 (4 pulsating) PG 1159 stars. A precise determination of the effective temperature should confine the blue and red edge of the GW Vir strip. This analysis is under way, however, we noticed a prominent difference between the stable and pulsating PG 1159 stars prior to detailed analyses. All cooler ($\sim$100 kK) pulsating stars clearly show the resonance doublet of N V while the stable stars do not. Whether N is relevant for the driving of pulsations has to be investigated by the pulsation experts. I any case, it is important for the evolution, since it shows that mixing processes must occur in order to produce N in the C and O rich layers (see discussion above).

The last aspect we want to investigate is rotation and mass loss in PG 1159 stars. Our high resolution Keck spectra revealed that instrumental and Stark broadening alone can not reproduce the observed line width. This could be caused by rotation (~70 km/sec). It would, however, be quite surprising since DA WDs are slow rotators (Koester & Herrero 1987, Heber et al. 1997) with upper limits of ~40 km/sec. However, the broadening could also be caused by mass loss (macroturbulence). In order to distinguish the two hypotheses we determined mass loss rates for all PG 1159 stars showing this additional broadening. The analysis of the very recently obtained Keck spectrum of PG 1159-035 will give further clues since it has an asteroseismologically determined rotation period.

Acknowledgement: This work is supported by the DFG by several travel grants for observation campaigns and the participation in this meeting.

References

Blöcker T. Schönberner D. 1997, IAU Symp 177, Kluwer in press
Bradley P. A., Dziembowski W.A. 1996, ApJ 462, 376
Dreizler S., Werner K. 1993, A&A 278, 199
Dreizler S., Werner K. 1996, A&A 314, 217
Dreizler S., Werner K. Heber U. 1995, Lecture Notes in Physics 443, 160
Dreizler S., Werner K., Heber U., Engels D. 1996, A&A 309, 820
Gautschy A., 1997, A&A, in press
Green R.F., Schmidt M., Liebert J. 1986, ApJS 61, 305
Handler G., Kanaan A., Montgomery M.H. 1996, IAU Symposium 180, Kluwer, in press
Heber U., Napiwotzki R. Reid I.N. 1997, A&A, in press
Iben I. Jr. 1984, ApJ 277, 333
Iben I. Jr., MacDonald J. 1995, Lecture Notes in Physics 443, 48
Iben I. Jr., Kaler J.B., Truran J.W., Renzini A. 1983, ApJ 264, 605
Kawaler S.D., Bradley P.A. 1994, ApJ 427, 415
Kawaler S.D., et al. 1995, ApJ 450, 350
Koester D., Herrero A. 1987, ApJ 332, 910
Koesterke L., Hamann W.-R. 1997, A&A, in press
McGraw J.T., Starrfield S.G., Liebert J., Green R. 1979, IAU Coll. 53, p. 377
Napiwotzki R., Schönberner D. 1991, A&A 249, L16
Rauch T. 1997, A&A, in press
Rauch T., Dreizler S., Werner K. 1997, in *White Dwarfs*, Kluwer, in press
Saio H. 1996, The ASP Conference Series 96, 361
Silvotti R., 1996, A&A 309, L23
Stanghellini L., Cox A.N., Starrfield S., 1991, ApJ 383, 766
Starrfield S., Cox A.N., Kidman R.B., Pesnell W.D. 1985, ApJ 293, L23
Werner K. 1996, A&A 309, 861
Werner K., Heber U., Hunger K. 1991, A&A 244, 437 (WHH)
Werner K., Dreizler S., Heber U., Rauch T. 1996, The ASP Conference Series 96, 267
Werner K., Dreizler S., Heber U., Rauch T. 1997, in *White Dwarfs*, Kluwer, in press
Wesemael F., Green R.F., Liebert J. 1985, ApJS 58, 379
Winget D.E. et al. 1991, ApJ 378, 326

THE ASYMPTOTIC GIANT BRANCH

J. C. LATTANZIO AND C. A. FROST
Department of Mathematics, Monash University, Australia

1. Introduction

For stars with masses between about 1 and 8 $M_{\odot}$ the ascent of the Asymptotic Giant Branch (AGB) is the last nuclear powered evolutionary stage. Following this the stars eject a planetary nebula and fade as white dwarfs. Although only short in duration, the AGB is very important due to the nucleosynthesis which occurs. Here we briefly review the species produced and the mechanism for their production. We then discuss the current uncertainties in quantitative predictions of this nucleosynthesis, which centre on our determination of convective boundaries.

2. AGB Evolution

The details of AGB evolution have been the subject of much theoretical work, and are, in one sense, quite well understood. The reader is referred to Iben & Renzini (1983), Frost & Lattanzio (1996a) and Lattanzio *et al.*(1996) for details. Briefly, an AGB star has the structure shown in Figure 1. The C-O core is the result of He burning, and will become the final white-dwarf remnant. Just above this is the He-shell. This is thermally unstable, and burns vigorously during shell flashes (or thermal pulses) but is essentially extinguished between them. Above the He-shell is the intershell region, so-called because it is between the He and H-shells. Above the H-shell is the convective envelope. During a thermal pulse, the He-shell will deliver some $10^7 L_{\odot}$ for a brief period, and this enormous energy production results in the formation of a convective zone. This "flash-driven intershell convection" extends over the region shown in Figure 1, and thus mixes the products of (partial) helium burning throughout this region. The approximate composition of this zone is 25% carbon and 75% helium. Following the pulse, the helium luminosity decreases and the star expands. This essentially extinguishes the H-shell,

T.R. Bedding et al. (eds.),
***Fundamental Stellar Properties: The Interaction between Observation and Theory*, 373–380.**

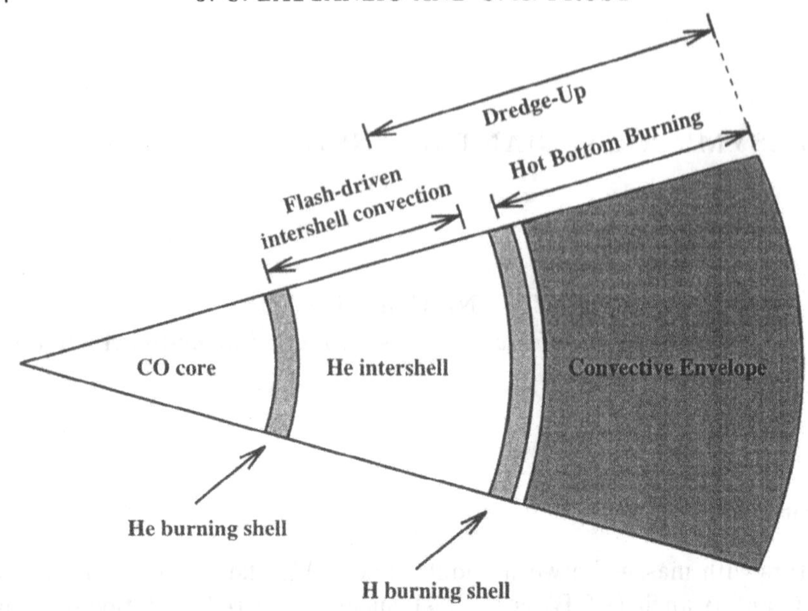

Figure 1. Schematic structure of an AGB star. During a thermal pulse the intershell convection extends over the region marked by "flash-driven intershell convection". During the dredge-up phase, the convective envelope moves inward to the depth marked by "dredge-up". For massive stars, during the interpulse phase the convective envelope penetrates the top of the H-shell, so that envelope convection extends down to the region marked "hot bottom burning".

and the bottom of the convective envelope moves inwards in mass. After a small number of pulses, this convection penetrates the maximum outward extent of the flash-driven convective zone and this results in the mixing of freshly produced carbon to the stellar surface. As the star begins to contract back to its normal configuration, the H-shell is re-ignited and provides all of the energy during the next interpulse phase, until the following pulse.

For more massive stars (above about $4M_\odot$) the bottom of the convective envelope penetrates into the top of the H-shell, and some nuclear reactions take place at the bottom of the convective envelope. This is known as "hot bottom burning" (hereafter HBB), and is shown schematically in Figure 1. The termination of the AGB phase is controlled by the mass loss. When the envelope becomes sufficiently thin, the star leaves the AGB. In the calculations presented below we used the formula of Vassilliadis & Wood (1993).

3. AGB Nucleosynthesis

3.1. NEUTRON-CAPTURE

The production of post-iron elements in AGB stars is well known, both observationally and theoretically. We now believe that the neutron source

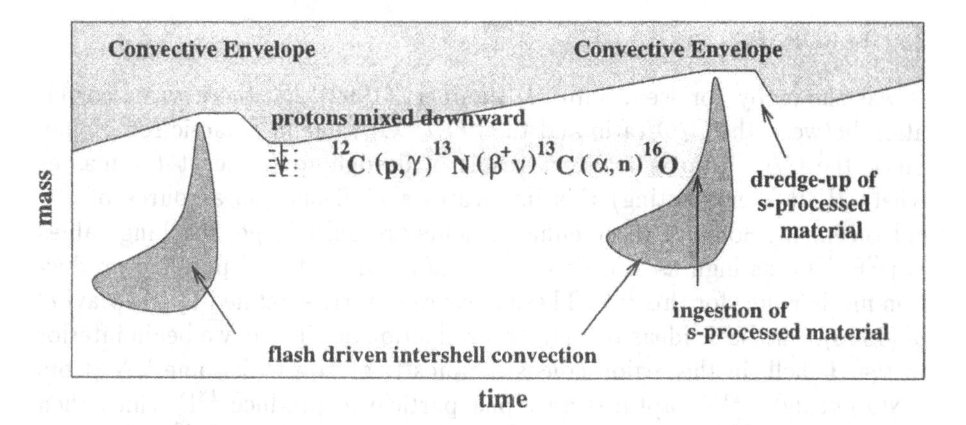

Figure 2. Schematic structure of two consecutive thermal pulses, showing how the downward mixing of hydrogen leads to the production of ^{13}C and then *s*-processing. Note that this *s*-processed material is later dredged to the surface of the star.

active in these stars (at least the lower masses) is ^{13}C. But how exactly does this ^{13}C arise? This remains a serious problem for the models. While it is true that some ^{13}C is produced by CNO cycling in the H-shell, this is nowhere near enough to produce the neutron exposures inferred from observations of these stars. We will return to this problem later, but for now we will assume that some kind of extra-mixing takes place at the bottom of the convective envelope during the dredge-up phase. This has indeed been found by Iben & Renzini (1982a,b), but has not been reproduced consistently. If this mixing occurs, then small amounts of hydrogen are mixed into a region which is relatively rich in ^{12}C. During the subsequent interpulse phase, these regions heat and the protons are captured by the ^{12}C to produce ^{13}C. It is crucial that there is not too much hydrogen in this region, or the CN cycle will progress further to ^{14}N, destroying the all-important ^{13}C. This "^{13}C-pocket" was believed to sit in the star and wait until the next thermal pulse, when it would be engulfed by the intershell convection. The high temperatures present would then release the neutrons and the *s*-processing would occur in the convective intershell region. But it has been shown recently (Straniero *et al.*1995; Lattanzio *et al.*1996; Mowlavi *et al.*1996) that the temperatures in the intershell are sufficiently high, and/or the interpulse duration sufficiently long, that the ^{13}C burns through to ^{16}O, thus releasing the neutrons *in situ* where the *s*-processing will then take place. Later, when this region is mixed into the flash-driven convection zone, the results of the neutron captures are also mixed into the convective zone. This situation is shown schematically in Figure 2. The details of the *s*-processing are not our concern here: the interested reader is referred to Gallino & Busso (1997).

3.2. FLUORINE

It was shown by Jorissen, Smith & Lambert (1992) that there was a correlation between the C/O ratio and the $[^{19}F/^{16}O]$ value in galactic red-giants. Since the C/O ratio is a direct result of dredge-up (at least for masses where HBB is not acting) this implicates shell-flashes as a source of ^{19}F enhancement. Some of these enhancements are quite large, reaching values of $[^{19}F/^{16}O]$ as high as 1.5. Forestini *et al.*(1992) outlined possible production mechanisms for this ^{19}F. These ideas were further refined by Mowlavi *et al.*(1996). The basic ideas behind the production of ^{19}F are: we begin interior to the H-shell, in the region consisting mostly of ^{4}He with some ^{14}N (from CNO cycling). ^{14}N captures an alpha particle to produce ^{18}F which then beta decays to ^{18}O. Assuming a source of protons (see below!) ^{18}O then produces ^{15}N via $^{18}O(p,\alpha)^{15}N$. This ^{15}N can then capture another ^{4}He nucleus to produce ^{19}F. But we need a proton source for this chain of reactions to take place. There are two likely sources of protons. ^{13}C suffers (α,n) reactions to produce ^{16}O and neutrons. These neutrons are used to produce the enhancements in *s*-process elements seen in these stars, but they are also available for ^{14}N and ^{26}Al to capture. Each undergoes (n,p) reactions and hence produces protons. The details of this process depend primarily on the temperature of the intershell region, both during a pulse and during the interpulse phase. The temperature, of course, also depends on the metallicity of the star as well as the core-mass, and details are provided in Mowlavi *et al.*(1996). An important point for our purposes is that once ^{19}F is dredged to the surface, it is destroyed by any HBB present via $^{19}F(p,\alpha)^{16}O$. Thus stars with enhanced ^{19}F are presumably those of lower masses, where HBB does not take place. For further details see Mowlavi *et al.*(1996). Finally, we note that to produce the largest observed enhancements of ^{19}F seems to require a substantial source of ^{13}C, just as is needed for the *s*-process abundances. The exact mechanism responsible for producing this ^{13}C remains unknown (see below).

3.3. HOT BOTTOM BURNING

The maximum envelope temperature for a $6M_{\odot}$ model with $Z = 0.02$ was found to rise with each pulses to a peak of 70 $\times 10^{6}$K. Toward the end of the evolution mass-loss results in a decreasing envelope mass, and hence the temperature at the base of the envelope decreases, and HBB is terminated just before the star leaves the AGB. This HBB burns H via the CNO, Ne-Na and Mg-Al cycles. But the first thing to happen is a significant increase in the ^{7}Li content of the stellar envelope. Cameron & Fowler (1971) suggested a mechanism for the production of ^{7}Li which required HBB. Boothroyd & Sackmann (1992) showed quantitatively that such a scenario can work in the required stars, and we refer to them for details.

But HBB does more than produce ^{7}Li. Any carbon added to the envelope by the repeated third dredge-up is converted into both ^{13}C and ^{14}N. In fact, the timescales are such that the entire envelope is mixed through this high-temperature region many times, and the resulting ratio of $^{12}C/^{13}C$ is about 3.5, appropriate to the equilibrium operation of the CN cycle. Further processing burns the ^{13}C into ^{14}N, which is able to prevent the star from becoming a Carbon star (see also Wood *et al.*1983, Boothroyd *et al.*1993).

A detailed analysis of the nucleosynthesis occurring during HBB is not the intention of this paper, but we mention here some of the more noteworthy changes. Firstly there is efficient destruction of the ^{19}F produced by thermal pulses! Hence, as stated earlier, we would not expect to see relatively massive stars with enhanced ^{19}F. Secondly, there is a dramatic destruction of ^{18}O as first discussed by Boothroyd *et al.*(1995). Finally, we also see a production of ^{26}Al from the Mg-Al cycle. This is even more pronounced in models of lower metallicity (Lattanzio *et al.*1997a,b). This is very important for the interpretation of the isotopic ratios measured in grains recovered from meteorites (see below).

4. New "Observations": Meteorite Grains

We have recently seen the exploitation of a new source of information about the composition of stellar material, provided by measurements of isotopic ratios in individual grains found in meteorites. A significant advantage of these measurements is that they can provide information about many elements for each grain, and since each grain has condensed in the outflow from a single star, we obtain much compositional information from a *single* stellar source.

The grains of interest to us are the silicon carbide (SiC) and oxide grains, especially corundum (Al_2O_3). Because the SiC grains must form in a carbon-rich environment it is believed that these grains originated in the envelopes of Carbon stars. The interest in corundum these is due to the oxygen and aluminium isotopic ratios, which show evidence of the three dredge-up episodes as well as HBB. It appears that a satisfactory explanation of all these grains requires us to consider stars of varying initial masses and a spread in the initial oxygen isotopic ratios, as discussed by Boothroyd *et al.* (1994). The effect of HBB on oxygen isotopic ratios has been calculated by Boothroyd *et al.* (1995), with particular reference to meteoritic data. Many of the oxide grains of Nittler *et al.* (1994) have been analysed for Al-Mg and show excesses of ^{26}Mg but have normal $^{25}Mg/^{24}Mg$. This indicates that live ^{26}Al has decayed *in situ* to produce the ^{26}Mg. The inferred initial $^{26}Al/^{27}Al$ ratios are as large as 0.016, quite consistent with the calculations reported in Lattanzio *et al.* (1997a,b). There is also a wealth of data available for other species

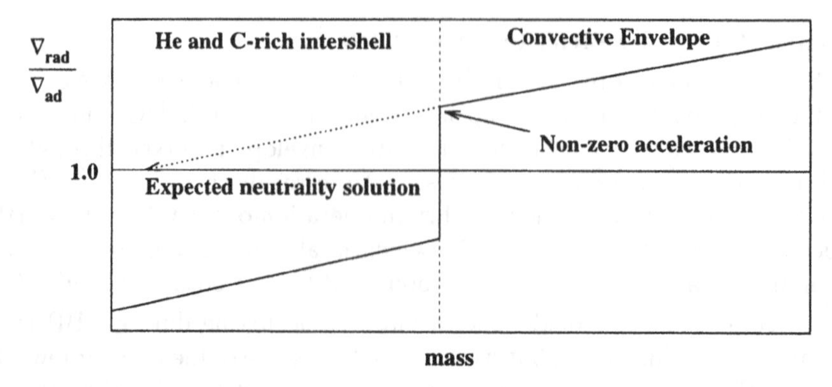

Figure 3. Schematic diagram showing the behaviour of the important temperature gradients at the bottom of the convective envelope during dredge-up in intermediate mass stars.

(see, for example, Gallino *et al.* 1994) but this will not be addressed here, except to remind us that we ignore this new source of highly accurate and wonderful data at our peril.

5. Model Limitations

Synthetic evolutionary calculations (*e.g.* Groenewegen & deJong 1993) using the best theoretical estimates for AGB evolution are unable to match the observed LMC Carbon star distribution. The observations require dredge-up to begin earlier in the evolution (*i.e.* at lower values of the core-mass) and to penetrate deeper into the star than current models predict. Both of these problems are related to determining the edge of a convective zone under complicated physical conditions.

There is, of course, a composition discontinuity at the bottom of the convective envelope during dredge-up: in the envelope the material is H-rich, but the envelope is growing into material which was previously part of the flash-driven intershell convective zone, and hence is about 75% He and about 25%C (see Figure 3). The Schwarzschild criterion for determining the convective boundary relies on finding the position where the acceleration is zero, which is where $\nabla_{rad} = \nabla_{ad}$. Convective eddies will still have a non-zero momentum when they reach this boundary, and hence they will penetrate into the radiatively stable region where they are decelerated to zero velocity. This is what is usually referred to in the literature as "convective overshoot". But for third dredge-up, the boundary is even more prone to extra-mixing, since the acceleration (as well as the momentum) is non-zero at the convective boundary. Exactly how the star will mix, to achieve the expected convective neutrality, is uncertain. It is, of course, a hydrodynamical problem. Somehow we expect the convective region to grow into the intershell region, until the

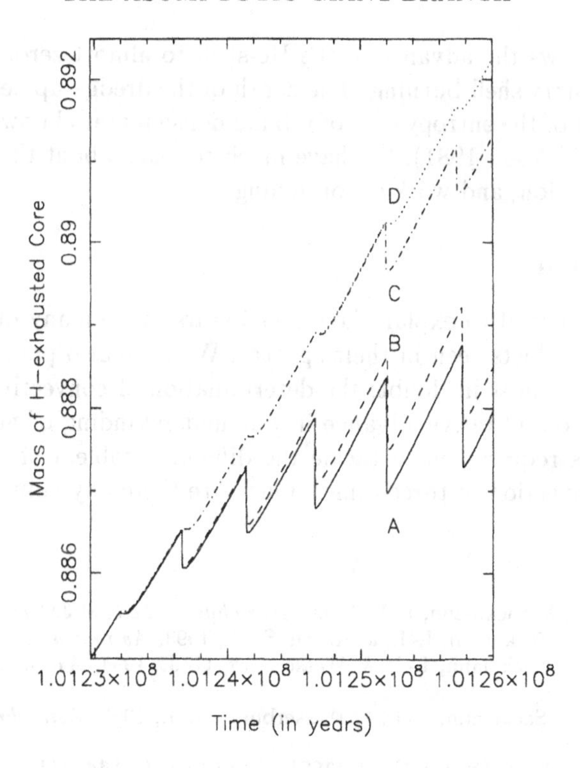

Figure 4. **Four different algorithms for mixing, and four different amounts of dredge-up. See Frost & Lattanzio (1996b) for details.**

gradients smoothly approach each other (see the dotted line in Figure 3). From this configuration we would still expect the usual overshoot.

This situation has been investigated in some detail by Frost & Lattanzio (1996b) who found that the depth of dredge-up depended critically on assumptions made at the boundary of the convective region, as well as the way in which the mixing was handled within the evolutionary calculation (*eg.* if the mixing is performed after each iteration, or only after a model has converged). For example, Figure 4 shows the mass of the H-exhausted core as a function of time for a $5M_{\odot}$ model with $Z = 0.004$. Each of the four evolutionary sequences A, B, C and D began with the same model; each uses a somewhat different algorithm for dealing with the convective boundary and associated mixing. The effect on dredge-up is dramatic, varying from none in case D to very deep in case A (see Frost & Lattanzio 1996b for details). Clearly this situation is not satisfactory, and much theoretical work must be done to clarify this picture.

Current work in progress has shown that the depth of dredge-up dramatically alters the evolution of the star. Deep dredge-up cools the intershell region, and slows the advance of the He-shell to almost zero. Thus we find almost stationary shell burning. The depth of the dredge-up depends also on the treatment of the entropy cost of mixing dense material upward in a gravitational field (Wood 1981). We have much to learn about this complicated phase of evolution, and work is continuing.

6. Conclusion

We have attempted to explain how AGB stars make many of the interesting nuclear products seen in their spectra. We have also pointed out where the models are most in doubt: the determination of convective boundaries. A significant quantitative advance in our understanding of nucleosynthesis in AGB stars requires us to tackle the difficult problem of hydrodynamic turbulent convection in three dimensions. Are there any volunteers?

References

Boothroyd, A. I. & Sackmann, I.-J., 1992, *Astrophys. J. Lett*, **393**, L21.
Boothroyd, A. I., Sackmann, I.-J., & Ahern, S. C., 1993, *Astrophys. J.*, **416**, 762.
Boothroyd, A. I., Sackmann, I.-J., & Wasserburg, G. J., 1994, *Astrophys. J. Lett*, **430**, L77.
Boothroyd, A. I., Sackmann, I.-J., & Wasserburg, G. J., 1995, *Astrophys. J. Lett*, **442**, L21.
Cameron, A. G. W., & Fowler, W. A., 1971, *Astrophys. J.*, **164**, 111.
Forestini, M., Goriely, S., Jorissen, & Arnould, M., 1992, *Astron. Astrophys.*, **261**, 157.
Frost, C.A. & Lattanzio, J. C., 1996a in *Stellar Evolution: What Should Be Done?*, Proceedings of the 32nd Liege Colloquium, Eds A. Noels et al, 307.
Frost, C. A¿, & Lattanzio, J. C., 1996b, *Astrophys. J.*, **473**, 383.
Gallino, R., *et al.*1994, *Astrophys. J.*, **430** , 858.
Gallino, R. & Busso, M., 1997, in Proceedings of *Astrophysical Implications of Laboratory Study of Interstellar Grains*, in press.
Groenewegen, M. A. T., & de Jong, T., 1993, *Astron. Astrophys.*, **267**, 410.
Iben, I., Jr., & Renzini, A., 1982a, *Astrophys. J. Lett*, **259**, L791.
Iben, I., Jr., & Renzini, A., 1982a, *Astrophys. J. Lett*, **263**, L231.
Iben, I., Jr., & Renzini, A., 1983, *Ann. Rev. Astr. Ap.*, **21**, 271.
Jorissen, A., Smith, V. V., & Lambert, D. L., 1992, *Astron. Astrophys.*, **261**, 164.
Lattanzio, J. C., Frost, C. A., Cannon, R. C., & Wood, P. R., 1996, *Mem. Astro. Soc. Italia*, in press.
Lattanzio, J. C., Frost, C. A., Cannon, R. C., & Wood, P. R., 1997a, "Nucleosynthesis in Intermediate Mass Stars" in *The Carbon Star Phenomenon*, Proceedings of IAU Symposium 177, Ed R. F. Wing, in press.
Lattanzio, J. C., Frost, C. A., Cannon, R. C., & Wood, P. R., 1997b, in preparation.
Mowlavi, N., Jorissen, A., & Arnould, M., 1996, *Astron. Astrophys.*, **311**, 803.
Nittler, L. R., Alexander, C. M. O'D., Gao, X., Walker, R., & Zinner, E. K., 1994, Nature, 370, 443.
Straniero, O., Gallino, R., Busso, M., Chieffi, A., Limongi, M., and Salaris, M., 1995, *Astrophys. J. Lett*, **440**, L85.
Vassilliadis, E. & Wood, P. R., 1993, *Astrophys. J.*, 413, 641.
Wood, P. R., 1981, *Astrophys. J.*, **248**, 311.
Wood, P. R., Bessell, M. S., & Fox, M. W., 1983, *Astrophys. J.*, **272**, 99.

COOL WHITE DWARFS : COOLING THEORY AND GALACTIC IMPLICATIONS

G. CHABRIER
C.R.A.L., Ecole Normale Supérieure, 69364 Lyon Cedex 07, France
chabrier@ens-lyon.fr

1. Introduction

The understanding of the physics of cool white dwarfs (WD) bears important consequences for Galactic evolution and cosmological implications. The observed cutoff in the disk WD luminosity function (WDLF) yields the determination of the age of the Galactic disk, as suggested initially by Winget et al. (1987). The recent microlensing observations toward the LMC (Alcock et al., 1996) suggest that WDs might provide a substantial fraction of the halo dark matter (Chabrier, Segretain & Méra, 1996; Adams & Laughlin, 1996). The correct analysis of these applications implies a correct WD cooling theory and reliable photometric predictions, which in turn require accurate interior and atmosphere models. Important improvement in this latter domain has been accomplished recently by Bergeron, Saumon & Wesemael (1995) and Bergeron, Wesemael & Beauchamp (1995), which yields the determination of photometric color indices and bolometric corrections down to 4000 K (see Leggett, these proceedings). In this paper, we review the most recent improvement in WD interior and cooling theory.

2. Internal structure. Cooling theory

The theory of WD cooling was first outlined by Mestel & Ruderman (1967) who identified the dichotomic properties of WDs, where degenerate electrons provide the pressure support but do not contribute significantly to the heat capacity ($c_V/k_B = \pi^2/2(T/T_F)$ for a fermion gas, where c_V is the specific heat per unit mass and $T_F \sim 10^9$ K is the electron Fermi temperature for WD central densities) while the ions contribute negligible pressure but

T.R. Bedding et al. (eds.),
Fundamental Stellar Properties: The Interaction between Observation and Theory, 381–388.

provide most of the thermal energy. Abrikosov (1960) and Salpeter (1961) independently pointed out the possible onset of crystallization in cool WDs. VanHorn (1968) first developed a consistent theory of WD crystallization and Lamb & VanHorn (1975) first calculated the evolution of a pure carbon crystallizing WD. These calculations have been extended to C/O mixtures by Wood (1992) who examined extensively the importance of the various parameters entering WD evolution, e.g. the core composition and the atmosphere structure. A further significant breakthrough in WD cooling theory is due to Stevenson (1980) who first pointed out the importance of the crystallization diagram in a *two-component* (e.g. C/O) plasma and of the difference of chemical composition in the fluid and in the solid phase. This motivated numerous calculations for the characterization of the phase diagram (Barrat, Hansen & Mochkovitch, 1988; Ichimaru, Iyetomi & Ogata, 1988), the effect on WD cooling (Mochkovitch, 1983) and on the WDLF (García-Berro, Hernanz, Mochkovitch & Isern, 1988). More recently Segretain & Chabrier (1993) characterized the evolution of the crystallization diagram of stellar plasmas for arbitrary binary mixtures as a function of the charge ratio. The chemical differentiation at crystallization calculated with these diagrams was shown to produce an extra source of energy ΔE in the WD (Segretain et al., 1994), which in turn leads to a substantial increase in the age of crystallized WDs for a given luminosity ($\Delta t \propto \Delta E/L$), an important issue for a correct determination of the age of the Galactic disk (Hernanz al., 1994).

In this section, we derive an analytical theory for the evolution of cool, crystallizing WDs, based on first principles of thermodynamics (Landau & Lifschitz, 1980), aimed at describing the main physical effects in terms of simple physics (see also Isern et al., 1997). These calculations include all the afore-mentioned processes and yield a reasonable estimate of the gravitational energy release and time delay induced by chemical fractionation at crystallization, a question of strong debate among the community. This provides useful guidelines to verify the validity of complete, numerical calculations.

The first laws of thermodynamics yield for white dwarf cooling

$$L = -\int_0^M \frac{dq}{dt} dm - \int_0^M \epsilon_\nu dm = -\int_0^M \frac{du + d\Omega}{dt} dm - \int_0^M \epsilon_\nu dm \qquad (1)$$

where ϵ_ν is the neutrino rate and dq/dt is the heat rate per unit mass along the change of an equilibrium state. du and $d\Omega$ are the change of specific internal energy and gravitational energy, respectively. The first one reads :

$$du = c_V dT + ((T\frac{dP}{dT})_V - P)dv + \Delta u_{crys} + [\Sigma_i(\mu_i dN_i)_l + \Sigma_i(\mu_i dN_i)_s]_{V,T}$$
$$= Tds - Pdv + T\Delta s + [\Sigma_i(\mu_i dN_i)_l + \Sigma_i(\mu_i dN_i)_s]_{V,T} \quad (2)$$

The μ_i denote the chemical potentials, s the specific entropy and $v = 1/\rho$ is the volume per unit mass. The dN_i are the variations of carbon and oxygen nuclei in the fluid and the solid phase due to the change of composition at crystallization. They can be calculated with the thermodynamics lever rule. When the central crystal grows[1], there is a thin C-enriched surrounding fluid layer where *locally* the variation of nuclei $d(\delta N_{i_l})$ is *not* equal to $-dN_{i_s}$ (no *local* mass conservation) [2]. How this carbon excess in the fluid is redistributed homogeneously will be examined below. No variation of composition yields of course $\delta N_i = 0$ and thus no extra internal energy.

The first two terms on the r.h.s. of (2) have been recognized originally by Mestel & Ruderman (1967), the third term was first introduced by Van Horn (1968) and is the crystallization latent heat $l = -\Delta u_{crys} = T(s_{sol} - s_{liq})$.

The condition of hydrostatic equilibrium yields for the variation of gravitational energy :

$$\delta\Omega = \int_0^M P\delta v\delta m \approx < \frac{P_e}{\rho} > M \quad (3)$$

since the electron pressure largely dominates the ionic pressure.

Eqns. (1)-(3) can be rewritten :

$$L + L_\nu = -\int_0^M c_V \frac{dT}{dt} dm - \int_0^M (T\frac{dP}{dT})_V \frac{dv}{dt} dm + l\frac{dm_S}{dt} + \delta u \frac{dm_S}{dt} \quad (4)$$

where $\delta u = \Sigma_i(\int \mu_i d(\delta N_i))$ and m_S is the mass crystallized. Note that the last two terms in eqns. (2) and (4) are evaluated at *constant volume* and thus *do not stem from a contraction work, but from the change of composition at crystallization.*

[1]Note that this fractionation process is described sometimes in the literature as drowning O-*flakes*. This is an erroneous picture, based on Stevenson's old *eutectic* diagram, which corresponds to an *inhomogeneous* solid. The correct *spindle* diagram yields a homogeneous solid. It is easy to verify that the solid is denser than the liquid, so that the crystal grows at the center of the star. Were WDs made of water, the story would be different because of the volume expansion at crystallization

[2]This is similar to e.g. silicium deposition for semi-conductor devices in a silicium+impurity liquid, which yields a silicium concentration gradient and eventually a silicium-rich layer. We have in fact a distillation process in the WD.

The contribution of the first three terms of eqn. (4) can be estimated easily, as done initially by Mestel & Ruderman (1967) and Lamb & Van Horn (1975) :

$$\begin{aligned}\delta U_{th} &= \int_0^M c_V dT\, dm \approx \delta\Omega \text{ from the virial theorem}\\ \delta U_{grav} &= \int_0^M T\frac{dP}{dT}dvdm \approx \int_0^M [T\frac{dP_i}{dT} + o(T/T_F)^2]dvdm \approx < \frac{P_i^{th}}{\rho} > M\end{aligned} \tag{5}$$

where $P_i^{th} = \rho\mathcal{R}T/\mu_0$ (where μ_0 is the mean ionic molecular weigth) is the thermal (non electrostatic) ionic pressure.

Equations (3) and (5) yield $\delta U_{grav} \sim \frac{<P_i^{th}>}{<P_e>}\delta\Omega$. WD characteristic central density $\rho \sim 10^6$ g.cm^{-3} and central temperature $T \sim 10^6$ K yield $P_i^{th}/P_e \sim 10^{-3}/\bar{Z}$. Thus only a negligible fraction of the energy due to gravitational contraction is radiated. Most of the work is expended in raising the electron Fermi energy, as first noted by Lamb & Van Horn (1975). The latent heat contribution can be estimated from the differences between the solid and liquid ionic entropies (VanHorn, 1968) : $l \sim -kT_c/AH$, where T_c is the crystallization temperature ($\sim 3 \times 10^6$ for C/O, see Segretain & Chabrier, 1993), A is the mean atomic mass and $H = 1$ a.m.u.$= 1.66\times10^{-24}$ g. This yields an energy release $U_{latent\,heat} \sim 10^{47}$ erg $\sim 10^{-2}\,\Omega$, where $\Omega = GM^2/R \sim 10^{49}$ erg is the WD gravitational energy. The negative sign indicates that the energy is *emitted* at crystallization.

The last term can be estimated as follows :

$$\delta u = (\frac{\partial u}{\partial X})_{v,T}\delta X \approx (\frac{\partial u_i}{\partial X})_{v,T}\Delta X \approx \Delta u_i \tag{6}$$

where X is the mass fraction of one of the components (say carbon), $\Delta X = X_l - X_s$ and $\Delta u_i = u_{i_l} - u_{i_s}$ is the difference of Madelung energy in the C/O plasma between the fluid and the solid phase[3]. The Madelung energy (per unit mass) of the mixture reads :

$$u_i/kT = \alpha\Sigma_k\frac{X_k}{A_k}\Gamma_k = \alpha\Gamma_e\{X\frac{Z_1^{5/3}}{A_1} + (1-X)\frac{Z_2^{5/3}}{A_2}\} \tag{7}$$

where α denotes the Madelung constant in the fluid or in the solid phase ($\alpha_s = -0.9, \alpha_l = -0.899$), $\Gamma_e = e^2/a_ekT$ (a_e is the mean inter-electronic distance) and the index k denotes each ionic species (C^{6+}, O^{8+}). This yields:

[3]In fact δu is only a fraction of Δu_i since it stems from the difference w.r.t. to the average energy over the C-enriched layer. This is not consequential for the present estimate

$$\Delta u_i/kT \approx -0.9\Gamma_e \Delta X(Z_1^{5/3}/A_1 - Z_2^{5/3}/A_2) \tag{8}$$

The virial theorem $P_i/\rho = \frac{1}{3}u_i$ yields :

$$\delta U/\delta\Omega = \frac{\int \delta u \, dm_s}{\delta\Omega} \sim \frac{< \Delta P_i > M_s}{< P_e > M} = \frac{\Delta P_i}{P_i}\frac{P_i}{P_e}\frac{M_s}{M} \tag{9}$$

Note that P_i is now the ionic electrostatic pressure. For $Z_1 = 6, Z_2 = 8$, $x_1{=}x_2{=}1/2$, we get $P_i/P_e \sim 10^{-2}{-}10^{-1}$, $\Delta P_i/P_i \sim 2\Delta x\,(Z_1^{5/3}{-}Z_2^{5/3})/(Z_1^{5/3}{+}Z_2^{5/3}) \sim -0.5 \times \Delta x$, where $\Delta x \sim 0.1 - 0.3$ is the difference of carbon number concentration between the solid and the fluid phase at crystallization (Segretain & Chabrier, 1993). With $M_s/M \sim 0.1$, this yields:

$$\delta U/\delta\Omega \sim 10^{-4} - 10^{-3} \tag{10}$$

in agreement with the detailed numerical calculations (see Figure 5 of Segretain et al., 1994).

Chemical differentiation at crystallization thus provides an additional source of energy wich remains much smaller than the gravitational energy. But, as shown below, the release of this quantity at a low-luminosity phase of the evolution has a significant effect upon the lifetime of the star at these stages.

$$\Delta t = \int_0^M \frac{\delta u(T)}{L(T)} dm \approx \frac{\delta U(T)}{L(T)} \approx \frac{\Delta P_i}{P_i}\frac{P_i}{P_e}\frac{M_S}{M}\frac{\Omega}{L} \tag{11}$$

With the afore-mentioned values we get $\Delta t \sim 5 \times 10^8$ yr *at the begining of crystallization*, $M_s/M \sim 10\%$ and $L = 10^{-3.5}\,L_\odot$ and $\Delta t \sim 2 \times 10^9$ yr at $L = 10^{-4.5}\,L_\odot$, $M_s/M \sim 80\%$, the observed cutoff luminosity. These simple calculations show the importance of the time delay induced by chemical fractionation at crystallization for cool (faint) WDs, even though the corresponding energy is small compared to the binding energy.

An other issue concerns the redistribution of the excess of carbon in the fluid at crystallization since the solid core is O-enriched, as obtained from the phase diagram. Since the fluid C-enriched layer around the crystal is lighter than the surrounding medium, a Rayleigh-Taylor instability develops locally, due to the variation of molecular weight. This problem has been considered in detail by Mochkovitch (1983) who showed that the typical crystallization time (for a $0.6M_\odot$ WD, the crystallization velocity at the begining of crystallization is $v_c \sim 10^{-2} M_{WD}/7 \times 10^7 yr \sim 10^{-8}$ cm.s^{-1}) is significantly larger than the convection time so that the liquid is likely to be rehomogeneized rapidly as crystallization goes on.

3. Galactic implications

The first application of these calculations concerns the age of the faintest WD ever observed, $ESO\,439-26$ (Ruiz et al., 1995). The trigonometric-parallax determination of this object yields an absolute magnitude $M_V = 17.4 \pm 0.3$, $M_{bol} = 17.1 \pm 0.1$ (Bergeron et al., 1997) about 1 mag faintward of the observed cut-off of the WDLF of the Galactic disk (Liebert et al., 1988). The location of this object in the M_V vs $V-I$ diagram, and a comparison with the photometric sequences of cool white dwarfs recently derived by Bergeron et al. (1995) yields the interpretation that it is a cool ($T_{eff} \sim 4500$ K), massive WD ($m \sim 1.2\,M_\odot$). Comparison with (pure carbon) evolutionary models of Wood (1992) yields an age determination for this WD, $t \sim 6.5$ Gyr, substantially below the lower limit for the age of the Galactic disk determined by detailed WD cooling theory (Hernanz et al., 1994). The Wood sequences do not include the afore-mentioned release of gravitational energy due to C/O differentiation at crystallization. Figure 1 displays several isochrones of a C/O WD for several masses (WDs more massive than $1.2\,M_\odot$ have a different O/Ne/Mg internal composition), obtained with a WD cooling theory including (solid line) and neglecting (dashed line) the fractionation process. The dotted lines display the luminosity of ESO 439-26 and its mass, assuming either a H-rich ($\sim 1.1\,M_\odot$) or a He-rich ($\sim 1.2\,M_\odot$) atmosphere (see Ruiz et al., 1995). As shown, ESO 439-26 is compatible with an age t$\sim$ 10 Gyr, in good agreement with the most recent determination of the age of the disk, whereas neglecting differentiation would yield $t < 9$ Gyr.

An other important galactic application is the age of the disk inferred from the comparison between the observed and the theoretical WDLF. As shown by Hernanz et al. (1994) and by the present analysis, chemical fractionation yields 1 to 2 Gyr older ages (i.e. 10-20% increase), depending on the initial C/O profile, w.r.t. estimates which do not include this process. Preliminary calculations along the present lines based on the recently observed WDLF (Oswalt et al., 1996) yield an estimate for the age of the cutoff at $\log L/L_\odot = -4.5$, i.e. an age for the Galactic disk, $t_D \sim 12$ Gyr.

4. Uncertainties in the theory

An important uncertainty in the models is the initial C/O composition and stratification. If the initial WD is already stratified, with oxygen accumulated near the center, as suggested by Mazzitelli & D'Antona (1986),

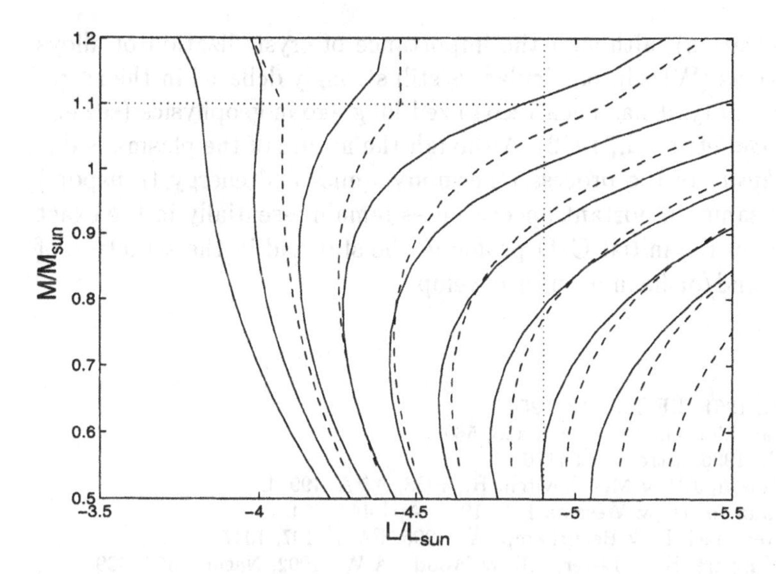

Figure 1. Isochrones for different masses of crystallized WDs ($\log L/L_{\odot} < -3.5$) in a mass-luminosity diagram. The solid lines include C/O differentiation at crystallization whereas the dashed lines do not. The isochrones are the *same in both cases* and go from $t = 7$ Gyr to $t = 15$ Gyr from left to right. The vertical dotted line is the inferred luminosity for $ESO\,439-26$ (Bergeron et al., 1997), while the horizontal dotted lines are the inferred mass for a H-rich ($1.1\ M_{\odot}$) and He-rich ($1.2\ M_{\odot}$) atmosphere. Crystallization starts earlier for massive (and thus rapidly evolving) WDs, which then enter the rapid Debye cooling regime, which causes the bending in the isochrones.

the energy release by differentiation, and the related time delay are reduced (see Segretain et al, 1994, Tables 3-4). The other main source of uncertainty concerns the model atmospheres and most importantly for the age of the disk, the thickness of the helium layer wich regulates the emergent heat flux. The thicker the layer the younger the age (Wood, 1992). These uncertainties translate in a $\sim 1-2$ Gyr uncertainty on the age of the disk. Futher progress in this direction is certainly essential for a better determination of the cooling and the observational properties of cool WDs.

5. Conclusion

We have shown in this review that substantial improvement in the theory of cool WDs has been accomplished within the past few years for the atmosphere as well as for the internal structure. We have shown that chemical fractionation at crystallization, although it liberates a negligible amount of gravitational energy, modifies substantially the cooling history of the star and yields an important time-delay for faint WDs. This process cannot be ignored in accurate WD cooling theory. As noted by Mochkovitch (private communication), although the importance of crystallization of alloys in stellar plasmas (WDs in particular) is still strongly debated in the astrophysical community, it has been recognized long ago in geophysics (see e.g. Loper, 1984; Buffett et al., 1992). Although the nature of the plasma is different, the *physics* of the process (thermodynamics and energy transport) is exactly the same. Important uncertainties remain essentially in the exact determination of the initial C/O profile in the star and in the structure of the hydrogen and/or helium outer envelope.

References

Abrikosov, A.A., 1961, J.E.T.P., 12, 1254
Adams, F. & Laughlin, G., 1996, ApJ, 468, 586
Alcock, C. et al., 1996, astroph-9604176
Barrat, J.L., Hansen, J.P. & Mochkovitch, R., 1988, A&A, 199, L15
Bergeron, P., Saumon, D., & Wesemael, F, 1995, ApJ 443, 764
Bergeron, P., Wesemael, F. & Beauchamp, A., 1995, PASP, 107, 1047
Buffett, B.A., Huppert, H.E., Lister, J.R. & Woods, A.W., 1992, Nature, 356, 329
Chabrier, G., Segretain, L. & Méra, D., 1996, ApJ, 468, L21
Dubin, D., 1990, Phys. Rev. A, 42, 4972
García-Berro, E., Hernanz, M., Isern, J. & Mochkovitch, R., 1988, A&A, 193, 141
Hernanz, M., García-Berro, E., Isern, J., Mochkovitch, R., Segretain, L. & Chabrier, G., 1994, ApJ, 434, 652
Ichimaru, S., Iyetomi, H. & Ogata, S., 1988, ApJ, 334, L17
Isern, J., Mochkovitch, R., Hernanz, M., García-Berro, E., 1997, ApJ, submitted
Lamb, D.Q. & VanHorn, H.M., 1975, ApJ, 200, 306
Landau, L. & Lifshitz, E., 1980, *Statistical Physics*, Pergamon Press
Loper, D.E., 1984, Adances in Geophysics, 26, 1
Liebert, J., Dahn, C.C.& Monet, D.G., 1988, ApJ, 332, 891
Mazzitelli, I. & D'Antona, F., 1986, 308, 706
Mestel, L. & Ruderman, M.A., 1967, M.N.R.A.S., 136, 27
Mochkovitch, R., 1983, A&A, 122, 212
Oswalt, T.D., Smith, J.A., Wood, M.A. & Hintzen, P., 1996, Nature, 382, 692
Ruiz, M.T., Bergeron, P., Leggett, S., & Anguita, C, 1995, ApJ, 455, L159
Salpeter, E.E., 1961, ApJ, 134, 669
Segretain, L. & Chabrier, G., 1993, A&A, 271, L13
Segretain, L., Chabrier, G., Hernanz, M., García-Berro, E., Isern, J. & Mochkovitch, R., 1994, ApJ, 429, 641
Stevenson, D., 1980, J. Physique Sup., 41, C2-61
VanHorn, H.M., 1968, ApJ, 151, 227
Winget, D. et al., 1987, ApJ, 315, L77
Wood, M. A., 1992, ApJ, 386, 536

DISCUSSION

ROBERT KURUCZ: What about the effect of rotation on solidification?

GILLES CHABRIER: It is not clear how rotation will affect the crystal growth, but there is no reason for it to *prevent* crystallization, which must occur below a certain temperature. The effect of rotation on the C-redistribution in the surrounding layer has been considered in detail by Mochkovitch (1983).

THEORY, OBSERVATION AND EXPERIMENT: STELLAR HYDRODYNAMICS

A Different Perspective

DAVID ARNETT
Steward Observatory,
University of Arizona,
Tucson AZ 85721, USA

Abstract. Computer technology now allows two dimensional (2D) simulations, with complex microphysics, of stellar hydrodynamics and evolutionary sequences, and holds the promise for 3D. Careful validation of astrophysical methods, by laboratory experiment, by critical comparison of numerical and analytical methods, and by observation are necessary for the development of simulation methods with reliable predictive capability. Recent and surprising results from isotopic patterns in pre-solar grains, 2D hydrodynamic simulations of stellar evolution, and laser tests and computer simulations of Richtmeyer-Meshkov and Rayleigh-Taylor instabilities will be discussed, and related to stellar evolution and supernovae.

1. Introduction

It is a personal pleasure to speak at this meeting in honor of Prof. Hanbury-Brown; this discussion may be seen as my colored reflection of his approach to making astrophysics a quantitative science.

A fundamental property of stars is their composition, and its possible variation throughout their structure. Stars are thermonuclear reactors, so that the change of abundances both drives the evolution and provides a diagnostic of that process. With the wealth of new quantitative data becoming available, theorists have an obligation to take a critical look at the conceptual framework within which predictions of nuclear abundance anomalies are made.

The investigation of rapidly evolving late stages of the lives of stars—by use of direct multidimensional hydrodynamic simulations—brings into question assumptions generally used in the theory of stellar evolution.

T.R. Bedding et al. (eds.),
Fundamental Stellar Properties: The Interaction between Observation and Theory, 389–394.

Quantitative predictions of nucleosynthesis yields and pre-supernova characteristics will require attention to these new physical effects.

For at least the last three decades there has been a steady and dramatic increase in computer performance. This has been accompanied by a comparable (probably greater) increase in the quality and efficiency of algorithms. How may we validate these increasingly complex simulations, to convince ourselves that the results are real and not just virtual reality?

2. Toward a Predictive Theory: Tests with the NOVA Laser

Ultimately simulations must be well resolved in three spatial dimensions. One of the great assets of computers is their ability to represent complex geometries. If we can implement realistic representations of the *essential* physics, then simulations should become tools to predict—not "postdict"—phenomena. An essential step toward that goal is the testing of computer simulations against reality in the form of experiment(Remington, Weber, Marinak, et al., 1995). This is a venue in which we can alter conditions (unlike astronomical phenomena), and thereby understand the reasons for particular results. Experiments are intrinsically three dimensional, with two dimensional symmetry available with some effort, so that they provide a convenient way to assess the effects of dimensionality.

For Rayleigh-Taylor instabilities, the NOVA experiments not only sample temperatures similar to those in the helium layer of a supernova, but hydrodynamically scale to the supernova as well(Kane, Arnett, & Remington, 1997). The NOVA laser is physically imposing. The building is larger in area than an American football field; the lasers concentrate their beams on a target about the size a BB (or a small ball bearing). This enormous change in scale brings home just how high these energy densities are. Preliminary results show that the astrophysics code (PROMETHEUS) and the standard inertial confinement fusion code (CALE) both give qualitative agreement with the experiment. For example, the velocities of the spikes and bubbles are both in agreement with experiment, and analytic theory which is applicable in this experimental configuration(Kane, Arnett, & Remington, 1997). The two codes give similar, but not identical results. These differences will require new, more precise experiments to determine which is most nearly correct.

3. Applications to Stellar Hydrodynamics

In discussion of stellar evolution, one encounters the topics of rotation, convection, pulsation, mass loss, microturbulence, sound waves, shocks, and instabilities—to name a few—which are all just hydrodynamics. However, direct simulation of stellar hydrodynamics is limited by causality. In analogy to light cones in relativity, in hydrodynamics one may define space-time regions in which communication can occur by the motion of sound waves. To correctly simulate a wave traveling through a grid, the size of the time step must be small enough so that sound waves cannot "jump" zones. Thus the simulation is restricted to short time steps—an awkward problem if stellar evolution is desired. While simulations of the solar convection zone are feasible, the simulation time would be of order hours instead of the billions of years required for hydrogen burning. For the latter, a stellar evolution code is used, which damps out the hydrodynamic motion, obviating the need for the time step restriction. Any presumed hydrodynamic motion is then replaced by an algorithm (such as adiabatic structure and complete mixing in formally convective regions). Thus, *stellar evolution* deals with the long, slow phenomena, and *stellar hydrodynamics* has dealt with the short term.

However, the stages of evolution prior to a supernova explosion are fast and eventful. Here direct simulation is feasible(Bazan & Arnett, 1997). A key region for nucleosynthesis is the oxygen burning shell in a pre-supernova star. Besides producing nuclei from Si through Fe prior to and during the explosive event, it is the site at which the radioactive ^{56}Ni is made and is mixed. The conventional picture of this region relies upon the notion of thermal balance between nuclear heating and neutrino cooling in the context of complete microscopic mixing by convective motions.

This is usually treated by the mixing length scenario for convection, which assumes statistical (well developed) turbulence, random walk of convective blobs approximated by diffusion, subsonic motions, and almost adiabatic flow. These approximations are further constrained by a simplistic treatment of the boundaries of the convective region.

The timescales for the oxygen burning shell are unusual. The evolutionary time is $\tau_{evol} \approx 4 \times 10^3$ s. The convective "turnover" time is $\tau_{conv} \approx \Delta r / v_{conv} \approx \tau_{evol}/10$, while the sound travel time across the convective region is $\tau_{sound} \approx \Delta r / v_{sound} \approx \tau_{conv}/100$. The burning time is $\tau_{burn} = E/\varepsilon \leq \tau_{conv}$. Obviously the approximations of subsonic flow, well developed turbulence, complete microscopic mixing, and almost adiabatic flow are suspect.

These time scales are rapid enough to make the oxygen shell a feasible target for direct numerical simulations, and an extensive discussion is about to appear(Bazan & Arnett, 1997). The two dimensional simulations show qualitative differences from the previous one dimensional ones. The oxygen shell is not well mixed, but heterogeneous in coordinates θ and ϕ as well as r. The burning is episodic, localized in time and space, occurring in flashes rather than as a steady flame. The burning is strongly coupled to hydrodynamic motion of individual blobs, but the blobs are more loosely coupled to each other.

Acoustic and kinetic luminosity are not negligible, contrary to the assumptions of mixing length theory. The flow is only mildly subsonic, with Mach numbers of tens of percent. This gives non-spherical perturbations in density and temperature of several percent, especially at the boundaries of the convective region.

At the edges of convective regions, the convective motions couple to gravity waves, giving a slow mixing beyond the formally unstable region. The convective regions are not so well separated as in the one dimensional simulations; "rogue blobs" cross formally stable regions. A carbon rich blob became entrained in the oxygen convective shell, and underwent a violent flash, briefly outshining the oxygen shell itself by a factor of 100. Significant variations in neutron excess occur throughout the oxygen shell. Because of the localized and episodic burning, the typical burning conditions are systematically hotter than in one dimensional simulations, sufficiently so that details of the nucleosynthesis yields will be affected.

The two dimensional simulations are computationally demanding. Our radial zoning is comparable to that used in one dimensional simulations, to which we add several hundred angular zones, giving a computational demand several hundred times higher. This has limited us to about a quarter of the final oxygen shell burning in a SN1987A progenitor model. Given the dramatic differences from one dimensional simulations, it is important to pursue the evolutionary effects to see exactly how nucleosynthesis yields, pre-supernova structures, collapsing core masses, entropies, and neutron excesses will be changed. It may be that hydrostatic and thermal equilibrium on average, and the temperature sensitivity of the different burning stages, taken together, tend to give a rough layering in composition, even if the details of how this happens are quite different.

4. Mixing

We sometimes forget that stars are really very large. Let us make a order of magnitude estimate of diffusion time scales in a dense stellar plasma. It is the nuclei, not the electrons which define the composition. The coulomb cross section for pulling ions past each other is of order $\sigma \approx 10^{-16}$ cm^2. For a number density $N \approx 10^{24}$ cm^{-3}, this implies a mean free path $\lambda = 1/\sigma N \approx 10^{-8}$ cm. For a particle velocity $v_d \approx 10^8$ cm/s, this gives a diffusion time $\tau_d = (\Delta r)^2/\lambda v_d \approx (\Delta r)^2$ s cm^{-2}. For a linear dimension of stellar size, $\Delta r = 10^{11}$ cm, $\tau_d = 3 \times 10^{14}$ y, or 3,000 Hubble times! While one may quibble about the exact numbers used, it is clear that pure diffusion is ineffective for mixing stars.

Actually, we all know from common experience—such as stirring cream into coffee (tea)—that this is incomplete. To diffusion must be added *advection*, or stirring. Stars may be stirred too. For example, rotation may induce currents, as may accretion, and perturbations from a binary companion. However, the prime mechanism for stirring that is used in stellar evolutionary calculations is thermally induced convection. The idea is that convective motions will stir the heterogeneous matter, reducing the typical length scale Δr to a value small enough that diffusion can insure microscopic mixing. For our stellar example above, this would require a reduction in scale of $(\lambda/\Delta r)^{1/2} \approx 10^{-8}$. Convection is not perfectly efficient, so that the actual mixing time would still be finite. Given that such a limit exists, we must examine rapid evolutionary stages to see if microscopic mixing is a valid approximation. For pre-supernovae, the approximation is almost certainly not correct, so that these stars are not layered in uniform spherical shells as conventionally assumed, but heterogeneous in angle as well as radius.

5. Pre-solar Grains

This new view of the nucleosynthesis process in a supernova, both prior to and during the actual explosion, has implications for any attempt to identify pre-solar grains with such events. For example, on the basis of their enrichments in ^{28}Si and the presence of ^{44}Ti at the time of grain formation, type II supernovae have been suggested as the most likely sources of the X grains(Amari, et al., 1996). Using the one dimensional view of nucleosynthesis gives serious difficulties in detail(Hoppe, et al., 1996). The higher burning temperatures in the two dimensional simulations shifts the detailed constraints on yields. The heterogeneity in neutron excess, and the possibility of rogue blobs, further relaxes these constraints. The probability of macroscopic mixing prior to explosion, but with ashes of high temperature burning by various processes, gives still other options.

This leads us to suggest two different views of what might be occurring.

1. **The Layered Mixing Model.** This view has been explored in some detail(Hoppe, et al., 1996) for the X grains, and found to have problems. Inner layers, containing ^{44}Ti must be brought out to coincide with outer layers, containing ^{26}Al. This must occur without any microscopic mixing with intermediate layers which are oxygen rich, and could prevent SiC grain formation. However, just before grain formation, these layers must be microscopically mixed.
2. **The Path Integral Model.** It may be more accurate to think of the anomalies in the X grains as the result of a "path integral" over the trajectory of a blob, sampling a variety of conditions. The blob would be affected by burning prior to explosion as well as after. In general,

the blob might have all the anomalies produced with no mixing, or it might be moved out to the region with which it should mix, prior to the explosion, to be microscopically mixed with adjacent matter upon expansion to lower density.

The anomalies in the X grains have also been identified with explosive helium burning in ^{14}N-rich matter(Clayton, et al., 1997) in type I supernovae. He-rich blobs in type II supernovae might be possible sources as well.

6. Conclusions

- Multidimensional simulations provide new perspectives of direct relevance to the understanding of stellar evolution, and the interaction between theory and experiment promises to become even more exciting.
- Experimental tests of astrophysical simulation methods are feasible, and preliminary results from the NOVA laser are promising.
- The slow stirring outside the formally unstable regions, if it extends to more slowly evolving stages, may imply important and needed modifications for stellar evolutionary theory.
- The qualitative nature of the pre-supernova evolutionary stages is changed when direct hydrodynamic simulation techniques are used. This promises to modify our understanding of supernovae and nucleosynthesis. It is particularly applicable to the interpretation of presolar grains in meteorites.

This work has been supported in part by the National Aeronautics and Space Administration grant NAGW-2798 and National Science Foundation grant AST94-17346.

References

Amari, S., Hoppe, P., Zinner, E., Lewis, R. S. *Astrophys. J.***394**, L43, 1992.
Arnett, D., *Supernovae and Nucleosynthesis*, Princeton: Princeton University Press, 1996.
Bazan, G. & Arnett, D., *Astrophys. J.*, 1997, in press.
Clayton, D. D., Arnett, D., Kane, J., & Meyer, B. S., 1997, *Astrophys. J. Letters*, in press.
Hoppe, P., Strebel, R., Eberhardt, P., et al., *Science***272**, 1314, 1996.
Kane, J., Arnett, D., Remington, B., et al., 1997, *Astrophys. J. Letters*, in press.
Landau, L. D., & Lifshitz, E. M., *Fluid Mechanics*, Reading, MA: Addison-Wesley, 1959, ch. VI.
Remington, B. A., Weber, S. V., Marinak, M. M., et al., *Phys. Plasmas* **2**, 241, 1995.

Discussion of this paper appears at the end of these Proceedings.

SUMMARY AND REVIEW: INTERACTION OF OBSERVATIONS AND THEORY OF STELLAR INTERIORS

STEVEN D. KAWALER
Department of Physics and Astronomy
Iowa State University, Ames, IA 50011 USA

"The star is never wrong." — *Leslie Kawaler, to Ed Nather*

1. Introduction: Observational Approaches to Studying Stellar Interiors

As we move into the next century of stellar astronomy, it is helpful to consider the modes by which stellar astronomy has advanced in the modern era. At this conference, our attention has been focussed on precision measurement of stellar fundamental properties such as mass, luminosity, and radius. This mode of inquiry is one of several that have borne fruit. In this review, I will discuss some of the topics that have arisen here with respect to how our understanding of stellar interiors progresses through interaction between theory and observation.

At some level, stars are relatively simple. However, sufficiently detailed study of individual stars, in particular our Sun, reveals they can be extremely complex. One of the issues that we constantly struggle with is in defining which observed properties are important to understand and which properties might be simply idiosyncratic. Study of unusual, prominent, or pathological stars sometimes proves to be meaningful for studies of more average subjects, while at other times these prominent stars can be misleading. Within this category fall some "famous" stars such as Mira, Betelgeuse, and Sirius. Here too we find objects whose fame is rooted in historical as well as in modern studies: the Crab pulsar, SN 1987A, and so on. Lesser known stars have served as the prototypes for important classes of objects, including (again) Mira, β Pictoris, and PG 1159-035.

In contrast to the prominent examples, there are several prototypical objects that are unremarkable save for their proximity to Earth and the ease with which we can study them. The principal example is of course our Sun; an otherwise (we hope) ordinary G2V star that we can study in

T.R. Bedding et al. (eds.),
Fundamental Stellar Properties: The Interaction between Observation and Theory, 395–404.

incredible detail. It is justifiably the star where all stellar interior theory begins...if you can't get the Sun right, then don't bother with globular cluster isochrones!

Beyond prominent or average "individuals", focus groups are also important areas of study. Because of the long time scales over which star change the ensemble properties of samples of stars are needed to learn about stellar evolution. The importance of star clusters for understanding stellar evolution is clear; examples of clusters that have been, and will continue to be, key objects include 47 Tuc, Ω Cen, and the Hyades. These objects have grown in importance over the years because understanding them in the greatest possible detail reveals important new insights into the physics of all stars. For example, matching the morphology of cluster color–magnitude diagrams has led to the discovery of the likely effects of convective overshoot in stellar cores (as reviewed by Cesare Ciosi) and continual progress in determining stellar ages (as reviewed by Don Vandenberg).

Additional important constraints come from more extensive samples, such as all–sky surveys, luminosity functions, and so on. The collection of all white dwarf stars in our part of the Galaxy, for example, provides important data on the star formation history of our neighborhood. As discussed by Gilles Chabrier at this meeting, the white dwarf luminosity function also plays an important role in stimulating studies of the physics of white dwarf interiors — including the process of crystallization and phase separation in degenerate matter. The collective properties of the variable stars uncovered by the MACHO and OGLE surveys are extremely illuminating for the understanding of stellar pulsation, as illustrated by Kem Cook and Dante Minniti at this meeting.

2. "Fundamental" Properties

Fundamental observed properties of stars have always played crucial roles in theoretical investigation of stellar structure and evolution. As is evident from the discussions at this meeting, they will continue to provide technical challenges, the solutions of which will push the envelope of technology. As a faculty member in a department of Physics and Astronomy, I've recently been exposed to heavy ion collider physics through our nuclear physics group. They are part of the RHIC (Relative Heavy Ion Collider) experiment at Brookhaven National Laboratory. One of the techniques that RHIC is employing to study the head–on collision of accelerated gold nuclei is, of all things, Hanbury-Brown Twiss interferometry for localization of interaction products in the search for the quark–gluon plasma. I expect that continued efforts in fundamental stellar astronomy will continue to develop techniques with broad application.

The revolution in stellar astronomy truly began with the first determination of stellar distances in the middle of last century. With distances known, the energy output of the stars beyond the Sun could be determined, and stellar masses calibrated. The assembly of catalogs of stellar distances enabled the "invention" of the modern H–R diagram, providing systematic relations amongst the fundamental stellar parameters for theoretical investigation. Rapid progress included the mass–luminosity relation, and its explanation on theoretical grounds, and quantitative exploration of the direction and rate of stellar evolution.

This is a continuing history, as made apparent during this meeting. Calibration of the cosmic distance scale relies upon the first steps in the chain. Precise masses and luminosities of Cepheid variables, RR Lyra and other horizontal branch stars are universal concern. The imminent release of the Hipparcos results, as previewed here by C. Turon, is eagerly anticipated. Progress in development of CCD astrometry is allowing ground–based parallaxes of faint (and relatively distant) stars (see the review in this volume by C. Dahn) that we could not have hoped for a generation ago.

Beyond mass and luminosity, what other "fundamental properties" do stars possess? Principal competition for another quantity is between some measure of the temperature, and some measure of the radius. Both temperature and radius demand further qualification. By radius does one mean the position of last scattering? If so, at what wavelength? If defined as where the density drops to zero, how does one measure such a thing? Similar ambiguities confront the use of temperature as a fundamental quantity. Is "effective temperature," as determined via the Stefan–Boltzmann law, sufficiently fundamental? Even if so, how does one measure such a quantity? If the star is not perfectly spherical, what then? As illustrated in Bob Kurucz's review, stars have temperature gradients along their surfaces caused by convective motions, and active regions. Spatial temperature variations also change with time.

One might argue that a truer "fundamental quantity" is stellar surface area! After all, it is the surface area that links the measured fluxes with stellar luminosity, independent (to high order in most round-ish stars) of star shape. That is, of course, if one knows the distance to the stars (or their angular diameters).

Stellar evolution theorists typically discuss models of given masses, luminosities and effective temperatures. But comparison with observations requires transformations of these ideal quantities into "observables" such as colors and broadband magnitudes. The difficulties of these transformations, described by many at this meeting, remain with us. It is rare that determination of $T_{\rm eff}$ to 1% or less is of great importance within this climate. However, for some stars the determination of $T_{\rm eff}$ to 10% or more remains a challenge to the observers. It is in these difficult cases that progress in the study of stellar interiors depends on the observational constraints on these fundamental quantities.

As a relatively recent example, in 1987 the temperatures of hot white dwarfs such as PG 1159–035 were very poorly known. But with the work of Werner and Dreizler on NLTE atmospheric models in the late 1980s and early 1990s, these uncertainties have been reduced to well below 10%. With this advance, seismological studies of these stars, described in the next section, became possible. In the future, more accurate determination of the fundamental parameters of a wide variety of pulsating stars, including the pulsating sdB stars described above, the rapidly oscillating Ap stars, and δ Scuti stars will provide a key to unlocking their seismic secrets.

Another fundamental property of stars is angular momentum. All stars rotate. Observationally, the amount of rotation varies significantly between otherwise identical stars. The influence of rotation ranges from mechanical (through partial hydrostatic support in rapidly rotating stars as well as rotationally-induced mass loss) to thermal (through horizontal heat transport in circulating material) and compositional (via rotationally driven mixing in stellar interiors).

At this meeting, we heard about several of the effects of stellar rotation. Contrary to what many of us learned in being introduced to stellar interiors, the effects of stellar rotation are frequently of 0th order, not first or second. Certainly during star formation and pre–main sequence evolution, angular momentum and its redistribution (and loss) is a controlling factor. Coupled to the angular momentum problem is the issue of mass loss, which, as demonstrated here by Andre Maeder's discussion, clouds the whole issue of the main sequence for massive stars. During main sequence evolution and beyond, rotation significantly alters the evolutionary tracks of stars, as shown in N. Langer's review.

Redistribution of angular momentum must occur as a consequence of the onset of rotational instabilities in regions of steep angular velocity gradients. Such redistribution can result in the redistribution of material. This, in turn, causes mixing of composition, and provides observational signatures as "peculiar" surface abundances in stars. The process of rotational mixing and composition transport was discussed by a variety of people at this meeting, including Andre Maeder, Corinne Charbonnel, and Charles Proffitt.

3. A Case Study: Stellar Seismology

The brightness of stars at various colors is not necessarily constant with time. Classical variable stars have long been evidence of this, but new techniques are showing that many stars are multiperiodic variables at a very small level, both photometrically and spectroscopically. These subtle variations are the signal for asteroseismology, which has the potential to revolutionize the study of stellar interiors. Asteroseismology involves long observing runs using large telescopes to look at relatively bright stars. Recent success in this area is very promising. Here, I briefly mention a few areas of progress in asteroseismology. There are others, including the newly-discovered multiperiodic sdB stars, the rapidly oscillating Ap stars, δ Scuti stars, etc.

Jørgen Christensen-Dalsgaard reviewed some of the progress in helioseismology in this meeting, and due to lack of time could only provide a few illustrative examples. One remarkable result is of broad impact in physics and astrophysics. The measured solar oscillation frequencies determine, through inversion, the sound speed in the solar interior. Comparison with the current "best" solar model shows agreement to remarkable precision — much better than 1% everywhere within the model. The independently-constructed solar model to which this inversion is compared includes the most accurately known input physics (equation-of-state, opacities, nuclear reaction cross-sections). This result is the essence of interaction between observations and theory. As he notes, this places stringent constraints on solutions of the solar neutrino problem, placing the burden firmly on particle physicists.

Other helioseismological results made possible by ground based (GONG, BISON, LOWL, etc.) and space-based (SOHO) observations include measurement of the solar rotation as a function of depth and latitude, the subsurface structure in the vicinity of sunspots and other active regions, mapping of large scale convection, etc. All of these results challenge current theoretical pictures of the Sun. By extension, our view of stellar interiors and evolution will never be the same.

Helioseismic results describe a single star at a single time in its history. Modifications of our understanding of the equation-of-state of stellar matter are appropriate for the line described by the Sun in the (ρ, T, μ) volume of phase space; we all know what fraction of a volume is covered by a line. Study of stars in a comparable level of detail would certainly broaden our understanding of all aspects of stellar physics.

Compared to the stars, the Sun is "easy pickings" for seismological study. We get many more photons from the Sun, and we can resolve its surface. This allows high signal-to-noise observations that reveal very small amplitude variations, and spatial modulation allows unambiguous mode identification. To even detect solar-like oscillations on solar-type stars, a combination of cleverness, benevolent telescope allocation committees, and a degree of luck (beyond clear skies) is needed.

Hans Kjeldsen gave compelling justification for undertaking the challenge to detect and measure these modes. Since convection plays a central role as a mechanical stimulator of the oscillation modes, a wide area of the H–R diagram may be populated by multimode nonradial pulsators analogous to our Sun. Kjeldsen and his collaborators, and many other groups, have attempted to detect oscillations of solar–type stars. As he reviewed at this meeting, this is indeed a difficult observational problem. Possible detection of oscillations in η Bootes await confirmation (Brown et al. 1997). Kjeldsen's progress report on α Cen is tantalizing, and illustrates again the technical challenges of these important observations.

Probably the most successful investigations in asteroseismology involve the pulsating white dwarf stars. Seismological analysis of these stars allows unprecedented access to their interiors. The pulsation modes in these stars, nonradial g-modes, provide a sensitive indicator of the mass of the star and can also be used to determine depth below the surface where steep composition gradients occur (Kawaler & Bradley 1994), among other things. Papers from recent ground based studies (for example, the papers by the Whole Earth Telescope collaboration) describe this procedure in some detail.

Observational results from white dwarf seismology pose a significant challenges to theoretical models. In one example, the thickness of the helium–rich surface layer of the hottest white dwarfs (Kawaler & Bradley 1994) was measured to be three orders of magnitude larger than that in cooler pulsating helium–rich white dwarfs (Bradley & Winget 1994). This challenged the notion that they were two links of the same evolutionary change until Dehner (1996, see also Dehner & Kawaler 1995) showed that diffusive purification of the surface helium layer produced a composition transition zone that reached the correct position at the time the models reached the cooler pulsational instability.

Interaction between observation and theory, and between different modes of observation, is an important facet in white dwarf pulsations. Spectroscopic determinations of the abundances, effective temperatures, and gravities (as reviewed by Dreizler at this meeting) provide essential constraints on seismological models (i.e. O'Brien et al. 1996, Kawaler et al. 1995), while independent determination of these quantities using both techniques foster confidence in both methods (Kawaler & Bradley 1994).

An additional benefit of seismological studies of white dwarf stars is that they can yield distance determinations. Asteroseismic distances to two stars have been independently confirmed. Bradley and Winget (1994) published an asteroseismic distance for the DB star GD358 that matched parallax measurements, and Kawaler et al. (1995) determined the distance to the PG1159 star PG2131 that was confirmed via the spectroscopic parallax of its K-dwarf companion by Howard Bond (1995, private communication).

4. The Road Ahead... AGB Thermal Pulse Dynamics as a Selected Example

As discussed by several at this conference (i.e. Lattanzio, Boothroyd, Charbonnel) the chemical abundances in AGB stars, and indeed the carbon-star phenomenon, is not easily solved with standard hydrostatic evolutionary models. Perhaps an understanding of the dynamical aspects of helium shell flashes in AGB stars will allow solution of this classic problem?

One of the fundamentals that we learn in the early days of stellar evolution courses is that there is a "hierarchy of time scales" that is usually obeyed within stars. That is, $\tau_{dyn} \ll \tau_{therm} < \tau_{nuc}$ during most phases of stellar evolution. It is this hierarchy that allows us to compute the evolution of stars as a quasi-static process. With the thermal and dynamical time scales shorter than the time scale for nuclear transformations, we can "freeze" the composition of the stellar material to follow thermodynamic transformations generated by a composition change. Similarly, if the time scale for heat flow is long compared to that for dynamical (hydrodynamical) adjustments, then we can assume hydrostatic equilibrium is maintained for the purposes of thermodynamic adjustments to a changing energy flux. By taking time steps that are comparable to the thermal time scale, the assumption of hydrostatic equilibrium is accurate to high accuracy. Th time step is sufficiently short that nuclear processing results in only small changes to the abundance of relevant elements over the time step.

If, however, the time scale for dynamic adjustments is comparable to the time scale for heat flow, then these assumptions are no longer valid. The equations solved by a standard hydrostatic stellar evolution code do not describe the stellar material in this circumstance. Such a standard stellar evolution code may in fact produce models that look nice and converge quickly, but such models are solutions to equations that are not relevant to the real situation that they are supposed to be modeling. Similarly, if nuclear transformation occur at a rate that is fast (i.e. comparable to the time scale for heat flow) the hydrostatic code will produce irrelevant results.

In thermally pulsing AGB stars, the worst of all situations can occur. During a thermal pulse, the helium-burning shell drives a convective region. In the convective region, the thermal and dynamical time scales are comparable. But this convective region rides atop the nuclear burning shell, which is experiencing a mild runaway. The coupling results in all three time scales being comparable in the shell-burning region.

To fully resolve the temporal behavior of the star during the pulse, and to "accurately" track nucleosynthesis through the pulse, time steps are required to be extremely short ... years or less. However, the dynamical time scale for these stars is of this order. Therefore, the evolutionary calculations are implicitly demanding hydrostatic evolution when, in fact the envelope of a real AGB star is free to respond on the same time scales in a dynamic way. Also, within the burning region, the luminosity can be enormous, and work out to be larger than the (local) Eddington luminosity. Under such conditions, the material there is dynamically unstable.

Clearly, we are making mistakes when we model the evolution of AGB stars. Perhaps when these mistakes are corrected, the observed abundances of AGB stars will be less mysterious.

5. Conclusions

The theoretical study of stellar interiors continues to rely on observational constraints, as it did at the turn of the 20th century and earlier. However, this reliance has now bifurcated. On the one hand, more detailed observations are pushing hard on the theoretical assumptions of spherical symmetry, global thermal and hydrostatic equilibrium, lack of importance of rotation and magnetic fields, etc. As observers pin down the abundances and fundamental properties of stars, theorists will have to work hard to produce realistic models that explain these observations in all of their detail.

On the other hand, there remain significant regions of the H–R diagram where parameters of the stars are not precisely known. These include, on the cool side, the formation and evolution of very low–mass stars, the pulsation/convection/mass loss interaction in AGB stars, and the production of planetary nebulae. On the hot side, the evolution of massive stars with rotation and mass loss remains an exciting and controversial research area.

In the early days of helioseismology, an important result was the measurement of the depth of the solar convection zone. While "modern" helioseismology has provided a large number of other important discoveries, such a measurement of the convection zone boundaries in other stars would profoundly affect our understanding of stellar interiors. To do so requires effort as a community to pursue asteroseismological observations. The only way for this to happen is for the stellar astronomers to educate themselves and the broader astronomical community about the potential benefits. Only then will telescope allocation committees and others facilitate the development of a new observational paradigm for stellar astronomy.

All this, and we barely discussed the generation, modulation, and physical effects of stellar magnetic fields! There is much to be done for observers and theorists alike.

Acknowledgements

The author gratefully acknowledges support from the Organizing Committees, NSF Young Investigator Award AST-9257049, and the NASA Astrophysics Theory Program.

References

Bradley, P. & Winget,D. (1994), *Ap. J.*, **430**, 850.
Brown, T., Kennelly, E., Korzennik, G., Nisenson, P., Noyes, R., & Horner, S. (1997), *Ap. J.*, **475**, 322.
Dehner, B.T. (1996), Ph.D. Dissertation, Iowa State University.
Dehner, B.T. & Kawaler, S.D. (1995), *Ap. J. Lett*, **445**, L141.
Kawaler, S.D. & Bradley, P.A. (1994), *Ap. J.*, **427**, 415.
Kawaler, S.D. et al. (the WET collaboration) (1995), *Ap. J.*, **450**, 350.
O'Brien, M.S., Clemens, J.C., Kawaler, S.D., & Dehner, B.T. (1996), *Ap. J.*, **467**, 397.

DISCUSSION

ARNOLD BOOTHROYD: Rotation is also important on the RGB (see, e.g., Charbonnel) to provide a mixing mechanism for the observed abundance anomalies. When a star becomes a giant, the core contracts and the envelope expands. There will be shear at the core boundary and we expect mixing. This may also have an effect when an HB star becomes and AGB star; rotational mixing may also affect He-shell flashes (thermal pulses). Since $\tau_{\rm nuc} \sim \tau_{\rm conv} \sim \tau_{\rm dyn}$ in the He shell flash, it would not be surprising if effects of non-sphericity were significant.

ANDRE MAEDER: I agree that the treatment of shears is essential. In particular, shears and steep μ-gradients occur at the same place and thus any mixing in a region with a steep μ-gradient would considerably affect the evolution.

NORBERT LANGER: I agree that the mixing of elements and angular momentum through the strong μ- and entropy gradients produced by nuclear shell sources is a fundamental problem. There may be a simple way to check this, i.e., to compare the specific angular momenta of white dwarfs and main-sequence stars. This would tell us how much angular momentum gets sucked out of the core into the envelope during the post-main-sequence evolution.

ROBERT KURUCZ: In your statement about getting observing time for asteroseismology you show a misapprehension of our basic knowledge. We do not know the abundances in the Sun to 10%. In other stars it is worse. The opacities that go into seismology models depend on these abundances. One has to get high- resolution, high-signal-to-noise spectra to fix the stellar parameters. It takes a long time to make the observations and a long time to analyze them.

STEVE KAWALER: It depends on what one is after. Seismology can allow us to determine the positions of convective zone boundaries (core and envelope); we have only theory right now. To me, it is much more exciting to measure, for example, the convective-zone mass in Procyon than to know better the solar nickel abundance. We can do so much with seismology, even allowing for uncertainty in composition, that not to do so would be a shame.

TIM BEDDING (also responding to Kurucz): In open clusters you can assume (perhaps wrongly) that all stars have the same metallicity.

JØRGEN CHRISTENSEN-DALSGAARD: In fact, we probably all agree. For asteroseismic analysis is it crucially important to have as accurate information as possible on the 'classical' stellar parameters, such as mass, luminosity, effective temperature and composition, from traditional observations. Only with such information is it truly possible to use the oscillation frequencies to probe details of the deep interior structure, or detect problems in the assumed physics. Without such information, parts of the precious seismic data must be used to constrain the overall stellar parameters.

PIERRE MAXTED: I want to mention my experience of CM Draconis, a star that has been mentioned before. This star determines the effective temperature scale for low mass stars, provides tests of models of stellar structure through its accurately known mass, radius and luminosity, and also allows limb darkening to be directly measured. Yet, I only heard of this star one year ago. I think this demonstrates the need for more interaction between observers and theoreticians, so that observers can observe the stars that best constrain the theoreticians' models.

BENGT GUSTAFFSON: A philosophical remark on your philosophical introduction to what is worthwhile and important to concentrate on and what is 'weather.' For a long time we have concentrated on fundamental principles, which I guess is a heritage from physics. This is alright, but there is a growing interest in natural sciences as regards complexity and formation of structure. Now, stellar physics offers a plethora of complex structure, both in atmospheres and, as we begin to understand, in their interiors. This has its own beauty that many of us love, while others get disappointed. However, one should note that stars, as compared to many other systems in nature, are still reasonably simple and admit understanding. They seem, therefore, to offer examples of structure formation that are suitable as case studies in a much wider context.

13. STELLAR CHEMICAL EVOLUTION

STELLAR CHEMICAL EVOLUTION

JOHN E. NORRIS
Mount Stromlo & Siding Spring Observatories,
The Australian National University,
Private Bag, Woden P.O., ACT 2611, Australia
email: jen@mso.anu.edu.au

1. Introduction

One of the major achievements of astrophysics has been the demonstration that most of the chemical elements have been synthesized in stars: nucleosynthesis calculations of homogeneous and inhomogeneous big bang cosmologies show that, in comparison with the most metal-poor stars currently known, essentially no elements heavier than B existed at the era of decoupling (see e.g. Wagoner, Fowler, & Hoyle 1967; Kajino, Mathews, & Fuller 1990). Following the pioneering work on stellar nucleosynthesis by Hoyle (1946), the basic precepts and the role of stars was set down in the classic papers of Burbidge et al. (1957) and Cameron (1957), and the ensuing decades have produced a vast body of theoretical and observational effort to more fully understand the details of the process.

Extensive reviews of the subject exist in the literature. Some, which the reader may find useful, are those of Arnett (1996), Trimble (1975), Truran (1984), and Wheeler, Sneden, & Truran (1989).

The nature of the first heavy-element-producing objects remains the subject of conjecture. Authors such as Couchman & Rees (1986) argue that the first zero-heavy-element-abundance stars formed as the result of cooling via the hydrogen molecule in clusters of mass $\sim 10^5$ $M_{\odot}$. While the mass function of this first stellar generation remains elusive its evolution will have two important effects. The first supernovae would halt any ongoing collapse of material to form further stars, and more important for the present discussion, the ejecta from these objects would chemically enhance the remaining gaseous material: Couchman & Rees advocate possible enrichment up to Population II values (i.e. $Z/Z_{\odot} \sim 0.01$).

It is interesting then to note that the low column density Lyα clouds inferred from the spectra of quasars appear to have abundances near this

T.R. Bedding et al. (eds.),
Fundamental Stellar Properties: The Interaction between Observation and Theory, 407–416.

value (Cowie et al. 1995), and that at redshifts of ~ 3 a similar result is found for the damped Lyα systems (Pettini et al. 1994). In the Galaxy one can look at things a little more closely: for the halo globular clusters $\langle$[Fe/H]$\rangle = -1.6$, with the lowest values found at ~ -2.3 (Zinn 1985). The halo stars yield more detail because of their greater number: to first approximation their abundance distribution is similar to that of the simple closed-box model of chemical enrichment, with normal stars of lowest abundance having [Fe/H] $= -4.0$ (Ryan & Norris 1991; McWilliam et al. 1995; Ryan, Norris, & Beers 1996). At the other abundance extreme, the most metal-rich material in the Galaxy for which high quality abundances exist occurs in the bulge, with [Fe/H] ~ 0.5 (McWilliam & Rich 1994).

A large body of observational material exists to test and guide theoretical studies. The aim here is to present a comparison between the results of the two approaches. Following a brief section on each, comparison will be made between observed and predicted relative abundances, [X/Fe], as a function of [Fe/H]. Broad general agreement is found, but important exceptions exist, and an attempt will be made to highlight these differences, with conjectures as to how the situation might be improved.

2. Theory

The production of the elements during stellar evolution has been the subject of intense theoretical study during the past five decades, and the reader is referred to Arnett (1996), and references therein, for a thorough discussion of the topic. Suffice it here to say that following relatively quiescent evolution stars possess a series of concentric regions containing the ashes of various phases of nuclear burning. Major uncertainties remain concerning the manner in which material is undoubtedly mixed within stellar interiors over and above the predictions of standard theories and in which it is lost by winds during the relatively quiescent stages. The expulsion and burning of material in the final explosive phases of the more massive stars remain the subject of great activity.

The picture has emerged of a large fraction of the elements having been produced in massive stars, in the range $10 < M/M_{\odot} < 30$. In particular, these relatively short-lived objects have produced some C, most of the O and elements in the range Ne–Ca, the iron-peak elements, and some of the heavier neutron-capture elements via the r-process. Large uncertainties remain, however, in the modelling. The recent supernova simulations, for example, of Woosley & Weaver (1995) show a strong sensitivity to the assumed energy release: for $M = 25\ M_{\odot}$, $Z = 0$ higher energy explosions result in iron-peak element production with a neutron star remnant, while less energetic ones produce no iron and result in a black hole. In the latter case more C is produced than O, in contradistinction to what normally happens. The complexity of the situation suggests that advances will only be made by a strong interaction between theory and observation, as has been emphasized by Arnett (1996).

Less massive objects, in the range $1 < M/M_{\odot} < 10$, are believed to be responsible for much of the C and N, and for most of the heavy neutron-capture elements via the s-process. Uncertainties remain: the C/N ratio depends critically of the treatment of convection (see Renzini & Voli 1981), and the neutron source for the production of the heaviest elements, once believed to be the ^{22}Ne(α,n)^{25}Mg reaction in stars of mass 4–8 $M_{\odot}$ (Iben & Truran 1978) is now thought, following the revision of nuclear cross sections, to be due to ^{13}C(α,n)^{16}O in stars of lower mass (Busso et al. 1988).

Type Ia supernovae, believed to involve the deflagration of accreting white dwarfs in binary systems, are invoked to produce large amounts of Fe at later times (see e.g. Nomoto et al. 1984).

In is important to note, however, that in spite of the large uncertainties a judicious choice of the free parameters can lead to a good reproduction of observed abundance patterns. Timmes, Woosley, & Weaver (1995; their Figure 5), for example, present theoretical results for solar neighborhood material which reproduce the observed relative abundances for elements H–Co to within a factor 2. Given that the absolute abundances range over a factor of 8 dex, this is no mean achievement.

3. Observation

Improvements in the theory of stellar atmospheres and in detector technology in the past two decades have resulted in observed abundances of very high internal precision for relatively large numbers of stars. Two examples of this are the works of Edvardsson et al. (1993) and McWilliam et al. (1995), where internal precisions of $\sim$ 0.05–0.15 dex are claimed, and indicated by the internal consistency of the material presented.

Here too, however, the treatment of convection in the models is the subject of concern, as had been emphasized by Kurucz in this volume. At its simplest this may lead to small systematic differences when using model atmospheres from different sources (see e.g. Ryan et al. [1996] for an example of differences of $\sim$ 0.10 dex), while basic inadequacies in the current treatments may compromise all current results. To the present author, at least, the internal consistency of abundances reported by Boesgaard & Friel (1990) for the dwarfs in the Hyades and Pleiades, and in the case of the former their similarity to values given for its giants by Lambert & Ries (1981) suggest that the current techniques lead to reasonable results, at least at the 0.1 dex level.

4. Comparison between Observation and Theory

A detailed comparison for all elements lies outside the limitations of the present discussion. In what follows attention will be restricted to CNO, the α elements Mg, Si, Ca, and Ti, the odd-numbered light elements Na and Al, the iron-peak elements, and the heavy neutron-capture elements, which are most amenable to observation and representative of the problem.

4.1. CARBON, NITROGEN, AND OXYGEN

The agreement for the CNO group as a function of metallicity is a mixed bag. A useful comparison of theory with observation is presented by Timmes et al. (1995; Figures 11, 13, and 14). In the view of the author the agreement is good for O, needs improvement for C, and is very poor for N.

For O the more recent data of King (1994) together with the results of Timmes et al. show excellent agreement with the prediction of $[O/Fe] \sim 0.4$ for $[Fe/H] < -1.0$ from Type II supernovae. The data of Edvardsson et al. (1993) show a well-defined decrease in [O/Fe] as [Fe/H] increases from -1.0 to solar, as expected at later times following enrichment by the O-poor, Fe-rich ejecta of deflagration supernovae.

The situation for C is not so clear. Timmes et al. predict that $[C/Fe] \sim 0.0$ for $-3.0 < [Fe/H] < 0.0$. In comparison Wheeler et al. (1989) noted that over the range $-2.5 < [Fe/H] < -1.5$ [C/Fe] appeared to rise slightly towards lower abundance, and the results of McWilliam et al. (1995) and Norris, Ryan, & Beers (1997) are suggestive of the effect continuing at even lower abundance. (Note that internal mixing effects such as the C depletions seen in globular cluster stars act in the opposite direction.) If C overabundances at lowest abundances are confirmed by future observation, they will provide a challenge to the model builders to determine whether the problem lies with the theoretical yields of C in the 10–30 $M_{\odot}$ objects or a more exotic source must be sought for this element.

The comparison of Timmes et al. for N shows quite clearly that, observationally, it is produced as a primary element in the most metal-poor stars, with the data suggesting $[N/Fe] \sim 0.0$ for $-3.0 < [Fe/H] < 0.0$, while the models quite starkly insist that it is secondary. This provides perhaps the clearest demonstration that at high mass, as has been evident for some time for low-mass objects, existing theory is quite unable to mix N produced in stellar interiors into more exterior regions. Several mechanisms suggest themselves to overcome the problem. Timmes et al. produce primary N by increasing convective overshoot in their models; Langer, in this volume, suggests that rotation may lead to greater production of N; and a more exotic solution might be the 500 $M_{\odot}$, $Z = 0$ hypernovae of Woosley & Weaver (1982), which produce copious amounts of N.

4.2. THE α ELEMENTS

A comparison between observation and theory for the four α elements Mg, Si, Ca, and Ti is shown in Figure 1, which has been taken from Ryan et al. (1996). On the left are the observations, where different symbols show the results of different workers. On the right the data are presented again as small symbols. Here the thin lines are robust estimators of the abundance trends, with the central one being the running mid-mean. The theoretical results for the Galactic enrichment model of Timmes et al.[1] (1995) are shown by thick dashed lines, where the lower refers to their basic calculations and the upper to their preferred result when the Fe from the model supernovae is arbitrarily reduced by a factor 2. With the possible exception of Ti there is good agreement between observation and theory.

The data of Edvardsson et al. (1993) for these elements at [Fe/H] > -1.0 show very clearly the transition from the enhanced Population II values to solar as the result of Type Ia supernova production of Fe-enriched, α-element-poor material at later times. Somewhat puzzling, however, are the results of McWilliam & Rich (1994) for giants in the Galactic bulge. Their stars have [Fe/H] > -1.0, but different values of [Mg/Fe] and [Ti/Fe], on the one hand, and [Si/Fe] and [Ca/Fe], on the other. As noted by them, this is difficult to understand from a theoretical point of view. More work, both theoretical and observational (since the result is based on a relatively small number of faint objects), is needed to resolve this important issue.

4.3. AL, NA

The odd-atomic numbered elements Al and Na are discussed here to highlight a couple of problems. Theory predicts that they will be underproduced relative to neighboring even-numbered elements in low abundance environments, and this appears to be borne out by observation, at least in principle if not in degree. In particular Ryan et al. (1996) show that for [Fe/H] < -2.0 [Al/Fe] ~ -0.8, while Timmes et al. (1995) predict a value in the range -0.4 to 0.0. As Ryan et al. note, however, their abundances were determined assuming LTE, while the preliminary work of Baumueller and Gehren shows that large non-LTE effects, sufficient to explain the observed difference, become important at these low abundances.

As discussed by Da Costa in this volume, Al and Na, together with Mg (see also the paper by Shetrone herein), show very peculiar abundances in globular clusters, not predicted by standard stellar evolution calculations,

[1] As emphasized by Gibson (1997) different workers sometimes produce different yields for the same input model parameters. While comparison is given here with Timmes et al., one should bear in mind that comparison with other workers might give somewhat different results.

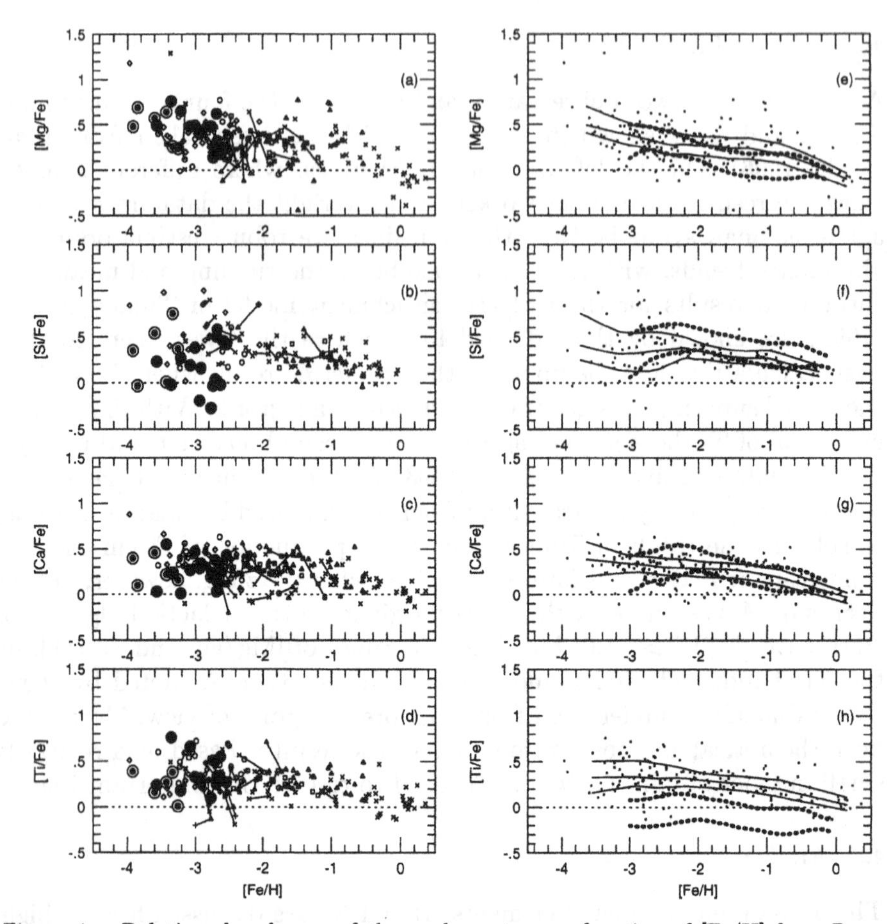

Figure 1. Relative abundances of the α elements as a function of [Fe/H] from Ryan et al. (1996). On the left different symbols present results from different workers, while on the right the data are again presented as small symbols. The thin lines represent an averaging of the data, while the dashed lines come from the Galactic enrichment model of Timmes et al. (1995). See text for discussion.

but thought nevertheless to originate from (p,γ) reactions in or near the hydrogen burning shells of clusters subgiants. The report by Briley et al. (1996), however, of Na abundance variations at the main sequence turnoff of 47 Tuc offers a challenge to this hypothesis and, unless the variations existed in the cluster at the formation of the Na-enhanced stars, suggests an important confrontation between theory and observation.

4.4. THE IRON-PEAK ELEMENTS

Figure 2 presents a comparison between observation and theory for the abundances of the iron-peak elements as a function of [Fe/H]. This demon-

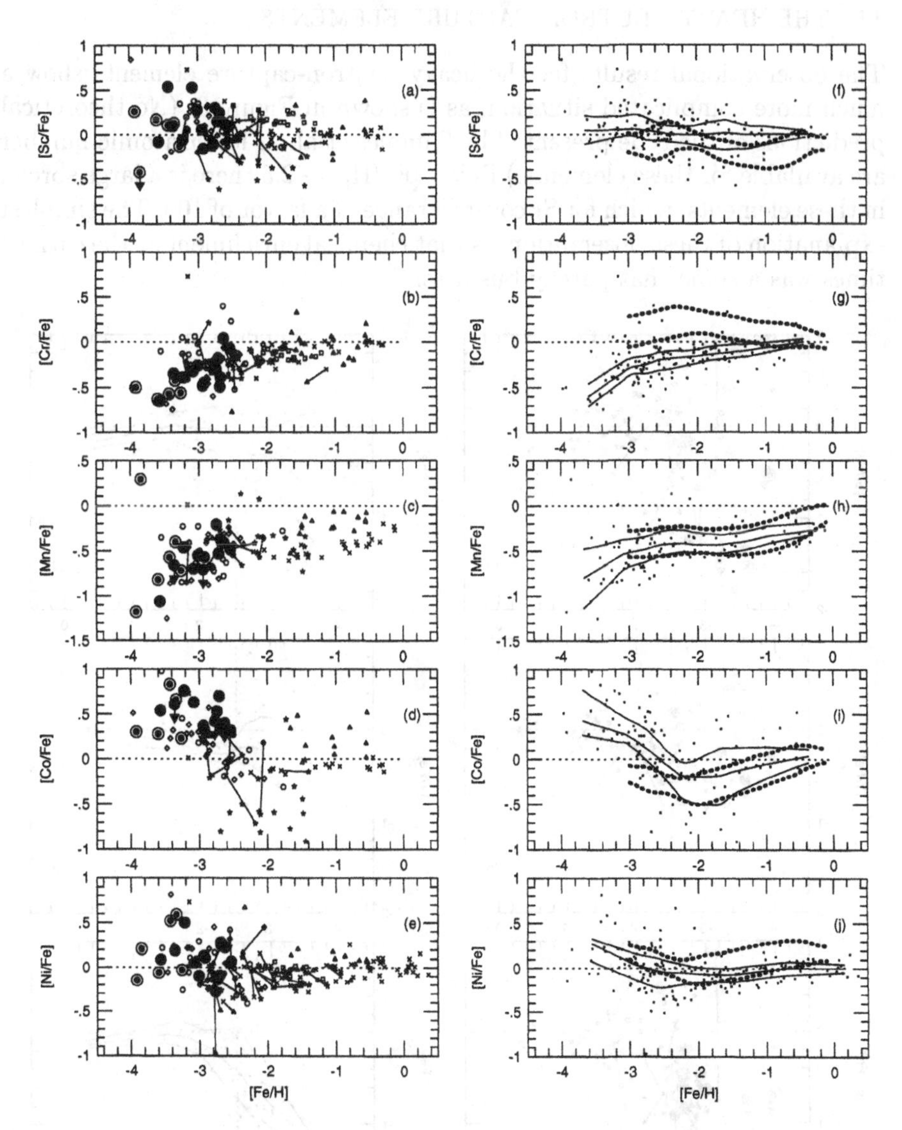

Figure 2. Relative abundances of the iron-peak elements as a function of [Fe/H] from Ryan et al. (1996). The presentation is the same as described for Figure 1.

strates the discovery of McWilliam et al. (1995) that Cr and Mn, which lie below Fe in the Periodic Table, are underproduced relative to solar values at [Fe/H] < –3.0, while Co, lying above Fe, is overproduced. These effects are not predicted by standard supernova enrichment models, and provide an important challenge for future work.

4.5. THE HEAVY NEUTRON-CAPTURE ELEMENTS

The observational results for the heavy neutron-capture elements show a much more complicated situation, as is shown in Figure 3. (No theoretical predictions of the type presented by Timmes et al. at lower atomic number are available for these elements.) Below [Fe/H] ~ -2.5 there is a large spread in these elements, which for Sr covers a range of a factor of 100. The simplest explanation of these observations is that chemical enrichment at the earliest times was a somewhat patchy business.

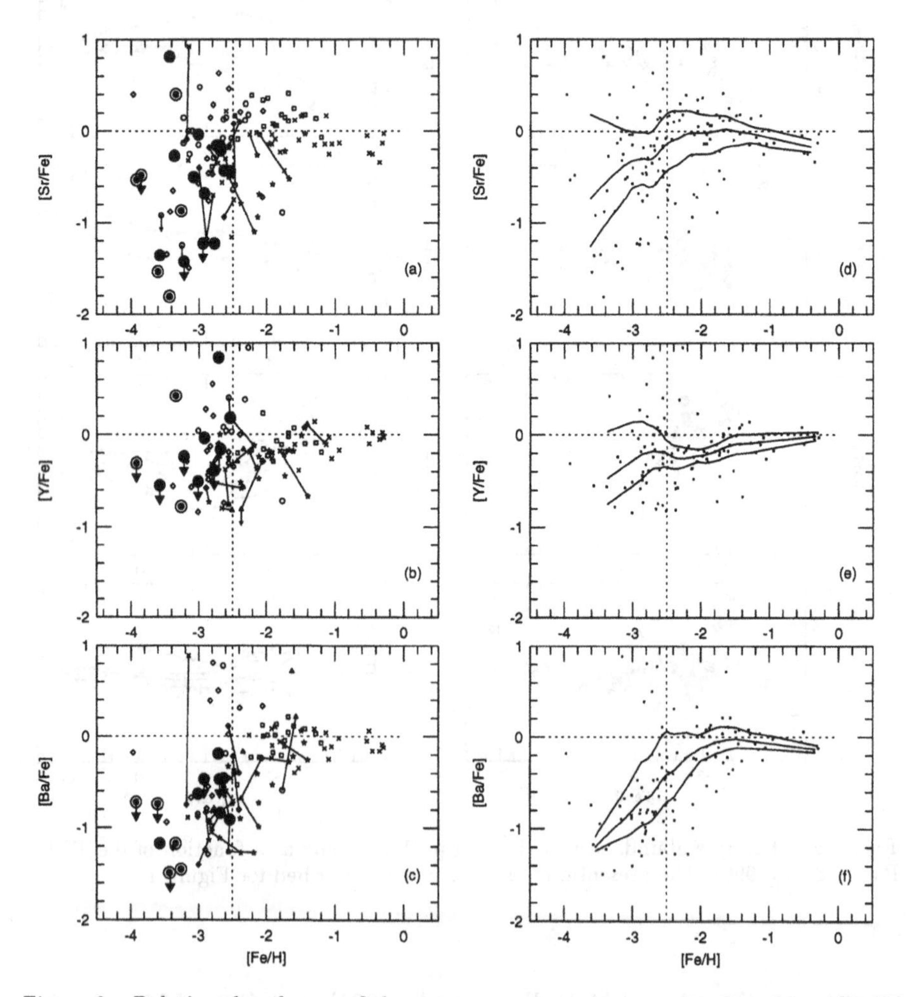

Figure 3. Relative abundances of the neutron-capture elements as a function of [Fe/H] from Ryan et al. (1996). The presentation is similar to that described for Figure 1.

One of the most important aspects of the abundance patterns of the heavy elements is that while at high abundance stars such as the Ba II giants clearly possess the signature of the s-process and identify $^{13}C(\alpha,n)^{16}O$ as the neutron source (see e.g. Tomkin & Lambert 1979, 1983), one finds at [Fe/H] < –2.0 the distribution is an r-process one (see e.g. McWilliam et al. 1995). The former receives its explanation in production of the heaviest elements in low- to intermediate-mass stars, while the latter presumably results from production in high-mass ones at the earliest times, perhaps in the formation of neutron stars as described by Woosley et al. (1994).

Of particular interest is the work of Sneden et al. (1994, 1996) on the metal-poor star CS 22892–052 ([Fe/H] = –3.0), which displays relative overabundances of the neutron-capture elements by $\sim$ 1 dex, in an r-process pattern. The overabundances are so large that the Th/Eu ratio may be determined, leading to an age estimate (and thus of the Galaxy) of 17 $\pm$ 4 Gyr (Cowan et al. 1996). Another important aspect is that it has [C/Fe] = 1.0. As far as the author is aware none of the current concepts of stellar nucleosynthesis simultaneously produces both C- and r-process enrichment (see Norris et al. 1997 for further discussion of this point).

5. Summary

Theoretical calculations of stellar nucleosynthesis provide a basic understanding of most of the chemical abundance trends observed in stars. Important inadequacies, however, remain. Of these:

- Some are no doubt associated with problems inherent in the difficulty of modelling supernova explosions (e.g. absolute yields, Cr, Mn, Co, and r-process heavy element production at low [Fe/H]).
- Others are suggestive of mixing phenomena (some driven perhaps by rotation?) not included or treated only approximately in the models (e.g. N at earliest times, CNO, and AlMgNa in globular clusters).
- Some may indicate basic problems in the modelling (e.g. C-rich/ r-process rich metal-poor stars such as CS 22892–052, Na variations on the main sequences of globular clusters, α elements in bulge giants).
- Some indicate a need for better model atmosphere abundance determination (e.g. non-LTE effects for Al in metal-weak stars).
- Some indicate a need for more and better observational data (e.g. α elements in bulge giants).

References

Arnett, D. 1996, Supernovae and Nucleosynthesis (Princeton : Princeton University Press)
Boesgaard, A. M. & Friel, E. D. 1990, ApJ, 351, 467
Briley, M. M., Smith, V. V., Suntzeff, N. B., Lambert, D. L., Bell, R. A., & Hesser, J. E. 1996, Nature, 383, 604
Burbidge, E. M., Burbidge, G. R., Fowler, W. A., & Hoyle, F. 1957, Rev.Mod.Phys., 29, 547

Busso, M., Picchio, G., Gallino, R., & Chieffi, A. 1988, ApJ, 326, 196
Cameron, A. G. W. 1957, PASP, 69, 201
Couchman, H. M. P. & Rees, M. J. 1986, MNRAS, 221, 53
Cowan, J. J., McWilliam, A., Sneden, C., & Burris, D. L. 1996, preprint
Cowie, L. L., Songaila, A., Kim, T. -S., & Hu, E. M. 1995, AJ, 109, 1522
Edvardsson, B., Andersen, J., Gustafsson, B., Lambert, D. L., Nissen, P. E., & Tomkin, J. 1993, A&A, 275, 101
Gibson, B. K. 1997, MNRAS, submitted
Hoyle, F. 1946, MNRAS, 106, 343
Iben, I. Jr. & Truran, J. W. 1978, ApJ, 220, 980
Kajino, T., Mathews, G. J., & Fuller, G. M. 1990, ApJ, 364, 7
King, J. 1994, AJ, 107, 350
Lambert, D. L. & Ries, L. M. 1981, ApJ, 248, 228
McWilliam, A., Preston, G. W., Sneden, C., & Searle, L. 1995, AJ, 109, 2757
McWilliam, A. & Rich, R. M. 1994, ApJS, 91, 749
Nomoto, K., Thielemann, F. -K., & Yokoi, K. 1984, ApJ, 286, 644
Norris, J. E., Ryan, S. G., & Beers, T. C. 1997, ApJ, submitted
Pettini, M., Smith, L. J., Hunstead, R. W., & King, D. L. 1994, ApJ, 426, 79
Renzini, A. & Voli, M. 1981, A&A, 94, 175
Ryan, S. G. & Norris, J. E. 1991, AJ, 101, 1865
Ryan, S. G., Norris, J. E., & Beers, T. C. 1996, ApJ, 471, 254
Sneden, C., McWilliam, A., Preston, G. W., Cowan, J. J., Burris, D. L., & Armosky, B. J. 1996, ApJ, 467, 819
Sneden, C., Preston, G. W., McWilliam, A., & Searle, L. 1994, ApJ, 431, L27
Tomkin, J. & Lambert, D. L. 1979, ApJ, 227, 209
Tomkin, J. & Lambert, D. L. 1983, ApJ, 273, 722
Timmes, F. X., Woosley, S. E., & Weaver, T. A. 1995, ApJS, 98, 617
Trimble, V. 1975, Rev.Mod.Phys., 47, 877
Truran, J. W. 1984, ARN&PS, 34, 53
Wagoner, R. V., Fowler, W. A., & Hoyle, F. 1967, ApJ, 148, 3
Wheeler, J. C., Sneden, C., & Truran, J. W. Jr. 1989, ARA&A, 27, 279
Woosley, S. E. & Weaver, T. A. 1982, in Supernovae: A Survey of Current Research, eds. M.J. Rees & R.J. Stoneham, (Dordrecht : Reidel), 79
Woosley, S. E. & Weaver, T. A. 1995, ApJS, 101, 181
Woosley, S. E., Wilson, J. R., Mathews, G. J., Hoffman, R. D., & Meyer, B. S. 1994, ApJ, 433, 229
Zinn, R. 1985, ApJ, 293, 424

Discussion

S. Balachandran: I thought your C results at low metallicities very interesting. Can you tell me which C transition you used?

Norris: At lowest abundance only the G band is useful.

N. Langer: The scatter in [r/Fe] you find at very low metallicity is really amazing. Can you disentangle whether mainly the scatter in the r-process abundances or in iron is responsible; i.e. if you would plot [r/O] vs. [O/H], and [Fe/O] vs. [O/H], which curve would have the strong scatter.

Norris: Unfortunately O abundances are not available at lowest abundance because of the weakness of the features. If, however, one were to use the α elements as proxy the scatter would be in the [r/α] vs. [α/H] plane rather than in the [Fe/α] vs. [α/H] one.

CHEMICAL EVOLUTION OF GALAXIES — CHALLENGES FOR STELLAR ASTRONOMY

ELAINE M. SADLER
School of Physics, University of Sydney
NSW 2006, Australia

Abstract. I discuss several current problems in understanding the chemical evolution of galaxies which have particular relevance for fundamental stellar astronomy.

1. Introduction

Stellar astronomy is the foundation on which we build our understanding of the evolution of galaxies and their stellar populations. A new generation of ground–based and space telescopes are now providing data of unprecedented quality for distant galaxies, so that it is becoming possible to investigate the stellar populations of these galaxies from spectroscopic studies of their integrated light (e.g. O'Connell 1995; Steidel *et al.* 1996).

Here, I will discuss three current problems in the chemical evolution of galaxies which relate closely to fundamental stellar astronomy. They are:

- Hot, UV–bright stars in old, metal–rich populations
- α–element abundances, Mg/Fe and the "iron discrepancy "
- Dating elliptical galaxies, and the age–metallicity relation

There is a unifying theme to this discussion, which is reflected in the assertion below.

The Galactic bulge is the 'new frontier' for stellar astronomy. It is being opened up by a new generation of large telescopes such as Keck and the ESO VLT, which will allow old, metal–rich stars to be studied in the kind of detail previously reserved for solar–neighbourhood disk stars. Such studies will also help improve our understanding of the evolution of distant galaxies.

T.R. Bedding et al. (eds.),
Fundamental Stellar Properties: The Interaction between Observation and Theory, 417–426.

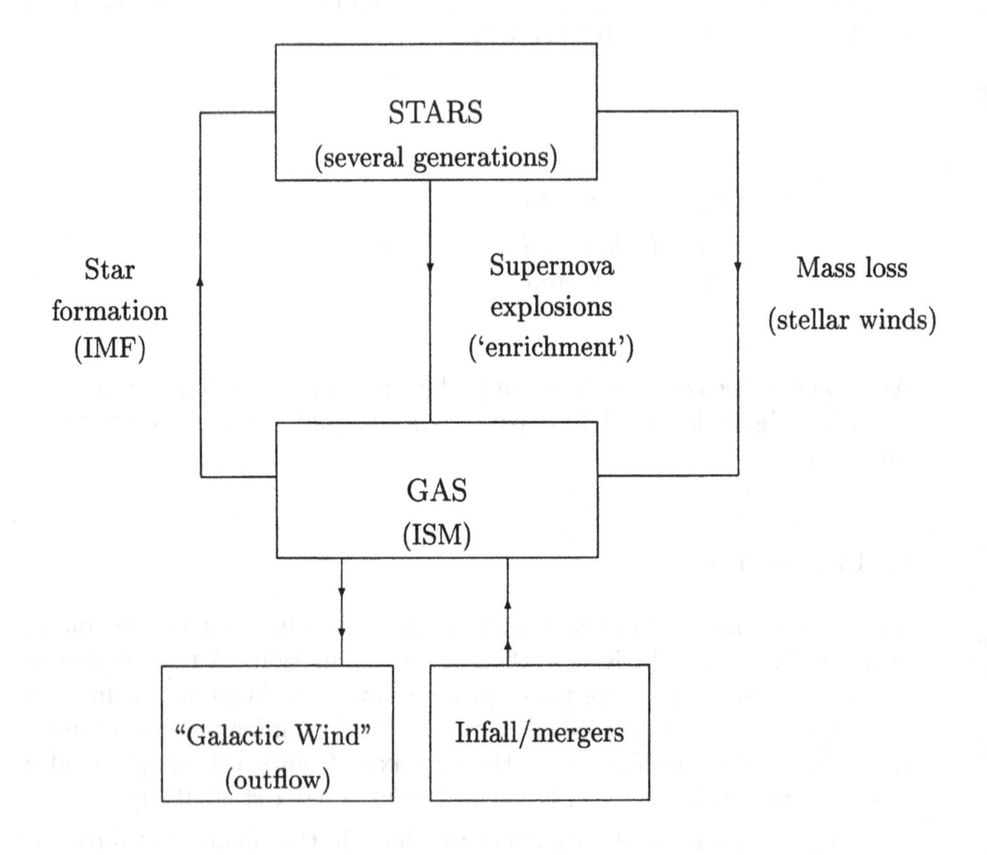

Figure 1. Chemical evolution of a galaxy

2. The chemical evolution of a galaxy

We can think of a galaxy as an ecosystem whose main components are stars and gas. As shown schematically in Figure 1, stars can lose mass to the interstellar medium (ISM) through mass loss in a stellar wind or, more spectacularly, in a supernova explosion. In either case, the material lost will chemically enrich the ISM. Gas in the ISM can in turn form new stars when conditions are suitable. If this is all that happens, then the galaxy is a closed system. Many galaxies may not be closed systems, however — gas may be lost in an outflow (or 'galactic wind') if star formation becomes too vigorous; and new gas (and even stars) can be added by the infall or merger of smaller galaxies.

Models for the chemical evolution of galaxies (e.g. Arimoto & Yoshii 1987; Matteucci 1992; Tantalo *et al.* 1996) attempt to reproduce the observed properties of nearby galaxies by simulating the collapse of a galaxy made of two components (luminous and dark matter) and allowing star formation to proceed in a well–determined way with consequent chemical enrichment of the ISM. They include the effects of galactic winds powered by supernova explosions and stellar winds from massive stars. Such models have been successful in reproducing many of the observed photometric and chemical properties of nearby galaxies, and suggest that the onset of a galactic wind is a key physical process in determining the chemical evolution of the most massive galaxies.

2.1. STELLAR POPULATIONS ALONG THE HUBBLE SEQUENCE

The outward appearance (Hubble type) of a galaxy depends on many factors, including kinematics, galaxy size, the relative amounts of gas and stars, the current star formation rate and the star formation history. In general, however, as we move along the Hubble sequence in the direction E–S0–Sa–Sb–Sc–Sd–Irr we are moving from systems which have already turned most of their available gas into stars to systems which still have a relatively high gas fraction and active star formation. Hence we might expect that galaxies of similar Hubble type have also had a similar star formation and chemical enrichment history.

TABLE 1. Stellar populations along the Hubble sequence

Class	Dominant population	Closest Templates
Dwarf	Varies/mixed	LMC, SMC
		Local Group dwarfs
Spiral	Halo: old, metal–poor	Galaxy, M 31
	Disk: mixed, young	
Elliptical	Old, metal–rich	Galactic bulge?

The problems I have chosen to discuss here relate mainly to giant elliptical galaxies. This is not because I think that giant ellipticals are the most interesting galaxies (although I do); but because disentangling the history of their chemical evolution is more difficult for these galaxies than for spiral and dwarf galaxies, where some members of the class are sufficiently nearby that we can resolve them into individual stars.

The only giant elliptical galaxy in which individual stars have been resolved is NGC 5128, for which Soria *et al.* (1996) were able to detect the tip of the halo red giant branch with HST. O'Connell (1986) points out that HST easily resolves the brightest members of young stellar populations at the distance of the Virgo cluster, but that the brightest members of an old population with [Fe/H]>0 cannot be resolved outside the Local Group.

We are left with two possibilities. Either we try to disentangle the chemical evolution history of elliptical galaxies from their integrated light alone (either through spectra or broad–band colours), or we seek additional help from the only old and metal–rich population in which we CAN observe individual members — the stars of the Galactic bulge.

2.2. MEASURING LINE–STRENGTH INDICES

There are two problems in trying to disentangle the star formation history of a galaxy from the spectrum of its integrated light. Not only do we see the superposition of many spectra; but the velocity dispersion of the stars (due to their individual motions within the galaxy potential) limits the spectral resolution as well. For a giant elliptical galaxy, the internal velocity dispersion can be up to 300 km/s (corresponding to line widths of several Å). Thus weak lines are blended or smeared out, and only strong lines can be measured. Since the Mg lines near 5200Å are among the strongest absorption lines, they have long been used as an abundance indicator for elliptical galaxies (e.g. Mould 1978; Gulati *et al.* 1991; Buzzoni *et al.* 1992; Barbuy 1994; Chavez *et al.* 1995).

The Lick line–strength indices (Faber *et al.* 1985) provide an easy way to quantify the absorption–line spectra of elliptical galaxies. The most commonly–used indices are Mg_2 (covering the Mg*b* line and MgH molecular band), <Fe> (the mean of two Fe absorption lines) and Hβ. Much effort has gone in to providing empirical abundance calibrations for the Lick indices, both for single stars and for composite populations (Burstein *et al.* 1984; Faber *et al.* 1985; Gorgas *et al.* 1993; Worthey *et al.* 1994). These measurements give us a broad, qualitative picture of the evolution of giant elliptical galaxies, as outlined below. Nevertheless, there remain some problems which will be discussed later.

2.3. THE FORMATION OF ELLIPTICAL GALAXIES

The first thing we can deduce from the integrated spectra of elliptical galaxies is that K–giant stars dominate the light, and hence that the dominant stellar population is old. Since there appears to have been little or no recent star formation in these galaxies, it is plausible that the abundance distribution (and especially the radial abundance gradient) provides a fossil record of the galaxy's formation.

Models in which elliptical galaxies formed by the dissipative collapse of a protogalaxy (Larson 1976; Carlberg 1985) predict that these galaxies should have steep radial abundance gradients (i.e. a decrease of 0.5 to 1.0 dex in [Fe/H] over a decade in radius). Although abundance gradients are observed, however, (Faber 1977; Gorgas *et al.* 1990; Davies *et al.* 1993; Carollo & Danziger 1994), they are much shallower (typically 0.2 dex per decade in radius) and uncorrelated with galaxy luminosity. This strongly favours mergers rather than dissipative collapse as the dominant formation mechanism for elliptical galaxies (White 1980), and the current view is that elliptical galaxies were 'assembled' from smaller subcomponents.

2.4. POPULATION SYNTHESIS MODELS FOR THE INTEGRATED LIGHT

O'Connell (1986) discusses two possible approaches to the spectral synthesis of an integrated galaxy spectrum. In both cases, the aim is to combine individual spectra (or spectral energy distributions) to produce a model composite. Recently, most workers have used the "evolutionary population synthesis" approach (Tinsley 1978; Bruzual & Charlot 1993), in which they predict the spectral energy distribution (SED) resulting from a single–burst population and evolve this with time. A composite population can then be modelled by adding several single–burst models of different ages. The main variables are the stellar IMF and star formation rate, and it is clearly important to use a stellar library which contains all the stellar evolutionary stages which are present in the galaxy being modelled.

Charlot *et al.* (1996) compare several evolutionary synthesis models for elliptical galaxies, and discuss some of the limitations associated with them. The most serious of these are the treatment of stars in post–main sequence evolutionary stages; the poorly–determined temperature scale for the red giant branch and cool stars; and the lack of complete and accurate libraries of stellar spectra, especially for cool stars and non-solar metallicities. These problems relate directly to the topics discussed at this meeting.

2.5. THE GALACTIC BULGE AS A TEMPLATE FOR ELLIPTICAL GALAXIES

The Galactic bulge comprises the inner 1–2 kpc of our Galaxy. Its light is dominated by old, metal–rich stars (Arp 1965) with a wide abundance spread (Rich 1988). The mean abundance is slightly below solar (McWilliam & Rich 1994; Sadler *et al.* 1996), with some stars possibly reaching 2–3 times solar abundance. Whitford (1978) first suggested that the stars of the Galactic bulge could be valuable templates for understanding the stellar populations of elliptical galaxies. The main arguments are that most of the bulge stars are old (>10 Gyr; Terndrup 1988), and that the abundances of the most metal–rich ones overlap the metallicity range deduced for elliptical galaxies.

The high (and patchy) reddening towards the Galactic bulge, along with the faintness of even the brightest bulge giants (V = 15–17 mag.) and the problems of disk–star contamination and crowding make high–resolution spectroscopy difficult with 4 m class telescopes. In the near future, however, the new generation of 8–10 m telescopes will make it possible to carry out abundance analyses for large samples for metal–rich bulge giants (see Castro *et al.* 1996 for an early Keck spectrum of the bulge K–giant BW IV–167). These studies should both clarify the chemical evolution history of the bulge itself and provide improved stellar libraries for synthesis modelling of more distant galaxies.

3. Problem 1: The UV upturn in elliptical galaxies

Elliptical galaxies and the bulges of spirals show an upturn in their UV spectra at λ <2500Å, and their observed far–UV flux well exceeds that expected from the old stellar population which dominates the visible and IR light (e.g. Code & Welch 1979; Burstein *et al.* 1988). In elliptical galaxies, the UV upturn increases with metallicity (Burstein *et al.* 1988), which cannot be explained by traditional stellar population models because increasing temperature ought to shift the red giant branch and asymptotic giant branch to cooler temperatures. The most plausible explanation appears to be that old, metal–rich stars can go through brief hot phases in their evolution (Greggio & Renzini 1990).

Stellar evolution theory suggests several possible hot phases for old stars, including extreme horizontal branch (EHB) stars, AGB–manqué stars (or 'failed' AGB stars), and post–AGB (PAGB) stars. All of these are likely to be rare, because they represent short evolutionary phases, and only a few have been observed in globular clusters. The Galactic bulge promises to be a valuable laboratory for their study because it has so many stars that even short–lived phases can be observed in abundance. The main problem is to sift through large numbers of hot stars and separate the bulge stars from those in the foreground disk, so multi–object spectrographs (like the 2dF on the Anglo–Australian Telescope) are ideal for this work.

Figure 2 shows the expected location of the various hot–star candidates in the Galactic bulge CMD (RGB marks the bulge giant branch). Bertelli *et al.* (1996) have recently reported the likely signature of hot horizontal–branch stars in a bulge field near the Galactic Centre.

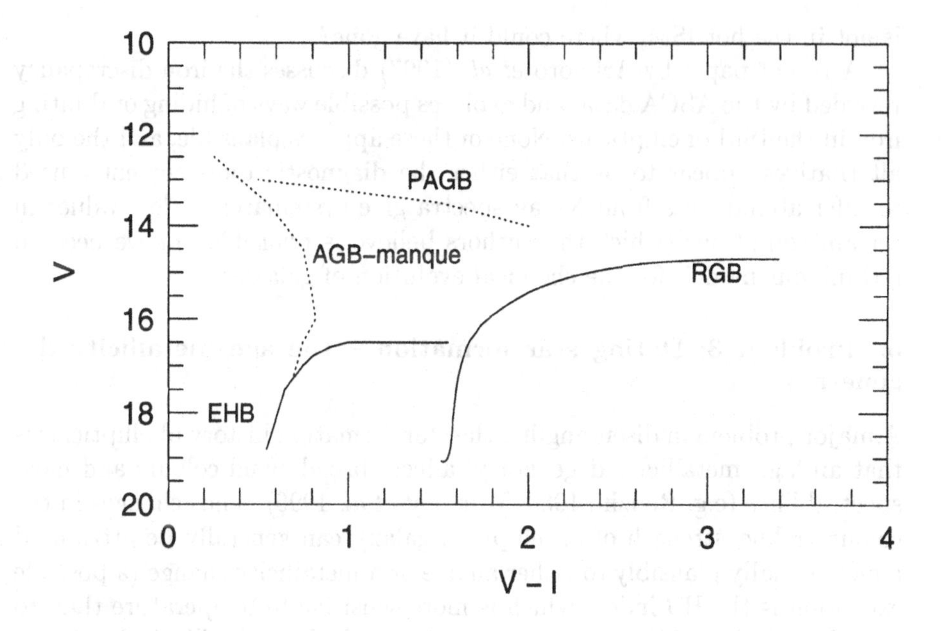

Figure 2. Schematic CMD for candidate UV–bright stars in the Galactic bulge

4. Problem 2: Enhanced Mg/Fe in ellipticals, and the fate of iron from SNeI

From the relative behaviour of the Mg_2 and <Fe> indices, Worthey *et al.* (1992) suggest that magnesium is overabundant relative to iron in elliptical galaxies, with [Mg/Fe] = +0.2 to 0.3 (McWilliam & Rich (1994) derive a similar enhancement for stars in the Galactic bulge). Enhanced Mg relative to Fe could plausibly arise if most of the stars in elliptical galaxies formed very quickly (i.e. in <1 Gyr), with very little star formation thereafter. In this case, there would be rapid chemical enrichment from type II supernovae, with a corresponding enhancement of α–elements such as magnesium.

While this explains what we observe in the integrated stellar light, nucleosynthesis doesn't stop when star formation stops. As noted in Figure 1, the interstellar medium (ISM) will continue to be enriched by stellar mass loss and by Type Ia supernovae, which should contribute substantial amounts of iron. Surprisingly, though, the ASCA satellite measured iron abundances in the hot ISM of giant elliptical galaxies which are LOWER than those in the stellar galaxy (e.g. Awaki *et al.* 1994). This poses two problems: (i) the hot ISM is believed to be shed by stars, yet it has a lower iron abundance than the stars from which it arose; and (ii) if the iron from type Ia supernovae has not been recycled into new generations of stars, and is not in the hot ISM, where could it have gone?

A recent paper by Arimoto *et al.* (1997) discusses the iron discrepancy revealed by the ASCA data, and explores possible ways of hiding or diluting iron in the ISM of ellipticals. None of these appears plausible, and the only alternatives appear to be that either the diagnostic tools currently used to infer abundances from X–ray spectra give systematically low values in certain conditions (which the authors believe is plausible), or we need to rethink our models for the chemical evolution of galaxies.

5. Problem 3: Dating star formation – the age–metallicity degeneracy

A major problem in disentangling the star formation history of ellipticals is that an 'age–metallicity degeneracy' affects broad–band colours and most spectral lines (e.g. Renzini 1986; Worthey *et al.* 1996). Thus changes in the colour or line–strength of an elliptical galaxy can generally be attributed almost equally plausibly to either an age or a metallicity change (a possible exception is the Hβ index, which is more sensitive to temperature than to abundance). For this reason, the epoch at which most elliptical galaxies formed most of their stars remains uncertain. Some of the arguments for 'old' and 'young' ages are:

Arguments that the bulk of the stellar population is old (>12 Gyr), from Renzini (1995):

- Clusters of galaxies at redshifts z$\sim$1 contain bright, red galaxies whose colours are consistent with passive evolution since z>2.
- The tightness of the fundamental plane for nearby galaxies implies either a very small age dispersion or formation at z>2.
- QSOs at z>3 show strong metal enrichment, suggesting that abundances in the central regions were already high 15 Gyr ago.
- If smaller ellipticals are much younger than bigger ones, this is hard to reconcile with a picture in which (as discussed earlier) larger ellipticals are "assembled" from smaller ones.

Arguments that the mean stellar ages of E nuclei are young (3–12 Gyr), from Faber et al. (1995):

- The Hβ line can be used as a temperature (and hence age) indicator, independent of metallicity.
- In the nuclei of elliptical galaxies, Hβ is stronger than expected from single–burst models of an old population.
- The flat Hβ gradients seen in most elliptical galaxies imply that the outer parts of these galaxies are both older and more metal–poor than the central regions.

A possible reconciliation of these views is offered by Bressan *et al.* (1996), who point out that the strength of the UV upturn (1550−V colour) may be a powerful age indicator for elliptical galaxies. They suggest that all ellipticals have old (13–15 Gyr) populations, but that their star formation histories differ, with star formation continuing for much longer in some galaxies than others.

We now have good-quality spectra of galaxies and increasingly sophisticated models, but it is unfortunate that the empirical calibration of the line-strength indices for metal-rich stars still rests on a handful of solar neighbourhood stars.

6. Summary

I have given a brief introduction to a few problems in extragalactic astronomy which have direct links to the topic of this meeting. We could make progress with several of these if we understood more about the late evolution of stars of solar metallicity and above. The message I would like to deliver is a simple one: studies of the formation and evolution of galaxies would be on firmer ground if the old, metal-rich stellar population of the Galactic bulge were as well observed, and as well understood, as the stellar populations of the Galactic bulge and halo.

I thank Don Terndrup for providing the diagram used in Figure 2.

References

Arimoto, N. and Yoshii, Y., 1987. *A&A*, **173**, 23.
Arimoto, N., Matsushita, K., Ishimaru, Y., Ohashi, T. and Renzini, A., 1997. *ApJ*, **477**, 128.
Arp, H., 1965. *ApJ*, **141**, 43.
Awaki, H., *et al.*, 1994. *PASJ*, **46**, L65.
Barbuy, B., 1994. *AJ*, **430**, 218.
Bertelli, G., Bressan, A., Chiosi, C. and Ng, Y.K., 1996. *A&A*, **310**, 115.
Bressan, A., Chiosi, C. and Tantalo, R., 1996. *A&A*, **311**, 425.
Buzzoni, A., Gariboldo, G. and Mantegazza, L., 1992. *AJ*, **103**, 1814.
Bruzual, A.G. and Charlot, S., 1993. *ApJ*, **405**, 538.
Burstein, D., Faber, S.M., Gaskell, C.M. and Krumm, N., 1984. *ApJ*, **287**, 586.
Burstein, D., Bertola, F., Buson, L., Faber, S.M. and Lauer, T.R., 1988. *ApJ*, **328**, 440.
Carlberg, R.. 1984. *ApJ*, **286**, 404.
Carollo, C.M. and Danziger, I.J., 1994. *MNRAS*, **270**, 523.
Castro, S., Rich, R.M., McWilliam, A., Ho, L.C., Spinrad, H., Filippenko, A.V., and Bell, R.A., 1996. *AJ*, **111**, 2439.
Chavez, M., Malagnini, M.L. and Morossi, C., 1995. *ApJ*, **440**, 210.
Charlot, S., Worthey, G. and Bressan, A., 1996. *ApJ*, **457**, 625.
Code, A.D. and Welch, G.A., 1979. *ApJ*, **228**, 95.
Davies, R.L., Sadler, E.M. and Peletier, R., 1993. *MNRAS*, **262**, 650.
Faber, S.M., 1977. In *The Evolution of Galaxies and Stellar Populations*, Yale University Press, p. 157.

Faber, S.M., Friel, E.D., Burstein, D. and Gaskell, C.M., 1985. *ApJS*, **57**, 711.
Faber, S.M., Trager, S.C., Gonzalez, J.J. and Worthey, G., 1995. In *Stellar populations*, IAU Symposium 164, p. 249.
Gorgas, J., Efstathiou, G. and Aragon Salamanca, A., 1990. *MNRAS*, **245**, 217.
Gorgas, J., Faber, S.M., Burstein, D., Gonzalez, J.J., Courteau, S. and Prosser, C., 1993. *ApJS*, **86**, 153.
Greggio, L. and Renzini, A., 1990. *ApJ*, **364**, 35.
Gulati, R.K., Malagnini, M.L. and Morossi, C., 1991. *A&A*, **247**, 447.
Larson, R.B., 1976. *MNRAS*, **176**, 31.
Matteucci, F., 1991. *ApJ*, **397**, 32.
McWilliam, A. and Rich, R.M., 1994. *ApJS*, **91**, 749.
Mould, J.R., 1978. *ApJ*, **220**, 434.
O'Connell, R.W., 1986. In *Stellar populations*, Cambridge University Press , p. 167.
O'Connell, R.W., 1995. In *Stellar populations*, IAU Symposium 164, p. 301.
Renzini, A., 1986. In *Stellar populations*, Cambridge University Press , p. 213.
Renzini, A., 1995. In *Stellar populations*, IAU Symposium 164, p. 325.
Rich, R.M., 1988. *AJ*, **95**, 828.
Sadler, E.M., Rich, R.M. and Terndrup, D.M., 1996. *AJ*, **112**, 171.
Soria, R., *et al.*, 1996. *ApJ*, **465**, 79.
Steidel, C.C., Giavalisco, M., Pettini, M., Dickinson, M. and Adelberger, K.L., 1996. *ApJ*, **462**, L17.
Tantalo, R., Chiosi, C., Bressan, A. and Fagotto, F., 1996. *A&A*, **311**, 361.
Terndrup, D.M., 1988. *AJ*, **96**, 884.
Tinsley, B.M., 1978. *ApJ*, **222**, 14.
White, S.D.M., 1980. *MNRAS*, **191**, 1P.
Whitford, A.E., 1978. *ApJ*, **226**, 777.
Worthey, G., Faber, S.M. and Gonzalez, J.J., 1992. *ApJ*, **398**, 69.
Worthey, G., Faber, S.M., Gonzalez, J.J. and Burstein, D. 1994. *ApJS*, **94**, 687.
Worthey, G., Trager, S.C. and Faber, S.M., 1996. In *Fresh Views of Elliptical Galaxies*, ASP Conf. Series, **86**, 203.

DISCUSSION

ROGER BELL: Houdashelt and I have calculated galaxy spectra, using synthetic spectra and isochrones. We can calculate Lick-type indices from these spectra without making use of the fitting functions. Do you know if there are galaxy spectra, rather than indices, which we could compare with to study the possible [Mg/Fe] versus [Fe/H] correlation?

ELAINE SADLER: I agree that the approach you describe is an excellent one, though at present I would worry a little about the synthetic spectra for the coolest stars. The Lick indices are attractive to observers because they are robust - for example, they don't require that the galaxy spectra be flux-calibrated, so data from non-photometric nights can still be used. Nevertheless, libraries of good-quality, flux-calibrated spectra do exist for nearby galaxies (e.g. Kennicutt 1992; ApJS 79, 255).

14. STELLAR AGES

NEW MODEL ATMOSPHERE ANALYSES OF COOL WHITE DWARFS: A REVISED LUMINOSITY FUNCTION AND CONSTRAINTS ON THE AGE OF THE GALAXY

S.K. LEGGETT
UKIRT
Joint Astronomy Centre, Hilo Hawaii

P. BERGERON
Lockheed Martin Canada

AND
MARIA TERESA RUIZ
Departamento de Astronomia, Universidad de Chile

1. Abstract

We have obtained new photometric and spectroscopic data for a large sample of cool white dwarfs. These data have been analysed with state-of-the-art model atmospheres and effective temperatures and atmospheric compositions have been determined (Bergeron, Ruiz & Leggett 1997). Radii and masses have also been obtained for those stars with accurate parallax measurements. These high quality data and models allow us to produce an improved cool white dwarf luminosity function based on the Liebert, Dahn & Monet (1988) proper motion sample. The turn-over seen at the faint end of this luminosity function, combined with theoretical cooling sequences, enable us to constrain the age of the local region of the Galaxy.

2. Introduction

The majority of stars will evolve into electron-degenerate white dwarfs and then cool to invisibility. The accurate trigonometric parallaxes determined by the Naval Observatory Flagstaff Station (described earlier in these proceedings) produce a well-defined degenerate sequence in color-magnitude diagrams, with a sharp terminus (Figure 10 of Monet *et al.* 1992). White

T.R. Bedding et al. (eds.),
***Fundamental Stellar Properties: The Interaction between Observation and Theory*, 429–432.**

dwarfs are most likely to be composed of carbon/oxygen cores, with surface envelopes of helium and sometimes hydrogen. Trace metals can be present. They cool at approximately constant radius while their surface composition can evolve in ways which are currently not fully understood.

Models of the cooling scales of white dwarfs show that even stars aged 10 Gyr will be visible, implying that the observed luminosity function of white dwarfs can be used to constrain the age of the local region of the disk. Liebert, Dahn & Monet (1988, hereafter LDM) produced an observational luminosity function that was used by Wood (1992) to derive an age for the disk of 8-11 Gyr. Observational uncertainties made up nearly half of the uncertainty in this age. The LDM sample suffered from uncertainties due to the unknown chemical composition of the coolest stars in the sample, as well as defiencies in the model atmospheres available at that time. The few models that were available did a poor job of reproducing the observed flux distributions (e.g. Leggett 1989).

3. Improvements to Model Atmospheres and Observations

Since 1989 my collaborators and I have been obtaining high quality optical and infrared data for a large sample of cool white dwarfs (Bergeron, Ruiz & Leggett 1997, hereafter BRL), while significant improvements have also been made to the model atmospheres (Bergeron *et al.* 1995). The improvements to the models include improved calculations of the collision-induced H_2 opacity, improved treatment of H_3^+ (important for H^- opacity), modelling of pressure ionization of He (important for He^- opacity) and more accurate line broadening calculations for Hα. Comparison with observation now shows excellent agreement (see BRL) although there are still some deficiencies. These are: theoretical treatment of pressure ionization of hydrogen; missing sources of UV opacity for the coolest hydrogen-rich white dwarfs; and an apparent 5% discrepancy in the absolute flux calibration.

4. Derived Parameters

The observed energy distributions are compared to model fluxes and T_{eff} and the angular diameter are derived, with an initial assumption that the surface gravity $logg = 8.0$. If the parallax is known, the radius R is determined from the angular diameter. Knowing R and T_{eff} the evolutionary models of Wood (1995) are used to derive the mass M. The implied value of $g = GM/R^2$ is compared to the initial value and the process iterated until these agree. If the parallax is unknown then $logg = 8.0$ is adopted.

The uncertainties in the derived parameters are typically: 150K for T_{eff}, 0.06 $M/M_\odot$ in mass, and 0.10dex in $logg$. Varying the mass of the surface envelope used by the evolutionary models leads to Δ M $\sim$ 0.03 $M/M_\odot$ and

$\Delta logg \sim 0.02$dex. For H-rich atmospheres the photometry and Hα profiles constrain the amount of He present to $< 50\%$, which has a negligible effect on the derived parameters. For He-rich atmospheres we can constrain the amount of H present to $< 1\%$ however this has a large effect on the derived parameters: this amount of H would lower T_{eff} by 250K, and lower *logg* by 0.15dex, with an implied reduction in M of $0.10 M/M_{\odot}$.

5. Implications for White Dwarf Evolution and Disk Age

A study of the chemical composition of our sample as a function of T_{eff} (or age) shows that for $6100K > T_{eff} > 5100K$ there is a significant lack of non-DA white dwarfs (i.e. stars not showing H features). We speculate that some combination of convective mixing of the (variable depth) outer layers, together with accretion of H onto a high-pressure He atmosphere (in such a way that H transitions are quenched), could produce this result.

We also find that the cooling curves of Wood, with a C/O core and thin He envelope, imply cooling ages for our sample of < 8 Gyr. To this can be added pre-PN ages of <2 Gyr. Uncertainties due to core composition and thickness of the outer envelope are ~2 Gyr.

We are able to improve the luminosity function (LF) derived by LDM as we have determined accurate stellar parameters for the stars in that sample. The earlier cooling curves by Wood, combined with the 1988 LDM results, implied an age for the disk of 8-11 Gyr. New models by Wood (with C/O cores, thick surface layers and new opacities; Oswalt *et al.* 1996) compared to the 1988 LDM results imply a younger age of 7-9 Gyr. These models have also been compared to an observational LF based on white dwarfs in binaries (Oswalt *et al.*) to derive an age of ~9.5 Gyr. However this sample suffers from severe incompleteness and the data is insufficient to determine composition for the coolest stars. It is also difficult to obtain accurate photometry for small-separation binaries, such as the three faintest stars in their sample, on which the disk age hinges.

Figure 1 shows the LF derived for hot white dwarfs by Fleming *et al.* (1986, slightly revised by LDM), our redetermination of the LDM function, and a comparison to the latest theoretical LF by Wood. The error bars are derived by assuming that the uncertainty for each star equals the size of its contribution to the space density. We can constrain the disk age to 8 ± 0.5 Gyr. Systematic uncertainties due to core composition and mass of the outer layers amount to 1–2 Gyr. Also, as described by Chabrier in these proceedings, there is the possibility that C/O separation occurs on crystallization of the older massive stars, releasing gravitational energy, and extending cooling times by a further 1–2 Gyr.

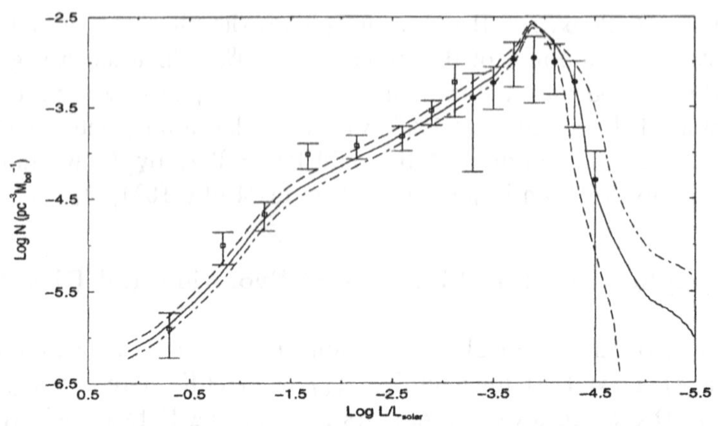

Figure 1. New luminosity function for the 8-tenths sample of LDM and theoretical curves from Wood (1996) for disk ages 7, 8 and 9 Gyr (left to right)

6. Conclusion

As the aim of this meeting is to discuss outstanding problems in stellar astrophysics, we close with a list of areas where further work is needed.

Problem	Solution?
UV opacity for $T_{eff} < 5300$K DA's	Lyman edge pseudo-continuum
Trace H in high pressure atmospheres	H occupation probability
Chemical evolution	convective mixing mechanisms
Core composition, surface layer mass	pulsation studies
Observational improvements	data, parallaxes, flux calibration

7. References

Bergeron, Saumon & Wesemael 1995, Ap.J. 443, 764
Bergeron, Ruiz & Leggett 1997, Ap.J.Suppl. in press
Fleming, Liebert & Green 1986, Ap.J. 308, 176
Leggett 1989, A&A 208, 141
Liebert, Dahn & Monet 1988, Ap.J. 332, 891
Monet *et al.* 1992, AJ 103, 638
Oswalt *et al.* 1996, Nature 382, 692
Wood 1992, Ap.J. 386, 539
Wood 1995, 9th European Workshop on White Dwarfs, NATO ASI Series, ed. Koester & Werner (Berlin: Springer), 41

Discussion of this paper appears at the end of these Proceedings.

HIPPARCOS SUBDWARFS AND GLOBULAR CLUSTER AGES: TOWARDS RELIABLE ABSOLUTE AGES

F. PONT AND M. MAYOR
Geneva Observatory, Sauverny, Switzerland

AND

C. TURON
Observatoire de Paris, Meudon, France

Abstract.

The maximum age of galactic globular clusters provides the best observational constraint on the minimum age of the Universe. One of the main "missing link" in the globular cluster age determination has been the lack of a precise calibration, with local subdwarfs, of the position of the subdwarf sequence at different [Fe/H].

Hipparcos data may change this situation. As many precise parallaxes become available for local subdwarfs, the distance to globular clusters can be estimated directly from ZAMS fitting to the subdwarf locus. The ages can then be inferred from the turnoff luminosity (a robust prediction of stellar evolution models), rather than using secondary indicators such as Horizontal-Branch position, or indicators depending on the uncertain colour scale such as turnoff colour.

Combining Hipparcos parallaxes with [Fe/H] values determined with the CORAVEL spectrometer, we are studying the position of the subdwarfs in the Colour-Magnitude Diagram from a sample of more than 900 subdwarf candidates. Preliminary results are presented here. It is shown that the distances of many subdwarfs had been underestimated in previous studies, mainly because a large fraction of them is in fact evolved off the main sequence into the turnoff or the subgiant branch.

T.R. Bedding et al. (eds.),
Fundamental Stellar Properties: The Interaction between Observation and Theory, 433–438.

1. The Hipparcos Subdwarfs

We are using parallax data for more than 900 metal-deficient stars measured by the Hipparcos astrometric satellite. This sample is the reunion of two different groups: 332 stars previously known or suspected to be subdwarfs, predominantly F stars, and about 600 stars shown by the Hipparcos parallax data to be situated well below the Main Sequence on the Colour-Magnitude Diagram (CMD).

2. The "Historical" Subdwarfs

Before the Hipparcos data, less than a dozen subdwarfs possessed well-known parallaxes. They are included in our sample, and it appears at once (Fig. 2) that most of the previous parallax estimates were inaccurate. Distances were in fact systematically underestimated, often by a large amount.

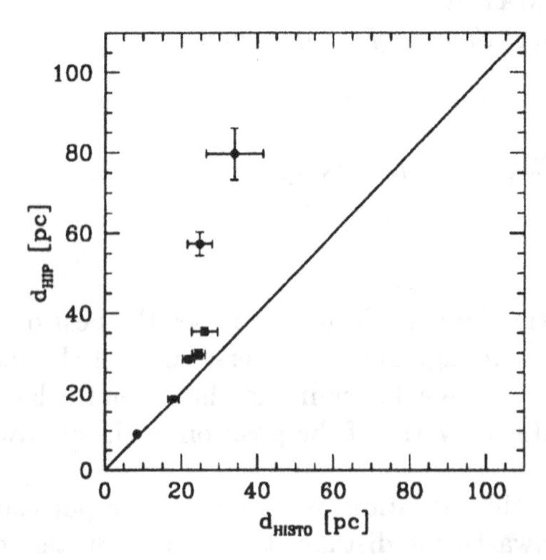

Figure 1. Distance for seven "historical" subdwarfs from the literature (Zinn&West 1981) and from Hipparcos. Most distances were importantly underestimated.

3. [Fe/H] determination with CORAVEL

The CORAVEL radial velocity spectrometer has been known for a long time to permit precise estimates of the metallicity of dwarf stars via the surface of the cross-correlation function. The cross-correlation function surface, **W**, representing the average surface of more than 1'500 weak metallic lines, is a sensitive function of [Fe/H] and of temperature. As early as 1980 a first calibration was made (Mayor 1980). But only recently have both precise, homogeneous metallicity studies of dwarfs (Edvardsson et al. 1993) and large CORAVEL samples (F-G dwarfs survey, radial velocity survey of Hipparcos stars, cluster data) become available that provide an accurate calibration of the method for F to K dwarfs.

The calibration is made using Edvardsson et al. (1993, **E93**) for F-dwarfs, and extrapolation based on the Hyades cluster for G-dwarfs, and the Carney et al. (1994, **C94**) surveys for subdwarfs. The residual dispersion of [Fe/H] is 0.08 dex with E93 and 0.14 dex with C94. This calibration will be refined as more measurements for subdwarfs and Hyades late-type stars are realized. Corrections for faint stars appear necessary and also have to be calibrated.

The CORAVEL determination is spectroscopic, thus much less sensitive to small anomalies in the colours or binarity than photometric indices, and has the crucial advantages of being available with little observing time in a homogeneous way for large sample of stars, including many of the 60'000 stars already measured with CORAVEL.

The sample of Hipparcos subdwarfs is now being measured with CORAVEL. The aim is a homogeneous sample of [Fe/H] determinations for all Hipparcos stars with accurate parallaxes and situated below the solar-metallicity ZAMS, in order to allow an optimal treatment of the biases. In the sample selection, a wide margin has been included below the main sequence, so that contamination by solar-metallicity stars can be controlled closely.

4. Position of the subdwarfs in the CMD

With accurate parallaxes and [Fe/H] determinations, colour-magnitude diagrams (CMD) can be constructed for different metallicities. Fig. 2 shows for instance the CMD for stars in the -2.4 to -2.0 dex range in [Fe/H]. For the first time, a *field* subdwarf sequence is defined.

The important number of evolved stars in the sample is immediately apparent. Indeed such an abundant presence of evolved stars among subdwarfs was not quite expected in recent subdwarfs studies (C94, Ryan & Norris 1991). C94 for instance estimate the contamination from evolved stars to be about 10% in their sample. The true value may be much higher, as is made clear in Fig. 3, thus modifying the distance and space velocity estimates. The distances to most F-type subdwarfs were consequently underestimated in such studies (meaning that the transverse velocity from proper motion was correspondingly underestimated).

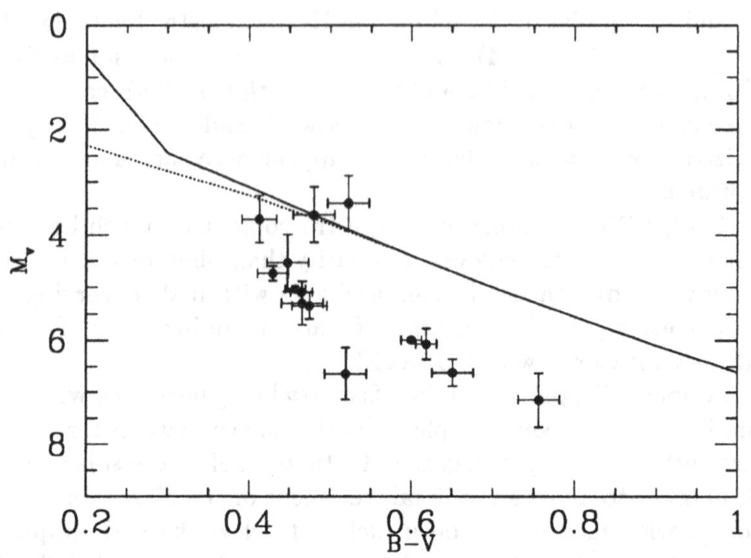

Figure 2. Position in the CMD of the subdwarfs with a [Fe/H] between -2.4 and -2.0 dex. The line is the Hyades sequence (observed – line, zero-age – dots). For the first time, we are able to see the turnoff and subgiant region in field subdwarfs, almost as clearly as in globular clusters !

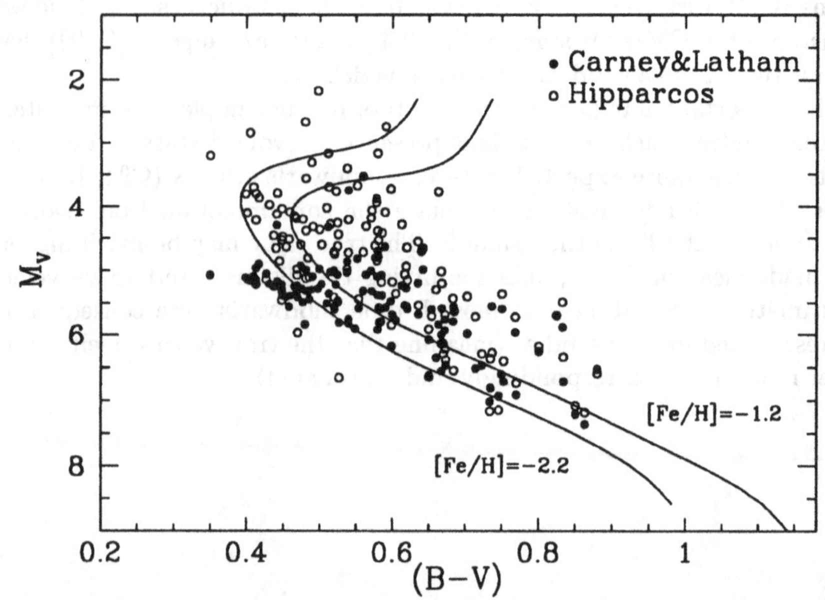

Figure 3. Comparison of the position in the CMD for some stars common to the C94 programme and to Hipparcos. 15 Gyr models isochrones from d'Antona et al. (1996) are superimposed. The unsuspected high number of evolved early type subdwarfs is clear.

5. Metal-deficient ZAMS

The displacement of the ZAMS in the CMD for various metallicities can be calibrated directly using the subdwarfs. The calibrators must be unevolved, and detected or suspected binaries, as well as objets with an uncertain parallax or large colour excess, are excluded. This reduces considerably the number of suitable calibrators. We shall increase this number by gathering more metallicity measurements, while refining the CORAVEL [Fe/H] calibration for G and deficient F stars.

6. Globular Cluster distances

The great usefulness of subdwarf luminosity calibrations is that the Globular Cluster (GC) distances can then be directly determined via main-sequence fitting to the subdwarf sequences of same [Fe/H]. No theoretical prediction is needed, except the assumption that field and GC stars of the same metallicity are similar. The reddening and metallicity of GCs must be known independently.

7. Globular Cluster ages

Recently, a considerable effort has been put into acquiring good CMD for Globular Clusters down to the lower main sequence. From these and the Hipparcos subdwarfs, reliable distances can be obtained. Then the turnoff luminosity, which is known to be a very robust prediction of stellar evolution and a sensitive function of age, can be compared *directly* to the model predictions, without any assumption on the poorly known colour and temperature calibrations. In practice, to avoid the uncertainty caused by the vertical tangent at the turnoff, we use the magnitude of the two points 0.04 mag redder than the turnoff.

The effect of errors on the reddening or metallicity of the clusters on this method is relatively severe (Δage[Gyr]$\simeq E(B-V)\cdot 10^{+2}$ [mag], Δage[Gyr]$\simeq$ $12 \cdot \Delta[Fe/H]$). For *relative* age estimations, indirect comparative methods are to be preferred.

Absolute ages, however, are another matter, and the subdwarf method is the most accurate. Pending some remaining uncertainties on stellar evolution models, the average ages of the oldest GC provided by this method can be considered as a very tight constraint on the lower limit of the age of the Universe.

Once good distances are known for a sample of GC, the position of other indicators can be calibrated, such as the HB luminosity, or the [Fe/H] dependence of RR Lyrae luminosity.

The subdwarf calibration will also permit a direct evaluation of the age of field subdwarfs relative to the globular clusters, a possible clue to the formation history of the Galaxy.

8. Towards reliable globular cluster ages

As more subdwarf CORAVEL measurements get added to the sample, the [Fe/H] calibration will be refined, and the number of suitable unevolved calibrators will increase. Once precise subdwarf sequences are determined, we shall try, in collaboration with stellar evolution modelists, to explore the changes needed in the models to fit the position and slope of the observed sequences, thereby improving the turnoff luminosity predictions. Finally, the effect of diffusion on the ages predicted by the models will be studied. The results will be presented in the Hipparcos meeting in Venice (May 1997).

References

Carney B.W., Latham D.W., Laird J.B. et al. 1994, AJ 107, 2240 (**C94**)
d'Antona F., Mazzitelli I., 1996, ApJ 456, 329
Edvardsson B., Andersen J., Gustafsson B. et al., 1993, A&A 275, 101 (**E93**)
Mayor M., 1980, A&A 87, L1
Ryan S.G., Norris J.E., 1991, AJ 101, 1835
Zinn R., West M.J., 1984, ApJSS 55, 45

GLOBULAR CLUSTER AGES: ARE THEY CONVERGING?

DON A. VANDENBERG
Dept. of Physics & Astronomy, University of Victoria
P.O. Box 3055, Victoria, B.C., Canada V8W 3P6

Abstract. Since the early 1970's, the consensus "best estimate" of the age of the Galactic globular cluster (GC) system has been 15–16 ±3 Gyr. However, a number of recent studies, which are briefly reviewed herein, have suggested that this estimate is too high by $\gtrsim$ 15%. Based on a consideration of new stellar models and the latest developments concerning the cluster distance scale, the present paper lends support to the notion that the first globulars began to form in the Galaxy 13–14 Gyr ago. In addition, our analysis of absolute GC ages adds to the mounting evidence that the dispersion in age among clusters having similar metallicities is probably quite small. As a consequence, age can hardly be the primary factor causing the wide diversity in their horizontal-branch morphologies.

1. Introduction

In a reasonably comprehensive review, VandenBerg, Bolte, & Stetson (1996; VBS96) concluded that the most metal-deficient (presumably the oldest) globular clusters in the Galaxy have ages of 15^{-3}_{+5} Gyr, where the estimated uncertainties were considered to be approximate 2σ limits. Virtually the same conclusion was reached by Chaboyer et al. (1996a) from a detailed Monte Carlo analysis of isochrone fits to observed C-M diagrams. However, a number of more recent investigations (Salaris, Degl'Innocenti, & Weiss 1996; Jimenez et al. 1996; and D'Antona, Caloi, & Mazzitelli 1997) have favored ages of $\lesssim$ 13 Gyr for the oldest clusters and, moreover, they have suggested that a further reduction of 1–2 Gyr is well within the realm of possibility. Given the far-reaching implications for cosmology of such "young" ages, it is clearly important to examine how they were obtained and to assess their reliability.

T.R. Bedding et al. (eds.),
Fundamental Stellar Properties: The Interaction between Observation and Theory, 439–448.

Salaris et al. (1996) have remarked that, if the Buser & Kurucz (1978, 1992; hereafter BK78, BK92) bolometric corrections are used to convert luminosity to M_V, instead of those tabulated by VandenBerg & Bell (1985; VB85), about a 2 Gyr reduction in the predicted ages for the most metal-poor globulars will result. On closer inspection, this assertion is not entirely correct because, in a relative sense, the BK92 BC_V's agree very well with those reported by VB85: their zero points do differ, but if both scales are normalized to some preferred value for the Sun, then the systematic variation of the BC_V values with $T_{\rm eff}$, gravity, or [Fe/H] is small. (This should not be too suprising given that both studies used essentially the same model atmospheres.) Be that as it may, there *are* systematic differences between the BK78 BC_V's, on the one hand, and those given by VB85 and BK92, on the other. Furthermore, the BK78 bolometric corrections appear to be in better agreement with both the Kurucz (1992) scale and the latest predictions by R. Bell (1996) in the sense that, at temperatures and gravities characteristic of turnoff stars, all three predict a much larger variation in BC_V as a function of [Fe/H] than one obtains from either VB85 or BK92 (see VandenBerg 1997a). This dependence is such as to imply a decrease in GC ages amounting to roughly $\delta(\text{age}) = 0.75[\text{Fe/H}]$ Gyr compared with estimates based on VB85 or BK92 BC_V's. Since the recent Kurucz and Bell bolometric corrections are based on improved model atmospheres, they should be superior to earlier predictions; consequently, the VBS96 estimate of 15 Gyr for the oldest GCs is arguably too high by ~ 1.5 Gyr.

Jimenez et al. (1996; JTJMP) tried a new approach to derive GC ages. Specifically, they attempted to determine the masses (and hence the ages) of stars presently evolving on the giant branches of 8 globular clusters from the average mass on the horizontal branch (HB), the average mass-loss efficiency $\langle\eta\rangle$, where η is the free parameter in the Reimers (1975) mass-loss formula, and its dispersion. To accomplish this, they adopted weighted averages of the cluster [Fe/H], $E(B-V)$, and $(m-M)_V$ values from the scientific literature along with the Kurucz (1992) color transformations. Published theoretical models were then used to infer the observed HB mass distributions — a risky procedure since the location of an HB star on the H-R diagram is known to be very sensitive to many parameters (cf. Tornambé 1988), and because the evolutionary state of a given HB star can often be ambiguous, thereby confusing the mass determination. Moreover, because masses tend to "pile up" at the red end, small errors in color/$T_{\rm eff}$ will necessarily translate to large errors in the estimated masses of red horizontal-branch stars in clusters possessing them.

By requiring agreement between the "observed" HB mass distribution in each GC and that predicted from numerical integrations of the mass loss along the upper RGB, $\langle\eta\rangle$ was determined. JTJMP appear to have either assumed or concluded from their analysis that the reddest star on an observed HB did not undergo any prior mass loss: they state (p. 938) that "$\langle\eta\rangle$ has been calculated from the value needed to reproduce both the point in the HB where $\eta = 0$ and the mean HB mass". This is highly questionable given the growing evidence (e.g., see the review by Stetson, VandenBerg, & Bolte 1996) against the hypothesis that age is *the* second parameter. For instance, M3 and M13 appear to be nearly coeval (see Ferraro et al. 1997); hence, for some yet unknown reason (differences in stellar rotation?), the HB stars in M13, which are almost entirely to the blue of the instability strip, must have undergone much more extensive mass loss than their counterparts in M3, which extend to very red colors. According to JTJMP, variations in HB morphology are due to differences in age.

Some other concerns with the JTJMP paper include the fact that their adopted cluster distances and derived ages are not consistent with turnoff age–luminosity relations. For instance, their derived age of 13.2 Gyr for M92 can only be obtained from an isochrone fit to the turnoff photometry if $(m-M)_V > 14.65$ (see §2) — assuming best estimates for the chemical abundances — whereas JTJMP adopted $(m-M)_V = 14.45$ (for which the turnoff age has to be > 16 Gyr). They suggest that a precise age cannot be accurately determined from the turnoff, even if the cluster distance were known, in part because of complications due to the mixing-length parameter. Granted, it is frequently quite difficult to define the turnoff point in an observed C-M diagram to within ± 0.1 mag; however, in practice (see §2), it is from the coincidence of the predicted and observed subgiant branch that the cluster age is inferred. The associated *internal* error is certainly $< \pm 0.5$ Gyr (cf. Chaboyer et al. 1996b). Both the distance scale problem and the small numbers of upper RGB stars in their samples (as few as 25 within 2 mag of the RGB tip) are further cause for skepticism. The approach used by JTJMP to derive GC ages, and their results, appear to be much more uncertain than they have acknowledged[1].

Finally, D'Antona et al. (1997) have suggested that the most metal-poor GCs may be as young as 12 Gyr. The 2–3 Gyr reduction in age from canonical estimates is due to two factors: (1) the use of the Canuto & Mazzitelli (1991) theory of convection in their calculations instead of the

[1]More recently, Jimenez & Padoan (1996) have obtained an age of 16.4 Gyr for M68 from an analysis of its luminosity function, on the assumption of nearly the same distance modulus that was adopted by JTJMP, who found an age of 12.7 Gyr for this cluster. No explanation was offered as to the cause of the discrepant results. In this particular case, the adopted distance, $(m-M)_V = 15.3$, simply cannot be reconciled with an age as high as 16 Gyr (see §3). The Jimenez–Padoan study is somehow flawed, though perhaps only in something as basic as their assumption that the relative lifetimes of stars in different evolutionary phases are accurately predicted by standard models — for which there is some evidence to the contrary (Bolte 1994, and references therein).

usual mixing-length theory, and (2) the adoption of increased distances to the globulars to be consistent with the prediction of their HB models. In an earlier paper, Mazzitelli, D'Antona, & Caloi (1995) demonstrated quite convincingly that turnoff age–luminosity relations do depend on how convection is treated, particularly those for low-metallicity stars that have very thin convective envelopes. VBS96 have argued that this uncertainty is alleviated to some extent by inferring cluster ages from the luminosities of subgiant-branch stars, which have much deeper surface convection zones. Even so, it remains a real possibility that the treatment of convection can affect GC age determinations at the $\sim \pm 1$ Gyr level.

As far as predicted HB luminosities are concerned, it has not been possible to reach a consensus, neither theoretically nor empirically (see VBS96), on what the true relation is between M_V(HB) and [Fe/H]. The horizontal-branch models that D'Antona et al. have used in their analysis appear to be brighter than those used in this investigation (§2) because they have somewhat larger helium core masses, but the cause of the differences in this quantity is presently unknown. Fortunately, Hipparcos data (see the previous paper in these proceedings), the determination of white dwarf distances to GCs (e.g., Renzini et al. 1996), and the observation of HB stars in M31 globular clusters (e.g., Fusi Pecci et al. 1996) promise to resolve these issues. First indications are (see below) that these constraints favor fainter HB luminosities (and older GC ages) than D'Antona et al. have advocated.

2. Absolute Globular Cluster Ages

VandenBerg et al. (1997) have computed an extensive new grid of evolutionary tracks and isochrones for ages from 8 to 18 Gyr, as well as fully-consistent zero-age horizontal-branch (ZAHB) loci, assuming $[\alpha/\mathrm{Fe}] = 0.0$, 0.3, and 0.6 for each of more than a dozen [Fe/H] values between -2.3 and -0.3. All α elements, including O, Ne, Mg, Si, S, etc., were assumed to vary together in abundance, and OPAL opacities (cf. Rogers & Iglesias 1992) were calculated for the adopted mixes. Furthermore, an improved equation of state allowing for Coulomb interactions and other non-ideal effects was employed, along with the rates for H-burning reactions described by Bahcall & Pinsonneault (1992). The transformation of the models from the theoretical to various C-M planes was carried using an empirically-constrained set of color transformations (VandenBerg 1997b) and the Kurucz (1992) bolometric corrections, adjusted by a constant amount to ensure consistency between the adopted M_{bol} and observed M_V values for the Sun.

The left-hand panel in Figure 1 shows that if the distance to M92 is set by matching the predicted and observed ZAHBs — yielding $(m-M)_V =$

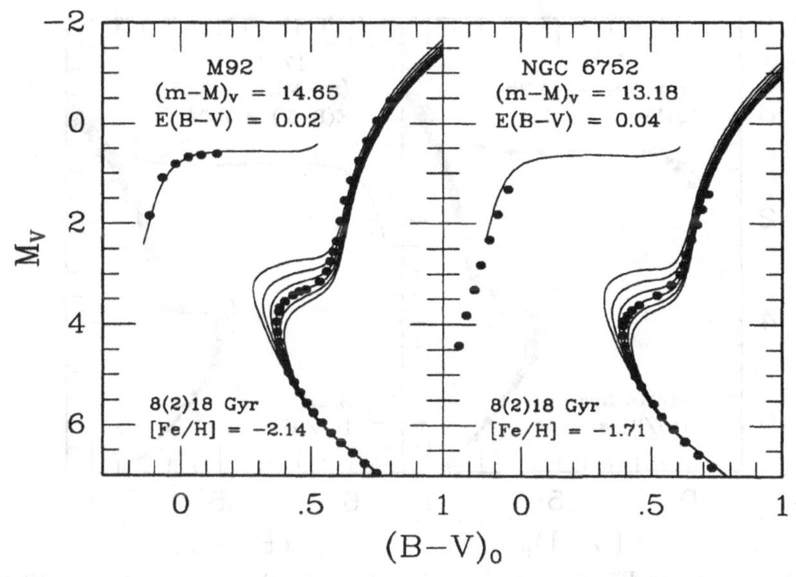

Figure 1. Comparisons of isochrones for the specified ages and [Fe/H] values with fiducial sequences for M92 (Stetson & Harris 1988, Bolte 1997) and NGC 6752 (as given by VBS90, based on photometry by Penny & Dickens 1986) for the indicated reddenings and distances. The position of the blue ZAHB in NGC 6752 was derived from an eye-estimated fit to the lower bound of the observations plotted by Caloi et al. (1986; their Fig. 1). Small color adjustments (< 0.015 mag) were applied to the isochrones, as necessary, in order to ensure that the predicted and observed lower main sequences coincided. In this, and all other plots to follow $[\alpha/\mathrm{Fe}] = 0.3$ has been assumed: see Carney (1996) and Ryan, Norris, & Beers (1996) for justification of this choice.

14.65, which is identical to the value adopted by VBS96 from a consideration of the classical Population II subdwarfs — an age near 14 Gyr is obtained. (We emphasize that the same isochrones were used in both studies and that differences in the derived age are entirely due to the BC_V scale revisions adopted here.) The right-hand panel indicates a very similar age for NGC 6752 on the assumption of the distance modulus that Renzini et al. (1996) have derived by fitting their *HST* observations of cluster white dwarfs to local white dwarf sequences. They suggest that this distance determination should be accurate to $\sim 10\%$. Interestingly, in spite of having an extremely blue HB, NGC 6752 does not appear to have an extreme age.

From the close similarity of the M3 and NGC 6752 C-M diagrams between the lower main sequence and the lower RGB, VandenBerg, Bolte, & Stetson (1990; VBS90) argued that the two clusters must be nearly coeval, despite their very different HB morphologies (also see Richer et al. 1996). And, indeed, when ZAHB luminosities are used to infer the distance to M3, an isochrone fit to the turnoff photometry indicates an age of 13–14

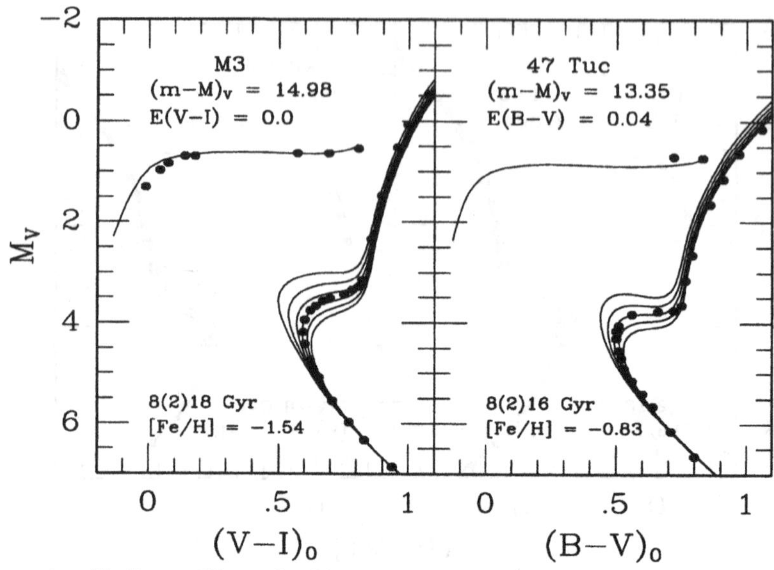

Figure 2. Similar to Fig. 1; in this case, isochrones for the specified ages and [Fe/H] values are compared with fiducial sequences for M3 (Johnson & Bolte 1997) and 47 Tucanae (Hesser et al. 1987).

Gyr (see the left-hand panel of Figure 2). The very encouraging agreement between the relative cluster ages as derived from a distance-independent method (VBS90) and from an evaluation of their absolute ages using different standard candles to determine the cluster distances suggests that the adopted moduli and the different approaches to the data are quite reliable. Although the uncertainties are still large enough not to preclude a 1–2 Gyr difference in age between M3 and NGC 6752, a variation of this size is at least a factor of 3 too small to explain the difference in their HB populations in terms of age. Thus the present analysis provides yet another (rather compelling) argument that age cannot be *the* second parameter. It is also worth mentioning that the relation between M_V(HB) and [Fe/H] predicted by VandenBerg et al. (1997), namely $M_V(\mathrm{HB}) = 0.19[\mathrm{Fe/H}] + 0.96$ (evaluated at $\log T_{\mathrm{eff}} \approx 3.84$ and assuming $[\alpha/\mathrm{Fe}] = 0.3$), agrees well with those predicted by other workers (e.g., Lee 1990) and that derived from *HST* observations of M31 globular clusters (Fusi Pecci et al. 1996). However, Baade-Wesselink and statistical parallax studies of field RR Lyraes appear to favor fainter HB luminosities (see the VBS96 review) and hence older ages for the globular clusters (e.g., see Carney 1996).

As illustrated in the right-hand panel of Fig. 2, a turnoff age of $\approx$ 12 Gyr is predicted for 47 Tuc if its apparent distance modulus is 13.35 — and the cluster stars have [Fe/H] = −0.83 (e.g., Brown, Wallerstein, & Oke 1990), $[\alpha/\mathrm{Fe}] = 0.3$ (cf. Carney 1996), and $Y = 0.24$ (Dorman, VandenBerg, &

Laskarides 1989). It may be recalled that Hesser et al. (1987) derived an age of 13.5 Gyr for this cluster, assuming $(m - M)_V = 13.40$ and isochrones for [Fe/H] = −0.65, [O/Fe] = 0.3, and $Y = 0.24$. While it may initially seem a bit disconcerting that a 1.5 Gyr younger age is obtained on the assumption of a slightly reduced distance (which would normally imply an increased age), there is no inconsistency. The cumulative effect of the revisions to the BC_V scale, the adoption of [α/Fe] = 0.3 instead of [O/Fe] = 0.3, and the treatment of Coulomb interactions in the equation of state used by VandenBerg et al (1997) is to reduce the age by $\gtrsim$ 1.5 Gyr at a given turnoff luminosity. Although Figs. 1 and 2, taken together, intimate that the Galactic GCs obey an age–metallicity relation, which they may well do, the sample of clusters considered here is obviously too small and the uncertainties still too large to allow one to regard these findings as more than suggestive.

3. On the Relative Ages of M68 and M92

VBS90 found that, when the C-M diagrams of M30, M68, and NGC 6397 were individually shifted both vertically and horizontally by the amounts needed to register them to the turnoff magnitude and color of M92, there was no perceptible difference in the location of their respective RGBs. Since a similar registration of theoretical isochrones relevant to these clusters indicated that the color difference between the turnoff and the lower giant branch changes at the rate $\Delta(B-V)_{\rm TO,RGB} \approx -0.012$ mag/Gyr, VBS90 concluded that the 4 clusters were "extremely uniform in age, with no convincing evidence for differences as great as 0.5 Gyr". Richer et al. (1996) reached basically the same conclusion, using the same approach, for a much larger sample of very metal-poor GCs. In stark contrast with this, Chaboyer, Demarque, & Sarajedini (1996c; CDS96) argued from measurements of the magnitude difference between the main-sequence turnoff and the HB, $\Delta V_{\rm TO}^{\rm HB}$, that there is a range of $\gtrsim$ 5 Gyr amongst these same clusters. Who is right?

In principle, the $\Delta(B-V)_{\rm TO,RGB}$ method is capable of much higher precision largely for the reason that the turnoff to lower-RGB color difference can be determined much more accurately than $\Delta V_{\rm TO}^{\rm HB}$ (see the extensive discussion of the advantages and disadvantages of the two techniques given by Stetson et al. 1996). Because, by definition, a C-M diagram is vertical at the turnoff, the turnoff magnitude cannot be determined more precisely than about ±0.05–0.1 mag, even from superb photometry. Since the rate at which $\Delta V_{\rm TO}^{\rm HB}$ changes with age is approximately +0.072 mag/Gyr — the estimated slope of turnoff M_V versus age relations at a representative age of 14 Gyr (VandenBerg et al. 1997) — the $\Delta V_{\rm TO}^{\rm HB}$ method as it has

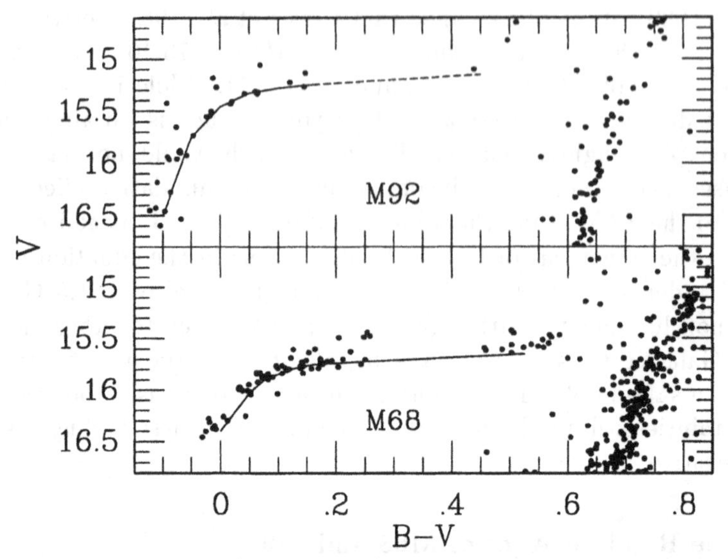

Figure 3. Plot of Bolte (1997) and Walker (1994) M92 and M68 photometry, respectively, for RGB and non-variable HB stars having $14.6 \leq V \leq 16.8$. The solid curves give eye-estimated ZAHB loci. The dashed curve extrapolates the M92 locus to redder colors: its slope is taken to be the same as that of the M68 ZAHB through the instability strip.

traditionally been used cannot reliably detect a difference in age smaller than ~ 1 Gyr. Furthermore, many GCs have few (or no) HB stars at the color of the turnoff, and even when they do, there could well be cluster-to-cluster differences in the mean evolutionary state of those stars, further complicating the evaluation of the $\Delta V_{\rm TO}^{\rm HB}$ parameter.

According to CDS96, the $\Delta V_{\rm TO}^{\rm HB}$ values for M30, M68, NGC 6397, and M92 are 3.62 ± 0.14, 3.42 ± 0.10, 3.74 ± 0.12, and 3.74 ± 0.14, respectively. Only in the case of M68 can one plausibly argue for a difference in age. But do the $\Delta V_{\rm TO}^{\rm HB}$ values for M68 and M92 really differ by 0.32 mag? Although based on relatively few stars, McClure et al. (1987) found no obvious difference in the location of the HB populations of M68 and M15 when the turnoffs of the two clusters were aligned: CDS96 give $\Delta V_{\rm TO}^{\rm HB} = 3.63 \pm 0.16$ for M15. Subsequently, Carney, Storm, & Jones (1992, see their Fig. 22) argued that there was no difference in the relative ages of M68 and M92 using either the $\Delta V_{\rm TO}^{\rm HB}$ or $\Delta(B-V)_{\rm TO,RGB}$ methods. And, if one defines ZAHB loci for M68 and M92 from the latest available photometry (see Figure 3), and compares them once the cluster turnoffs have been registered to one another (thereby removing most of the uncertainty in the definition of $V_{\rm TO}$), the result is Figure 4. The difference in the $\Delta V_{\rm TO}^{\rm HB}$ values for the two clusters is only 0.08 mag, suggesting rather similar ages. Perhaps the greatest concern with this analysis is that the data are taken

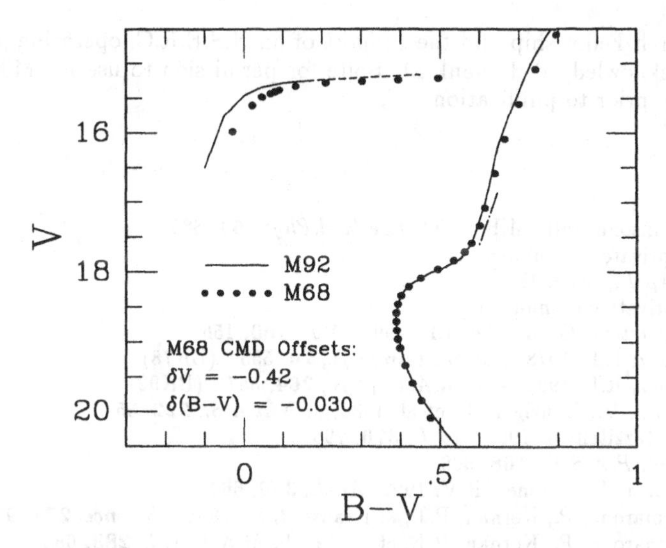

Figure 4. Using the indicated offsets and the ZAHB loci from Fig. 3, the fiducial sequences for M68 (McClure et al. 1987) and M92 (Stetson & Harris 1988) are intercompared. The dot-dashed line shows where the lower RGB of M68 would be located if this cluster were 2 Gyr younger than M92.

from several sources, inviting the possibility that small differences exist in the respective photometric zero-points which could have some ramifications for the measured ΔV_{TO}^{HB} values. (This is a potential problem for much of the CDS96 dataset as well.) But, until more definitive studies are carried out, we conclude that there is no inconsistency in the results of the ΔV_{TO}^{HB} and $\Delta(B-V)_{TO,RGB}$ methods as applied to M68 and M92. Indeed, relative GC age estimates based on the color-difference technique are arguably much more secure than those based on the ΔV_{TO}^{HB} method.

4. Summary

Due to steady improvements to the stellar models and tightening constraints on the cluster distances, some convergence of GC ages is certainly occurring. This study suggests that (1) the most metal-poor GCs formed about 14 Gyr ago, (2) the globulars obey an age-metallicity relation with [Fe/H] ~ -0.8 clusters like 47 Tuc being about 2 Gyr younger than very metal-deficient systems, (3) M3 and NGC 6752 have similar ages despite huge differences in their HB morphologies, and (4) M68 and M92 are close to being coeval in spite of recent claims to the contrary. (The absolute ages derived here should probably be reduced by ~ 1 Gyr to allow for the effects of He diffusion, which was not treated in the present models.)

A Killam Research Fellowship and the support of an N.S.E.R.C. operating grant are gratefully acknowledged. I thank M. Bolte for permission to use his M3 and M92 photometry prior to publication.

References

Bahcall, J.N. & Pinsonneault, M.H. 1992, *Rev.Mod.Phys.*, **64**, 885
Bell, R.A. 1996, private communication
Bolte, M. 1994, *Ap.J.*, **431**, 223
Bolte, M. 1997, private communication
Brown, J.A., Wallerstein, G., & Oke, J.B. 1990, *A.J.*, **100**, 1561
Buser, R. & Kurucz, R.L. 1978, *Astron.Astrophys.*, **70**, 555 (BK78)
Buser, R. & Kurucz, R.L. 1992, *Astron.Astrophys.*, **264**, 557 (BK92)
Caloi, V., Castellani, V., Danziger, J., et al. 1986, *M.N.R.A.S.*, **222**, 55
Canuto, V.M. & Mazzitelli, I. 1991, *Ap.J.*, **370**, 295
Carney, B.W. 1996, *P.A.S.P.*, **108**, 900
Carney, B.W., Storm, J., & Jones, R.V. 1992, *Ap.J.*, **386**, 663
Chaboyer, B., Demarque, P., Kernan, P.J., & Krauss, L.M. 1996a, *Science*, **271**, 957
Chaboyer, B., Demarque, P., Kernan, P.J., et al. 1996b, *M.N.R.A.S.*, **283**, 683
Chaboyer, B., Demarque, P., & Sarajedini, A. 1996c, *Ap.J.*, **459**, 558 (CDS96)
D'Antona, F., Caloi, V., & Mazzitelli, I. 1997, *Astron.Astrophys.*, in press
Dorman, B., VandenBerg, D.A., & Laskarides, P.G. 1989, *Ap.J.*, **343**, 750
Ferraro, F.R., Paltrinieri, B., Fusi Pecci, F., et al. 1997, *Ap.J.Lett.*, submitted
Fusi Pecci, F., Buonanno, R., Cacciari, C., et al. 1996, *A.J.*, **112**, 1461
Hesser, J.E., Harris, W.E., VandenBerg, D.A., et al. 1987, *P.A.S.P.*, **99**, 739
Jimenez, R. & Padoan, P. 1996, *Ap.J.Lett.*, **463**, L17
Jimenez, R., Thejll, P., Jorgensen, U.G., MacDonald, J., & Pagel, B. 1996, *M.N.R.A.S.*, **282**, 926 (JTJMP)
Johnson, J. & Bolte, M. 1997, in preparation
Kurucz, R.L. 1992, *Solar Abundance Model Atmospheres* (CD-ROM 19)
Lee, Y.-W. 1990, *Ap.J.*, **363**, 159
Mazzitelli, I., D'Antona, F., & Caloi, V. 1995, *Astron.Astrophys.*, **302**, 382
McClure, R.D., VandenBerg, D.A., Bell, R.A., et al. 1987, *A.J.*, **93**, 1144
Penny, A.J. & Dickens, R.J. 1986, *M.N.R.A.S.*, **220**, 845
Reimers, D. 1975, *Mem.Soc.Roy.Sci. Liège*, **8**, 369
Renzini, A., Bragaglia, A., Ferraro, F.R., et al. 1996, *Ap.J.Lett.*, **465**, L23
Richer, H.B., Harris, W.E., Fahlman, G.G., et al. 1996, *Ap.J.*, **463**, 602
Rogers, F.J. & Iglesias, C.A. 1992, *Ap.J.Suppl.*, **79**, 507
Ryan, S.G., Norris, J.E., & Beers, T.C. 1996, *Ap.J.*, **471**, 254
Salaris, M., Degl'Innocenti, S., & Weiss, A. 1996, preprint
Stetson, P.B. & Harris, W.E. 1988, *A.J.*, **96**, 909
Stetson, P.B., VandenBerg, D.A., & Bolte, M. 1996, *P.A.S.P.*, **108**, 560
Tornambé, A. 1987, in *ESO Workshop on Stellar Evolution and Dynamics in the Outer Halo of the Galaxy*, ed. M. Azzopardi & F. Matteucci (Garching: ESO), p. 307
VandenBerg, D.A. 1997a, in *Advances in Stellar Evolution*, ed. R.T. Rood & A. Renzini (Cambridge: Cambridge U. Press), in press
VandenBerg, D.A. 1997b, in preparation
VandenBerg, D.A. & Bell, R.A. 1985, *Ap.J.Suppl.*, **58**, 561 (VB85)
VandenBerg, D.A., Bolte, M., & Stetson, P.B. 1990, *A.J.*, **100**, 445 (VBS90)
VandenBerg, D.A., Bolte, M., & Stetson, P.B. 1996, *Ann.Rev.Astron.Astrophys.*, **34**, 461 (VBS96)
VandenBerg, D.A., Swenson, F.J., Rogers, F.J., Iglesias, C.A., & Alexander, D.R. 1997, in preparation
Walker, A.R. 1994, *A.J.*, **108**, 555

15. CONCLUSION

ACHIEVEMENTS AND PROSPECTS

L.E. CRAM
School of Physics
University of Sydney, NSW 2006
Australia

1. Introduction

The scientific study of the topic of this Symposium, "Fundamental Stellar Properties," was in its infancy at the turn of the Century. As illustrated at this meeting, there has been significant progress in many directions over the past 100 years. Despite this progress, there are however many unsolved problems that will be addressed by theoretical and observational techniques that are emerging only now.

Some astronomers would argue that the investigation of fundamental stellar properties is a mature field, one in which only a few significant discoveries remain to be made. Indeed, there can be little doubt that new technologies are likely to stimulate rapid progress in other fields where exploration is particularly difficult (such as cosmology or in situ planetology). However, the central role of fundamental stellar properties in astrophysical research, combined with a number of significant unsolved problems, will ensure that the subject remains vigorous well into the next Century.

The presentations of many participants in this Symposium point to two particularly pressing classes of problem that will receive much attention in the future. The first can be described by the term 'stellar gas dynamics,' the study of (magneto-) gas-dynamic processes in and around stars. The second can be described by the term 'stellar populations,' the study of the relationship between stars as individuals and stars in groups sharing common characteristics.

In the first area, the study of stellar gas dynamic processes such as accretion and mass loss must be undertaken if we are to understand problems in late and early stages of evolution. These include the stellar initial mass function, stellar multiplicity, and the origin of (baryonic) dark matter, white dwarfs, neutron stars and black holes. For stars that are neither young nor old, the behavior of waves and convection in the stellar interior

T.R. Bedding et al. (eds.),
Fundamental Stellar Properties: The Interaction between Observation and Theory, 451–457.

and atmosphere can influence structure and complicate spectroscopic diagnostics. Studies of waves may also open new ways to seek to verify theories of stellar structure. Although ever more powerful computers are enabling the investigation of hitherto intractable gas dynamic problems, there seem to be profound difficulties in establishing connections between gas-dynamic theory and spectroscopic observations of phenomena that take place on scales much smaller than the stellar radius.

In the second area, the classification of stars and their arrangement into populations is a subject with a long history. However, despite significant progress with our understanding of stellar populations and physical associations, we are far from a satisfactory understanding of the relationships between the properties of individual stars, and the collective behavior of groups of stars. The acquisition and exploitation of large-scale spectroscopic, photometric and dynamical data bases is likely to be a particularly important route to understanding the origins of stellar groups on scales ranging from associations to clusters of galaxies. A foundation of such research is a clear understanding of the relationship between theory and observation of fundamental stellar properties.

As we are approaching the end of a century of research on fundamental stellar properties, it seems to be opportune to use this closing address to review some of the achievements over this period. This will help me indicate some of the direction that the subject might take over the next few decades. Of necessity the survey will be brief and hence incomplete, but nevertheless we will be able to see how threads from the past lead us in certain directions to the future. It will be convenient to divide the review into three roughly equal time spans.

2. 1900–1935

At the turn of the Century, almost nothing was known about the source of stellar luminosity. Speculations regarding the possibility that nuclear reactions (or radioactivity) could power the stars were advanced as early as 1915 (by Eddington). However, it was only during the latter part of the 1930s that theories for nuclear fueling similar to those current today were developed. Without these theories it was not possible to discover the modern interpretation of the Hertzsprung-Russell diagram, and consequently their development was a major turning point in stellar astrophysics. Nevertheless, in the three decades leading up to this development, many remarkably resilient discoveries were made about the structure of stars, and about many of the basic features of stellar spectroscopic and photometric diagnostics.

Prior to the development of quantum mechanics and Saha's theory of ionisation ($\sim$1920), understanding of the effective temperatures of stars was relatively undeveloped despite the availability of Planck's essentially complete theory of radiation thermodynamics. However, it was recognised that the relationship between the sequence of spectral types and stellar colours (determined by photographic photometry and spectroscopy) provided an uncalibrated sequence ordered by effective temperature. The available refined techniques for measuring and conducting statistical error analyses of parallax measurements also allowed the inference of absolute magnitudes. The Hertzsprung-Russell diagram could thus be constructed. As the importance of this system of classification became widely appreciated (∼1912–15), the rather subtle differences of spectral line shapes between giants and dwarfs were used to provide 'spectroscopic parallaxes' greatly extending the scope of the diagram. The astrophysical significance of this method, as well as the general behaviour of spectral lines in different spectral types, was explained by Saha's theory (∼1920). By this time, the angular diameters of a few stars had also been measured, confirming the existence of the giant stars implied by the Hertzsprung-Russell diagram.

Although the application of Planck's radiation theory to temperature determinations required Saha's theory to be fully effective, the theory could be applied to the problem of radiation transfer through a star. This commenced in ∼1905 with the work of Schuster and Schwarzschild. One thrust of this work was directed at the determination of stellar structure under the condition of radiative equilibrium, while another was directed at classical diagnostic problems such as the theory of solar limb darkening, the shape of the continuous spectrum, and the interpretation of the strengths and shapes of spectral lines. As a result of the long-running but erroneous underestimate of the hydrogen abundance (due mainly to incomplete understanding of the nuclear energy source and of the role of H^- as an opacity source) the quantitative results of spectroscopic diagnostic were often quite wrong. However, the foundations of such important procedures as curve-of-growth analysis and line profile fitting were laid at that time by Milne, Minneart, Menzel and others.

3. 1935–1965

As noted above, the discovery of the energy source of the stars in 1935–37 was a major advance shaping the future course of stellar astrophysics. Major research themes over the subsequent three decades (i.e, until the advent of a new generation of 4-metre class telescopes, solid-state panoramic detectors, space observatories and powerful computers) were naturally related to this discovery, particularly in relation to the abundances of the elements, the nature of stellar populations and the evolution of the stars.

Throughout the period 1935–65, the exploitation of photo-multipliers and other electronic devices opened up possibilities for observations to be made with improved precision and/or of fainter sources. Some of the most fruitful advances in instrumentation were applied to investigations of the Sun, which was studied from X-ray to radio wavelengths, and with even greater discernment of details in the optical wavebands. There was a vigorous exchange of people and ideas between stellar and solar astrophysics, providing insights and analogies that were particularly illuminating. Unfortunately, increased specialisation seems to have reduced the opportunities for such exchanges.

Discoveries made by solar physicists in this period included the convective origin of granulation and super-granulation, the five-minute oscillations, the high temperature of the corona, the compact nature of the chromosphere, and the ubiquity of magnetic flux tubes in the solar photosphere. Applications of the curve-of-growth and of line profile fitting provided estimates of the abundances of many elements, few of which have needed to be significantly revised. The techniques and concepts of non-LTE spectroscopic diagnostics were developed for solar applications at this time. Towards the end of the period, most of these areas of solar research were being directed also towards the stars.

Stellar astrophysicists discovered micro-and macro-turbulence (presumably the analogues of solar granulation and oscillations), began to explore magnetic stars and stellar chromospheres, and identified and provided preliminary accounts of the two great stellar populations. Significant advances were made in the absolute and relative measurement of stellar spectra and in the theory of stellar atmospheres. Together, these allowed a reliable link to be established between observations of the spectroscopic and photometric properties of stars and the fundamental quantities predicted and used by theory, such as effective temperature, luminosity and surface gravity.

The period 1935–65 saw not only the major nuclear reaction chains elucidated, but also the rapid development of methods for solving the equations of stellar structure. The importance of including convective energy transport was also discovered. With the resulting theory of stellar structure and evolution, it was possible to account for the relative abundances of many elements in various kinds of stars and stellar populations, and to explain the Hertzsprung-Russell diagram of field and cluster stars. The period also saw rapid developments in the theory of variable stars, including the theory of the Cepheid instability strip.

4. 1965–1996

New technology has stimulated rapid advances in the study of fundamental stellar properties over the past three decades. The construction of several

4-metre class telescopes combined with the application of panoramic solid-state optical and IR detectors has improved greatly the quality of ground-based observations. Space experiments, from those of the early days of Copernicus and Skylab to the recent spectacular results from Hipparcos and HST, opened up new spectral bands for accurate spectroscopy and photometry. This has provided a view of the stars as being far more 'dynamic' than hitherto recognised, stimulating many fruitful lines of research on fundamental stellar properties. During the same period, the increasing power of computers has allowed ever-more challenging problem in stellar astrophysics to be addressed.

The oral, poster and written contributions at this Symposium have provided an impressive picture of many of the most significant outcomes of this research. For example, we have seen that there has been great progress in measuring stellar parallaxes and proper motions in the galaxy. The technical barriers to interferometric determinations of stellar angular diameters are finally breaking down. These measurements, combined with a raft of new determinations of stellar masses from refined studies of binary systems, will provide new and tight constraints on models of stellar structure and evolution. Calibrations of absolute flux measurements in the near UV, visual and IR spectral regions are approaching 1% uncertainty, providing significant challenges to the most refined models of stars. Spectroscopic observations interpreted using refined diagnostic techniques are providing relative and differential abundance estimates with uncertainties believed to be smaller than a factor of two. Observations of stellar oscillations and pulsations are providing new ways of probing the internal structure of stars.

As our understanding of 'normal' stars improves, so attention turns increasingly to the early and late stages of stellar evolution where both theory and observation is more difficult. The genesis of stars and of any planetary systems that they might possess is increasingly amenable to observational and theoretical investigation. The origin and evolution of stellar chromospheres and coronae, and of magnetic phenomena in general, are becoming increasingly clear. The importance of mass-loss, especially in the late stages of stellar evolution, is readily acknowledged, and there has been steady progress in understanding the phenomenon.

On the theoretical side, powerful and flexible codes now include millions of absorbing species and quite complex models of important physical processes to yield remarkably refined predictions of stellar spectra. Models of stellar evolution are based on ever-sounder determinations of data on nuclear reaction rates and opacities, and methods are available to deal with evolution so rapid that hydrodynamic effects become important. Significant progress has been made with theories that combine gas-dynamic models of radiation pressure driven mass-loss with self-consistent non-LTE predictions of spectra, allowing new interpretations of observations of stars at critical stages of their evolution. Increasing attention is also being directed at sophisticated models of internal gas-dynamics, including the (non-linear) interaction between convection, rotation, pulsation and magnetic fields. These are providing new and valuable insights that help to identify the processes that control the observed differences in structure and evolution from star to star.

5. Prospects

There are a number of profound questions about the formation and structure of the universe that turn on topics that lie in the province of fundamental stellar astrophysics. For example, it is widely acknowledged that there is the need for a great deal more work on the relationships between age, kinematic behaviour and abundances of stars and star groupings in the Galaxy to elucidate some of the basic processes involved in galaxy formation and structure. As another example, despite quite rapid recent progress the relationship between the nature of dark matter and the outcomes of stellar evolution remains poorly understood. On the very large scale, the interpretation of the spectra of external galaxies rests squarely on the understanding of the spectra of the individual stars (and populations) that contribute to the integrated light, and many crucial linkages still remain to be made in this field. On a much smaller (but no less interesting) scale is the desire and improving capacity to explore the environs of stars, to understand better the formation and properties of companions, winds, planets and so forth.

These topics, and many others, will be the subject of future research in the field of fundamental stellar astrophysics. However, while it seems clear that the field offers important and tractable research opportunities, several speakers at this Symposium have displayed caution and concern about the future of the subject. Much of this appears to have its origins in a problem shared by some other fields of research: a recognition that progress seems to demand increasing attention to the details, combined with an uncomfortable feeling that the subject could become constipated through preoccupation with just these details.

There is, I believe, no easy answer to this dilemma. It can be exemplified by an important theme emerging from this Symposium, namely the prospects for major developments in computer-based theories of convection and waves in stellar atmospheres. We recognise that these are likely to represent significant advances in our understanding of micro-and macroturbulence and other motions in stellar atmospheres, which are certainly unsatisfactory aspects of contemporary diagnostic practice. Furthermore, we suspect that temperature inhomogeneities and temperature/velocity correlations could be leading to significant systematic errors in determinations of abundances and line shifts. However, it is extremely difficult to see how the wealth of details and the multitude of parameters that will characterise refined hydrodynamic modelling are going to be incorporated into a theory of the gross behaviour of stars. Presumably, researchers should aim to concentrate on those details that do indeed influence (or probe) the gross properties of stars, and to forge linkages between theory and observation so that observable consequences can be identified and critical tests can be devised.

This Symposium has provided a timely and wide ranging overview of a branch of science whose birth coincided with the turn of the Century, and which has witnessed remarkable progress of the past 100 years. Its prospects for the next few decades took equally intriguing.

DISCUSSION

Di Benedetto (p. [illegible])

PIERRE MAXTED: [illegible] distance estimates to Cepheids based on period-luminosity relations, the metallicity contribution to the [illegible] via the correction for differential reddening [illegible] the [illegible] the Baade-Wesselink methods.

PAOLO DI BENEDETTO: The metallicity corrections affecting the actual BW distances [illegible] Cepheids are found to be quite negligible, according to the overall [illegible] term. For instance, the correction to the BW distance to [illegible] Galaxy is a few hundredths of [illegible] and such a small correction cannot [illegible] the significant reduction of a factor three [illegible] the final quoted error with respect to that affecting the distance derived by period-luminosity relations.

Kunk (p. [illegible])

ANDREAS [illegible]: Do you have any explanation of discrepancies concerning your UV flux distribution with the observed ones in the visible?

JEFFREY [illegible]: There are [illegible] discrepancies when the calibration is derived from [illegible] while [illegible] by the magnitude [illegible] extrapolating [illegible] the [illegible] are [illegible] low when compared with ground-based [illegible] atmosphere [illegible] would be [illegible] their [illegible] calibration.

[illegible] Bessell (p. 73)

ROBERT KURUCZ: Why are you willing to settle for 4% in the [illegible] it is [illegible] to have [illegible] standards.

BRUCE [illegible]: I [illegible] have made myself [illegible] I am not happy with 4%. [illegible] simply saying that [illegible] care to do better. When you start with [illegible] of the calibration standards, [illegible] these [illegible] your [illegible] programs stars, and [illegible] your results with those of [illegible] observed [illegible] better than 3-4%. [illegible] we should always try to improve.

Di Benedetto (p. 25)

PIERRE MAXTED: In distance estimates to Cepheids based on period-luminosity relations, the metalicity contributes to the uncertainty via the correction for differential reddening. How does diffferential reddening affect the Baade-Wesselink method?

PAOLO DI BENEDETTO: The reddening corrections affecting the actual BW distances to extragalactic Cepheids are found to be quite negligible, according to the overall absorption term. For instance, the correction for BW distance to M 100 Galaxy is few hundredths of magnitude and such a small contibution causes the significant reduction of a factor three in the final quoted error with respect to that affecting the distance dcerived by period-luminosity relations.

Kruk (p. 67)

ANDREAS KELZ: Do you have any problems or discrepancies connecting your UV flux distribution with the observed ones in the visible?

JEFFREY KRUK: There are no discrepancies when the calibration is derived by normalizing the white dwarf models by the V magnitude and extrapolating to the FUV. IUE fluxes are 6% low when compared with ground-based spectrophotometry at the atmospheric cutoff, which is what would be expected from their choice of normalization.

Bless & Percival (p. 73)

ROBERT KURUCZ: Why are you willing to settle for 4% in the visible? It is a disgrace to have such low standards.

ROBERT BLESS: I must not have made myself clear. I am not happy with 4%. I'm simply saying that it is hard to do better. When you start with the uncertainty in the calibration standards, transfer these to your own program stars, and then compare your results with those of another observer, it is very difficult to do better than 3-4%. Obviously, however, we should always try to improve.

Bell (p. 159)

MARTIN COHEN: I am not convinced that the few mountain-top near-IR measurements of Vega fluxes that lie above the Kurucz Vega model are indicative of a problem. None lies more than 2 sigma from the model spectrum and Donald Blackwell (priv. comm.) has suggested that the great difficulty associated with comparison of a star with a standard lamp on a distant site means that formal published uncertainties on the measurements are probably underestimated, perhaps by a factor of two.

We have two other methods to check the Kurucz spectra of Vega and Sirius, however. One is a parallel programme that's been running for about five years, in which we try to compare Kuiper Airborne Observatory spectra of bright cool stars like α Tau and α Boo with a specially commissioned absolute blackbody furnace made by NPL, UK. This method eliminates dependence on any adopted models but, because we have already published absolute spectra of these cool giants calibrated through Sirius and Vega, we can indirectly test these hot stellar spectra. We are just completing a year of diffraction calculations necessary to make the direct comparisons of KAO 3-30 μm spectra of K-M giants with the pre- and post-flight ground-based blackbody calibrations. It may be very difficult to recover the absolute *level* of stellar spectra but we should get the *shape*.

The second technique comes from the MSX satellite, currently in orbit, which carries a series of IR radiometers between 4 and 20 μm. This satellite also carried a series of small 'emissive spheres' that were ejected and observed repeatedly by the radiometers. These spheres have been modeled very precisely by their creators (MIT Lincoln Lab.) so that their thermal behaviour in space is highly predictable. MSX then made observations of some of our cool giant calibrators that can be compared directly with measurements of the emissive spheres. Again, all our K-M giant spectra have been assembled from pieces calibrated against Sirius and Vega, so a test of these cool stellar spectra is implicitly a test of these hot stellar models.

Nissen (p. 171)

MONIQUE SPITE: In the cool stars of the Small Magellanic Cloud an unexplained deficiency of nickel has been also found (V. Hill 1997).

POUL ERIK NISSEN: This is interesting and supports our suggestion that the halo stars, which are deficient in the alpha-elements, Na and Ni, have been accreted from a dwarf galaxy.

ROBERT KURUCZ: There is a systematic error that you did not point out in presenting the solar abundances. Abundances are usually from a minor stage of ionization, so the number density is proportional to electron number. Most of the determinations are independent. But the electron density varies with Si or Fe abundance, so changing Si or Fe changes all the abundances. Rigorously, everything should be determined simultaneously.

POUL ERIK NISSEN: I agree. The Holweger-Müller model was computed for $A_{\rm Fe} = 7.67$, which is inconsistent with the meteoritic value of 7.51.

DAVID ARNETT: Why should we believe that the overshoot region is spherically symmetric? The assumption should affect both the models of evolution and of spectra, shouldn't it?

POUL ERIK NISSEN: Yes, but that is what the 3D hydrodynamic models of the solar photosphere try to take into account.

BENGT GUSTAFSSON: The ad-hoc recipe to handle convective overshoot in the Kurucz ATLAS9 models leads to two effects which both increase the estimated abundances: (1) the effective temperature is increased and (2) the temperature gradient is decreased in the line-forming region. There is no clear proof that real convective overshoot would lead to these effects. The only way to clarify this is to carry out model simulations of Nordlund's type for stars with different metallicity and calculate lines for these models. This could and should be done.

Spite (p. 185)

JOHN NORRIS: You report very small scatter on the Spite Plateau. Could you comment on the result of Deliyannis, Boesgaard and King that a real lithium abundance scatter exists near the main sequence turn off of the metal poor cluster M92?

MONIQUE SPITE: Deliyannis et al. found different lithium abundances for stars having the same $B-V$. First, the S/N ratio of the spectra is not very good and thus the accuracy of the equivalent widths is not excellent. Moreover, the $(B-V)$ colors in a crowded globular cluster are not very accurate, and it is possible that the stars have not exactly the same temperature - I will add that the stars observed by Deliyannis et al. are not exactly at the turn off; they are subgiants and dilution could explain the 'lithium-poor' stars. One of the stars is, on the contrary, very 'lithium-rich,' but it has been found also very 'magnesium-rich.' The star is, as a consequence, very peculiar and a detailed analysis would be useful.

BENGT GUSTAFSSON: How would you explain a decrease of [O/Fe] from +0.65 to +0.40 within the halo? As a result of different O/Fe yields for SNe II of different masses or of different metal abundances, or is it too early to speculate about that?

MONIQUE SPITE: Up to now, with the ATLAS9 (without overshooting) or the NMARCS models and Teff/colour relations of Carney or Nissen, the oxygen abundance has been determined or estimated in the halo for four stars, one with a metallicity of about -1 and three with a metallicity of about -2 dex. For HD 103095 ([Fe/H] about -1), computations give a good agreement between the different systems of the oxygen lines and the ratio [O/Fe] has been found to be about 0.3 dex.

For the three other stars with [Fe/H] about -2, I could only estimate the change of the [O/Fe] ratio due to the change of the temperature. Thus it is really too early to speculate about a possible tilt of the ratio [O/Fe] versus metallicity in the halo. At least, before doing it, it is necessary to redo the computations of the OH molecular lines in the stars of the Nissen et al. (1994) paper with the new temperatures.

ROGER BELL: The high O abundances are disturbing in terms of results from globular cluster giants. Would you expect to see CO bands in these sub-dwarfs?

MONIQUE SPITE: These stars are hotter than the globular cluster giants and I am not sure that the CO band can be measured.

Sasselov (p. 253)

JOHANNES ANDERSEN: Your results seem to demolish most of the assumptions underlying the simple-minded applications of the Baade-Wesselink method, in particular the determination of radial-velocity curves from spectra averaging lots of lines of different origins. Is there any hope of getting useful results from these existing observations, or will they all have to be thrown out and replaced by new data from selected sets of spectral lines?

DIMITAR SASSELOV: The Baade-Wesselink method is attractive because it has the potential to provide accurate one-step distances to nearby galaxies. The simple-minded application of the method suffers from systematic errors which puts its reliability in question and makes it less competitive with other methods. One way to improve the quality of BW solutions is with the use high-resolution spectra. Unfortunately, the necessary information on line profiles and strengths cannot be recovered from existing observations.

MICHAEL SCHOLZ: Perhaps, the situation is different in Cepheids, but in case of the Miras the structure of the deep atmospheric layers (= upper interior 'envelope') changes little as you replace the simple grey atmosphere by a more sophisticated non-grey atmosphere.

DIMITAR SASSELOV: Yes, the atmospheres of Miras are so extended that regions in them may be completely decoupled and behave locally with the corresponding molecular opacitites.

BENGT GUSTAFSSON: Concerning the effects of blanketing - are they, in addition to being significant for colour calculations, etc., also significant for the structure and dynamics of the upper layers?

DIMITAR SASSELOV: Yes, I suspect they may be significant. We will not know how significant until a consistent calculation is made.

Kjeldsen & Bedding (p. 279)

ROBERT KURUCZ: I have a philosophical comment; do not take it personally. We do not know the spectrum of a single star. We would learn a lot more if you would spend some of your observing time taking a high-resolution, high signal-to-noise spectrum.

HANS KJELDSEN: I agree that we would learn a lot from such a spectrum. Firstly, however, it would not take very much time to do this for Procyon. In fact, at $R \approx 200\,000$ you can reach S/N of 1000 in 10 s and S/N of 10 000 in 15 minutes on a 4-m telescope. Secondly, I disagree that such a spectrum would provide more information than would detections of p modes. The solar oscillations provide a huge amount of information and tell us almost all that we know about the Sun. This will never be the case for a high S/N spectrum.

JØRGEN CHRISTENSEN-DALSGAARD (also responding to Kurucz): Observations of the stellar radiative spectrum, however precise, would not give information about the structure of the deep stellar interior. One might perhaps be forgiven for regarding such information as being more fundamental than fine details of the surface properties and composition.

Christensen-Dalsgaard (p. 285)

GIUSEPPE BONO: What is the dependency of nonradial polar modes on the assumption of the adiabatic approximation?

JØRGEN CHRISTENSEN-DALSGAARD: The error made in assuming adiabatic oscillations is part of a larger set of errors coming from an uncertain understanding of the near-surface regions, including effects of turbulent pressure, etc.

However, since in this region the modes propagate almost vertically, the effects are independent of degree, when properly scaled. This allows the errors to be eliminated in the analysis of the data, when inverting for the internal structure.

JOHANNES ANDERSEN: How does your solar model perform as regards neutrinos?

JØRGEN CHRISTENSEN-DALSGAARD: The computed neutrino fluxes for the model essentially agree with other standard solar models, such as that of Bahcall & Pinsonneault. Hence they are much higher than measurements. However, it must be emphasised that helioseismology does not directly constrain the computed neutrino flux. What is measured is squared sound speed $c^2 \propto T/\mu$ (T is temperature and μ is mean molecular weight). Thus T and μ are not constained individually. In principle it is possible to make models with low neutrino fluxes that agree with helioseismology, through carefully tuned mixing and, e.g., changes in opacity and nuclear parameters. In practice, such a solution seems unlikely.

DON VANDENBERG: Can you say something about the internal rotation in the Sun and how well models that allow for rotation do in matching the observed oscillations?

JØRGEN CHRISTENSEN-DALSGAARD: Solar internal rotation, as a function of radius and latitude, can be inferred from splitting of the frequencies according to m. The results show the surface differential rotation to persist through the convection zone, with a sharp transistion to nearly constant rotation in the radiative interior; there is somewhat controversial evidence for slow core rotation. This profile is quite inconsistent with the Yale rotation model, which predicts rather rapid rotations in the interior.

Chiosi (p. 323)

JØRGEN CHRISTENSEN-DALSGAARD: Could you elaborate on the difference between the normal overshoot model and the diffusive scheme?

CESARE CHIOSI: In a normal overshoot scheme, the extension of the overshoot region is customarily derived from the velocity law of convective elements as a function of the radial distance (cf. Bressan et al. 1981). Within this region, the temperature gradient is either adiabatic or radiative, but what is more relevant, mixing is assumed to be instantaneous and complete. In the diffusive-overshoot scheme, the same picture holds with the only difference that in the overshoot region mixing is thought to take place over a finite time scale (this may be a function of local properties) and a suitable prescription is sought for the diffusion coefficient (or equivalently time scale of mixing). See for instance the algorithm by Deng et al. (1996). In an ideal model both the overshoot distance and mixing time scale should be derived from a complete physical description of this phenomenon (see for instance the formulation by Grossman et al. 1993). The problem is with the complexity of these models which has so far hampered the practical usage of these (more physically sounded) theories in stellar model calculations.

STEVE KAWALER: Can you include overshooting during the convective phase of a thermal pulse on the AGB ?

CESARE CHIOSI: In principle yes. Marigo is indeed working on it. Our plan is to adopt the same diffusive-overshoot scheme as in Deng et al. (1996). The overshoot distance should result from the velocity profile (layer at which the velocity of convective elements vanishes) and mixing in it should be governed by diffusion at a suitable rate.

BENGT GUSTAFSSON: Does your work with Marigo produce different distribution of C/O ratios for a population of metal-poor carbon stars, such as in the LMC, as compared with galactic ones ? If so, which are the results ?

CESARE CHIOSI: We must distinguish between stars without and with envelope burning, i.e., stars lighter and heavier than about $4M_{\odot}$. In the first group, at given mass of the star and assuming constant the parameters governing the *3rd dredge-up*, owing to the longer AGB lifetime (larger number of pulses) the ratio C/O is higher in low metallicity stars. In stars of the second group the situation is more complicated. Even if envelope burning is more efficient in low metallicity stars, the interplay between number of pulses and nuclear burning may lead to the case in which the ratio C/O is on the average higher in low than in high metallicity stars.

Langer (p. 343)

JOHANNES ANDERSEN: Three comments and a question:

1. Wouldn't a radial velocity survey be a more direct way to exclude (post-mass-exchange) binaries?
2. Comparison of slowly rotating (synchronized) B-type binary stars with models seems to indicate the kind of convective core enlargement commonly labeled by the nickname 'overshooting.'
3. Because of orbital synchronization, the M-L relation derived from binary data will be for slowly rotating stars (mostly).

Q. Could fast interior rotation decoupled from the surface make any of these comparisons inaccurate/invalid?

NORBERT LANGER:

1. That is done and most appear to be single stars. Still, they could have been in a binary system before.
2. Other mixing mechanisms might act in these close binaries, e.g., tidal mixing. I would be careful to conclude anything about mixing processes in single stars from them.
3. It might only apply to components of close binaries (see point 2).

Q. No. It has been shown by J.P. Zahn that the timescale of angular momentum transport in massive main sequence stars is too short to allow a decoupling of core and envelope.

Proffitt (p. 355)

GILLES CHABRIER: In the collision integrals, the potential is screened by the ions and by the electrons. How do you take that into account?

CHARLES PROFFITT: Yes, all charged particles are taken into account when calculating the Debye screening length, which is the relevant screening length for the Sun and the metal-poor turn-off stars.

Arnett (p. 389)

BENGT GUSTAFSSON: Could you speculate concerning how your results would change qualitatively when you will get 3D hydro simulations?

DAVID ARNETT: With 2D, the angular momentum vectors must be perpendicular to the plane, so that vortices are 'pinned.' This tends to exaggerate them. In 3D, they would interact in a more complex way and break apart more easily. The NOVA experiments are really 3D, of course, and we are beginning a study of the 2D/3D differences for the Rayleigh-Taylor/Richtmeyer-Meshkov instability.

ROBERT KURUCZ: Norris quoted you as saying that most supernova calculations are duds. That reminded me that most supernovae in nature might be duds. We see only the successes. Modellers force their calculations to give the answers they expect (I do). If your calculations produce duds, they might be real. For example, wouldn't your asymmetric convection yield asymmetric explosions?

DAVID ARNETT: Our understanding of the event rates of supernovae, supernova remnants and pulsars has shown steady improvement over the years. While your question would have had some force some decades past, it appears now that 'silent supernovae' are not so common. We have long expected that some of the duds became black holes. As for asymmetric explosions - they tend to become spherical as they expand, that is, toward a Hubble law for velocity. For me, the point of the asymmetries caused by convection is that they should give rise to a breaking of symmetry in the explosion.

Leggett et al. (p. 429)

JOHANNES ANDERSEN: You referred to your sample as a 'local' one, but the galactic orbits of 'local' F dwarfs show that they originate from a wide range of galacto-centric distances. Can you do a similar analysis for the white dwarfs?

SANDY LEGGETT: Unfortunately we do not have radial velocities and so we cannot calculate UVW velocities. The tangential velocities only show that they are old disk stars.

INDEXES

Author Index

Subject Index

Object Index